Pyrolysis

THEORY AND INDUSTRIAL PRACTICE

Pyrolysis

THEORY AND INDUSTRIAL PRACTICE

EDITED BY

Lyle F. Albright

School of Chemical Engineering
Purdue University
West Lafayette, Indiana

Billy L. Crynes

School of Chemical Engineering
Oklahoma State University
Stillwater, Oklahoma

William H. Corcoran

Department of Chemical Engineering
California Institute of Technology
Pasadena, California

1983

ACADEMIC PRESS
A Subsidiary of Harcourt Brace Jovanovich, Publishers
New York London
Paris San Diego San Francisco São Paulo Sydney Tokyo Toronto

ACADEMIC PRESS, INC.
111 Fifth Avenue, New York, New York 10003

United Kingdom Edition published by
ACADEMIC PRESS, INC. (LONDON) LTD.
24/28 Oval Road, London NW1 7DX

Library of Congress Cataloging in Publication Data

Main entry under title:

Pyrolysis, theory and industrial practice.

Includes bibliographies and index.
1. Pyrolysis. I. Albright, Lyle Frederick, Date . II. Crynes, Billy L., Date . III. Corcoran, William Harrison, Date .
TP156.P9P95 660.2'96 82-6838
ISBN 0-12-048880-9 AACR2

PRINTED IN THE UNITED STATES OF AMERICA

83 84 85 86 9 8 7 6 5 4 3 2 1

DEDICATED TO

William H. Corcoran

1920–1982

During his long and illustrious career, Bill Corcoran accepted many challenges—handling all with skill and great wisdom. To mention just a few high points: As an eminent educator at the California Institute of Technology, he was Institute Professor of Chemical Engineering and for 10 years Vice President for Institute Relations. He left an indelible mark on the engineering profession and served over a period of many years in many capacities (ranging from inspector to chairman of key committees) dealing with the accreditation of engineering curricula. He served in several national offices of the American Institute of Chemical Engineers, and was its president in 1978. He was also a member of the National Academy of Engineering.

Among his many research interests Bill enjoyed an international reputation in pyrolysis. As his coeditors we are honored to have worked with him on the preparation of this book. It is most fitting that it should be dedicated to him.

In addition to being a top-notch engineer, Bill Corcoran was always a true gentleman. He has many friends who will miss him dearly. The world is a better place because of Bill.

Contents

3. Pyrolysis of *n*-Butane

WILLIAM H. CORCORAN

4. Pyrolysis of Heavy Hydrocarbons

CHARLES REBICK

5. Thermal Reaction of Olefins and Diolefins, and Formation of Aromatics

TOMOYA SAKAI

6. Pyrolysis: Hydrogenolysis of Toluene

MUNEYOSHI YAMADA and AKIRA AMANO

7. Mathematical Modeling of Hydrocarbon Pyrolysis Reactions

MARIO E. DENTE and ELISEO M. RANZI

8. Nonpetroleum Feedstocks

OTIS C. DERMER and BILLY L. CRYNES

9. Fundamental Aspects of the Formation and Gasification of Coke

DAVID L. TRIMM

Contributors

Numbers in parentheses indicate the pages on which the authors' contributions begin.

Lyle F. Albright (233), School of Chemical Engineering, Purdue University, West Lafayette, Indiana 47907

Akira Amano (117), Department of Applied Chemistry, Tohoku University, Aoba-Aramaki, Sendai 980, Japan

Margaret H. Back (1), Department of Chemistry, University of Ottawa, Ottawa, Ontario, Canada

Robert A. Back (1), Division of Chemistry, National Research Council of Canada, Ottawa, Ontario, Canada

Guy-Marie Côme (255), Département de Chimie-Physique des Réactions, Institut National Polytechnique de Lorraine et Université de Nancy I, 54042 Nancy Cedex, France

William H. Corcoran[1] (47), Department of Chemical Engineering, California Institute of Technology, Pasadena, California 91125

Billy L. Crynes (177), School of Chemical Engineering, Oklahoma State University, Stillwater, Oklahoma 74078

Mario E. Dente (133), Dipartimento de Chimica Industriale, e Ingegneria Chimica, Politecnico de Milano, 20133 Milano, Italy

Otis C. Dermer (177), Department of Chemistry, Oklahoma State University, Stillwater, Oklahoma 74078

Hubert Günschel (277), Academy of Sciences of the GDR, Zentralinstitut für Organische Chemie, 1199 Berlin, German Democratic Republic

Billy D. Head (25), Dow Chemical Company, Freeport, Texas 77541

[1]Deceased.

J. C. Heiman (365), Graff Engineering Corporation, Houston, Texas 77006

S. Ibarra (427), Gulf Research and Development Company, Pittsburgh, Pennsylvania 15230

Christopher F. McConnell[2] (25), Dow Chemical Company, Moorestown, New Jersey 08057

Alfred Mol (451), AM & Associates, 2596 CV Den Haag, The Netherlands

Siegfried Nowak (277), Academy of Sciences of the GDR, Zentralinstitut für Organische Chemie, 1199 Berlin, German Democratic Republic

Eliseo M. Ranzi (133), Dipartimento di Chimica Industriale, e Ingegneria Chimica, Politecnico di Milano, 20133 Milano, Italy

Charles Rebick[3] (69), Exxon Research and Development Laboratories, Baton Rouge, Louisiana 70821

Lawrence L. Ross (327), C. F. Braun and Company, Alhambra, California 91802

Tomoya Sakai (89), Department of Chemical Reaction Engineering, Nagoya City University, 3-1 Tanabedori, Mizuhoku, Nagoya 467, Japan

David L. Trimm (203), School of Chemical Engineering and Industrial Chemistry, University of New South Wales, Kensington, New South Wales, Australia 2033

Tom Chung-Hu Tsai (233), Bechtel Group, Inc., Houston, Texas 77001

Frank W. Tsai (437), Department of Pyrolysis Processes, KTI Corporation, Pasadena, California 91101

Muneyoshi Yamada (117), Department of Applied Chemistry, Tohoku University, Aoba-Aramaki, Sendai 980, Japan

Stanley B. Zdonik (377), Stone and Webster Engineering Corporation, Boston, Massachusetts 02107

[2]*Present address:* 2 Soldier's Field Park, Apt. 620, Boston, Massachusetts 02163.

[3]*Present address:* Corporate Research, Exxon Research and Engineering Company, P.O. Box 45, Linden, New Jersey 07036.

Preface

The objectives of this book are to collect and to organize the latest results and thinking on both mechanistic and industrial aspects of pyrolysis reactions and processes. The main emphasis is on the production of light olefins, although important by-products frequently include diolefins, aromatics, pyrolysis gasoline, and even in some cases acetylenic compounds.

The first step in organizing the book was to prepare a table of contents covering what we considered to be the most important features of pyrolysis. Advice and counsel on this outline were then sought from several experts. Finally, we obtained writers who were also experts for the various chapters. In choosing the authors we sought a blend of individuals with various backgrounds. The authors range from theoreticians to highly successful industrial representatives; they bring a truly international perspective to this book. A total of eight countries is represented.

The initial chapters mainly consider mechanistic and fundamental aspects of pyrolysis (also designated as thermal cracking or simply, cracking). Several chapters are devoted to feedstocks ranging from methane through naphthas, gas oils, and nonpetroleum feedstocks. Modeling of complicated pyrolysis reactions is also discussed, and three chapters cover various aspects of surface phenomena in pyrolysis units, including coking, carburization, corrosion, erosion, decoking, surface oxidation, and surface sulfidation.

Reactor designs for both laboratory and industrial units are considered and compared. Key aspects of planning, designing, and operating an olefin unit are reported and discussed in considerable detail. Economic problems for olefin production in the 1980s are many, but, as discussed in this book, so are the opportunities. The chapter by East German researchers on pyrolysis of naphthas and heavier hydrocarbons presents interesting data and a perspective not readily available elsewhere.

In selecting authors for the various chapters our goal was to obtain the best overall coverage of the field, even though we realized that there would not be complete unanimity on all subjects. We believe that one strength of the book is the examination of the differences, with emphasis on the point that there is need for more research to develop further understanding of pyrolysis. Some areas of controversy are as follows:

1. Agreement on reaction mechanisms, especially for lighter feedstocks, has not been realized. Different researchers may interpret experimental data in different ways. One complicating factor, as discussed next, is the role of the surface.
2. The role and significance of surface reactions in both industrial and laboratory units have not been clearly established. Surface effects are usually of greater consequence in laboratory units than in industrial units because of the larger surface-to-volume ratio of the former. The key point is the recognition that the predominantly homogeneous gas-phase reactions may be influenced by heterogeneous effects.
3. Coke is formed both by surface reactions, for which metal surfaces are often catalysts, and by gas-phase reactions. With regard to the latter source, there is some evidence for two conflicting mechanisms. It is not yet clear which is correct or whether both may occur under the proper conditions. In one mechanism, coke formed in the gas phase collects on the surfaces of the reactor vessel or coil; in the other, tar droplets formed in the gas phase collect on the solid surfaces, where they then decompose to coke.
4. The nature of velocity profiles in reactor systems with laminar or turbulent flow in tubular coils is not certain. Is the flow at low Reynolds numbers in small-diameter laboratory reactors truly laminar with no-plug-type characteristics? The analysis by Cleland and Wilhelm[1] indicates that rapid diffusion at pyrolysis temperatures results in plug flow. Further, not everyone agrees that surface effects are really eliminated in the so-called wall-less type of reactor.
5. As would be expected, there are differences in various techniques and flowsheets for industrial plants in areas such as separation and recovery of products, hydrogenation of acetylene, and best reactor design.

[1]Cleland, F. A., and Wilhelm, R. H. (1956). *AIChE* **2**, 489.

6. What are the best feedstocks for pyrolysis? The emphasis in the different chapters varies in this matter.

Although the main emphasis of the book is on ethylene or olefin production, the information presented should be most helpful in understanding other thermal processes. For example, the same coke precursors occur during the dehydrochlorination of 1,2-dichloroethane (used for industrial production of vinyl chloride) as in production of ethylene and other olefins. Yet there certainly are differences to be expected relative to the surface reactions in a vinyl chloride unit as compared to an olefin unit.

We hope that this book will serve as a logical starting point to study all key aspects of the process. Most chapters discuss material that has been carefully reviewed at some time. We believe, however, that significant updating of material has been realized in all cases, and that coverage and discussion on surface reactions and transfer-line exchangers are broader and more complete than in previous reports. Neither the editors nor the authors planned an exhaustive or all-inclusive literature survey for any chapter. Instead, the intent was to reference only the more pertinent publications and hence to provide optimal coverage within a reasonable number of pages.

Finally, we note that in the field of pyrolysis, as with many other fields, communication between the theoretician or research scientist and the applied scientist or engineer working on the industrial unit is not always optimal. We hope this book will help to bridge that gap. The editors, all of whom have had frequent contacts in the last 15–20 years with both the theoretician and the applied scientist in pyrolysis, believe the gap can and should be closed.

The editors are indebted to a large number of individuals. The cooperation and dedication of the authors in meeting deadlines and considering revisions, etc., were outstanding. We are also most appreciative of the tutorial assistance, guidance, advice, and/or encouragement provided us in the past relative to pyrolysis. For that help, we would especially like to recognize R. T. K. Baker, H. G. Davis, A. G. Oblad, K. F. Frech, W. A. McGill, M. J. Weinbaum, F. M. Wall, A. J. Apiecionek, J. A. Kivalen, O. L. Hollis, and C. F. King. Finally, our thanks go to our outstandingly helpful secretaries: Phyllis Beck at Purdue University, Terri Tackett and Norma White at Oklahoma State University, and Donna Johnson at the California Institute of Technology.

1

Thermal Decomposition and Reactions of Methane

MARGARET H. BACK and ROBERT A. BACK

I. INTRODUCTION

Methane has not been generally used as a feedstock in simple thermal cracking processes because the decomposition temperature is too high and yield of useful products too low. Nevertheless, much effort has been expended, particularly in the Soviet Union, toward the maximization of

Pyrolysis: Theory and Industrial Practice

ISBN 0-12-048880-9

yields of ethylene and acetylene (Stepanov *et al.*, 1971), and the pyrolysis of methane has been used for the production of acetylene and carbon black (Tesner, 1978). Diamonds can be formed during methane pyrolysis under suitable conditions (Deryaguin and Fedoseev, 1973). Methane is also important as a source of hydrogen in several industrial processes (Burollet, 1980). A good understanding of the thermal properties and reactions of methane is basic to the improvement and modification of established processes and in the search for new applications and uses.

There is an enormous literature on the thermal decomposition and reactions of methane, and the present chapter makes no attempt to be comprehensive. We have chosen instead to emphasize the more fundamental studies of the decomposition, especially those dealing with the mechanism and the quantitative measurement of the rates of elementary processes. The early stages of the decomposition are now quite well understood in this regard, but the details of the later stages, particularly the autocatalysis and the formation of carbon, are not yet clear, and we have tried to point out some of the areas of uncertainty where further careful quantitative work is needed. All oxidation reactions, including oxidative pyrolysis, have been excluded.

II. GENERAL FEATURES OF THE DECOMPOSITION

The thermal reactions of methane, the simplest hydrocarbon, have been extensively studied for many years, and the subject has been reviewed on a number of occasions (Steacie, 1954; Kramer and Happel, 1955; Khan and Crynes, 1970; Laidler and Loucks, 1971).

By the 1930s it had been established that the pyrolysis of methane largely followed a stepwise dehydrogenation sequence:

$$2\,CH_4 \xrightarrow{-H_2} C_2H_6 \xrightarrow{-H_2} C_2H_4 \xrightarrow{-H_2} C_2H_2 \xrightarrow{-H_2} 2\,C$$

Because the C_2 hydrocarbons are much less thermally stable than CH_4 and are readily attacked by CH_3 radicals formed in the decomposition of methane, secondary reactions of these products set in at very early stages of the decomposition. These and higher hydrocarbon products formed in secondary reactions rapidly attain steady states at very low concentrations, typically less than 1% of the CH_4 concentration and often much less, and the overall decomposition can then be described to a good approximation as $CH_4 \rightarrow 2\,H_2 + C$. However, each of the four dehydrogenation steps is itself complex, involving a variety of free-radical and molecular intermediates. The final step from acetylene to carbon is particularly complicated, probably proceeding through the formation of polyunsaturated and polyaromatic hydrocarbons, followed by polymerization, condensation, and loss of hydrogen. In such a complex system of reactions, which can show under various conditions induction periods, autocatalysis, or self-inhibition, it is not surprising that the measurement of overall rates of reaction based on pressure change, carbon or hydrogen formation, or disappearance of CH_4 led to a wide range of Arrhenius parameters and reaction orders as the pyrolysis was studied over a range of temperatures and pressures in static systems (Kassel, 1932, 1935; Odochian and Schneider, 1970), flow systems (Gordon, 1948; Germain and Vaniscotte, 1957; Schneider and Murgulescu, 1961; Schneider, 1962; Eisenberg and Bliss, 1967; Palmer and Hirt, 1963; Palmer *et al.*, 1968; Kaminski and Sobkowski, 1971; Gordon *et al.*, 1972), and high-temperature shock tubes (Skinner and Ruehrwein, 1959; Glick, 1959; Kevorkian *et al.*, 1960; Kozlov and Knorre, 1960; Hartig *et al.*, 1971; Napier and Subrahmanyam, 1972; Gardiner *et al.*, 1974). It is only in recent years that improved analytical techniques, notably gas chromatography, have permitted accurate measurement of product yields in the initial and early stages of the decomposition (Chen *et al.*, 1976a). In this way, products have been clearly identified as primary, secondary, tertiary, etc., in the overall sequence, and rates of many of the individual elementary reactions have been determined. Such studies have confirmed the simple dehydrogenation sequence, but have also revealed the importance of methylation reactions and the participation of higher-molecular-weight products, at least at lower temperatures, in the route to the formation of carbon.

Typical measurements of product formation in the early stages of the reaction in a static system at temperatures around 1000°K are shown in Fig. 1 (Chen *et al.*, 1975). The course of the reaction may be resolved into three stages, all occurring within a total conversion of methane of $<2\%$. The first stage may be described as the primary reaction; ethane and hydrogen are the only products. In the second stage the rate of formation of ethane gradually falls off and its concentration approaches a plateau value, while ethylene is observed as a secondary product. In the third stage the tertiary products, acetylene and propylene, appear, and at about the same time a sharp increase in the rate of formation of ethane is observed and other products such as butadiene and benzene begin to become significant. Late in the third stage, deposition of carbon becomes apparent. In the first two stages a simple, homogeneous radical mechanism can describe the results satisfactorily. The third stage, which involves some autocatalysis and formation of carbon, is much more complex.

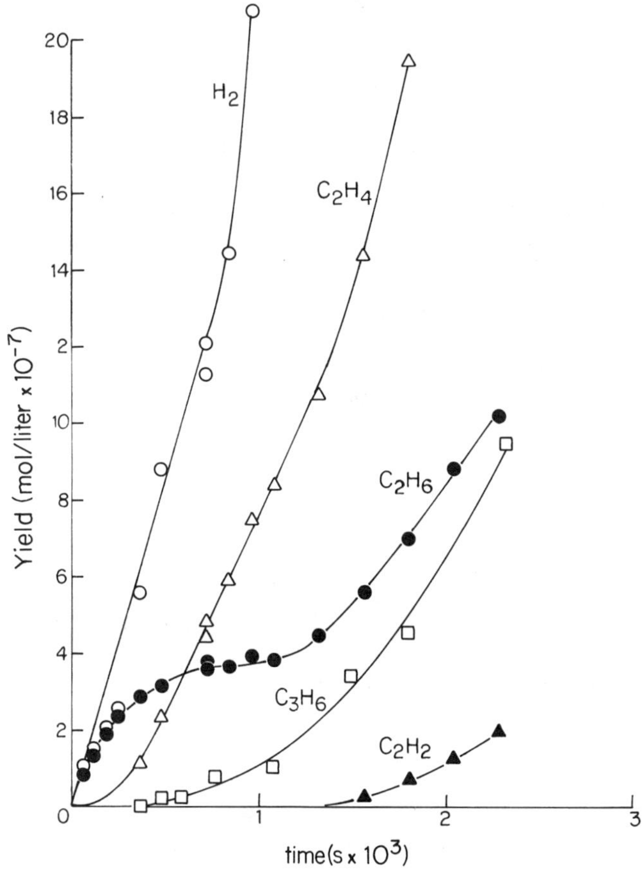

Fig. 1. Yields of products in the early stages of the pyrolysis of methane in a static system at 1038°K and 441 torr. (Taken from data of Chen *et al.*, 1975.)

III. REACTION MECHANISM AT 1000°K

The following mechanism for the early stages of methane decomposition accounts qualitatively for all the products observed in a static system at 1000°K and pressures above 50 torr, and accounts quantitatively for the yields of hydrogen, ethane, and ethylene at conversions up to about the attainment of the plateau in the ethane concentration (Fig. 1). It does not account for the subsequent acceleration in the decomposition or for the processes leading to carbon formation.

A. Stage 1: Primary Formation of Ethane and Hydrogen

$$CH_4 \longrightarrow CH_3 + H \qquad \Delta H = +104^* \qquad (1)$$

$$H + CH_4 \longrightarrow CH_3 + H_2 \qquad \Delta H = 0 \qquad (2)$$

$$2\,CH_3 \longrightarrow C_2H_6 \qquad \Delta H = -88.2 \qquad (3)$$

Net reaction: $$2\,CH_4 \longrightarrow C_2H_6 + H_2 \qquad \Delta H = +15.8$$

Reaction (1) is rate controlling and the only primary source of free radicals in the system. It is effectively always followed by reactions (2) and (3), which have much lower activation energies. Reaction (3) is apparently the only significant fate of CH_3 radicals at this stage of the decomposition. Note that stage 1 entails no pressure change, and only in stage 2 and subsequent stages does the pressure increase. In some static pyrolysis studies based on pressure change alone, short induction periods were noted that presumably correspond to stage 1.

B. Stage 2: Secondary Reactions of Ethane

(a) *Radical-chain dehydrogenation*:

$$CH_3 + C_2H_6 \longrightarrow CH_4 + C_2H_5 \qquad \Delta H = -6.0 \qquad (4)$$

$$C_2H_5 \longrightarrow C_2H_4 + H \qquad \Delta H = +38.9 \qquad (5)$$

$$H + CH_4 \longrightarrow CH_3 + H_2 \qquad \Delta H = 0 \qquad (2)$$

Net reaction: $$C_2H_6 \longrightarrow C_2H_4 + H_2 \qquad \Delta H = +32.9$$

(b) *Unimolecular decompostion*:

$$C_2H_6 \longrightarrow 2\,CH_3 \qquad \Delta H = +88.2 \qquad (3R)$$

Reactions (4), (5), and (2) constitute a chain reaction converting ethane to ethylene and hydrogen. Reaction (4) is rate controlling, and the reverse of 4 (4R) is largely negligible at pressures around 1 atm; so almost every C_2H_5 radical formed in reaction (4) decomposes. At much higher pressures of CH_4, (4R) may become significant and the dehydrogenation less efficient. Reaction (3R), the reverse of reaction (3), becomes an important secondary source of CH_3 radicals at higher temperatures and as C_2H_6 builds up.

* Values of ΔH throughout are in kcal/mol at 298°K, based on standard heats of formation (JANAF, 1970; Benson and O'Neal, 1970; Benson, 1968).

C. Stage 3: Secondary Reactions of Ethylene, Acetylene, and Propylene

Secondary Reactions of Ethylene

(a) *Radical-chain dehydrogenation:*

$$CH_3 + C_2H_4 \longrightarrow CH_4 + C_2H_3 \qquad \Delta H = -0.4 \qquad (6)$$

$$C_2H_3 \longrightarrow C_2H_2 + H \qquad \Delta H = +42.4 \qquad (7)$$

$$H + CH_4 \longrightarrow CH_3 + H_2 \qquad \Delta H = 0 \qquad (2)$$

Net reaction:

$$C_2H_4 \longrightarrow C_2H_2 + H_2 \qquad \Delta H = +42.0$$

(b) *Radical-chain methylation:*

$$CH_3 + C_2H_4 \longrightarrow n\text{-}C_3H_7 \qquad \Delta H = -25.8 \qquad (8)$$

$$n\text{-}C_3H_7 \longrightarrow C_3H_6 + H \qquad \Delta H = +36.2 \qquad (9)$$

$$H + CH_4 \longrightarrow CH_3 + H_2 \qquad \Delta H = 0 \qquad (2)$$

Net reaction:

$$C_2H_4 + CH_4 \longrightarrow C_3H_6 + H_2 \qquad \Delta H = +10.4$$

As ethylene formed from reaction (5) accumulates in the system, it begins to disappear in the reaction sequences (6), (7), and (2) and (8), (9), and (2), two parallel chain reactions that consume ethylene; they are propagated by methyl radicals, by their abstraction and addition reactions, respectively. Reaction (6) is rate controlling in the first sequence, and its reverse is probably negligible at 1000°K and 1 atm of CH_4. In the addition sequence, on the other hand, reaction (8R) is probably much faster than reaction (9), making the latter rate controlling at 1000°K. The ratio of propylene to acetylene in the products reflects the relative rates of the two chain sequences and will vary with temperature. At 1000°K, initial propylene yields were several times those of acetylene, but at 2000°K propylene was negligible compared to acetylene. Direct decomposition of ethylene to acetylene and rapid secondary loss of propylene are also factors at the higher temperatures.

Secondary Reactions of Acetylene

(a) *Radical-chain dehydrogenation:*

$$CH_3 + C_2H_2 \longrightarrow CH_4 + C_2H \qquad \Delta H = +7.8 \qquad (10)$$

$$C_2H \longrightarrow C_2 + H \qquad \Delta H = +139 \qquad (11)$$

$$H + CH_4 \longrightarrow H_2 + CH_3 \qquad \Delta H = 0 \qquad (2)$$

Net reaction:

$$C_2H_2 \longrightarrow C_2 + H_2 \qquad \Delta H = +146.8$$

This dehydrogenation sequence does *not* occur, because reaction (11) is much too endothermic to be of any importance at 1000°K, and (10R) is probably very fast. This sequence, therefore, cannot be a source of carbon in methane decomposition at these temperatures.

(b) *Radical-chain methylation*:

$$CH_3 + C_2H_2 \longrightarrow CH_3CHCH \qquad \Delta H = -34.5 \qquad (12)$$

$$CH_3CHCH \longrightarrow CH_3CCH + H \qquad \Delta H = +44.7 \qquad (13)$$

$$H + CH_4 \longrightarrow CH_3 + H_2 \qquad \Delta H = 0 \qquad (2)$$

Net reaction: $$C_2H_2 + CH_4 \longrightarrow CH_3CCH + H_2 \qquad \Delta H = +10.2$$

As with ethylene, the reverse of the addition reaction (12R) is probably much faster than reaction (13), so the latter is rate controlling.

Secondary Reactions of Propylene

(a) *Radical-chain dehydrogenation*:

$$CH_3 + C_3H_6 \longrightarrow CH_4 + C_3H_5 \qquad \Delta H = -17.7 \qquad (14)$$

$$C_3H_5 \longrightarrow CH_2CCH_2 + H \qquad \Delta H = +58.6 \qquad (15)$$

$$H + CH_4 \longrightarrow CH_3 + H_2 \qquad \Delta H = 0 \qquad (2)$$

Net reaction: $$C_3H_6 \longrightarrow CH_2CCH_2 + H_2 \qquad \Delta H = +40.9$$

(b) *Radical-chain methylation*:

$$CH_3 + C_3H_6 \longrightarrow CH_3CH_2CHCH_3 \qquad \Delta H = -26.2 \qquad (16)$$

$$CH_3CH_2CHCH_3 \longrightarrow CH_3CH_2CH{=}CH_2 + H \qquad \Delta H = +39.4 \qquad (17)$$

$$H + CH_4 \longrightarrow CH_3 + H_2 \qquad \Delta H = 0 \qquad (2)$$

Net reaction: $$C_3H_6 + CH_4 \longrightarrow CH_3CH_2CH{=}CH_2 + H_2 \qquad \Delta H = +13.2$$

(c) *Unimolecular decomposition*:

$$C_3H_6 \longrightarrow H + C_3H_5 \qquad \Delta H = +87.9 \qquad (18)$$

or $$C_3H_6 \longrightarrow CH_3 + C_2H_3 \qquad \Delta H = +93.2$$

Both addition and abstraction sequences can occur with propylene. Abstraction appears to predominate at 1000°K as yields of allene were much higher than those of butene, although rapid loss of the thermally less stable butene may also be a factor. Reaction (18) can be a significant source of radicals in the system when the propylene concentration becomes high enough.

D. Summary of the Mechanism

This rather simple mechanism just outlined depends on the following assumptions:

1. The CH_3 radical is the only radical present in significant concentrations.
2. Hydrocarbon products disappear rapidly, largely through reaction with methyl radicals in chain sequences with no net consumption of radicals. Dissociation reactions such as (3R) and (18) are of minor importance.
3. Higher hydrocarbon radicals dissociate so rapidly that they take almost no part in other reactions. An exception is C_2H, which instead is maintained at a very low concentration by reaction with methane (10R).
4. Hydrogen atoms always react with CH_4 by reaction (2) to produce hydrogen.

These assumptions appear to be largely true in methane decomposition at 1000°K. Their validity depends chiefly on the large excess of methane (even at high conversions hydrocarbon products amount to only a few percent of the methane) and on the high temperature relative to that of other hydrocarbon pyrolyses, which makes the higher radicals very short-lived with respect to decomposition.

The several chain-reaction sequences propagated by methyl radicals follow a common pattern, with abstraction or addition followed by radical decomposition, but there are notable differences. Thus with ethane, only abstraction is possible, but with ethylene both abstraction and addition sequences occur. With acetylene the abstraction sequence is blocked by the stability of the C_2H radical, so only addition is important; with propylene, however, abstraction of the allylic hydrogen is apparently much faster than addition, and it predominates. Finally, it should be noted that in this mechanism there is initially no radical-chain decomposition of methane. The secondary chain sequences propagated through abstraction also do not consume methane, and it is only with the methylation sequences propagated by addition reactions, beginning with ethylene, that methane is consumed. Even then, these chains are limited by the amount of olefin present, and probably at most might double the rate of disappearance of methane, that is, a chain length of two.

E. Later Stages of the Decomposition

Shortly after the appearance of C_2H_2 as a product, the decomposition of methane rapidly becomes much more complex. Higher unsaturated and aromatic hydrocarbons and finally carbon appear in surprisingly rapid succession, and at the same time a strong acceleration of the overall reaction is observed. A detailed mechanism has not been delineated for this stage of the decomposition. However, although the mechanism presented stops at reactions of propylene, it is obvious that similar dehydrogenation and methylation chain sequences, propagated by methyl radicals, will operate with the propyne, allene, and butene formed in reactions (13), (15), and (18), with other higher unsaturated hydrocarbons that progressively make their appearance as the decomposition proceeds. It is clear that such methyl radical-initiated reactions leading to increasing molecular weight and decreasing hydrogen content will lead ultimately to carbon formation, and must play a major part in this process, at least at moderate temperatures. New sources of free radicals are required in the system to account for the observed acceleration, and reactions at carbon surfaces may become important under some conditions; these aspects of the decomposition are not yet well understood.

IV. THE DECOMPOSITION AT HIGHER TEMPERATURES

With increasing temperature, the three stages of the decomposition observed at 1000°K become sharply compressed in time, and the lifetimes and steady-state concentrations of hydrocarbon products become progressively less. Thus at 1400°K (Trenwith, 1979), the half-life of C_2H_6 with respect to dissociation is about 1 ms, and concentrations of C_2H_6 and CH_3 become controlled by an equilibrium between reactions (3) and (3R). Ethylene and acetylene are more stable and reach higher concentrations, but also disappear in a rapid sequence of reactions. Experiments in static systems are limited to about 1100°K by the time taken to admit gas to the reaction vessel and remove it at low enough conversions to study the early stages of the reaction. Flow systems, with time resolution of 10^{-1} or exceptionally 10^{-2} s, extend the temperature range, and methane pyrolysis has been extensively explored in flow reactors at temperatures as high as 2100°K (Holman *et al.*, 1976). Resolution of the early stages of the reaction, however, is limited to much lower temperatures; at 1400°K, for example, conversion in a typical flow system at 200 ms was about 10% (Khan and Crynes, 1970), carbon was already a major product, and C_2 hydrocarbons were at low steady-state values. Carbon-14 labeling experiments in a comparable system showed that extensive free-radical equilibration reactions between products and reagent had occurred (Kaminski and Sobkowski, 1975, 1977). In studies at still higher temperatures in shock tubes, the much shorter reaction times that can be achieved again permit detailed resolution and measurements of the initial stages of the reaction. Figure 2

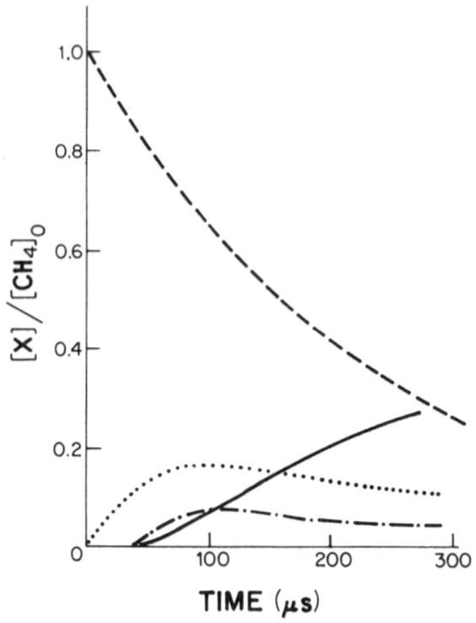

Fig. 2. Approximate time profiles of reagent and products, measured *in situ* by mass spectrometer in the decomposition of methane in a reflected shock wave at 2655°K, total initial pressure of 700 Pa and 0.4% CH_4, ----, CH_4; · · · ·, CH_3;–·–·–, C_2H_4;——, C_2H_2. (Taken from data of Gardiner *et al.*, 1974.)

shows methane and products, measured by a time-of-flight mass spectrometer, as a function of time in reflected shock experiments at 2655°K (Gardiner *et al.*, 1974). The only products of importance after 300 μs are CH_3, C_2H_4, and C_2H_2. Reaction (3) is almost totally reversed, with ethane only present at very low concentrations in equilibrium with CH_3 radicals. The very low concentration of ethane makes the methyl radical-dehydrogenation sequence, reactions (4), (5), and (2), ineffective, and ethylene probably arises from an alternate decomposition path:

$$2\,CH_3 \longrightarrow C_2H_6^* \longrightarrow C_2H_5 + H \qquad \Delta H = +9.8 \tag{19}$$

$$C_2H_5 \longrightarrow C_2H_4 + H \qquad \Delta H = +38.9 \tag{5}$$

where $C_2H_6^*$ represents a vibrationally excited molecule, or possibly by by the molecular dissociation:

$$2\,CH_3 \longrightarrow C_2H_6^* \longrightarrow C_2H_4 + H_2 \qquad \Delta H = +55.5 \tag{20}$$

These can be viewed as alternative channels either for methyl combination or for the dissociation of ethane, depending on whether or not the energy of formation of an individual molecule of ethane is retained long enough to contribute to its subsequent dissociation; if the (3)–(3R) system is at thermodynamic equilibrium, the two views become indistinguishable. The dissociation of ethane via reaction (19) or (20) is negligible compared to reaction (3R) in the normal thermal decomposition of ethane, and even at high temperatures (3R) is still much faster, but these reactions become important in methane decomposition at high temperature as the only effective escape routes from the (3)–(3R) equilibrium.

Reactions (3), (19), and (5), when followed by reaction (2), constitute a linear chain reaction propagated by methyl radicals and consuming methane.

$$2\,CH_3 \longrightarrow C_2H_4 + 2\,H \qquad \Delta H = +48.7 \qquad (3 + 19 + 5)$$

$$2\,H + 2\,CH_4 \longrightarrow 2\,H_2 + 2\,CH_3 \qquad \Delta H = 0 \qquad (2)$$

Net reaction: $$2\,CH_4 \longrightarrow C_2H_4 + 2\,H_2 \qquad \Delta H = +48.7$$

Reaction (20), following reaction (3), on the other hand, is a radical-termination process, the only effective one at these temperatures when reaction (3) is almost totally reversed. The shock-tube data do not distinguish clearly between reactions (19) and (20), although there was certainly no very long chain process occurring.

In a more recent shock-tube study at temperatures from 1948 to 2773°K, using a laser to measure density gradients after the shock-front, Tabayashi and Bauer (1979) found it necessary to include in their model the displacement reaction,

$$CH_3 + CH_4 \longrightarrow C_2H_6 + H \qquad \Delta H = 16.2 \qquad (21A)$$

and its variant,

$$CH_3 + CH_4 \longrightarrow C_2H_4 + H_2 + H \qquad \Delta H = 10.3 \qquad (21B)$$

in order to fit the experimental data. These reactions, followed by reactions (2) and (5), would also constitute a chain decomposition of methane, and will be discussed in more detail later.

Acetylene production in the high-temperature shock tube probably occurs by the same methyl-radical chain reaction of ethylene—reactions (6), (7), and (2)—postulated at lower temperatures, although the alternative molecular elimination reaction

$$C_2H_4 \longrightarrow C_2H_2 + H_2 \qquad \Delta H = +41.8 \qquad (22)$$

which has been observed in infrared laser pyrolysis experiments (Peterson *et al.*, 1978), might also be important. A trace of C_4H_2, presumably diacetylene, was also tentatively identified in the shock tube, probably formed by reaction of C_2H radicals.

$$C_2H + C_2H_2 \longrightarrow C_4H_3 \longrightarrow C_4H_2 + H \qquad (23)$$

This reaction would be favored by the relatively high concentration of C_2H_2 attained (Fig. 2) and by a resultant shift, also assisted by the high temperature, in the CH_3/C_2H ratio controlled by reactions (10) and (10R). No hydrocarbon products higher than C_4H_2 were observed in these shock-tube experiments.

V. ELEMENTARY REACTIONS IN THE EARLY STAGES

A. The Primary Dissociation

Two possible channels must be considered for the dissociation of methane:

$$CH_4 \longrightarrow CH_3 + H \qquad \Delta H = +104 \tag{1A}$$

$$CH_4 \longrightarrow CH_2 + H_2 \qquad \Delta H = +109 \tag{1B}$$

Arguments for one or the other of these two processes in the past have been based either on their energetics or on attempts to show the presence of CH_3 or CH_2 radicals in the system. Reaction (1B) gained favor for a while (Kevorkian *et al.*, 1960; Kassel, 1932, 1935; Kozlov and Knorre, 1960) when it was believed that its endothermicity was substantially less than that of (1A), and could therefore account for experimental activation energies that tended to fall in the range of 80–100 kcal/mol. Most recent evidence, however, indicates that the heat of formation of the $^3B_1$ ground state of the CH_2 radical is about 93 kcal/mol (Warneck, 1971; Chupka and Lifshitz, 1968), making reaction (1B) endothermic by 109 kcal/mol. Formation of the first excited singlet state of CH_2 would be even more endothermic by another 6–10 kcal/mol. It is not clear whether singlet or triplet CH_2 would be more readily formed in reaction (1B); production of the singlet conserves spin but the energy barrier is higher, whereas triplet formation has a lower energy requirement but would probably have a reduced frequency factor due to the required change of spin. By assuming an activation energy (E) of 109 kcal/mol, the endothermicity for *triplet* CH_2 formation, and a frequency factor (A) based on the measured rate of the reverse reaction of *singlet* CH_2 with H_2 (Braun *et al.*, 1970), it was estimated that at 1000°K reaction (1B) was 1000 times slower than the measured primary dissociation of methane (Chen *et al.*, 1975). This is obviously an artificial upper limit, based on the most favorable combination of A and E for the two CH_2 species, and the real rate must be much less; it seems safe to conclude that reaction (1A) is the only significant primary dissociation in the methane pyrolysis.

There is some direct experimental evidence supporting this conclusion. In early flow-tube experiments (Steacie, 1954), methyl radicals, but not methylene, were detected by tellurium mirrors and by mass spectrometry. More recently, in shock tubes, methyl radicals have been detected in the decomposition by mass spectrometry (Gardiner *et al.*, 1974), and hydrogen atoms have been measured by absorption of Lyman-α radiation at 121.5 nm (Roth and Just, 1975). Both CH_3 and H appeared to be primary products;

however, reaction (1B), followed by the very fast reaction,

$$CH_2 + CH_4 \longrightarrow C_2H_6^* \longrightarrow 2\ CH_3 \qquad \Delta H = -7.1 \tag{24}$$

could conceivably lead to an apparent primary formation of CH_3 (but not of H). Analysis of the isotopic composition of the hydrogen produced in the initial stages of the decomposition of a CH_4–CD_4 mixture was reported in a shock-tube study (Burcat and Lifshitz, 1970) and also in the pyrolysis in a static system (Chen *et al.*, 1975). In both cases the ratio $(HD)^2/(H_2)(D_2)$ was close to 4, as expected from an atomic mechanism, whereas reaction (1B) would lead to a value of 0. Studies in which the isotopic composition of the ethane was analyzed (Yano, 1973; Yampolskii *et al.*, 1974) led to the conclusion that CH_3 and CD_3 were the only precursors to ethane, that is, reaction (24) with or without stabilization of ethane, was not occurring; this argument is weakened, however, by the efficient methyl-radical exchange reactions known to occur in this system (Burcat and Lifshitz, 1970). In summary, all experimental results can be explained adequately by the occurrence of reaction (1A) only, and no positive evidence whatever exists for reaction (1B).

Most of the measurements reported for k_1, the first-order rate constant for the dissociation of methane, have been made in shock-tube studies in which it was argued that the initial rate could be measured, usually by the rate of disappearance of CH_4, with minimal complications from secondary reactions because of the high temperatures and short reaction times. This is approximately true, although Gardiner *et al.* (1974) have recently noted the possibility of a chain decomposition via reaction (19), discussed earlier. Although it can be predicted that the dissociation of a molecule as simple as methane should be pressure dependent at the temperatures achieved in shock tubes even at relatively high pressures, it was assumed in most shock-tube work that the reaction was in its high-pressure limit. Shock-tube data, with their typically large experimental scatter, are not well suited to the detection of pressure effects or to the accurate determination of activation energies, which would have shown the pressure dependence of k_1, and it was only in the work of Hartig *et al.* (1971) at 2000°K that the dependence was looked for and observed. Values of k_1 have also been estimated in flow systems, notably by Palmer and Hirt (1963), who followed the reaction by the deposition of carbon, and argued that secondary decomposition of methane was inhibited by removal of radicals at the carbon surface so that the rate of the primary dissociation could be measured. Probably the most accurate measurements of k_1 and its pressure dependence were made in a static system at temperatures from about 1000 to 1100°K and pressures from 25 to 700 torr, using methane carefully purified to remove traces of ethane (Chen *et al.*, 1975). First-order rate constants were obtained from initial rates of formation of ethane, which in the first stage of the reaction should be a direct measure of reaction (1), which is followed quantitatively by reactions (2) and (3). Values of k_1 obtained in this way at four temperatures are shown in Fig. 3 as a function of pressure and are seen to be strongly pressure dependent. The solid lines in this graph were calculated by fitting the results to the general expression for the unimolecular rate constant obtained by RRKM theory (Forst, 1973). The model for the activated complex involved reduction of the frequencies of the stretching and bending modes of the dissociating bond and was similar to that described by Placzek *et al.* (1965).

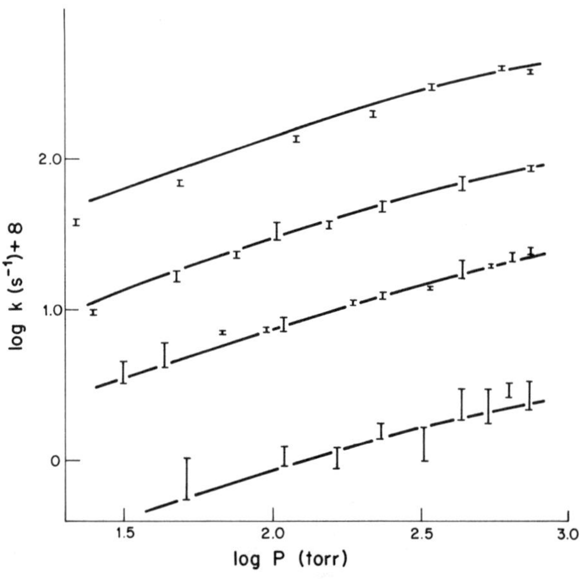

Fig. 3. Variation of k_1 with pressure in the static pyrolysis of methane. Solid lines are RRKM calculations, vertical bars are experimental measurements (Chen *et al.*, 1975, Fig. 3).

From the general expression for the rate constant, values of k_1^∞ and k_1^0, the limiting high- and low-pressure rate constants, were calculated, and from these, values of $A^\infty = 2.8 \times 10^{16}\ s^{-1}$ and $E^\infty = 107.6$ kcal/mol were obtained. These parameters are in good accord with expectations for a simple unimolecular decomposition in its pressure-dependent region. A notable feature is the relatively large difference between E^∞ and E^0, the critical energy, which was taken as 103 kcal/mol (close to the accepted endothermicity of the reaction), a value chosen to give the best fit with the measured values of k_1. In most unimolecular decompositions treated by RRKM theory, this difference is about 2 kcal/mol (Chen *et al.*, 1975). This large difference in the case of methane may be related to the relatively high activation energy of the dissociation with consequent high population of vibrational levels in the activated complex. A consequence of this effect is that in several shock-tube studies that reported activation energies of 102–104 kcal/mol, close to the value of ΔH, the reaction was assumed, incorrectly, to be in its high-pressure limit.

From the general equation for the unimolecular rate constant using the parameters obtained by a best fit to the data in Fig. 3, values of the rate constant k_1 were calculated for the wide range of pressures and temperatures studied in shock tubes and flow systems (Chen *et al.*, 1975). These values are shown as solid lines in Fig. 4, together with recent experimental values. Agreement between observed and calculated values is very good considering the combined uncertainties in the extrapolation of the calculations and in the experimental measurements at high temperatures. It is clear that all measurements of rate constants for the dissociation of methane above 1200°K have been made in the pressure-dependent region.

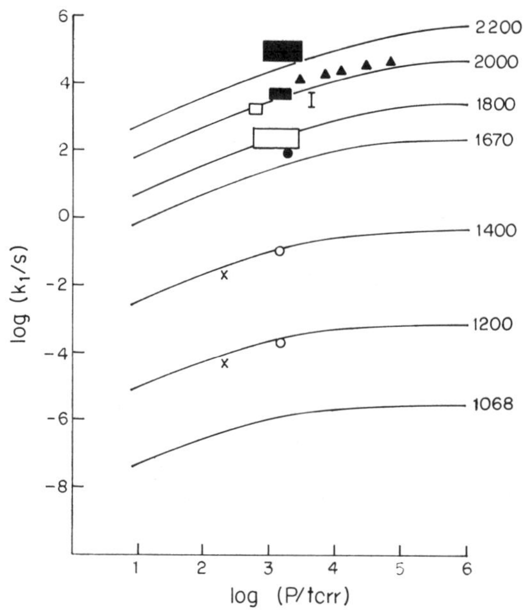

Fig. 4. Variation of k_1 with pressure in the pyrolysis of methane in shock tubes. ○, Skinner and Ruehrwein (1959); ●, Kevorkian *et al.* (1960); □, Kozlov and Knorre (1960); ■, Glick (1959); I, Napier and Subrahmanyam (1972); ▲, Hartig *et al.* (1971) and in a flow system; ×, Palmer and Hirt (1963). Solid curves are RRKM calculations based on pyrolysis in a static system at 995–1103°K (Chen *et al.*, 1975, Fig. 4).

B. Secondary Reactions in Stage 1

Although it has been shown that the pressure dependence exhibited by the methane decomposition is quantitatively explicable as a classical pressure dependence of a unimolecular reaction, an alternative explanation in terms of a complex radical-chain mechanism should be considered. The only possible chain-propagation step in the earliest stages of the methane decomposition is the inversion substitution reaction (21A), which if combined with reactions (1), (2), and (3) would lead, in a long-chain approximation with $k_{(3)} \gg k_{(21A)}$, to a reaction rate of $k_{(21A)}(k_{(1)}/k_{(3)})^{1/2}[CH_4]^{3/2}$ and an activation energy of $E_{(21A)} + \frac{1}{2}E_{(1)}$. This mechanism was first considered by Kassel (1935) as a possible way to reconcile the observation of CH_3 in fast-flow mirror experiments by Rice and Dooley (1934), with his own observation of an overall activation energy much less than ΔH for reaction (1A). Kassel rejected the mechanism, however, arguing that it would give neither first-order kinetics nor the observed activation energy, and concluded instead that reaction (1B) was occurring.

It is now clear that the low activation energies observed by Kassel and others were measured at conversions well into the autocatalytic region where the mechanism becomes complex, and cannot be related simply either to the primary dissociation or to the simple chain reaction suggested. There is in fact no direct evidence at the present time for the occurrence of reaction (21A) in the thermal decomposition of methane. As discussed elsewhere (Chen *et al.*, 1975), the dependence of the first-order rate constant on methane pressure is not that expected if the radical chain propagated by reaction (21A) were of any importance. Displacement reactions such as (21A) have never been observed in saturated hydrocarbons in the gas phase, and a rather high activation energy is expected on theoretical grounds, although this in itself is no impediment because $E_{(21A)}$ could be as high as 50 kcal/mol and still lead to an overall activation energy for the chain decomposition about equal to that observed. It was noted earlier that Tabayashi and Bauer (1979) have argued that reactions (21A) and (21B) are important in the shock pyrolysis of methane at high temperatures, and have estimated activation energies of 40 and 23 kcal/mol and frequency factors of 8×10^{10} and 1×10^{10} M^{-1} s^{-1}, respectively, for the two reactions. Based on these parameters, the rate of reaction (21A) in the static pyrolysis at 1038°K can be calculated as outlined previously (Chen *et al.*, 1975), and leads to the conclusion that at 440 torr, (21A) would have occurred at a rate about 20% that of reaction (1A). This would probably not have been easily detected in the kinetic studies of the pyrolysis (Chen *et al.*, 1975), and reaction (21A), therefore, cannot be ruled out from these experiments. Similar calculations of the rate of reaction (21B), however, show that a long-chain process comprising reactions (21B), (5), and (2) should have predominated in the static pyrolysis between 1000 and 1200°K, giving ethylene and H_2 as primary products, 3/2 order kinetics, and an overall activation energy of about 76 kcal/mol. This is totally incompatible with the experimental observations, and it is clear that reaction (21B) is much slower than suggested by Tabayashi and Bauer (1979). On the other hand, the occurrence of reaction (21A) with Arrhenius parameters close to those suggested remains a distinct possibility. It could be important both at very high temperatures, when the CH_3 radical concentration builds up due to the reversal of reaction (3), and at very low temperatures, at which it could be favored if the overall activation energy for the chain process (given by $\frac{1}{2}E_{(1A)} + E_{(21A)}$) is appreciably less than $E_{(1A)}$.

If reactions (21A) and (21B) are excluded, the only secondary reactions occurring in the first stage of the methane decomposition are reactions (2) and (3), which have both been studied independently in a variety of systems over a wide range of temperatures. Values of $k_{(2)}$ have been summarized by Roth and Just (1975), who themselves evaluated $k_{(2)}$ directly in the methane pyrolysis in a shock tube between 1700 and 2000°K by measuring the steady-state concentration of hydrogen atoms and its approach to the steady state. In common with many simple bimolecular abstraction reactions (Gardiner, 1977), reaction (2) shows marked Arrhenius curvature, with E varying from 11.9 kcal/mol at temperatures below 500°K to 15.1 kcal/mol at 1000 − 1700°K (Roth and Just, 1975).

Values of $k_{(3R)}$ have been measured directly in the pyrolysis of ethane over a range of conditions (Trenwith, 1979) and can also be calculated from values of $k_{(3)}$ determined in a number of methyl-radical combination studies (Pacey and Wimalasena, 1980). At the temperature of the methane decomposition, reaction (3R)—and of course (3)—is pressure dependent; at higher temperatures, as discussed earlier, reactions (19) and (20), yielding ethylene, effectively become alternative channels for the combination of methyl radicals.

C. Secondary Reactions of Ethane

As the concentration of ethane, formed in reaction (3), builds up in the system, it begins to be consumed in reactions (4) and (3R) in stage 2 of the decomposition. Around 1000°K, reaction (4) is the main loss process, and under some conditions ethane reaches a steady-state concentration before it turns upward again in the autocatalytic stage of the reaction (Fig. 1). Values of $k_{(4)}$ were estimated at several temperatures in the static pyrolysis system from this steady-state concentration (Chen *et al.*, 1976b), making a small correction for reaction (3R). These values confirm the curvature in the Arrhenius plot for this rate constant reported earlier by Clark *et al.*,

(1971), and by Pacey and Purnell (1972). Rate constants for a number of similar abstraction reactions have now been measured over a wide range of temperatures, and Arrhenius curvature seems to be the rule rather than the exception. This behavior has been discussed by Gardiner (1977) and by Clark and Dove (1973), who have shown that it can be explained in terms of the effect of temperature on the energy distribution in the reactants and the activated complex. Pacey (1979), on the other hand, has emphasized the importance of tunneling on the rate.

With increasing temperature, direct dissociation of ethane by reaction (3R) becomes dominant, and as noted earlier, effectively reverses reaction (3). The steady-state concentration of ethane becomes progressively smaller until it is a negligible product and its secondary reactions are unimportant. Reactions (3) and (3R) are strongly pressure dependent at all temperatures of the methane decomposition.

VI. SURFACE EFFECTS AND AUTOCATALYSIS

The reactions discussed to this point have been homogeneous gas-phase processes. As in most thermal reactions of hydrocarbons, the possibility that heterogeneous processes contribute to the decomposition must be considered, and some studies have examined this aspect (Germain and Vaniscotte, 1957; Iosif and Schneider, 1968). This question is important first in the measurement of k_1, discussed earlier, and the identification of this rate constant with the homogeneous dissociation reaction (1). To test the influence of the surface of the vessel on this reaction, experiments were performed in a vessel packed with quartz tubes making the surface-to-volume (S/V) ratio about 10 times that of the unpacked vessel (Chen *et al.*, 1975). The initial rate of formation of ethane was identical in the two vessels, and it was concluded that this rate was therefore a measure of the homogeneous dissociation of methane. Ethane and other products were also unchanged up to the ethane steady state, confirming the homogeneous mechanism postulated for these early stages. In the acceleration region a small surface enhancement was noted in all products, suggesting a possible new surface initiation process, but the mechanism in this autocatalytic stage is too complex to permit any firm conclusions.

The question of surface effects in the autocatalytic region is further complicated by the formation of carbon at relatively early stages in the decomposition. Little attention has been given to the effect of the accumulation of carbon on the rate of the overall reaction. Solid carbon introduces a greatly enhanced and active surface for heterogeneous processes and is undoubtedly a factor in some of the autocatalytic effects that have been reported in the methane reaction. Because the rate of formation of carbon is a function of many variables, such as pressure, temperature, reaction time, and condition of the vessel surface, and because its yield is not generally amenable to measurement as it is formed, its influence has been difficult to determine precisely. In most of the studies of the decomposition of methane, the presence of carbon has been shown to have an accelerating, although sometimes erratic, effect. One study, however, reported a decrease in rate following deposition of carbon on the reaction vessel (Odochian and Schneider, 1971, 1972). Several studies have attempted to deal quantitatively with the rate of decomposition on a carbon surface. Makarov and Pechik (1974) studied the reaction in vessels with different S/V ratios. They concluded that radicals are both formed and destroyed on a carbon surface and the result was a complex function of the pressure and dimensions of the reactor. In a rather different approach, Kuppers and Lydtin (1975) studied the decomposition of "cold" methane at low pressures on a hot carbon surface. The purpose of the work was an investigation of the efficiency of energy transfer rather than the chemical transformation at the surface.

It is thus expected that the carbon formed during the course of the decomposition of methane can affect the reaction; it is not clear at what stage of decomposition carbon becomes an important influence. One approach to this problem was the development of a method for the direct measurement of carbon deposited on the vessel surface during the course of the reaction in a static system around 1000°K, using a light-absorption technique (Chen and Back, 1979). The sensitivity of this method allowed the detection of carbon on the surface in amounts equivalent to less than a monolayer of atoms (the carbon in fact is probably not deposited uniformly). The earliest deposition of carbon observed in this way was well after the onset of autocatalysis (Fig. 5). It thus appeared that carbon did not cause the autocatalysis, but the possibility remained that a deposit of carbon in amounts below the limit of detection could occur early enough to do so. To test this possibility, a series of experiments was conducted in which, at regular intervals, the reagent and volatile products were removed and analyzed, a new charge of methane admitted, and the reaction continued (Chen *et al.*, 1976b). If carbon or any other nonvolatile deposit were formed on the surface during one pyrolysis interval, it should accumulate over the series of pyrolyses and the yields of products would be expected to increase steadily if the deposit were catalytic. Experiments were done under conditions identical to those illustrated in Fig. 1 for intervals of 6, 12, 24, and 36 min, and in all of these experiments the yields of the products were reproducible within the error limits of the analysis and no change with successive experiments was observed. Even for the 36-min interval, which

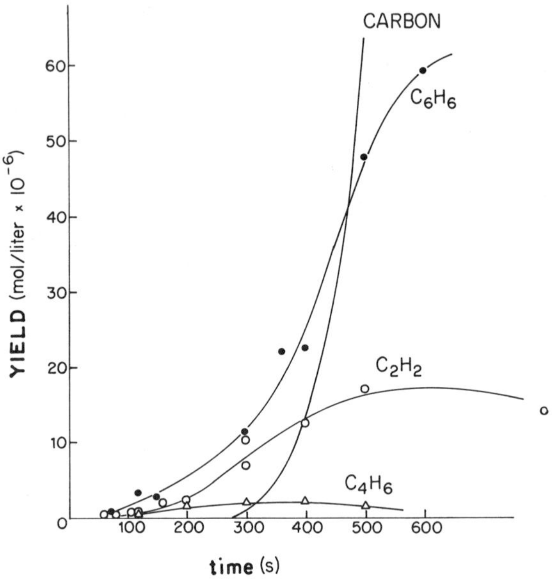

Fig. 5. Yields of later products from the pyrolysis of methane in a static system at 1118°K and 307 torr (Chen and Back, 1979).

was well into the autocatalytic region, there was no accumulation of carbon on the walls of the vessel in sufficient quantity to affect the rate of formation of the products.

These experiments show conclusively that under the conditions employed (static system, 1038°K, 400 torr CH_4), carbon on the surface of the vessel could not account for the autocatalysis. A final possibility remained that a "smoke" of very small carbon particles, which would have an enormous surface area, and which would have been pumped away like a volatile product in the interval experiments, might have caused the autocatalysis. Examination of a helium–neon laser beam passing through the vessel revealed no visible scattering by such particles; very small numbers of particles can be seen by this method, although the wavelength of the light puts a lower limit on the size of particle that can be detected.

The conclusion from all these experiments that carbon is probably not the primary cause of the autocatalysis was also reached from an entirely different type of experiment. Lahaye and Prado (1976) measured the size distribution of particles of carbon deposited from pyrolyzing hydrocarbons in a flow system and proposed a general theory of carbon deposition (Prado and Lahaye, 1978). From results on the number of particles deposited from various mixtures of methane and benzene, they concluded that the time required to form carbon could not be identified with the "induction period" (i.e., the time to the onset of autocatalysis) in the methane decomposition.

Because carbon does not seem to be responsible for the observed autocatalysis in the methane decomposition, and because surface does not seem to have a large effect, homogeneous reactions involving volatile reaction products must be considered. The fact that all the products are enhanced to more or less the same degree suggests that new radical-initiation reactions are simply feeding additional radicals into the system with little other change in mechanism. Various homogeneous processes can be suggested as new radical sources. The most thermally unstable product with respect to simple dissociation is propylene, but its rate of dissociation is probably too slow to be important. Reactions of unsaturated products with methane, or bimolecular reactions of unsaturates, are probably the most promising candidates. The bimolecular reactions combined with the sharply rising concentration of unsaturates could account for the observed rather sudden onset of autocatalysis (Chen *et al.*, 1976b). Estimates of the rates of these reactions suggest that they may be too slow, but there are few direct measurements of their rates and none at all at the temperatures of the methane decompositions; because these are essentially abstraction reactions, some Arrhenius curvature is not an unlikely possibility. Bimolecular reactions of unsaturates are at the present time the most probable explanation of homogeneous autocatalysis in the methane decomposition, although the mechanism remains in considerable doubt.

VII. FORMATION OF CARBON

Although the more general problem of coke formation in pyrolysis of hydrocarbons will be discussed in another chapter, some of the fundamental aspects of the formation of carbon from methane will be considered here. Several articles have discussed the mechanisms of conversion of organic compounds (Prado and Lahaye, 1978; Palmer and Cullis, 1965; Fitzer *et al.*, 1971) to carbon. The processes are complex, involving simultaneous decomposition and polymerization as well as phase changes from gas to liquid to solid. Industrially the problem is important from two opposite points of view. On the one hand, carbon black is an important commercial product and conditions for its most efficient production are required. On the other hand, it is an undesirable side product in hydrocarbon cracking processes, often causing poisioning of catalysts, and a minimum production is essential.

It is generally agreed that carbon is formed from high molecular weight compounds following polymerization, rather than by fragmentation and eventual production of C_2 vapor. In the thermal conversion of methane below 1200°K, polyaromatic hydrocarbons appear to be the most important intermediates (Lieberman and Noles, 1973; Oro and Han, 1966).

The mechanism of the formation of polyaromatic hydrocarbons probably involves both radical and molecular processes, and the relative importance of these is undoubtedly a function of temperature and pressure for any reactant. The kinetics of the formation and disappearance of the main intermediate products in the conversion of methane to carbon were measured at temperatures in the neighborhood of 1100°K by Chen and Back (1979) (Fig. 5). Although the low molecular weight products were undoubtedly formed by radical processes, it was shown that molecular addition processes, probably Diels–Alder type, could account for the rates of formation of large polyaromatic molecules. In this regard, 1,3-butadiene is a significant intermediate, appearing at relatively high conversions, somewhat after acetylene; it is also a major intermediate at high temperatures and conversions in flow systems. The mechanism of formation of butadiene is in some doubt, as there is no obvious sequence of methyl-radical reactions leading to it; dehydrogenation of 1-butene, for example, is not suitable because the latter is only a very minor product. Formation by reactions of vinyl radicals, as in the ethylene pyrolysis, is a more promising source:

$$C_2H_3 + C_2H_4 \longrightarrow C_4H_7 \longrightarrow C_4H_6 + H \qquad \Delta H = +11.6 \tag{31}$$

or

$$C_2H_3 + C_2H_2 \longrightarrow C_4H_5 \xrightarrow{(+CH_4)} C_4H_6 + CH_3 \qquad \Delta H = -48.3 \tag{32}$$

As ethylene and acetylene build up in the system, these reactions might begin to compete with loss of vinyl by reaction (7); they would also be enhanced by increased production of vinyl in reaction (6) and perhaps in reactions (26), (27), (29), and (30). Whatever its mechanism of formation, butadiene is a very probable intermediate in the formation of aromatic hydrocarbons, both through Diels–Alder reactions (Sakai *et al.*, 1970) and free-radical additions and polymerization. Further studies of the formation and disappearance of the various intermediates and products are needed before the details of these processes in methane decomposition can be established.

REFERENCES

Benson, S. W. (1968). "Thermochemical Kinetics." Wiley, New York.

Benson, S. W., and O'Neal, H. E. (1970). "Kinetic Data on Gas Phase Unimolecular Reactions," NSRDS–NBS No. 21. Natl. Bur. Stand., Washington, D.C.

Braun, W., Bass, A. M., and Pilling, M. (1970). *J. Chem. Phys.* **52**, 5131.

Burcat, A., and Lifshitz, A. (1970). *J. Chem. Phys.* **52**, 3613.

Burollet, P. F. (1980). *Rev. Inst. Fr. Pet.* **35**, 191.
Chen, C.-J., and Back, M. H. (1979). *Carbon* **17**, 175.
Chen, C.-J., Back, M. H., and Back, R. A. (1975). *Can. J. Chem.* **53**, 3580.
Chen, C.-J., Back, M. H., and Back, R. A. (1976a). *In* "Industrial and Laboratory Pyrolyses" (L. F. Albright and B. L. Crynes, eds.), ACS Symposium Series, **32**, p. 1. Am. Chem. Soc., Washington, D.C.
Chen, C.-J., Back, M. H., and Back, R. A. (1976b). *Can. J. Chem.* **54**, 3175.
Chupka, W. A., and Lifshitz, C. (1968). *J. Chem. Phys.* **48**, 1109.
Clark, T. C., and Dove, J. E. (1973). *Can. J. Chem.* **51**, 2147.
Clark, T. C., Izod, T. P. J., and Kistiakowski, G. B. (1971). *J. Chem. Phys.* **54**, 1295.
Deryaguin, B. V., and Fedoseev, D. V. (1973). *Carbon* **11**, 299.
Eisenberg, B., and Bliss, H. (1967). *AIChE Monogr. Ser.* **63**, 3.
Fitzer, E., Mueller, K., and Schaefer, W. (1971). *Chem. Phys. Carbon* **7**, 237.
Forst, W. (1973). "Theory of Unimolecular Reactions." Academic Press, New York.
Gardiner, W. C. (1977). *Acc. Chem. Res.* **10**, 326.
Gardiner, W. C., Owen, J. H., Clark, T. C., Dove, J. E., Bauer, S. H., Miller, J. A., and McLean, W. J. (1974). *Symp. [Int.] Combust [Proc.]* **15**, 857.
Germain, J. E., and Vaniscotte, C. (1957). *Bull. Soc. Chim. Fr.* p. 692.
Glick, H. S. (1959). *Symp. (Int.) Combust. [Proc.]* **7**, 98.
Gordon, A. S. (1948). *J. Am. Chem. Soc.* **70**, 395.
Gordon, M. D., Rumyantsev, A. N., and Lavrovskii, K. P. (1972). *Neftekhimiya* **12**, 37.
Hartig, R., Troe, J., and Wagner, H. G. (1971). *Symp. (Int.) Combust. [Proc.]* **13**, 147.
Holman, A., Rokstad, O. A., and Sobbakken, A. (1976). *Ind. Eng. Chem., Process Des. Dev.* **15**, 439.
Iosif, I., and Schneider, I. A. (1968). *Z. Phys. Chem. (Leipzig)* **237**, 81.
"JANAF Thermochemical Tables" (1970). 2nd ed., NSRDS–NBS No. 37. Natl. Bur. Stand., Washington, D.C.
Kaminski, A., and Sobkowski, J. (1971). *Rocz. Chem.* **45**, 649.
Kaminski, A. M., and Sobkowski, J. (1975). *Rocz. Chem.* **49**, 1367.
Kaminski, A. M., and Sobkowski, J. (1977). *Rocz. Chem.* **51**, 779, 1183.
Kassel, L. S. (1932). *J. Am. Chem. Soc.* **54**, 3949.
Kassel, L. S. (1935). *J. Am. Chem. Soc.* **57**, 833.
Kevorkian, V., Heath, C. E., and Boudart, M. (1960). *J. Phys. Chem.* **64**, 964.
Khan, M. S., and Crynes, B. L. (1970). *Ind. Eng. Chem.* **62**, 54.
Kozlov, G. I., and Knorre, V. G. (1960). *Combust. Flame* **6**, 253.
Kramer, L., and Happel, J. (1955). *In* "The Chemistry of Petroleum Hydrocarbons" (B. T. Brooks, S. S. Kurtz, C. E. Bond, and L. Schmerling, eds.), Vol. 2, p. 71. Reinhold, New York.
Kuppers, D., and Lydtin, H. (1975). *J. Cryst. Growth* **31**, 98.
Lahaye, J., and Prado, G. (1976). *C. R. Hebd. Seances Acad. Sci., Ser. C* **283**, 425.
Laidler, K. J., and Loucks, L. F. (1971). *In* "Comprehensive Chemical Kinetics" (C. H. Bamford and C. F. H. Tipper, eds.), Vol. 5, p. 1. Elsevier, Amsterdam.
Lieberman, M. L., and Noles, G. T. (1973). *Chem. Vap. Deposition, Int. Conf., 4th, Boston* p. 19.
Makarov, K. I., and Pechik, V. K. (1974). *Carbon* **12**, 391.
Napier, D. H., and Subrahmanyam, N. (1972). *J. Appl. Chem. Biotechnol.* **22**, 303.
Odochian, L., and Schneider, I. A. (1970). *Rev. Roum. Chim.* **15**, 1647.
Odochian, L., and Schneider, I. A. (1971). *Rev. Roum. Chim.* **16**, 1467.
Odochian, L., and Schneider, I. A. (1972). *Rev. Roum. Chim.* **17**, 441.
Oro, J., and Han, J. (1966). *Science* **153**, 1393.
Pacey, P. D. (1979). *J. Chem. Phys.* **71**, 2966.
Pacey, P. D., and Purnell, J. H. (1972). *J. C. S. Faraday Trans. I* **68**, 1462.
Pacey, P. D., and Wimalasena, J. H. (1980). *J. Phys. Chem.* **84**, 2221.
Palmer, H. B., and Cullis, C. F. (1965). *Chem. Phys. Carbon* **1**, 265.

Palmer, H. B., and Hirt, T. J. (1963). *J. Phys. Chem.* **67**, 709.
Palmer, H. B., Lahaye, J., and Hou, K. C. (1968). *J. Phys. Chem.* **72**, 348.
Peterson, N. C., Manning, R. G., and Braun, W. (1978). *J. Res. Natl. Bur. Stand.* **83**, 117.
Placzek, D. W., Rabinovitch, B. S., and Whitten, G. Z. (1965). *J. Chem. Phys.* **43**, 4071.
Prado, G., and Lahaye, J. (1978). *Chem. Phys. Carbon* **14**, 167.
Rice, F. O., and Dooley, M. D. (1934). *J. Am. Chem. Soc.* **56**, 2747.
Roth, P., and Just, T. (1975). *Ber. Bunsenges. Phys. Chem.* **79**, 682.
Sakai, T., Soma, K., Sasaki, Y., Tominaga, H., and Kunugi, T. (1970). *In* "Refining Petroleum for Chemicals," Advances in Chemistry Series, No. 97, p. 68. Am. Chem. Soc., Washington, D.C.
Schneider, I. A. (1962). *Z. Phys. Chem. (Leipzig)* **220**, 199.
Schneider, I. A., and Murgulescu, I. G. (1961). *Z. Phys. Chem. (Leipzig)* **218**, 338.
Skinner, G. B., and Ruehrwein, R. A. (1959). *J. Phys. Chem.* **63**, 1736.
Steacie, E. W. R. (1954). "Atomic and Free Radical Reactions," 2nd ed., Vol. 1. Reinhold, New York.
Stepanov, A. V., Chermalykh, G. N. and Stadnik, V. F. (1971). *Neftepererab. Neftekhim (Kiev)* No. 6, 85.
Tabayashi, K., and Bauer, S. H. (1979). *Combust. Flame* **34**, 63.
Tesner, P. A. (1978). *Gazov. Promst.* No. 10, 42.
Trenwith, A. B. (1979). *J. C. S. Faraday Trans. I* **75**, 614.
Warneck, P. (1971). *Z. Naturforsch., A* **26A**, 2047.
Yampolskii, Y. P., Rybin, V. M., and Molodykh, E. I. (1974). *Dokl. Akad. Nauk SSSR* **217**, 888.
Yano, T. (1973). *Bull. Chem. Soc. Jpn.* **46**, 1619.

2

Pyrolysis of Ethane and Propane

CHRISTOPHER F. McCONNELL and BILLY D. HEAD

I. INTRODUCTION

The growth of ethylene production since 1945 is one of the most extensive of any chemical product. Today, it is the single most basic and important building block in the petrochemical industry (*Chemical Engineering Progress*, 1966). The Texas Gulf Coast is the production center in the United States because of the availability of large quantities of ethane and propane from the nearby gas fields (Peters, 1966). The overall production of ethylene has continued to grow, with installed capacity from all feed-stocks reaching 26.5 billion pounds in 1977 (*Oil and Gas Journal*, 1977), 32.6 billion pounds as of June 30, 1979 (*Oil and Gas Journal*, 1979), and finally, 36.9 billion pounds as of June 30, 1980 (*Oil and Gas Journal*, 1980). The actual production levels were about 90% of capacity (*Oil and Gas Journal*, 1976).

Pyrolysis: Theory and Industrial Practice

ISBN 0-12-048880-9

The pyrolysis of ethane and propane is often the process of choice for production of ethylene. In the United States light paraffins have been, and continue to be, the predominant feedstocks. In Europe, the feedstock is more than 90% naphtha because of the unavailability of liquified petroleum gases (LPG) and the high demand for the heavier gas oils as heating fuels. Japan is similar to Europe in that primarily naphtha and gas oil are pyrolized there. Figure 1 shows the feedstock distribution for the United States in 1976 (*Oil and Gas Journal*, 1976). Fully 72% of the feedstock was ethane and propane, and this is also representative of years prior to 1976. It is projected that in the 1980s the pyrolysis of heavier feedstocks will increase such that the use of ethane, propane, and butane will account for less than 50% of the ethylene production in the United States, as will be discussed in more detail.

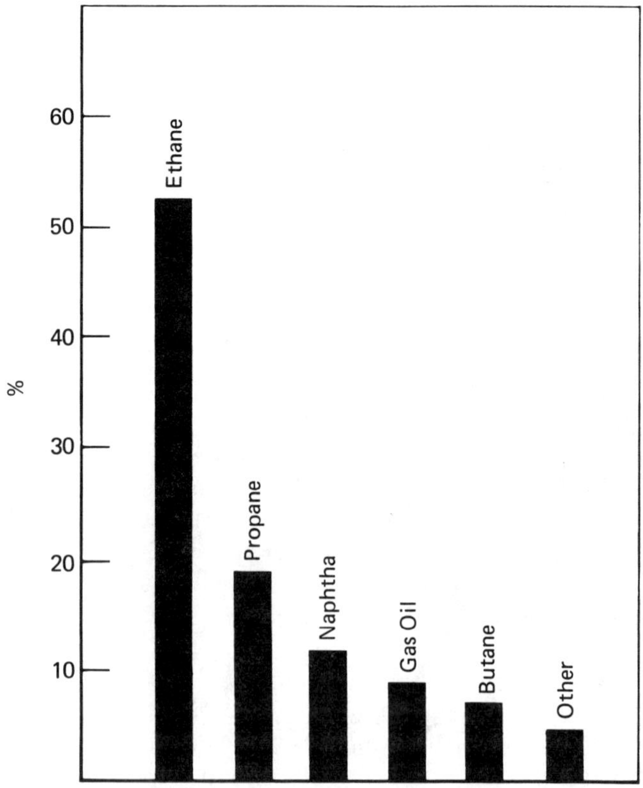

Fig. 1. Percentages of various feedstock types in use for ethylene production, United States, 1976.

II. COMMERCIAL CONSIDERATIONS

To understand why United States industry prefers to pyrolyze ethane and propane to produce ethylene, it is necessary to look at the factors that determine the feedstock for any chemical process. These include feedstock availability and price, by-product value, reaction conditions, complexity of the processing unit, and capital requirements. Because natural gas is the basis for the American energy industry, the supply of natural gas liquids has been abundant, and it has been attractive for this industry to separate and sell these liquids as petrochemical feedstocks, which command a premium value over fuel value. At the same time, the lack of a competitive market for ethane in other commercial areas has made it available to the petrochemical industry and helped keep the price attractive.

The pyrolysis reaction of choice when the primary goal is to produce ethylene is the one that minimizes the formation of methane, C_4^+ hydrocarbons, and even propylene to some extent. The reaction parameters that most affect this selectivity are temperature, hydrocarbon partial pressure, residence time, and feedstock composition. The most efficient feedstock for ethylene production is ethane. It requires the simplest processing unit with the lowest capital requirement. Marketing of the products from ethane pyrolysis is greatly simplified by the low yield of by-products.

Overall feedstock composition is one of the primary factors in determining the quantity and composition of the by-products. As the molecular weight of the feedstock increases, the amount of by-product production also increases. This can be seen from the data shown in Table I. The feedstock composition also determines the conditions under which the pyrolysis takes place. Ethane is the most refractory of the feedstocks and requires the highest coil outlet temperature. As the molecular weight of a feedstock increases, it becomes easier to crack, and the reaction severity as measured by the coil outlet temperature decreases. This is assuming that the residence time of the hydrocarbon in the coil reactor remains constant. Reaction conditions required by various feedstocks in a plant designed for a wide range of feeds are shown in Table II. By varying the residence time and/or dilution steam, it is possible to maintain a constant coil outlet temperature, and some units are operating in this manner.

Another factor in the choice of ethane and propane as the preferred feedstock for the production of ethylene is the size of the production plant. Pyrolysis of ethane and propane produces the lowest yield of by-products, which in turn minimizes the size of downstream units such as the depropanizer, debutanizer, and compressors. As heavier hydrocarbon

TABLE I

Pyrolysis Products from Various Feedstocks

Feedstock	Product (in pounds/100 pounds feedstock)				
	Ethylene	Propylene	Butadiene	BTX	Other
Ethane[a]	82.3	1.8	2.6	0.7	12.6
Propane[a]	43.7	21.2	4.1	4.8	26.2
Butane[a]	42.2	14.6	3.9	4.8	34.5
Light naphtha[a]	29.3	14.4	4.0	13.8	38.5
Full-range naphtha[a]	27.2	12.8	4.5	11.3	44.2
Gas oil[b]	25.0	12.4	4.8	11.2	46.6
Crude oil[c]	25.2	8.3	3.5	15.3	47.7

[a] From Dow Chemical Co. (1981).
[b] From *Hydrocarbon Processing* (1975).
[c] From Wing (1980).

TABLE II

Reaction Conditions for Various Feedstocks

Feedstock	Coil outlet temperature (°F)	Steam/hydrocarbon weight ratio	Residence time (s)
LPG[a]	1550	0.3	0.3
Light naphtha	1540	0.5	0.3
Heavy naphtha	1520	0.7	0.3
Gas oil	1480–1500	0.9	0.3

[a] Liquefied petroleum gases (mainly mixtures of ethane and propane).

fractions are utilized for the feedstock, the size of these units must increase dramatically. Furthermore, desulfurization units must be added for by-product purification, which adds to the overall plant complexity. Existing production facilities have very little flexibility, if any, to handle a wide range of feedstocks. The current trend is to design plants that do have this capability, but the added capital expense is very significant. In 1976 it was calculated that the capital required to construct a 500,000 MT/year ethylene plant on the American Gulf Coast was $104 million. This plant was designed to crack ethane and the values were for inside battery limits with a 1978 startup (W. Tucker and M. Abrahams, unpublished observations). The relative investment cost for heavier feedstocks was also calculated, and these values are shown in Table III.

TABLE III

Effect of Feedstock on Investment Cost

Feedstock	Relative investment cost
Ethane	1.00
Propane	1.15
n-Butane	1.20
Naphtha	1.45
Atmospheric gas oil	1.65
Vacuum gas oil	1.84

With an increasing rate of inflation, the cost of building a world-scale ethylene plant is increasing at an almost unbelievable pace. A recent study by The Dow Chemical Company has shown that a plant scheduled for startup in 1986 and slated to produce 1.2 billion pounds/year will require $225 million in capital. This plant was designed to handle a feedstock consisting of 80% ethane and 20% propane.

The question of primary interest is what will be the feedstock of the 1980s and 1990s. Early in the 1970s, forecasters began to predict the decline of ethane and propane as the major cracking feedstock, and the necessity of building production units in the United States capable of handling heavier material, such as naphtha and gas oil. Indeed, an analysis of the domestic supply of natural gas liquids made by the United States Bureau of Mines (see Fig. 2) shows a downward trend (*Chemical Week*, 1980) due to both the decline in proven natural gas reserves and the leveling off of the amount of ethane extracted from the gas. In 1974–1975, roughly 20% of the contained ethane was extracted. This will increase to 26% in 1980, and will probably level out at 30% in 1985 (Luger, 1976). Ethane recovery must be increased from this 30% level to 50% in order to supply sufficient feedstock for plants producing an additional 2 billion pounds/year of ethylene in the United States. There is some question that this will be done in view of the high value now placed on natural gas.

Based on predictions such as these, most of the new ethylene-production units constructed in the United States during the 1970s were designed to pyrolyze heavier feedstocks. However, recent developments in the petroleum industry have thrown the entire supply forecast into a state of turmoil. New discoveries of gas fields in western Canada and Mexico, along with the distinct possibility of pipeline deliveries from northern Alaska, should offset the downward trend in gas production in the continental United States. The international supply of natural gas liquids is now projected to triple because of additional recovery from the Middle East oil-producing nations. Much of the estimated 1.5 million bbl/day production will be

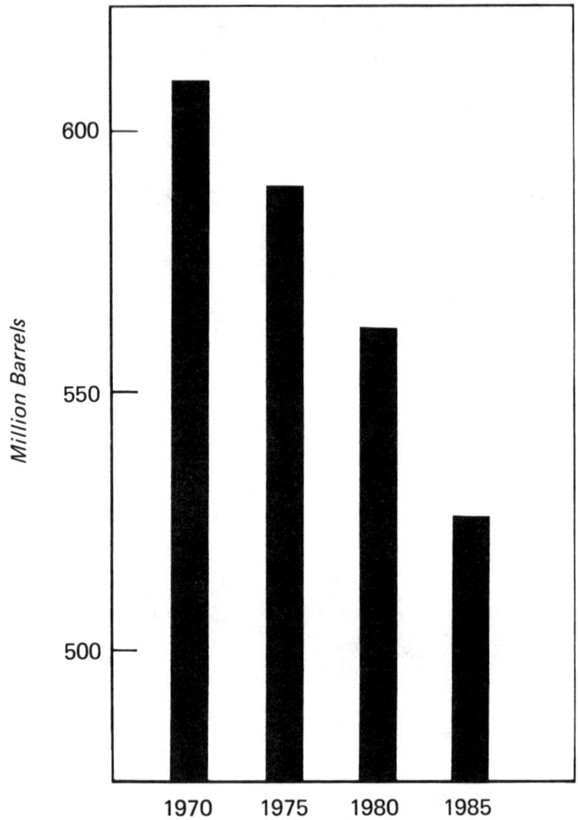

Fig. 2. Gas liquids production in the United States, in millions of barrels, for 4 representative years.

available for import to the United States. As a result, there is a good possibility that ethane and propane will be available to American chemical producers throughout the next two decades, and that the pyrolysis of these compounds will continue to be utilized throughout the ethylene industry.

In subsequent sections, the pyrolysis of ethane and propane will be reviewed from a theoretical standpoint.

III. GENERAL REACTION CHARACTERISTICS

The pyrolysis of light hydrocarbons can be considered as an approach to equilibrium. For ethane pyrolysis, a major portion of the reaction is described by a single stoichiometric equation.

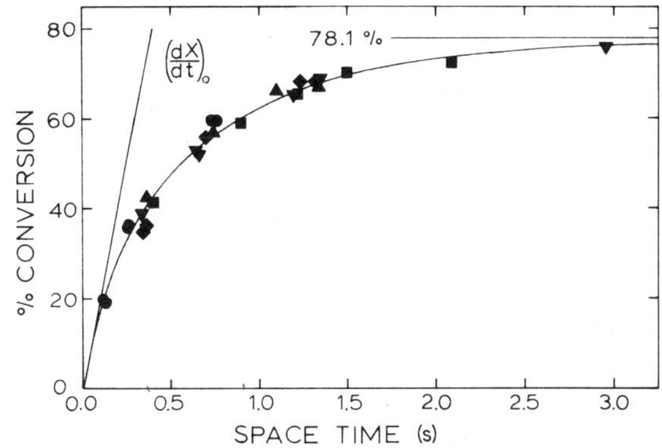

Fig. 3. Percentage ethane conversion with steam diluent (50 mol %) at 800°C and 1 atm, versus space time. Surfaces as follows: ●, Vycor glass; ■, stainless steel 410; ◆, stainless steel 304; ▲, Incoloy 800 steel; ▼, Hastelloy X. (Reprinted with permission from *Thermal Hydrocarbon Chemistry*. Copyright 1979 by the American Chemical Society.)

$$C_2H_6 \rightleftharpoons C_2H_4 + H_2 \tag{1}$$

Figure 3 shows that an approach to equilibrium does indeed occur. At 1073°K, the equilibrium constant is almost 1.0 (Kramer and Happel, 1955). With 50 mol % steam, Eq. (1) predicts conversion at equilibrium to be 78.1%.

Propane pyrolysis is similar, but in this case, two stoichiometric equations are important. The ethylene-forming reaction is strongly favored to the right.

$$C_3H_8 \rightleftharpoons C_2H_4 + CH_4 \tag{2}$$

$$C_3H_8 \rightleftharpoons C_3H_6 + H_2 \tag{3}$$

Equilibrium conversion at 1073°K and 50 mol % steam is over 90%.

Today, we know that light paraffin pyrolysis is not as simple as these equations portend. Many by-products are formed that are not explained. Rice and Herzfeld (1934) suggested that pyrolysis occurs through free-radical reactions. Martin and Niclause (1964) confirmed this intermediate theory in 1962. They showed that when a trace of oxygen is present, cracking is totally suppressed in high *S/V* ratio reactors coated with lead oxide.

Free-radical reactions are very complex. Literally dozens of reactions occur simultaneously. Analytical solutions are derived only by making free-radical steady-state assumptions. However, pyrolysis reactions are sensitive to free-radical concentration, and thus steady-state assumptions can lead to significant error. Computer analysis is also difficult. Free radicals are often 5 or more orders of magnitude less in concentration than their molecular counterparts. Numerical step sizes must be kept quite small to keep track of the radicals. Computational costs become large.

In itself, the optimization of an olefins plant is tedious. The furnaces, exchangers, compressors, and distillation equipment are all mathematically coupled. Any change requires resimulation of the cracking reactions. Consequently, industrial designers rely on proven empirical models using many stoichiometric equations similar to Eqs. (1), (2), and (3). Even though these models are admittedly not mechanistic, they are accurate and require much less computer time. Ross and Shu (1979) reviewed literature on stoichiometric modeling. An example of ethylene process optimization was given by Jinkerson and Gaddy (1979).

On the other hand, researchers who desire a priori understanding employ models based on the elementary free-radical reactions. Experimental data are used to confirm rate parameters and the basic mechanism. Because these free-radical elementary steps are mechanistic, they offer theoretical significance. The remainder of this chapter will be concerned with the elementary steps, and not stoichiometric equations.

An outstanding characteristic of free radicals is that their reactions are fast—sometimes close to gas-phase collision frequency. For this reason, a pseudoequilibrium concentration of free radicals is obtained very quickly. The level is mostly dependent on temperature. It is difficult to visualize any commercial furnace or transfer-line exchanger inducing a temperature change sharp enough to outpace free-radical pseudoequilibrium. The speed of approach has been observed in computer analysis (Lee and Yeh, 1979), as well as in experimental work (Tsuboi, 1978).

Free-radical pseudoequilibrium is dynamic, however, and it shifts with changes in the concentrations of molecular species. During the early stages of pyrolysis, pseudoequilibrium is shifting rather quickly; but in the late stages, concentrations of both molecular species and free radicals become relatively constant. Sundaram and Froment (1978) have used a large mechanistic model under conditions similar to industrial practice. Table IV shows the reactor-exit radical concentrations for ethane and propane pyrolysis. As a comparison, total molecular concentration is on the order of 10^{-2} mol/liter.

There are three general classifications of homogeneous reactions involving the various free radicals. Each group consists of forward and reverse reactions.

Scission and Coupling. Scission of a molecule introduces a pair of free radicals. High activation energy is characteristic, thus the pyrolysis requirement of high temperature. Coupling is the reverse reaction and zero activation energy is characteristic. To a large extent, these two types of

TABLE IV

Concentrations of Free Radicals at Reactor Exit[a]

Radical	Ethane pyrolysis[b] (concentration, $mol\,L^{-1}$)	Propane pyrolysis[c] (concentration, $mol\,L^{-1}$)
H	$10^{-8.3}$	$10^{-9.3}$
CH_3	$10^{-8.3}$	$10^{-7.7}$
C_2H_3	$10^{-6.6}$	$10^{-7.0}$
C_2H_5	$10^{-6.7}$	$10^{-7.7}$
C_3H_5	$10^{-7.9}$	$10^{-6.4}$
1-C_3H_7	N.S.[d]	$10^{-9.3}$
2-C_3H_7	N.S.	$10^{-8.0}$
C_4H_7	$10^{-6.6}$	$10^{-7.4}$
1-C_4H_9	$10^{-10.3}$	N.S.
2-C_4H_9	N.S.	$10^{-10.6}$

[a] Estimated from Sundaram and Froment (1978).
[b] Experimental conditions: 1123°K, 50 mol % steam.
[c] Experimental conditions: 1123°K, 40 mol % steam.
[d] N.S., not shown.

reactions dictate the total pseudoequilibrium radical *concentration* as a function of temperature.

Hydrogen Abstraction. Here, a free radical abstracts a hydrogen from a molecular species. The result is that the original free radical becomes a molecule, and the original molecule becomes a free radical. The reverse reaction is similar. Low, but not zero, activation energies are characteristic. These reactions adjust and maintain the pseudoequilibrium *distribution* of the various free radicals.

Decomposition and Addition. Unimolecular decomposition of a free radical results in an olefin and a smaller free radical, often a hydrogen radical. Activation energy is normally 30–40 kcal/mol. Addition of a small radical to an olefin results in a single larger free radical. Typical activation energy is 1–2 kcal/mol for hydrogen-radical addition, and 7–9 kcal/mol for methyl-radical addition. The forward decomposition reaction is responsible for conversion to olefin and the reverse addition reaction is responsible for inhibition.

Starting with one component feed, the above set of reactions leads to a broad range of products. Coupling reactions produce larger molecules. Free-radical analogs to these larger molecules are formed via hydrogen abstractions. The larger free radicals may decompose to form higher olefins, or couple to form even larger molecules. An array of reactions involving the various paraffins, free radicals, and olefins is necessary for a detailed description of the pyrolysis. Furthermore, heterogeneous wall reactions and gas-phase polymerizations complicate the picture. To review all of the possible reactions is nearly an impossible task. In subsequent sections, however, some of the more important and better understood reactions are discussed. Emphasis is placed on pairs of forward and reverse elementary steps.

IV. SCISSION AND COUPLING

The most intensively studied and best-understood pair of pyrolysis reactions is ethane scission and methyl-radical coupling.

$$C_2H_6 \rightleftarrows CH_3\cdot + CH_3\cdot \tag{4}$$

Researchers studied the scission reaction by extrapolating ethane pyrolysis data to zero reaction time. The coupling reaction was studied by photolysis and shock-tube experiments. Essentially no temperature dependence was observed (zero activation energy) for the coupling reaction. Thermochemists began constructing activated-complex models to unify the data, and today, much of our present understanding of unimolecular reaction theory can be attributed to the study of these two reactions.

Zaslonko and Smirnov (1979) reviewed the literature on methyl-radical coupling. Data from 16 recent studies show that the reaction rate decreases slightly with increasing temperature. Between 300 and 1750°K, the high-pressure limiting rate constant is described by

$$k_{\infty,\mathrm{rev}} = 10^{11.40}\, T^{-0.38}\ \mathrm{L\,mol^{-1}\,sec^{-1}} \tag{5}$$

where, for the bimolecular reaction, $d(CH_3)/dt = -2\, k_{\infty,\mathrm{rev}}\,(CH_3)^2$.

At low total pressure, methyl coupling is pressure dependent. The activated complex, containing excess energy, must be deactivated by third-body collision; otherwise, the excited ethane molecule may redissociate. Falloff curves for the rate constant have been constructed by Luther and Troe (1979).

The "high" value for methyl-radical heat of formation (McCulloh and Dibeler, 1976) confirmed thermochemical consistency to the forward and reverse rate parameters. Zaslonko and Smirnov (1979) found that, for ethane scission, the following equation best fits the available high-pressure limiting data.

$$k_{\infty,\mathrm{for}} = 10^{16.5} \exp(-88{,}160\ \mathrm{cal\,mol^{-1}}/RT)\ \mathrm{sec^{-1}} \tag{6}$$

Recently measured values are given by Trenwith (1979), who used computer-assisted, zero-time extrapolation of ethane pyrolysis data.

$$k_{\infty,\text{for}} = 10^{16.72\pm0.17} \exp(-88{,}850 \pm 680 \text{ cal mol}^{-1}/RT) \text{ sec}^{-1} \quad (7)$$

Trenwith's curves indicate that roughly 30% falloff occurs at 913°K and 1 atm. Falloff becomes more significant at higher temperature and lower pressure (Luther and Troe, 1979). Mechanistic models may have to incorporate corrections for falloff behavior.

Parameters for propane scission are not as well known.

$$C_3H_8 \rightleftarrows C_2H_5\cdot + CH_3\cdot \quad (8)$$

During propane pyrolysis, side reactions occur promptly so that the propane scission rate constant is difficult to obtain by zero-time extrapolation. No directly measured rate parameters were found. Sundaram and Froment (1978) studied propane pyrolysis extensively with mechanistic models. They used the following equation, which is reasonably consistent with the computer-fitted data given by Pratt and Rogers (1979a).

$$k_{\text{for}} = 10^{16.3} \exp(-84{,}500 \text{ cal mol}^{-1}/RT) \text{ sec}^{-1} \quad (9)$$

The methyl–ethyl coupling reaction poses experimental problems because nonidentical radicals are involved. Again, the expression of Sundaram and Froment may be the best.

$$k_{\text{rev}} = 10^{9.5} \text{ L mol}^{-1} \text{ sec}^{-1} \quad (10)$$

Lifshitz and Frenklach (1975) found a similar value of $10^{9.4}$ L mol^{-1} sec^{-1}. Falloff behavior probably occurs below industrial pressures because the propane-activated complex has several degrees of freedom to distribute excess energy.

Parameters for butane scission are again not well known, but those for the reverse reaction, ethyl–ethyl coupling, are well known.

$$C_4H_{10} \rightleftarrows C_2H_5\cdot + C_2H_5\cdot \quad (11)$$

Three recent measurements (Golden *et al.*, 1976; Parkes and Quinn, 1976; Walker, 1976) for the coupling rate constant confirm the "high" value.

$$k_{\text{rev}} = 10^{9.9} \text{ L mol}^{-1} \text{ sec}^{-1} \quad (12)$$

This reaction's falloff behavior also probably occurs at pressures below those of commercial interest (Golden *et al.*, 1976).

No direct measurements for the rate of butane scission were found in the literature. Pratt and Rogers (1979b) present recent computer-fitted data. Their values are reasonably consistent with the rate expression that follows, used by Powers and Corcoran (1974).

$$k_{\text{for}} = 10^{15.7} \exp(-81{,}000 \text{ cal mol}^{-1}/RT) \text{ sec}^{-1} \quad (13)$$

A minor increase in the preexponential factor would eliminate any discrepancy.

Other scission reactions are surely occurring during pyrolysis. C—C bond rupture is more likely than C—H bond rupture (Bradley, 1974), based on the high activation energy for methane dissociation (~104 kcal/mol). Large paraffin molecules tend to be more susceptible than the smaller, more refractory molecules. Single bonds in higher olefins may go through scission, and some molecules may dissociate in two or more ways.

Coupling reactions are also abundant. If 10 free-radical species are assumed present, then theoretically, 55 unique coupling reactions could be occurring. The literature indicates that rate constants are consistently high, and that no evidence has been found for significant temperature dependence (Walker, 1976). Because of their speed, these reactions strictly regulate the total radical concentration.

The scissions of ethane and propane are the important initiation steps during light paraffin pyrolysis. They are net producers of free radicals. Likewise, coupling reactions, the termination steps, are net consumers of free radicals. The other major groups of homogeneous pyrolysis reactions, hydrogen abstraction and decomposition/addition, always start and finish with a free radical. No net change occurs. A fourth group, however, also produces and consumes radicals. These are disproportionation reactions.

$$C_2H_4 + C_2H_4 \rightleftharpoons C_2H_3\cdot + C_2H_5\cdot \qquad (14)$$

$$C_2H_6 + C_2H_4 \rightleftharpoons C_2H_5\cdot + C_2H_5\cdot \qquad (15)$$

Such elementary steps are necessary to explain ethylene pyrolysis data and low-temperature ethyl–ethyl data. In light paraffin cracking, disproportionations are not as common as scission and coupling reactions, and will not be further discussed.

V. HYDROGEN ABSTRACTIONS

Abstraction reactions are unique in that over a wide temperature range, their Arrhenius plots show curvature. The reactions are relatively fast, as indicated by high frequency factors (measured at high temperature) and moderate activation energies. Arrhenius curvature corresponds to decreasing activation energy with decreasing temperature. Pressure dependence, associated with a third-body requirement, is not observed.

The most basic pair of hydrogen abstractions involves hydrogen, methane, and their free-radical analogs.

$$CH_4 + H\cdot \rightleftharpoons H_2 + CH_3\cdot \qquad (16)$$

Clark and Dove (1973a) fitted the forward reaction to a semiempirical equation for the curved Arrhenius plot.

$$k_{for} = 10^{1.35}\, T^{3.0} \exp(-8750 \text{ cal mol}^{-1}/RT) \text{ L mol}^{-1} \text{ sec}^{-1} \quad (17)$$

Above 950°K, the linear equation of Roth and Just (1979) adequately described the data.

$$k_{for} = 10^{11.86} \exp(-15{,}060 \text{ cal mol}^{-1}/RT) \text{ L mol}^{-1} \text{ sec}^{-1} \quad (18)$$

Clark and Dove (1973b) also examined the reverse reaction between molecular hydrogen and the methyl radical. The Arrhenius plot was curved, but above 1200°K, a straight line fit the data.

$$k_{rev} = 10^{10.19} \exp(-15{,}500 \text{ cal mol}^{-1}/RT) \text{ L mol}^{-1} \text{ sec}^{-1} \quad (19)$$

This reverse abstraction is more than an order of magnitude slower than the forward abstraction. Both reactions are relatively slow at industrial pyrolysis temperatures.

In the production of ethylene, an important chain-carrying reaction is hydrogen abstraction from ethane by a hydrogen free radical.

$$C_2H_6 + H\cdot \rightleftharpoons H_2 + C_2H_5\cdot \quad (20)$$

Slight Arrhenius plot curvature was noted by Clark and Dove (1973a).

$$k_{for} = 10^{-0.27}\, T^{3.5} \exp(-5200 \text{ cal mol}^{-1}/RT) \text{ L mol}^{-1} \text{ sec}^{-1} \quad (21)$$

Camilleri *et al.* (1974) found that a linear equation fit the data from 350 to 1250°K.

$$k_{for} = 10^{11.12} \exp(-9360 \text{ cal mol}^{-1}/RT) \text{ L mol}^{-1} \text{ sec}^{-1} \quad (22)$$

The ethyl radical, produced in this reaction, is the direct precursor to the valuable ethylene product.

There are limited data for the reverse reaction between molecular hydrogen and the ethyl radical. Kaminski and Sobkowski (1979) quoted a linear equation.

$$k_{rev} = 10^{8.47} \exp(-10{,}800 \text{ cal mol}^{-1}/RT) \text{ L mol}^{-1} \text{ sec}^{-1} \quad (23)$$

This reverse abstraction is more than 2 orders of magnitude slower than the forward abstraction. In general, small radicals abstract hydrogen faster than large radicals, and large molecules are more susceptible to abstraction than small molecules.

A similarly important reaction is hydrogen abstraction from ethane by a methyl radical.

$$C_2H_6 + CH_3\cdot \rightleftharpoons CH_4 + C_2H_5\cdot \quad (24)$$

Here there is little doubt about deviation from Arrhenius behavior. A semi-empirical equation was found by Clark and Dove (1973a).

$$k_{for} = 10^{-3.26}\,T^{4.0}\exp(-8280\ \mathrm{cal\,mol^{-1}}/RT)\ \mathrm{L\,mol^{-1}\,sec^{-1}} \tag{25}$$

Pacey and Purnell (1972) used a different form for their nonlinear expression.

$$k_{for} = 10^{11.85}\exp(-23{,}900\ \mathrm{cal\,mol^{-1}}/RT) + 10^{7.4}\exp(-9100\ \mathrm{cal\,mol^{-1}}/RT)\ \mathrm{L\,mol^{-1}\,sec^{-1}} \tag{26}$$

Both forms have appeal from a theoretical standpoint. Transition-state theory predicts a T^x temperature dependence as in Clark and Dove's equation. The possibility of a change in mechanism (e.g., the appearance of quantum tunneling or a different transition-state configuration) is in concord with Pacey and Purnell's asymptotic formula. Both equations are accurate to a large extent.

For the reverse reaction, abstraction from methane by the ethyl radical, Kaminski and Sobkowski (1979) quoted a linear equation.

$$k_{rev} = 10^{8.36}\exp(-11{,}400\ \mathrm{cal\,mol^{-1}}/RT)\ \mathrm{L\,mol^{-1}\,sec^{-1}} \tag{27}$$

Because both the forward and reverse reactions share the same activated complex, both should exhibit similar Arrhenius deviation. The low pre-exponential factors quoted by Kaminski and Sobkowski indicate that the values were measured at low temperature. They are probably applicable below 500°K.

The abstraction reactions that produce propyl radicals are not as well characterized as those that produce ethyl radicals. No rate parameters for reverse reactions could be found in recent literature. From 1210 to 1680°K, Bradley (1979) used a linear equation for abstraction of a hydrogen radical from propane.

$$C_3H_8 + H\cdot \rightleftharpoons H_2 + C_3H_7\cdot \tag{28}$$

$$k_{for} = 10^{10.60}\exp(-8190\ \mathrm{cal\,mol^{-1}}/RT)\ \mathrm{L\,mol^{-1}\,sec^{-1}} \tag{29}$$

For hydrogen abstraction from propane by a methyl radical, Bradley was able to computer-optimize rate parameters.

$$C_3H_8 + CH_3\cdot \rightleftharpoons CH_4 + C_3H_7\cdot \tag{30}$$

$$k_{for} = 10^{9.32}\exp(-11{,}700\ \mathrm{cal\,mol^{-1}}/RT)\ \mathrm{L\,mol^{-1}\,sec^{-1}} \tag{31}$$

This expression is reasonably consistent with several other studies (Pratt and Rogers, 1979a; Lifshitz and Frenklach, 1975; Camilleri *et al.*, 1975) between 650 and 1550°K.

In the preceding equations, no differentiation is made between *n*-propyl and *i*-propyl radicals. Hydrogen atoms of the methylene group are about 10 times more reactive than those of the methyl groups (Gray *et al.*, 1971); therefore, roughly 70% of the product propyl radicals are *i*-propyl radicals.

Acetylene is a by-product of high-severity ethane and propane cracking. The vinyl radical, acetylene's precursor, is formed via hydrogen abstraction from ethylene.

$$C_2H_4 + H\cdot \rightleftharpoons H_2 + C_2H_3\cdot \qquad (32)$$

Olsen *et al.* (1979) used a linear-rate model for this reaction during shock-tube (1300°–2500°K) pyrolysis of ethane.

$$k_{for} = 10^{11.84} \exp(-14{,}500 \text{ cal mol}^{-1}/RT) \text{ L mol}^{-1} \text{ sec}^{-1} \qquad (33)$$

No information was available on the reverse reaction.

There are many possible hydrogen abstractions that could occur during light hydrocarbon pyrolysis. If 10 free-radical species and 10 molecular species are assumed present, then 45 reaction pairs are conceivable. Throughout the literature, the forward abstractions, which produce larger free radicals, are found dominant over the reverse reactions. The key role of abstraction reactions is to quickly convert smaller free radicals, the products of scission and radical decomposition, into larger free radicals, the precursors to olefins.

VI. DECOMPOSITION AND ADDITION

A primary elementary step for producing ethylene is the unimolecular decomposition of the ethyl radical.

$$C_2H_5\cdot \rightleftharpoons C_2H_4 + H\cdot \qquad (34)$$

It is a pressure-dependent reaction. The high-pressure limiting rate constant was recently determined by Pratt and Rogers (1979a).

$$k_{\infty,for} = 10^{13.5} \exp(-41{,}800 \text{ cal mol}^{-1}/RT) \text{ sec}^{-1} \qquad (35)$$

The value of Benson and O'Neal (1970) has found wide application.

$$k_{\infty,for} = 10^{13.5} \exp(-40{,}700 \text{ cal mol}^{-1}/RT) \text{ sec}^{-1} \qquad (36)$$

Pressure corrections are normally employed with both expressions.

The reverse reaction, hydrogen-radical addition to ethylene, has been studied extensively at room temperature. Michael *et al.* (1973) reviewed the literature. They extrapolated to a high-pressure limiting rate constant at 300°K.

$$k_{\infty,rev} = 10^{9.0} \text{ L mol}^{-1} \text{ sec}^{-1} \qquad (37)$$

However, there is some temperature dependence, and Penzhorn and Darwent (1971) found $E_a = 1500$ cal/mol. This value can be combined with the preceding room-temperature result.

$$k_{\infty,\text{rev}} = 10^{10.1} \exp(-1500 \text{ cal mol}^{-1}/RT) \text{ L mol}^{-1} \text{ sec}^{-1} \quad (38)$$

Although high-temperature data are scarce, pressure corrections may also be necessary for this reverse reaction at industrial pyrolysis conditions.

The *n*-propyl radical decomposes mainly into ethylene and a methyl radical, whereas the *i*-propyl radical decomposes mainly into propylene and a hydrogen radical.

$$n\text{-}C_3H_7\cdot \rightleftharpoons C_2H_4 + CH_3\cdot \quad (39)$$

$$i\text{-}C_3H_7\cdot \rightleftharpoons C_3H_6 + H\cdot \quad (40)$$

The most recent forward-rate parameters were given by Camilleri *et al.* (1975).

$$k_{\text{for},n} = 10^{12.1} \exp(-32{,}500 \text{ cal mol}^{-1}/RT) \text{ sec}^{-1} \quad (41)$$

$$k_{\text{for},i} = 10^{13.4} \exp(-40{,}800 \text{ cal mol}^{-1}/RT) \text{ sec}^{-1} \quad (42)$$

No pressure dependence was observed, reflecting the stabilizing influence of higher degrees of freedom in the activated complex. Although the preexponential factor for *n*-propyl decomposition is lower than that found in previous studies, it was supported recently by the findings of Bradley (1979).

Using known thermodynamic data, Camilleri *et al.* (1975) also calculated rate constants for the corresponding reverse addition reactions.

$$k_{\text{rev},n} = 10^{7.5} \exp(-7900 \text{ cal mol}^{-1}/RT) \text{ L mol}^{-1} \text{ sec}^{-1} \quad (43)$$

$$k_{\text{rev},i} = 10^{9.8} \exp(-1700 \text{ cal mol}^{-1}/RT) \text{ L mol}^{-1} \text{ sec}^{-1} \quad (44)$$

Inhibition, so evident in propane pyrolysis, is caused by these addition reactions.

Acetylene is produced in high-severity cracking via decomposition of the vinyl radical.

$$C_2H_3\cdot \rightleftharpoons C_2H_2 + H\cdot \quad (45)$$

This reaction is also pressure dependent. No recent measurements were found in the literature for the forward step. Just *et al.* (1976) estimated a high-pressure limiting expression.

$$k_{\infty,\text{for}} \simeq 10^{15} \exp(-42{,}000 \text{ cal mol}^{-1}/RT) \text{ sec}^{-1} \quad (46)$$

Their preexponential factor is unusually large. Sundaram and Froment

(1978) used a much lower value for their mechanistic pyrolysis model (1000°K, 1.5 atm).

$$k_{for} = 10^{9.3} \exp(-31{,}500 \text{ cal mol}^{-1}/RT) \text{ sec}^{-1} \quad (47)$$

This latter expression was supported by Olsen *et al.* (1979), who modeled ethane pyrolysis at higher temperatures (1300–2500°K).

Payne and Stief (1976) measured the rate of hydrogen-radical addition to acetylene. An Arrhenius equation was determined for the high-pressure limiting rate constant.

$$k_{\infty,rev} = 10^{9.74} \exp(-2410 \text{ cal mol}^{-1}/RT) \text{ L mol}^{-1} \text{ sec}^{-1} \quad (48)$$

At 400°K, falloff occurred below 0.7 atm. Higher-temperature data were not available.

Butadiene is also formed in high-severity cracking. In the overall reaction scheme presented here, there are only two ways to increase molecular carbon number—coupling and addition. Vinyl–vinyl coupling may be one route to butadiene. Another might be methyl addition to propylene (or ethyl addition to ethylene), followed by decomposition to butene, secondary hydrogen abstraction to a butene radical, and further decomposition to butadiene. If butane is formed by methyl–propyl or ethyl–ethyl coupling, a series of abstractions and decompositions will also form butadiene.

With the exception of small radicals H· and $CH_3\cdot$, any free radical can presumably decompose into an olefin. With larger radicals, there may be more than one set of products. Similarly, there are a number of possible addition reactions that convert olefins to free radicals.

Light paraffin pyrolysis is very complex with seemingly endless permutations of elementary steps based on three general groups of forward and reverse reactions. The important point is that the maintenance of pseudo-equilibrium in free-radical concentrations is fast and dynamic: fast because several types of reactions occur at rates comparable to gas-collision frequency; dynamic, not only because the free-radical distribution changes with conversion, but also because this fleeting population churns, converting paraffin to olefin, and constantly readjusting.

The complexity does not stop here. Smaller yields of aromatic liquids, tar, and coke are also formed. These topics will be discussed next.

VII. BY-PRODUCT REACTIONS

A by-product from ethane and propane cracking is pyrolysis gasoline, which is a mixture of various aromatic hydrocarbons such as benzene and toluene, plus higher paraffins, olefins, and various ringed compounds. The linear and branched hydrocarbons are formed by coupling and addition reactions. Dehydrogenation occurs through abstraction followed by radical decomposition. All of these reactions are extensions of the elementary steps that have already been discussed.

Cyclizations may proceed via unimolecular reactions of large unsaturated free radicals. The forward elementary step can be regarded as an internal addition reaction. Nohara and Sakai (1980) studied the formation of 5-membered rings and found that the allyl radical is an important precursor. Decomposition of the substituted cyclopentyl radical yields a substituted cyclopentene. Alternatively, a 6-membered ring may be formed by isomerization and aromatization.

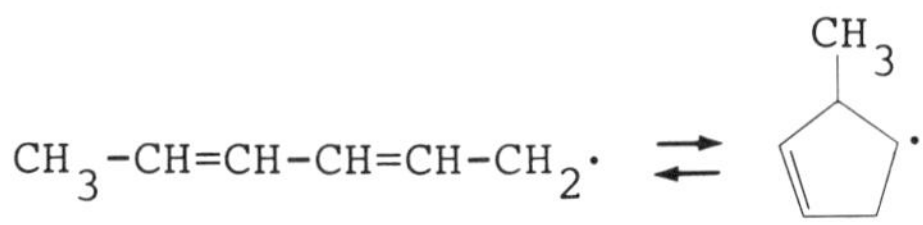

(49)

Six-membered rings may also be formed directly. In addition to radical cyclization, the molecular Diels–Alder mechanism yields 6-membered rings. In the pyrolysis of butadiene with ethylene, the selectivity for cycloaddition is nearly 100% (Sakai and Nohara, 1976).

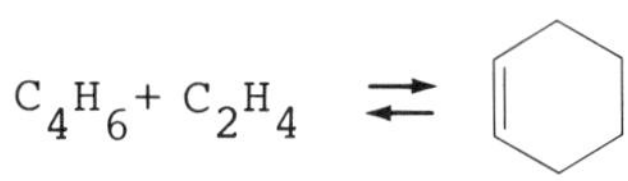

(50)

Tars are formed during ethane and propane cracking. These are generally high-molecular-weight polycyclic aromatic hydrocarbons. Cundall *et al.* (1979) noticed the spectra of polyacetylenes, H—C≡C—C≡C—C≡C—C≡C—C≡C—C≡C—H, prior to tar formation. They concluded that the polyacetylenes are intermediate to polycyclic aromatic formation; however, the cyclization step has not been fully elucidated.

By far the least desirable by-product of cracking is coke. Its formation causes heat-transfer surface fouling, high pressure drop, reduced yield, and eventually furnace downtime for decoking. In fact, there is recent evidence that costly tube failure is mostly due to coke deposition (Albright and McConnell, 1979).

One pathway to coke formation is condensation and dehydrogenation of the polycyclic aromatic tars formed in the gas phase. Globules of coke have been observed on Vycor glass, aluminized steel, and Incoloy 800 steel (Marek and Albright, 1981; Albright *et al.*, 1979). Using an electron microscope, Johnson and Anderson (1962) watched the transformations from polymer droplets to coke particles. This coking mechanism proceeds without the aid of metal catalyst and the coke product is relatively amorphous in structure. More details on coking are reported in Chapter 9 (Trimm) and Chapter 10 (Albright and Tsai).

Catalytic coking also occurs where light hydrocarbons decompose on metal surfaces. Whereas molecular species are absorbed and decomposed below 850°K, carbon-containing free radicals are perhaps adsorbed and decomposed above 850°K (McConnell, 1976). This is consistent with the temperature at which, for many light hydrocarbons, free radicals begin to appear. However, the radical-sensitive reactions occurring in the gas phase should be altered by such a significant quenching of free radicals. To maintain free-radical pseudoequilibrium and make up for lost radicals, either more scission reactions or fewer coupling reactions should occur. The consistent production of methane during ethane pyrolysis, even with different carbon-deposition rates, indicates that the scission reaction rate is little affected (Dunkleman and Albright, 1976). Ethane pyrolysis data (McConnell, 1976), however, indicate reduced yields of higher molecular weight products during periods of high coking. Fewer coupling reactions may indeed occur. A third possibility for maintaining the balance of gas-phase free radicals is that for each carbon-bearing radical that is adsorbed and decomposed, a hydrogen radical is released.

It should be pointed out that steam is nearly always used as a diluent in commercial light hydrocarbon cracking. Steam is essentially inert to the pyrolyzing gas, although it does improve olefin yield by reducing hydrocarbon partial pressure. The most important function of steam diluent is to partially remove coke from the reactor tube walls. This is accomplished through the water–gas/shift reactions, which gasify coke to produce carbon monoxide, carbon dioxide, and hydrogen.

VIII. SUMMARY

Ethylene is today's single most important hydrocarbon chemical. The process of choice for its manufacture is pyrolysis of light hydrocarbons. Ethane and propane feedstocks produce fewer by-products and have traditionally been favored. Naphtha and gas oil have recently gained in popularity, but developments indicate that ethane and propane pyrolysis will continue to be important. Other commercial-scale olefins, such as propylene and butadiene, are also produced in ethane and propane cracking.

Side reactions form the cyclic products in pyrolysis gasoline. More adverse side products are polycyclic aromatic tars and amorphous and catalytic cokes. Steam, used as diluent, interacts with the surface coke to form carbon oxides.

The simple combination of heat with ethane and/or propane results in a complex set of reactions, so complex that simulation strains the capabilities of computers. Basic elementary steps are well characterized, but there remain numerous elementary steps for which limited rate data are available. In fact, the mechanistic details of several side reactions are still not fully elucidated.

ACKNOWLEDGMENTS

We thank Mary Murphy and Diane Stevens of The Dow Chemical Company for their help in preparing this chapter.

REFERENCES

Albright, L. F., and McConnell, C. F. (1979). *In* "Thermal Hydrocarbon Chemistry" (A. G. Oblad, H. G. Davis, and R. T. Eddinger, eds.), Advances in Chemistry Series, No. 183, pp. 205–224. Am. Chem. Soc., Washington, D.C.

Albright, L. F., McConnell, C. F., and Welther, K. (1979). *In* "Thermal Hydrocarbon Chemistry" (A. G. Oblad, H. G. Davis, and R. T. Eddinger, eds.), Advances in Chemistry Series, No. 183, pp. 175–191. Am. Chem. Soc., Washington, D.C.

Benson, S. W., and O'Neal, H. E. (1970). "Kinetic Data on Gas Phase Unimolecular Reactions," NSRDS–NBS No. 21. Nat. Bur. Stand., Washington, D.C.

Bradley, J. N. (1974). *Proc. R. Soc. London, Ser. A* **337**, 199–216.

Bradley, J. N. (1979). *J.C.S. Faraday I* **75**, 2819–2826.

Camilleri, P., Marshall, R. M., and Purnell, J. H. (1974). *J.C.S. Faraday I* **70**, 1435–1444.

Camilleri, P., Marshall, R. M., and Purnell, J. H. (1975). *J.C.S. Faraday I* **71**, 1491–1502.

Chemical Engineering Progress (1966). **62**(6), 87.

Chemical Week (1980). July 16, 41–42.

Clark, T. C., and Dove, J. E. (1973a). *Can. J. Chem.* **51**, 2147–2154.

Clark, T. C., and Dove, J. E. (1973b). *Can. J. Chem.* **51**, 2155–2160.

Cundall, R. B., Fussey, D. E., Harrison, A. J., and Lampard, D. (1979). *J.C.S. Faraday I* **75**, 1390–1394.

Dow Chemical Co. (1981). Unpublished data.

Dunkleman, J. J., and Albright, L. F. (1976). *In* "Industrial and Laboratory Pyrolyses" (L. F. Albright and B. L. Crynes, eds.), ACS Symposium Series, No. 32, pp. 241–260. Am. Chem. Soc., Washington, D.C.

Golden, D. M., Choo, K. Y., Perona, M. J., and Piszkiewicz, L. W. (1976). *Int. J. Chem. Kinet.* **8**, 381–387.

Gray, P., Herod, A. A., and Jones, A. (1971). *Chem. Rev.* **71**, 247–291.

Hydrocarbon Processing (1975). **54**(11), 143.

Jinkerson, K. R., and Gaddy, J. L. (1979). *Ind. Eng. Chem., Process Des. Dev.* **18**, 579–587.

Johnson, G. C., and Anderson, R. C. (1962). *Proc. Conf. Carbon, Buffalo, 1961* **1**, 395.

Just, T., Roth, P., and Damm, R. (1976). *Symp. (Int.) Combust. [Proc.]* **16**, 961–969.

Kaminski, A. M., and Sobkowski, J. (1979). *Radiochem. Radionanal. Lett.* **39**, 209–220.
Kramer, L., and Happel, J. (1955). *In* "The Chemistry of Petroleum Hydrocarbons" (B. T. Brooks, S. S. Kurtz, C. E. Boord, and L. Schmerling, eds.), Vol. 21, pp. 146–147. Reinhold, New York.
Lee, W. M., and Yeh, C. T. (1979). *J. Phys. Chem.* **83**, 771–774.
Lifshitz, A., and Frenklach, M. (1975). *J. Phys. Chem.* **79**, 686–692.
Luger, C. R. (1976). *Hydrocarbon Process.* **55**(6), 105–106.
Luther, K., and Troe, J. (1979). *Symp. (Int.) Combust. [Proc.]* **17**, 535–542.
McConnell, C. F. (1976). M. S. Thesis, Purdue Univ., Lafayette, Indiana.
McCulloh, K. E., and Dibeler, V. H. (1976). *J. Chem. Phys.* **64**, 4445–4450.
Marek, J. C., and Albright, L. F. (1981). *182nd Am. Chem. Soc. Natl. Meet., New York*, Paper 53.
Martin, R., and Niclause, M. (1964). *J. Chim. Phys. Phys.-Chim. Biol.* **61**, 802.
Michael, J. V., Osborne, D. T., and Suess, G. N. (1973). *J. Chem. Phys.* **58**, 2800–2806.
Nohara, D., and Sakai, T. (1980). *Ind. Eng. Chem., Fundam.* **19**, 340–344.
Oil and Gas Journal (1976), **74** (47), 79.
Oil and Gas Journal (1977). **75** (9), 50.
Oil and Gas Journal (1979). **76** (36), 59–64.
Oil and Gas Journal (1980). **77** (35), 59–64.
Olsen, D. B., Tanzawa, T., and Gardiner, W. C. (1979). *Int. J. Chem. Kinet.* **11**, 23–44.
Pacey, P. D., and Purnell, J. H. (1972). *J.C.S. Faraday I* **68**, 1462–1473.
Parkes, D. A., and Quinn, C. P. (1976). *J.C.S. Faraday I* **72**, 1952–1971.
Payne, W. A., and Stief, L. J. (1976). *J. Chem. Phys.* **64**, 1150–1155.
Penzhorn, R. D., and Darwent, B. B. (1971). *J. Chem. Phys.* **55**, 1508–1511.
Peters, E. H. (1966). *Chem. Eng. Prog.* **62**(6), 87.
Powers, D. R., and Cocoran, W. H. (1974). *Ind. Eng. Chem., Fundam.* **13**, 351–355.
Pratt, G., and Rogers, D. (1979a). *J.C.S. Faraday I* **75**, 1089–1110.
Pratt, G. L., and Rogers, D. (1979b). *J.C.S. Faraday I* **75**, 2688–2693.
Rice, F. O., and Herzfeld, K. F. (1934). *J. Am. Chem. Soc.* **56**, 284.
Ross, L. L., and Shu, W. R. (1979). *In* "Thermal Hydrocarbon Chemistry" (A. G. Oblad, H. G. Davis, and R. T. Eddinger, eds.), Advances in Chemistry Series, No. 183, pp. 129–152. Am. Chem. Soc., Washington, D.C.
Roth, P., and Just, T. H. (1979). *Natl. Bur. Stand. (U.S.), Spec. Publ.* No. 561, 1339–1352.
Sakai, T., and Nohara, D. (1976). *In* "Industrial and Laboratory Pyrolysis" (L. F. Albright and B. L. Crynes, eds.), ACS Symposium Series, No. 32, pp. 152–177. Am. Chem. Soc., Washington, D.C.
Sundaram, K. M., and Froment, G. F. (1978). *Ind. Eng. Chem., Fundam.* **17**, 174–182.
Trenwith, A. B. (1979). *J.C.S. Faraday I* **75**, 614–620.
Tsuboi, T. (1978). *Jpn. J. Appl. Phys.* **17**, 709–716.
Walker, R. W. (1976). *In* "Gas Kinetics and Energy Transfer" (P. G. Ashmore and R. J. Donovan, eds.), Vol. 2, pp. 296–334. Chem. Soc., London.
Wing, M. S. (1980). *Chemtech* **10**, 20–22.
Zaslonko, I. S., and Smirnov, V. N. (1979). *Kinet. Katal.* **20**, 575–583.

3

Pyrolysis of *n*-Butane

WILLIAM H. CORCORAN

I. INTRODUCTION

Pyrolytic decomposition of *n*-butane has received significant attention in the past. Important among the studies have been those of Pease and Durgan (1930), Nevhaus and Marek (1932), Steacie and Puddington (1938), Crawford and Steacie (1953), Hepp and Frey (1953), Purnell and Quium (1961a,b, 1962, 1965), Wang *et al.* (1963), Zdonik *et al.* (1967), Blakemore and Corcoran (1969), Pacey and Purnell (1972a,b), Blakemore *et al.* (1973), Powers and Corcoran (1974), and Sundaram and Froment (1977). In the earlier work the studies were made on conversions of butane at less than about 10%. Rice (1931) first showed the free-radical nature of the pyrolytic decomposition of *n*-butane. In the work by Powers and Corcoran (1974), an effort was made to discuss the effects of primary and secondary butyl radicals and of secondary reactions. In all of the work the major products were methane, ethane, ethylene, and propylene. Minor products were hydrogen, *trans*-2-butene, 1-butene, and *cis*-2-butene. Methane and propylene were

Pyrolysis: Theory and Industrial Practice

ISBN 0-12-048880-9

produced in about equal molal quantities, and their relative fractions in the product were not strong functions of temperature of pyrolysis. The amount of ethylene increased with temperature, and the amount of ethane decreased. Figure 1 is a plot of a typical product distribution of the major products as a function of temperature with pyrolysis at atmospheric pressure.

In further discussions in this chapter, the goal will be to present detailed information on the nature of the free-radical reactions, to discuss the roles of primary and secondary butyl radicals and secondary reactions, to observe the effects of oxygen as an impurity on the pyrolysis of oxygen, to note the effects of surfaces on a seemingly homogeneous reaction, and finally, to show data for use in engineering design of reactors for pyrolysis of *n*-butane.

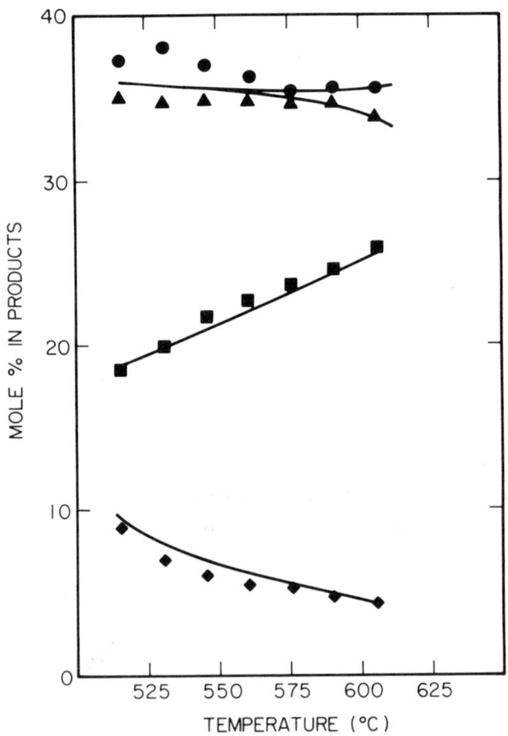

Fig. 1. Product distribution as a function of temperature at an initial butane concentration of 0.001 *M*. ●, Methane; ▲, propylene; □, ethylene; ◆, ethane. Lines, prediction by mechanism. [Reprinted with permission from D. R. Powers and W. H. Corcoran, Pyrolysis of *n*-butane—Explicit effects of primary and secondary butyl radicals and of secondary reactions, *Ind. Eng. Chem., Fundam.* **13**, 351–355 (1974). Copyright 1974, American Chemical Society.]

II. THE FREE-RADICAL NATURE OF n-BUTANE PYROLYSIS

Rice in 1931 was the first to suggest the free-radical nature of the pyrolysis reactions in the decomposition of n-butane. A reasonably simple mechanism is given by the following sets of equations as presented by Blakemore et al. (1973):

Initiation

$$C_4H_{10} \xrightarrow{k_1} C_2H_5\cdot + C_2H_5\cdot \tag{1}$$

$$C_4H_{10} \xrightarrow{k_2} C_3H_7\cdot + CH_3\cdot \tag{2}$$

Propagation

$$C_2H_5\cdot + C_4H_{10} \xrightarrow{k_3} C_2H_6 + C_4H_9\cdot \tag{3}$$

$$CH_3\cdot + C_4H_{10} \xrightarrow{k_4} CH_4 + C_4H_9\cdot \tag{4}$$

$$p\text{-}C_4H_9\cdot \xrightarrow{k_5} C_2H_4 + C_2H_5\cdot \tag{5}$$

$$s\text{-}C_4H_9\cdot \xrightarrow{k_6} C_3H_6 + CH_3\cdot \tag{6}$$

$$C_4H_9\cdot \xrightarrow{k_7} C_4H_8 + H\cdot \tag{7}$$

$$C_2H_5\cdot \xrightarrow{k_8} C_2H_4 + H\cdot \tag{8}$$

$$H\cdot + C_4H_{10} \xrightarrow{k_9} H_2 + C_4H_9\cdot \tag{9}$$

$$C_3H_7\cdot \xrightarrow{k_{10}} C_3H_6 + H\cdot \tag{10}$$

$$C_3H_7\cdot \xrightarrow{k_{11}} C_2H_4 + CH_3 \tag{11}$$

Termination

$$C_2H_5\cdot + C_2H_5\cdot \xrightarrow{k_{12}} C_2H_4 + C_2H_6 \tag{12}$$

$$C_2H_5\cdot + C_2H_5\cdot \xrightarrow{k_{13}} C_4H_{10} \tag{13}$$

In the experimental evaluation of the pyrolysis of n-butane, significant differences in activation energies and orders of reaction were observed in the reports from various investigators. Most probably the variation in results related to lack of control of the amount of oxygen present in the system and to variations in the surface present in the pyrolysis studies. In the work by Blakemore et al. (1973), a careful effort was made to control the amount of oxygen. A value of 1 ppm or less of oxygen in the gas stream was considered to be an oxygen-free system. In addition, the studies were carried out in a gold reactor. Gold had not been used extensively in kinetics

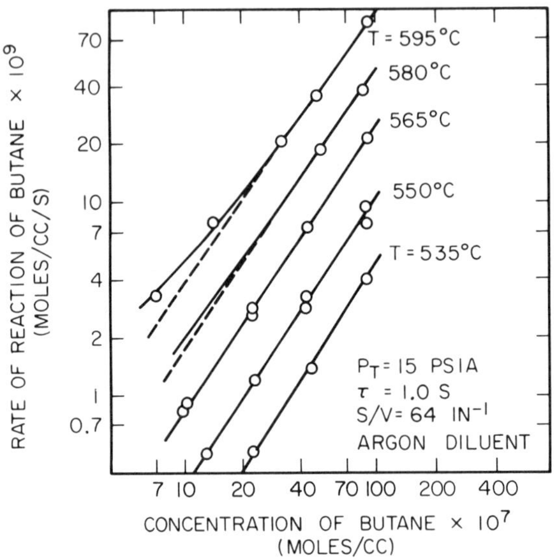

Fig. 2. Pyrolysis of butane as a function of temperature and concentration of butane in an untreated reactor (15 PSIA = 0.103×10^6 Pa and 64 in.$^{-1}$ = 25.2 cm^{-1}). [Reprinted with permission from J. E. Blakemore, J. R. Barker, and W. H. Corcoran, Pyrolysis of *n*-butane and the effects of trace quantities of oxygen, *Ind. Eng. Chem., Fundam.* **12**, 147–155 (1973). Copyright 1973, American Chemical Society.]

studies of pyrolytic reactions but was thought to have possibilities as a material that would give a low surface activity relative to the homogeneous reaction.

The role of the surface had not been unnoticed previously. Crynes and Albright (1969) observed an increased rate of pyrolysis when their stainless-steel reactor had been treated with oxygen. There was then a decrease in rate until a steady-state value was obtained. In their work a rapid return to the rates of reactions observed in untreated reactors occurred because of relatively high conversions of the propane that was being pyrolyzed. Their work then indicated that a carbonaceous layer had formed in steady state in the reactor.

Blakemore and Corcoran (1969) observed that higher rates of initial decomposition of butane were obtained in untreated gold reactors. Figure 2, for example, shows the rate of reaction of butane as a function of the concentration of butane for various temperatures in an untreated gold reactor. The steady-state nature of the surface seemed to be gold with a thin carbonaceous layer. After the untreated reactor had been exposed to hot oxygen for several hours, the rate of decomposition of butane was significantly decreased as shown in Fig. 3. Subsequent work with oxygen showed that it has a significant depressing effect on the rate of decompo-

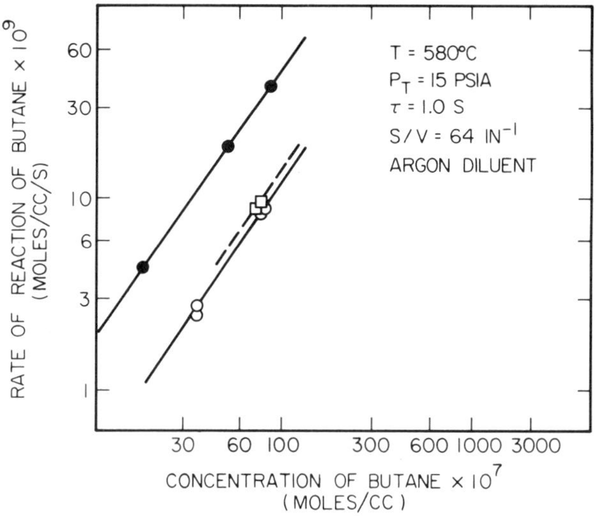

Fig. 3. A comparison of the rate of pyrolysis in an untreated reactor, before (●) and after exposure to "hot" oxygen (□); ○, after exposure to "hot" oxygen and vacuum of 10 μm. (15 PSIA = 0.103×10^6 Pa and 64 in.$^{-1}$ = 25.2 cm^{-1}). [Reprinted with permission from J. E. Blakemore, J. R. Barker, and W. H. Corcoran, Pyrolysis of n-butane and the effects of trace quantities of oxygen, *Ind. Eng. Chem., Fundam.* **12**, 147–155 (1973). Copyright 1973, American Chemical Society.]

sition of the butane. The issue will be discussed in more detail, but in the present comments it would suffice to say that residual oxygen was present in the surface of the gold. It was earlier thought that the oxygen would not chemisorb on the gold, but Kummer (1959) did report that oxygen would chemisorb on impurities in gold. Removal of the oxygen and impurities was thought possible by an etching of the gold surface with nitric acid and heating the surface to 500°C.

In consideration of Kummer's work and the depressed rate of reaction of butane after exposure of the gold surface to hot oxygen, the surface was then washed with nitric acid. After the acid wash, it was observed that the initial rate of pyrolysis came back to 50% of the pyrolysis rate in the untreated reactor. The rates of reactions observed in the acid-washed reactor did not return to the level of the untreated reactor until operation of the reactor for several days at low conversions of butane. Figure 4 shows the pyrolysis of butane as a function of temperature and concentration of butane for a gold reactor that had been treated with acid and then subsequently operated for several hours. Figure 4 does not indicate a full recovery to the level of operation shown in Fig. 2, but the full, steady-state operation of the reactor in the absence of oxygen would allow the data in Fig. 4 to finally achieve the level of performance shown in Fig. 2.

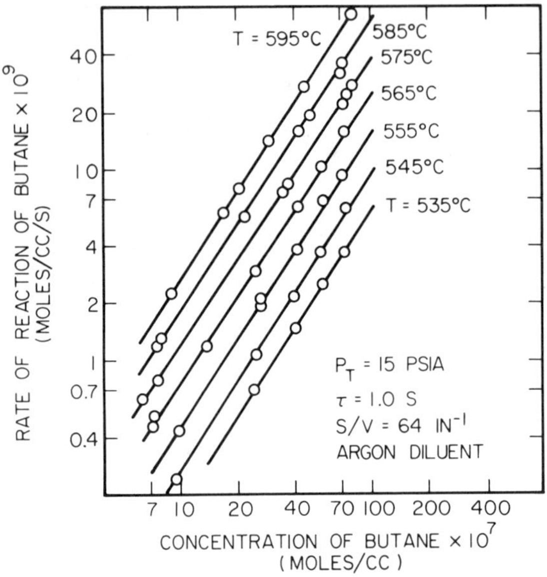

Fig. 4. Pyrolysis of butane as a function of temperature and concentration of butane is an acid-treated reactor (15 PSIA = 0.103×10^6 Pa and 64 in.$^{-1}$ = 25.2 cm^{-1}). [Reprinted with permission from J. E. Blakemore, J. R. Barker, and W. H. Corcoran, Pyrolysis of *n*-butane and the effects of trace quantities of oxygen, *Ind. Eng. Chem., Fundam.* **12**, 147–155 (1973). Copyright 1973, American Chemical Society.]

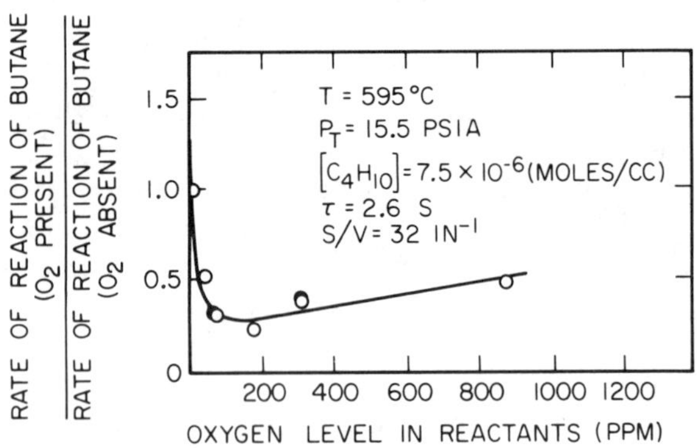

Fig. 5. Relative rate of decomposition of butane as a function of the oxygen level of the reactants at 595°C (15.5 PSIA = 0.107×10^6 Pa and 32 in.$^{-1}$ = 12.6 cm^{-1}). [Reprinted with permission from J. E. Blakemore, J. R. Barker, and W. H. Corcoran, Pyrolysis of *n*-butane and the effects of trace quantities of oxygen, *Ind. Eng. Chem., Fundam.* **12**, 147–155 (1973). Copyright 1973, American Chemical Society.]

As noted, not only does the surface of the reactor have an effect on the pyrolysis of n-butane, but trace amounts of oxygen in the feedstream have significant effects as well. Figure 5 shows the role of oxygen in affecting the rate of decomposition of n-butane. It can be seen that only a few parts per million of oxygen in the reactants gives a significant decrease in the rate of decomposition of the n-butane. A more detailed structure of that figure for the low concentrations of oxygen is shown in Fig. 6. Significant variations in reported activation energies for the thermal decomposition of butane are reasonably explained by noting the roles of surface effects in the reactors and of parts-per-million quantities of oxygen.

In the work by Blakemore *et al.* (1973), an assumption was made that the rate of disappearance of butane could be represented by a simple equation of the form

$$d(C_4H_{10})/dt = k(C_4H_{10})^m \tag{14}$$

with k being the regular specific rate constant and m being the experimentally observed order of reaction. For conversions of butane usually less than 1% it was found that the value of m was 1.5 ± 0.05 in the untreated gold reactor and 1.51 ± 0.03 in the acid reactor. An Arrhenius plot for the pyrolysis of n-butane is given in Fig. 7 for an acid-treated gold reactor. The value of the activation energy for the Arrhenius equation was determined to be 63.3 kcal/mol, and the Arrhenius frequency factor was 2.4×10^{16} $cc^{1/2}/(mol^{1/2}\ s)$. When the gold reactor had not been treated with acid, the activation energy was observed to be 66.6 kcal/mol, and the Arrhenius frequency factor was 1.8×10^{17} $cc^{1/2}/(mol^{1/2}\ s)$. The acid-treated and untreated reactors gave essentially the same results within

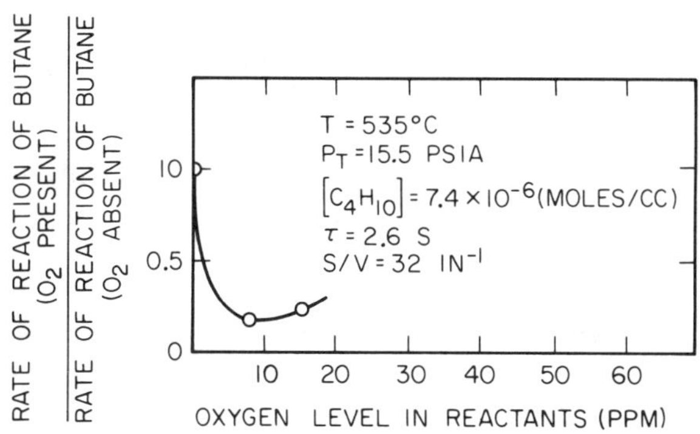

Fig. 6. Relative rate of decomposition of butane as a function of the oxygen level of the reactants at 535°C (15.5 PSIA = 0.107×10^6 Pa and 32 in.$^{-1}$ = 12.6 cm^{-1}). [Reprinted with permission from J. E. Blakemore, J. R. Barker, and W. H. Corcoran, Pyrolysis of n-butane and the effects of trace quantities of oxygen, *Ind. Eng. Chem., Fundam.* **12**, 147–155 (1973). Copyright 1973, American Chemical Society.]

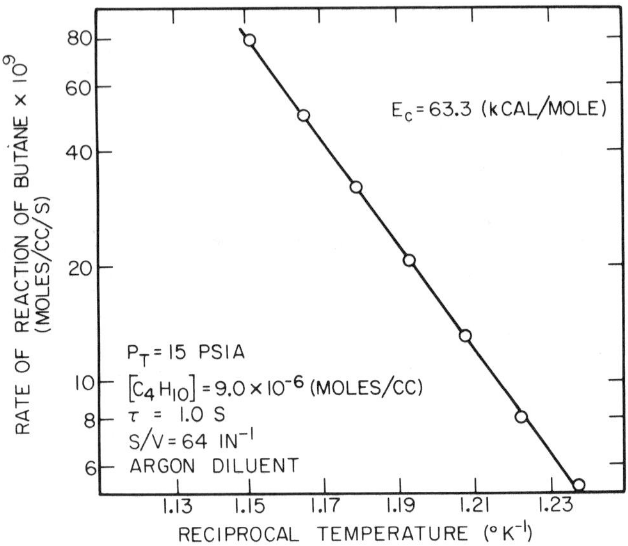

Fig. 7. Arrhenius plot for the pyrolysis in an acid-treated reactor (15 PSIA = 0.103×10^6 Pa and 64 in.$^{-1}$ = 25.2 cm^{-1}). [Reprinted with permission from J. E. Blakemore, J. R. Barker, and W. H. Corcoran, Pyrolysis of *n*-butane and the effects of trace quantities of oxygen, *Ind. Eng. Chem., Fundam.* **12**, 147–155 (1973). Copyright 1973, American Chemical Society.]

the limits of the experiment. To be reasonably sure that there was little effect of surface, a change in the S/V ratio in the reactor from 32 in.$^{-1}$ to 64 in.$^{-1}$ to 165 in.$^{-1}$ (12.6 cm^{-1} to 25.2 cm^{-1} to 65 cm^{-1}) did not affect the kinetics of the pyrolysis to any extent.

In these studies in the gold reactor, the primary products of the reaction were methane, propylene, ethylene, and ethane. Less than 2% of the products consisted of 1-butene, *cis*-2-butene, and *trans*-2-butene. Hydrogen was present but not easily measured in the decomposition of small amounts of butane. Measurements were not made for 1,3-butadiene, 1-pentene, or propane. Figures 8, 9, and 10 show distribution of major products in the pyrolysis of butane for temperatures of 595°C, 565°C, and 535°C, respectively. As previously observed, the percentages of methane and propylene did not change significantly with temperature. Ethane decreased with temperature, and ethylene significantly increased as temperature was increased in the pyrolysis reaction. Values of the rate constants for the simple equations in the mechanism for Eqs. (1–13) were obtained from Blakemore and Corcoran (1969), Kerr and Lloyd (1968), Purnell and Quinn (1962), and Sagert and Laidler (1963). The use of the rate constants for the individual simple equations showed that the 3/2 term for butane was the predominant term consistent with the experimentally observed result. If

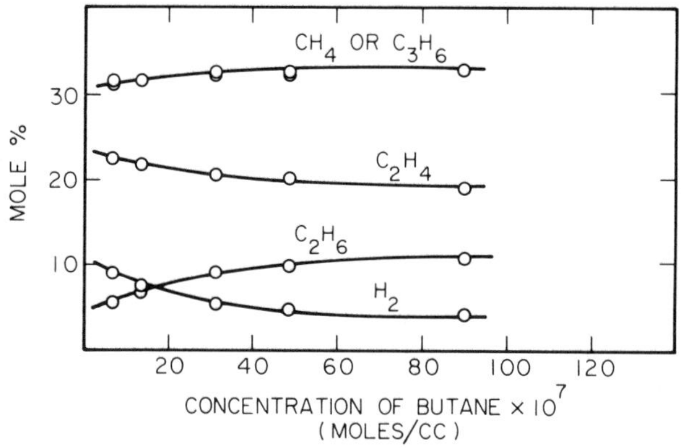

Fig. 8. Distribution of major products for the pyrolysis of butane with mole % referring to percentage of total products at 595°C. [Reprinted with permission from J. E. Blakemore, J. R. Barker, and W. H. Corcoran, Pyrolysis of *n*-butane and the effects of trace quantities of oxygen, *Ind. Eng. Chem., Fundam.* **12**, 147–155 (1973). Copyright 1973, American Chemical Society.]

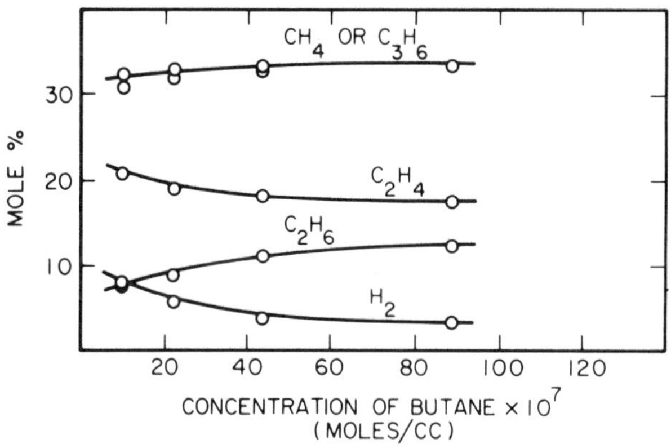

Fig. 9. Distribution of major products for the pyrolysis of butane with mole % referring to percentage of total products at 565°C. [Reprinted with permission from J. E. Blakemore, J. R. Barker, and W. H. Corcoran, Pyrolysis of *n*-butane and the effects of trace quantities of oxygen, *Ind. Eng. Chem., Fundam.* **12**, 147–155 (1973). Copyright 1973, American Chemical Society.]

the 3/2 order of butane is considered in Eq. (14), an apparent activation energy of 55 to 63 kcal/mol may be obtained depending on what values of the rate constants are used. That range of values is not too far removed from the experimentally observed values of 63.3 kcal/mol for the acid-treated reactor and 66.6 for the untreated reactor. Because work in previous investigations showed activation energies from approximately 38 to 75 kcal/mol, it is believed that the rather simple approach afforded by Eqs. (1–13) gives reasonable results when the predicted values are compared with the experimental values.

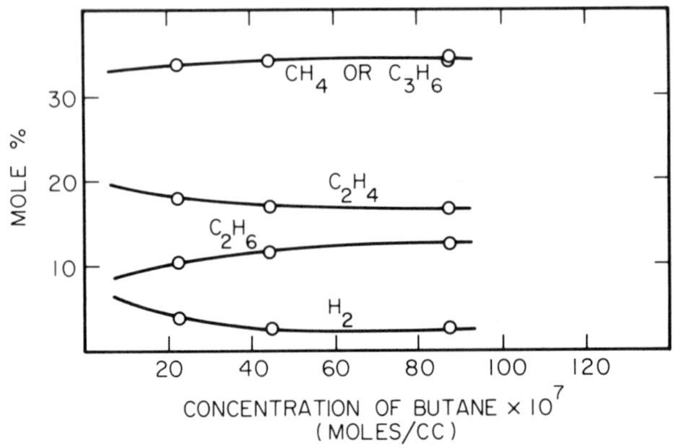

Fig. 10. Distribution of major products for the pyrolysis of butane with mole % referring to percentage of total products at 535°C. [Reprinted with permission from J. E. Blakemore, J. R. Barker, and W. H. Corcoran, Pyrolysis of *n*-butane and the effects of trace quantities of oxygen, *Ind. Eng. Chem., Fundam.* **12**, 147–155 (1973). Copyright 1973, American Chemical Society.]

III. THE ROLES OF PRIMARY AND SECONDARY BUTYL RADICALS AND OF SECONDARY REACTIONS IN THE PYROLYSIS OF *n*-BUTANE

In the discussions to this point, the pyrolysis of *n*-butane has been treated in reasonably simple style. The results reported were for very small percentages of decomposition of *n*-butane, so the effects of secondary reactions were negligible. In industrial pyrolysis, significantly greater percentage decompositions of *n*-butane than 1% must be acheived. Therefore, in a practical consideration of the rates and mechanisms in the pyrolysis of *n*-butane there must be consideration of secondary reactions and of explicit effects of the primary and secondary butyl radicals that can form. The work by Powers and Corcoran (1974) was directed to a development of a more detailed understanding of the pyrolysis of *n*-butane. Just as in the work by Blakemore *et al.* (1973), a gold reactor was used, and it was found that there could be some effect at the surface of gold. The rate of reaction was found to be independent of reactor pretreatment after a sufficient amount of conditioning. In particular, after the surface had been washed with nitric acid, there was a slow increase in the rate of decomposition of the *n*-butane in the reactor. That suggests the removal of oxygen from its chemisorption on the impurities of the gold and the development of the steady-state form of the gold surface with a slight carbonaceous layer. The work suggested that before the surface of the gold reactor was totally conditioned there was sufficient chain termination to reduce the rate of reaction of the *n*-butane. When the carbonaceous layer became sufficiently thick, the chain termination decreased, and the rate of decomposition of the butane acheived a steady state.

The product distribution for the work in the more detailed study of the kinetics of decomposition of *n*-butane has already been shown in Fig. 1. A comparison between Blakemore *et al.* (1973) and Powers and Corcoran (1974) may be made by observing the distribution of major products in Figs. 8, 9, and 10 from Blakemore *et al.* (1973) with similar information in Fig. 11 from Powers and Corcoran (1974). Any differences relate to the experimental procedures and end effects in the energy transfer to the reacting system. In the work by Powers and Corcoran (1974), the argon diluent was added to the *n*-butane well upstream from the reactor in contrast to addition immediately next to the reactor in the earlier work by Blakemore *et al.* (1973). Nevertheless, the qualitative results are essentially the same.

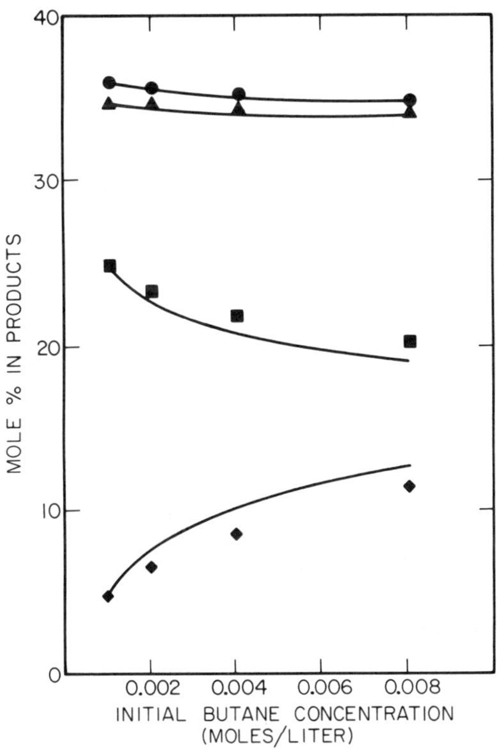

Fig. 11. Product distribution as a function of initial butane concentration at 590°C. ●, Methane; ▲, propylene; ■, ethylene; ◆, ethane. Lines, prediction by mechanism. [Reprinted with permission from D. R. Powers and W. H. Corcoran, Pyrolysis of *n*-butane—Explicit effects of primary and secondary butyl radicals and of secondary reactions, *Ind. Eng. Chem., Fundam.* **13**, 351–355 (1974). Copyright 1974, American Chemical Society.]

TABLE I

Proposed Mechanism for the Pyrolysis of *n*-Butane[a]

Elementary step[b]			Preexponential (s^{-1} or $M^{-1} s^{-1}$)	Activation energy (kcal/mol)
1. Butane	⟶	ET + ET	5.0E 15	81.0
2. Butane	⟶	ME + PR	5.0E 15	85.0
3. ET + butane	⟶	2-BU + ethane	3.0E 10	20.1
4. ET + butane	⟶	1-BU + ethane	9.0E 10	22.8
5. ME + butane	⟶	2-BU + methane	5.0E 10	18.2
6. ME + butane	⟶	1-BU + methane	1.0E 11	20.9
7. ET	⟶	ethylene + H	1.7E 13	40.5
8. H + butane	⟶	2-BU + H_2	1.0E 12	14.0
9. H + butane	⟶	1-BU + H_2	1.3E 12	16.7
10. PR	⟶	propylene + H	2.6E 14	37.8
11. PR	⟶	ethylene + ME	2.0E 13	28.0
12. 1-BU	⟶	ethylene + ET	2.0E 13	24.0
13. 1-BU	⟶	1-butene + H	2.6E 14	38.0
14. 2-BU	⟶	propylene + ME	2.0E 13	26.0
15. 2-BU	⟶	2-butene + H	2.6E 14	37.5
16. ET + ET	⟶	butane	1.6E 09	0.0
17. ET + ET	⟶	ethane + ethylene	2.5E 08	0.0
18. ME + ME	⟶	ethane	2.0E 09	0.0
19. H + ethylene	⟶	ET	1.0E 10	0.5
20. ET + ethylene	⟶	1-BU	5.0E 06	2.0
21. ME + ethylene	⟶	PR	1.0E 07	2.0
22. ME + propylene	⟶	2-BU	1.0E 07	2.0
23. H + propylene	⟶	PR	6.0E 09	0.5
24. ME + ET	⟶	propane	2.0E 09	0.0

[a] Reprinted with permission from D. R. Powers and W. H. Corcoran, Pyrolysis of *n*-butane—Explicit effects of primary and secondary butyl radicals and of secondary reactions, *Ind. Eng. Chem., Fundam.* **13**, 351–355 (1974). Copyright 1974, American Chemical Society.

[b] Radical designations: ME, methyl; ET, ethyl; PR, propyl; 1-BU, 1-butyl; 2-BU, 2-butyl.

Whereas the mechanism shown by Blakemore *et al.* (1973) involved some 13 steps, the mechanism including the effects of secondary reactions and primary and secondary butyl radicals amounts to some 24 steps as shown in Table I. Initial estimates of rate constants for the individual simple reactions in Table I were taken from the summation of Blakemore and Corcoran (1969). Values for rate constants for abstraction of hydrogen atoms from butane by methyl and ethyl radicals were obtained from the work by Pacey and Purnell (1972a,b). There was some adjustment of the rate constants, keeping within the limits of accepted values, in order to allow matching of predicted and experimental results. The values for the rate constants finally established in Table II find their basis mainly in the

TABLE II

Relative Importance of Individual Elementary Steps in the Pyrolysis of *n*-Butane[a]

Elementary step[b]			Temp. 515°C		Temp. 560°C		Temp. 605°C	
			Initial conc. of *n*-butane (*M*)					
			0.001	0.008	0.001	0.008	0.001	0.008
			Relative frequency					
1. Butane	⟶	ET + ET	0.00116	0.00056	0.00205	0.00109	0.00356	0.00199
2. Butane	⟶	ME + PR	0.00009	0.00004	0.00018	0.00010	0.00036	0.00020
3. ET + butane	⟶	2-BU + ethane	0.09601	0.16704	0.06905	0.15834	0.05022	0.14646
4. ET + butane	⟶	1-BU + ethane	0.05137	0.08937	0.04055	0.09298	0.03206	0.09350
5. ME + butane	⟶	2-BU + methane	0.53558	0.52430	0.51379	0.49844	0.50115	0.48079
6. ME + butane	⟶	1-BU + methane	0.19103	0.18701	0.20114	0.19513	0.21329	0.20463
7. ET	⟶	ethylene + H	0.11981	0.02606	0.17455	0.05015	0.24264	0.08973
8. H + butane	⟶	2-BU + H_2	0.10156	0.02593	0.13842	0.04338	0.15651	0.05742
9. H + butane	⟶	1-BU + H_2	0.02355	0.00601	0.03522	0.01104	0.04330	0.01588
10. PR	⟶	propylene + H	0.00003	0.00002	0.00026	0.00021	0.00165	0.00135
11. PR	⟶	ethylene + ME	0.00112	0.00096	0.00737	0.00590	0.03495	0.02848
12. 1-BU	⟶	ethylene + ET	0.26567	0.28243	0.27691	0.30085	0.29006	0.32318
13. 1-BU	⟶	1-butene + H	0.00045	0.00048	0.00076	0.00083	0.00124	0.00138
14. 2-BU	⟶	propylene + Me	0.72772	0.71264	0.71572	0.69744	0.70897	0.69529
15. 2-BU	⟶	2-butene + H	0.00612	0.00599	0.00895	0.00872	0.01266	0.01241
16. ET + ET	⟶	butane	0.00034	0.00027	0.00041	0.00051	0.00045	0.00087
17. ET + ET	⟶	ethane + ethylene	0.00005	0.00004	0.00006	0.00008	0.00007	0.00014
18. ME + ME	⟶	ethane	0.00042	0.00011	0.00103	0.00023	0.00228	0.00048
19. H + ethylene	⟶	ET	0.00061	0.00026	0.00558	0.00250	0.03254	0.01579
20. ET + ethylene	⟶	1-BU	0.00018	0.00053	0.00076	0.00253	0.00264	0.01055
21. ME + ethylene	⟶	PR	0.00036	0.00059	0.00215	0.00303	0.01041	0.01385
22. ME + propylene	⟶	2-BU	0.00069	0.00136	0.00340	0.00600	0.01376	0.02304
23. H + propylene	⟶	PR	0.00069	0.00036	0.00529	0.00298	0.02583	0.01578
24. ME + ET	⟶	propane	0.00043	0.00019	0.00073	0.00038	0.00113	0.00072

[a] Reprinted with permission from D. R. Powers and W. H. Corcoran, Pyrolysis of *n*-Butane—Explicit effects of primary and secondary butyl radicals and of secondary reactions, *Ind. Eng. Chem. Fundam.* **13**, 351–355 (1974). Copyright 1974, American Chemical Society.

[b] Radical designations: ME, methyl; ET, ethyl; PR, propyl; 1-BU, 1-butyl; 2-BU, 2-butyl.

experimental results, which agreed quite well with the predicted values of products.

Table II gives a tabulation of the relative importance of each of the elementary reactions cited in Table I. In the establishment of the numbers in the table, the sums of the rates of the steps involved in consumption of *n*-butane were normalized arbitrarily to a value of 1.000. A review of the relative frequency presented in Table II shows clearly that the pyrolysis of the *n*-butane evolved by way of propagating steps 3–6, 12, and 14. For

TABLE III

Comparison of Experimental Orders of Reaction and Activation Energies with Predicted Values[a,b]

A. Order of Reaction with Respect to *n*-Butane

	Methane	Ethane	Ethylene	Propylene	Butane
			530°C		
EXP	1.284	1.602	1.243	1.333	1.329
DRP	1.319	1.638	1.215	1.319	1.332
			560°C		
EXP	1.307	1.606	1.208	1.314	1.316
DRP	1.285	1.694	1.176	1.286	1.300
			590°C		
EXP	1.250	1.676	1.166	1.258	1.259
DRP	1.261	1.750	1.152	1.267	1.279
			605°C		
EXP	1.291	1.768	1.212	1.296	1.293
DRP	1.251	1.774	1.144	1.262	1.271

B. Activation Energy for the Formation of Each Product at Various Initial Concentrations of *n*-Butane

Conc. of butane (*M*)		Methane	Ethane	Ethylene	Propylene	Butane
0.001	EXP	63.52	55.04	69.77	64.39	64.99
	DRP	63.52	55.81	68.48	62.92	63.56
0.002	EXP	62.93	56.87	68.78	63.33	63.99
	DRP	62.99	58.18	68.12	62.41	63.22
0.004	EXP	62.32	58.69	67.81	62.28	63.02
	DRP	61.80	59.72	66.69	61.30	61.72
0.008	EXP	61.72	60.52	66.84	61.22	62.03
	DRP	60.54	60.57	64.93	60.15	60.97

[a] Reprinted with permission from D. R. Powers and W. H. Corcoran, Pyrolysis of *n*-butane—Explicit effects of primary and secondary butyl radicals and of secondary reactions, *Ind. Eng. Chem., Fundam.* **13**, 351–355 (1974). Copyright 1974, American Chemical Society.

[b] Key to abbreviations: EXP, experimental results of this paper; DRP, predictions by mechanism in Table I.

low concentrations of butane, and especially at high temperatures, step 7 increases in importance. Because the hydrogen radical is produced by step 7, then steps 8 and 9, which show hydrogen extraction from butane to form the secondary butyl radical and the primary butyl radical, respectively, are affected in direct proportion to changes in step 7.

Table III presents experimental orders of reaction and activation-energies in a comparison with predicted values using Powers' more detailed

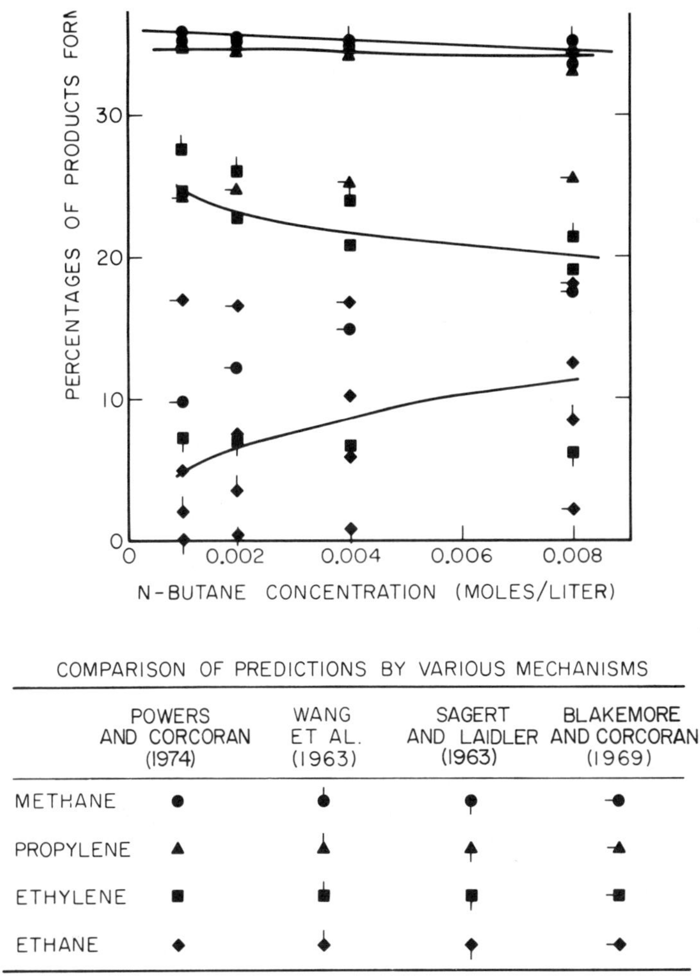

Fig. 12. Product distribution in comparison with the predictions of various mechanisms. See accompanying chart for explanation of symbols used in figure. Lines are experimental results in this work. [Reprinted with permission from D. R. Powers and W. H. Corcoran, Pyrolysis of n-butane—Explicit effects of primary and secondary butyl radicals and of secondary reactions, *Ind. Eng. Chem. Fundam.* **13**, 351–355 (1974). Copyright 1974, American Chemical Society.]

analysis. The variability between the experimentally observed values and predicted values is strikingly small.

A summation of work on the pyrolysis of *n*-butane may be observed in Fig. 12. The solid lines in the figure show experimental results obtained with the conditioned gold reactor in the work by Powers. The various indices for the plots are given in the legend. The figure shows values for the percentage of products formed as a function of the initial concentration of *n*-butane in the gas phase. The investigators prior to Powers and Corcoran (1974) did not take into account the effects of secondary reactions and of extraction of hydrogen to form secondary and primary butyl radicals. By accounting for these secondary effects it may be seen in Fig. 12 that a significant improvement in prediction of results is possible.

IV. DETAILS ON THE ROLE OF TRACE QUANTITIES OF OXYGEN IN THE PYROLYSIS OF *n*-BUTANE

Of significant interest in the applied chemistry of the pyrolysis of *n*-butane is the role of trace quantities of oxygen. As noted previously, Blakemore *et al.* (1973) examined the pyrolysis of *n*-butane for temperatures between 530° and 600°C and for conversion of butane less than 1%. In gold reactors and for S/V ratios of 35 to 125 in^{-1} (13.8 to 49.2 cm^{-1}), trace quantities of oxygen in the reactant system of amounts from 5 to 500 ppm had major effects on the pyrolysis. When oxygen was present in as small a quantity as 7 ppm, it decreased the amount of cracked products—namely, methane, ethane, ethylene, and propylene—by about 70% relative to the pyrolysis in the absence of oxygen. Figures 5 and 6 already presented show the decrease in rate of reaction of butane as a function of the oxygen concentration at low levels of oxygen. Although the relative distribution of the methane, ethane, ethylene, and propylene did not change significantly with increasing concentrations of oxygen, increasing amounts of 1-butene, *trans*-2-butene, *cis*-2-butene, 1,3-butadiene, and water were observed. Figure 13 shows the role of initial concentrations of butane in the distribution of products for 14.5 ppm of oxygen and a residence time of 2.6 s in the gold microreactor.

The distribution of methane, ethane, ethylene, and propylene was not too dependent on the initial concentration of oxygen at very low concentrations of oxygen. Figure 14 shows the effects on product distribution as the concentration of oxygen was increased at a temperature of 595°C and for a residence time of 2.6 s. With an increase in the concentration of oxygen, the figure shows that the amount of ethylene relative to the amount of ethane increased slightly. In the work reported by Blakemore *et al.* (1973) it was

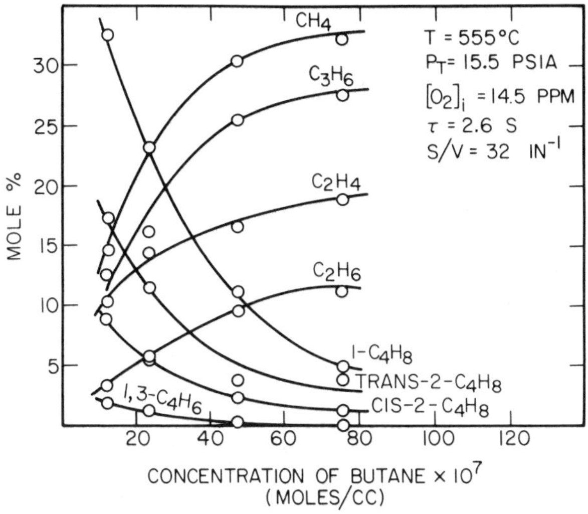

Fig. 13. Mole % of hydrocarbon products for moles of butane converted as a function of the concentration of butane for the pyrolysis in the presence of trace quantities of oxygen (15.5 PSIA = 0.107×10^6 Pa and 32 in.$^{-1}$ = 12.6 cm^{-1}). [Reprinted with permission from J. E. Blakemore, J. R. Barker, and W. H. Corcoran, Pyrolysis of n-butane and the effects of trace quantities of oxygen, *Ind. Eng. Chem., Fundam.* **12**, 147–155 (1973). Copyright 1973, American Chemical Society.]

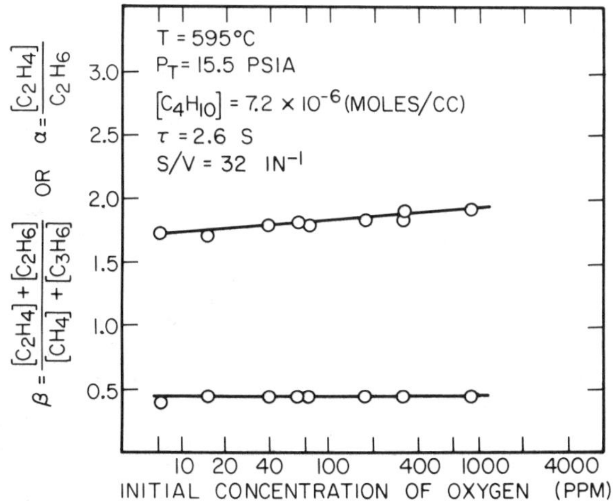

Fig. 14. Variation of the functions α (upper curve) and β (lower curve) at 595°C with initial levels of oxygen in the reactants (15.5 PSIA = 0.107×10^6 Pa and 32 in.$^{-1}$ = 12.6 cm^{-1}). [Reprinted with permission from J. E. Blakemore, J. R. Barker, and W. H. Corcoran, Pyrolysis of n-butane and the effects of trace quantities of oxygen, *Ind. Eng. Chem., Fundam.* **12**, 147–155 (1973). Copyright 1973, American Chemical Society.]

not clearly understood why there was an increased formation of butenes in the presence of oxygen. One suggested reaction was the interaction of oxygen with primary or secondary butyl radicals to yield 1-butene, *trans*-2-butene, or *cis*-2-butene and a hydroperoxide radical as follows:

$$C_4H_9\cdot + O_2 \longrightarrow C_4H_8 + HO_2\cdot \tag{15}$$

Formation of the 1,3-butadiene was then attributed possibly to a subsequent oxidation of 1-butene. Increase in the amount of ethylene at the expense of ethane was thought to come possibly from the reaction between oxygen and ethyl radicals in accord with the expression:

$$C_2H_5\cdot + O_2 \longrightarrow C_2H_4 + HO_2\cdot \tag{16}$$

Blakemore *et al.* (1973) examined the rates of disappearance of butane as a function of oxygen concentration. The order of the decomposition of the butane was approximately 1.5 for various oxygen concentrations. An examination of the rate, however, indicated an increase in activation energy to 81 kcal/mol versus the 63 to 66 kcal/mol noted in the absence of oxygen. The thought was presented that the presence of oxygen could influence initiation and termination steps in the chain reaction. Initiation of the chain reaction in the presence of oxygen was cited to be 30–35 kcal/mol more favorable than the homolytic decomposition of the butane, but for the temperatures studied in the pyrolysis and with small amounts of oxygen, the homolytic decomposition was thought to be relatively competitive to the formation of butyl and hydroperoxide radicals in the reaction between oxygen and butane. Nevertheless, there was a decrease in initial rate of decomposition of butane as oxygen was added to the system, and it was suggested that oxygen could react with alkyl radicals to terminate chains leading to the cracked products and that oxygenated free radicals could be destroyed on the surface of the reactor to reduce the overall rate of decomposition of butane.

Niclause *et al.* (1967) examined pyrolysis of propane, isobutane, *n*-butane, and isopentene in the presence of oxygen. In their work, oxygen inhibited or accelerated the rate of decomposition of the particular hydrocarbon as a function of the S/V ratio in the reactor. For a relatively low S/V ratio, oxygen accelerated the reaction, and at higher S/V ratios there was an inhibition of the reaction. The work by Blakemore *et al.* (1973) was at a high S/V ratio of 32 in.^{-1} (12.6 cm^{-1}), and inhibition was noted consistent with the observation by Niclause. Both sets of experiments are consistent with the idea that a reduction in the rate could relate to direct combination of alkyl radicals with oxygen and the resulting increase and ease of destruction of the peroxy radicals on the walls. The increased inhibition then with the high S/V ratios would be explained. Acceleration could come from the additional processes of formation of free radicals in the presence of oxygen over those formed in the absence of oxygen. The major difference between the work of Martin *et al.* (1976) and Blakemore *et al.* (1973) related to the presence of 1,3-butadiene in the Blakemore work and its absence around 500°C for low extent of reaction in the Martin work.

Table IV shows that the demethanation reaction of *n*-butane giving C_3H_6 and CH_4 increases relatively as the small amount of oxygen is increased from 0 to 0.13%. Martin *et al.* (1976) suggested that the relative increase in the demethanation related to the slower rate of oxidation of the methyl radical by the oxygen rather than oxidation of other chain carriers in the chain reaction for the pyrolysis of the *n*-butane.

Increased understanding of the pyrolysis of *n*-butane emphasizes the need to know more precisely the nature of the reacting systems. A significant amount of past literature in the pyrolysis of *n*-butane has obviously had significant error of interpretation as a result of lack of control of impurities such as oxygen and a lack of understanding of the interaction between the gas phase and the wall in the chain reaction. Although the mechanisms are not totally understood even now, a rational basis exists for thinking about the mechanism and rate of thermal decomposition of *n*-butane. The industrial challenge to produce quantities of unsaturated hydrocarbons of a given structure relates to a proper understanding of the applied chemistry

TABLE IV

Relative Importances (Percentages) of the Primary Stoichiometries of *n*-Butane Decomposition and Oxidation at Low Extent of Reaction[a,b]

	$(O_2)_0$ %	0	0.01	0.07	0.13
Decomposition	$n\text{-}C_4H_{10} = C_3H_6 + CH_4$	71	68.7	47.5	37
	$n\text{-}C_4H_{10} = C_2H_4 + C_2H_6$	21	10.7	1.5	$\simeq 0$
	$n\text{-}C_4H_{10} = 2\,C_2H_4 + H_2$	8	4.7	$\simeq 0$	$\simeq 0$
	$n\text{-}C_4H_{10} = n\text{-}C_4H_8 + H_2$	<1	$\simeq 0$	$\simeq 0$	$\simeq 0$
Oxidation	$n\text{-}C_4H_{10} + O_2$ (or $\frac{1}{2}O_2$) $= 2\,C_2H_4 + H_2O_2$ (or H_2O)	—	10.7	21	19.5
	$n\text{-}C_4H_{10} + O_2$ (or $\frac{1}{2}O_2$) $= n\text{-}C_4H_8 + H_2O_2$ (or H_2O)	—	5.2	30	43.5

Industrial and Laboratory Pyrolyses

[a] Refer to Martin *et al.* (1976).

[b] Experimental conditions: temperature, 494°C; $P_0^{n\text{-}C_4H_{10}} = 50$ mmHg; PbO-coated Pyrex vessel ($S/V \simeq 5\ cm^{-1}$).

of additives and of heterogeneity. If there is a clear control of the amount of oxygen present in the *n*-butane, then the data discussed in this chapter can be used in prediction of products and rates for the thermal decomposition of *n*-butane over reasonable temperature ranges and initial concentrations of *n*-butane. In addition to having a source of prediction of composition, and especially if reference is made to Table III, an engineering evaluation of the rate of reaction of the *n*-butane can lead to considering its decomposition as 3/2 order with respect to *n*-butane. A more precise ordering can come from the 24 elementary steps cited in Table III as presented by Powers and Corcoran (1974).

V. DATA FOR ENGINEERING DESIGN OF A REACTOR FOR PYROLYSIS OF *n*-BUTANE

Zdonik *et al.* (1967) presented an extensive discussion on how cracking proceeds in ethylene pyrolysis reactions. In that presentation a plot of reaction velocity constants for light hydrocarbons is shown and included here as Fig. 15. It shows a semilog plot of a first-order rate constant versus temperature obtained mainly by examination of data from commercial installations.

In development of first-order kinetics, Zdonik *et al.* (1967) used an equation of the form $K = Ae^{-e/RT}$. In that equation a value of A for *n*-butane was 1.80×10^{12}/s, and E was 56.308 kcal/mol. For a quick approach to the design of a butane reactor the first-order constant shown in Fig. 15 could be used.

Another approach to the problem, however, would come from the analysis of the type used by Powers and Corcoran (1974). The value of that approach, wherein 24 elementary steps were established and then used to determine the product distribution as shown in Fig. 12, is clear. Figure 12 shows a comparison of predictions by various authors, including Wang *et al.* (1963), Sagert and Laidler (1963), and Blakemore and Corcoran (1969). The solid curves in Fig. 12 are the predictions using the 24-step process, and the various indices show the results from the other sources of information. The calculation using the 24 steps gives a remarkably good representation of the experiment. One could question the temperature limitation in the work by Powers and Corcoran (1974), but in consideration of Fig. 15 from Zdonik *et al.* (1967), it would appear that there can be reasonable extrapolation of the data to temperatures of 1600°F (871°C) and beyond without too much difficulty.

In sum, a quick analysis of the butane reactor could be developed by using the first-order kinetics of Zdonik *et al.* (1967). A more thorough

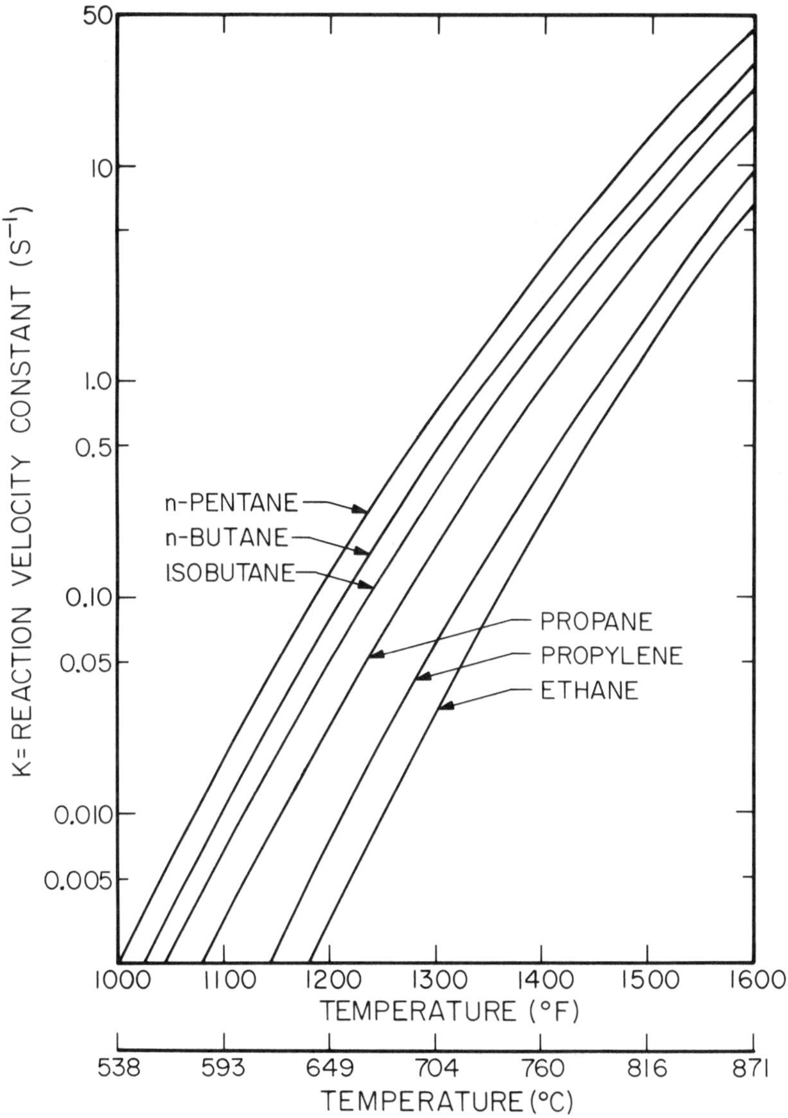

Fig. 15. First-order rate constants for pyrolysis of light hydrocarbons. [Reprinted with permission from S. B. Zdonik, E. J. Green, and L. P. Hallee, How cracking proceeds in the ethylene-pyrolysis reaction, *Oil Gas J.* June 26, 96–101 (1967).

analysis taking into account steps involving primary and secondary butyl radicals and secondary reactions gives a much more satisfactory result and is recommended for engineering design.

REFERENCES

Blakemore, J. E., and Corcoran, W. H. (1969). *Ind. Eng. Chem., Process Des. Dev.* **8**, 206–209.

Blakemore, J. E., Barker, J. R., and Corcoran, W. H. (1973). *Ind. Eng. Chem., Fundam.* **12**, 147–155.

Crawford, V. A., and Steacie, E. W. R. (1953). *Can. J. Chem.* **31**, 937–948.

Crynes, B. L., and Albright, L. F. (1969). *Ind. Eng. Chem., Process Des. Dev.* **8**, 25–31.

Hepp, H. J., and Frey, F. E. (1953). *Ind. Eng. Chem.* **45**, 410–415.

Kerr, J. A., and Lloyd, A. C. (1968). *Q. Rev., Chem. Soc.* **22**, 549–577.

Kummer, J. T. (1959). *J. Phys. Chem.* **63**, 460–463.

Martin, R., Niclause, M., and Scacchi, G. (1976). *In* "Industrial and Laboratory Pyrolyses" (L. F. Albright and B. L. Crynes, eds.), ACS Symposium Series, No. 32, p. 40. Am. Chem. Soc., Washington, D.C.

Neuhaus, M., and Marek, L. F. (1932). *Ind. Eng. Chem.* **24**, 400–402.

Niclause, M., Martin, R., Combes, A., Fusy, J., and Dzierzynski, M. (1967). *Ind. Chim. Belge* **32**, 674–679.

Pacey, P. D., and Purnell, J. H. (1972a). *Ind. Eng. Chem., Fundam.* **11**, 233–239.

Pacey, P. D., and Purnell, J. H. (1972b). *Int. J. Chem. Kinet.* **4**, 657–666.

Pease, R. N., and Durgan, E. S. (1930). *J. Am. Chem. Soc.* **52**, 1262–1267.

Powers, D. R., and Corcoran. W. H. (1974). *Ind. Eng. Chem., Fundam.* **13**, 351–355.

Purnell, J. H., and Quinn, C. P. (1961a). *Nature (London)* **189**, 656–658.

Purnell, J. H., and Quinn, C. P. (1961b). *J. Chem. Soc.* pp. 4128–4132.

Purnell, J. H., and Quinn, C. P. (1962). *Proc. R. Soc. London, Ser. A* **270**, 267–284.

Purnell, J. H., and Quinn, C. P. (1965). *Can. J. Chem.* **43**, 721–723.

Rice, F. O. (1931). *J. Am. Chem. Soc.* **53**, 1959–1972.

Sagert, N. H., and Laidler, K. J. (1963). *Can. J. Chem.* **41**, 838–847.

Steacie, E. W. R., and Puddington, I. E. (1938). *Can. J. Res., Sect. B* **16**, 260–272.

Sundaram, K. M., and Froment, G. F. (1977). *Chem. Eng. Sci.* **32**, 609–617.

Wang, Y. L., Rinker, R. G., and Corcoran, W. H. (1963). *Ind. Eng. Chem., Fundam.* **2**, 161–168.

Zdonik, S. B., Green, E. J., and Hallee, L. P. (1967). *Oil Gas J.* June 26, 96–101.

4

Pyrolysis of Heavy Hydrocarbons

CHARLES REBICK

I. INTRODUCTION

Previous chapters of this book have dealt with the cracking of relatively light saturated hydrocarbons—methane through butane. In this chapter, the pyrolysis of heavier paraffins will be discussed, and the discussions

Pyrolysis: Theory and Industrial Practice

ISBN 0-12-048880-9

will be extended to olefins, naphthenes, and alkyl aromatics. The cracking of these heavier compounds is of interest for several reasons: (1) A significant fraction of the ethylene and propylene produced commercially worldwide comes from the pyrolysis of heavier feedstocks—naphthas and gas oils. Current trends are toward the use of even heavier feeds including whole crudes (Zdonik *et al.*, 1975; Wing, 1980; Stinson, 1979). (2) Although most pyrolysis feeds are highly paraffinic, they contain significant amounts of naphthenes and aromatics whose behavior must also be understood. (3) Among the primary products of higher paraffin cracking are higher α-olefins. Their subsequent decomposition is thus a significant factor in high-conversion paraffin pyrolysis. In order to understand the pyrolysis of heavy feeds, and in particular to relate cracking rates and yields to feedstock properties, one needs to understand the response of the individual components to pyrolysis; hence the study of pure heavy compounds is important.

Although hydrocarbon thermal cracking has been a subject of study for over 100 years, heavier molecules were largely ignored until the past 20 years. This was primarily because thermal cracking is basically a nonselective process; thus the heavier the reactant, the more complex the mix of products. Before the development of modern analytical techniques, the pyrolysis of a heavy paraffin such as hexadecane presented a formidable product identification and quantification challenge. Hence most studies involved light reactants such as ethane, propane, and butane (e.g., Egloff *et al.*, 1936; Neuhaus and Marek, 1932). The first definitive study of heavier paraffins was that of Voge and Good (1949) on dodecanes and *n*-hexadecane. For the latter compound, they separated over 20 products using a combination of distillation, acid treating, bromination, and ultraviolet adsorption, a painstaking and time-consuming process. Since the advent of gas chromatography (GC), mass spectrometry (MS), and combined GC–MS, the analysis of many cracked-product mixtures is routine, and for others, most of the important products can be readily and rapidly identified. These analytical techniques have also spawned new pyrolysis approaches such as pyrolysis GC or MS, where the compound to be studied is actually pyrolyzed in the inlet of the analytical instrument (Greco, 1978). As a result of these developments, the past 20 years have seen a significant increase in the number of studies of the pyrolysis of heavy hydrocarbons.

In parallel with this increase in experimental capability and activity has been an even more rapid increase in the capabilities of computers to handle complex computations. This has allowed the development of computer models for the cracking of heavier molecules, and even complex mixtures of hydrocarbons. Such modeling will be discussed in Chapter 7. The combination of increased experimental activity and computer analysis has resulted in a great increase in the sophistication with which we understand the rates and mechanisms of pyrolysis reactions. Yet surprisingly, the generally accepted mechanism for paraffin cracking was first presented almost 40 years ago (Kossiakoff and Rice, 1943) and has survived intact with only a few adjustments. This Rice–Kossiakoff mechanism will be discussed in the next section, followed by descriptions of olefin and alkyl aromatic pyrolysis, the cracking of mixtures, and finally a discussion of the effects of feed additives and diluent gases.

II. PYROLYSIS OF PURE HEAVY PARAFFINS

A. Product Distributions at Low Conversions and Pressures

At low conversions and pressures of a few atmospheres and below, higher paraffins selectively crack to form olefins, methane, and ethane. Small amounts of hydrogen and propane are also observed. Among the olefins, the distribution is strongly weighted toward ethylene and propylene, although all possible olefinic products of molecular weights lighter than the reactant are observed. A typical product distribution for hexadecane cracking at 500°C is shown in Table I (Voge and Good, 1949). In general, straight-chain paraffins produce only straight-chain products, and all olefins produced are alpha-olefins. Branched paraffins yield some branched products and possibly internal olefins as primary products. No paraffins higher than propane are observed in significant amounts at low pressures. The reactions are generally first order in reactant, with observed activation energies of 60 ± 5 kcal/mol. The product distributions can be reconciled by the simple mechanism described in the next section.

B. Free-Radical Mechanism

It is now generally accepted that almost all hydrocarbon pyrolysis proceeds by a free-radical mechanism. The effects of inhibitors such as nitric oxide and propylene had led some investigators to propose a parallel molecular pathway (Stubbs *et al.*, 1952), but it was subsequently shown that such effects could also be adequately explained by alternate free-radical paths and that the molecular path hypothesis presented other problems. A free-radical mechanism allowed the most consistent explanation of experimental data. Such a mechanism involves several reaction

TABLE I

Products from Hexadecane Cracking[a,b]

Product	Yield (mol/100 mol cracked)		
	Experimental result	Predictions	
		Simple Rice	Rice–Kossiakoff
CH_4	53.0	50.7	61.2
C_2H_4	76.9	338.7	99.9
C_2H_6	52.6	49.4	38.8
C_3H_6	47.4	13.5	50.3
C_3H_8	13.0	0	0
C_4H_6	1.6	0	0
i-C_4H_8	1.6	0	0
n-C_4H_8	16.6	10.1	27.2
C_4H_{10}	2.8	0	0
C_5H_{10}	8.5	6.8	14.9
C_5H_{12}	0.1	0	0
C_6	24.3	6.8	16.5
C_7	16.2	6.8	14.0
C_8	13.4	6.8	12.3
C_9	10.1	6.8	10.9
C_{10}	10.9	6.8	9.9
C_{11}	9.3	6.8	8.9
C_{12}	7.3	6.8	7.8
C_{13}	7.4	6.8	7.1
C_{14}	4.9	6.8	7.0
C_{15}	—	3.4	3.5
Total hydrocarbon	378.2	533.8	390.2
H_2	16.7	0	0

[a] From Voge and Good (1949). Reproduced with permission from the *Journal of the American Chemical Society*. Copyright 1949 by the American Chemical Society.

[b] Experimental conditions: 500°C and 1 atm hydrocarbon partial pressure.

types: (1) *Initiation*—the introduction of free radicals into the reaction system. In pyrolysis, the reaction is initiated thermally; typically higher paraffin pyrolysis initiation involves homolytic cleavage of a carbon–carbon bond. (2) *Propagation*—a series of reactions that converts reactants to products while leaving the radical concentrations unchanged. Typical propagation steps might include radical decomposition, radical isomerization, hydrogen transfer, and radical addition. (3) *Termination*—the combination or disproportionation of radicals to give stable products.

A balance is rapidly set up between initiation and termination reactions such that the overall free-radical concentration changes very slowly with reaction time. [Over small increments of conversion, constant radical concentration can be assumed. This steady-state hypothesis allows the algebraic determination of radical concentrations and of the form of the rate equation (Laidler, 1965). Such an approach is useful for thermal cracking reactions at low conversions, although computer solution of the coupled differential elementary rate equations, when possible, provides a more accurate approach.]

This basic mechanism for paraffin pyrolysis was proposed by Rice and co-workers (1931, 1933, 1934). Specifically, they proposed that the propagation steps involved hydrogen abstraction by small free radicals from the reactant to form what might be termed a parent radical, and decomposition of this parent radical to form olefinic products and smaller free radicals. The radical decomposition involved scission of the carbon–carbon bond β to the radical site:

$$R{-}CH_2{-}CH_2{-}\dot{C}H_2 \longrightarrow R\dot{C}H_2 + CH_2{=}CH_2$$

or

$$R{-}CH_2\dot{C}H{-}CH_3 \longrightarrow R\cdot + CH_2{=}CH{-}CH_3$$

The hydrogen-abstraction reaction was assumed to be random, with all equivalent C—H bonds assumed to be equally likely to be broken. For paraffins, three types of C—H bonds are possible: primary, secondary, and tertiary. Rice found that the best fit with experimental data required activation energies for secondary and tertiary H abstraction to be 2 and 4 kcal/mol, respectively, less than for primary H abstraction. By statistically weighting according to these stabilities and the relative numbers of each type of C—H bond, the relative concentration of each possible parent radical could be predicted. These parent radicals then decomposed rapidly by β scission until there was no further C—C bond β to a radical site. Because such radicals (e.g., methyl, ethyl) had no ready decomposition pathway open, they were able to stabilize themselves by abstracting hydrogen from the reactant, thus continuing the chain.

This mechanism described the product distributions well for hexane and lighter paraffins but proved inadequate for heavier reactants. It predicted far too much ethylene and not enough higher olefin products. The model was subsequently modified by Kossiakoff and Rice (1943) to correct for this. They pointed out that alkyl radicals, in addition to decomposing by β scission, could isomerize by coiling back on themselves and transferring hydrogen internally. Such internal hydrogen transfer, it was assumed, could only take place between carbons 5 atoms apart on the chain, for

example,

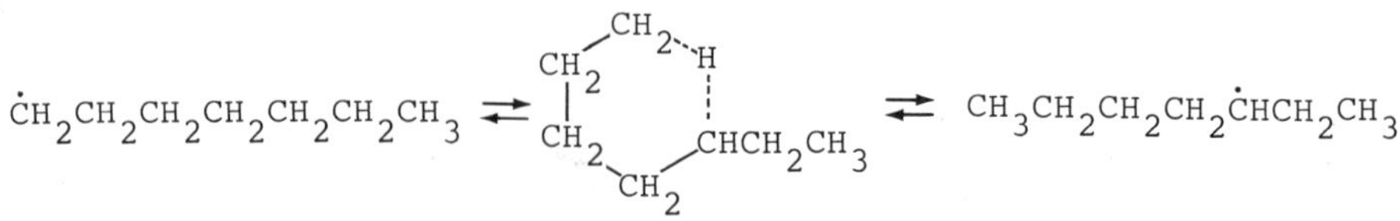

They further assumed that isomerization occurs on all radicals with chains of 6 carbons or more, and that isomerization brings all possible isomers into equilibrium. Rice and Kossiakoff showed that the model predicted the behavior of hexane, but the first real test for heavier paraffins was the study of Voge and Good (1949) mentioned earlier. Table I compares the prediction of the simple Rice and Rice–Kossiakoff mechanisms with their experimental results for hexadecane cracking at 1 atm. (Voge and Good modified the Rice–Kossiakoff isomerization hypothesis by allowing internal hydrogen transfer between carbons 4 *or more* positions apart on the chain. If isomerization to equilibrium is assumed, these hypotheses are almost equivalent.) Given the complexity of the product distribution, the Rice–Kossiakoff mechanism fits the experimental data reasonably well. Similar comparisons were made by the same authors for the experimentally observed and predicted products of isododecane and *n*-dodecane with good agreement.

Almost 50 years after it was proposed and 40 years after its initial experimental test for heavy paraffins, the Rice–Kossiakoff mechanism is still considered to be the best explanation of the product distributions from higher paraffin cracking. Minor modifications have been proposed over the years for the purpose of "fine tuning." These have been primarily concerned with the details of the isomerization step. Rice and Kossiakoff specifically mention only 1,5-hydrogen transfers in their mechanism. Both Voge and Good (1949) and Fabuss *et al.* (1962) extended this to include all H transfers greater than 1,5 as well. Murata *et al.* (1973) show that equilibrium among all possible isomers of higher radicals is necessary for good agreement, suggesting that at least 1,6-hydrogen transfer in addition to 1,5 is occurring. They also generalize the Rice–Kossiakoff (R–K) mechanism using a Markov-chain approach to simplify its use.

Rumyantsev *et al.* (1975) also propose modifications to the isomerization step. They suggest that primary radicals can isomerize to secondary through 5-, 6-, 7-, or 8-membered rings (1,4, 1,5, 1,6, and 1,7-hydrogen transfer, respectively), and that the relative probabilities for such isomerizations are 0.4, 1.0, 0.4, and 0.3, respectively. Isomerization among secondary radicals is assumed to proceed only by 6-membered rings. They show that this modification significantly improves the agreement with experiment for *n*-dodecane pyrolysis.

Despite these proposed improvements, the basic mechanism of radical isomerization, decomposition by β scission, and intermolecular hydrogen transfer involving methyl and ethyl radicals still stands. However, it has its limitations. It provides a good description of the product distribution only at low pressures and fairly low conversions. Also, because very few low-conversion studies have been done at temperatures much above 650°C, it is not known whether the mechanism will adequately describe pyrolyses at temperatures of current commercial interest. Finally, it fails to predict the formation of any molecular hydrogen because only C—C bond scissions are allowed. H_2 is observed in pyrolysis products even at moderate conversions and temperatures as low as 500°C (Voge and Good, 1949; Rebick, 1981).

The latter deficiency in the R–K mechanism can be largely corrected by allowing for ethyl-radical decomposition ($C_2H_5\cdot \rightarrow C_2H_4 + H\cdot$) as well as hydrogen transfer ($C_2H_5\cdot + RH \rightarrow C_2H_6 + R\cdot$), and by weighting these two reactions with reasonable relative activation energies. Murata and Saito (1974) have demonstrated the utility of this approach. It is unlikely that much hydrogen is formed by C—H scission in larger radicals.

The other problems with the mechanism will be dealt with in turn as the effects of conversion, pressure, and temperature on paraffin pyrolysis are considered.

C. Effects of Conversion

An increase in conversion at constant temperature results in two changes: (1) The rate constant for paraffin disappearance is reduced; and (2) the product distribution changes. The former is due to product inhibition. As noted earlier, olefins are the main primary products of paraffin pyrolysis. Olefins are also known to be potent inhibitors of free-radical decompositions due to their ability to react with radicals to form stable allylic species. As conversion increases, the concentration of olefins increases and self-inhibition sets in. This phenomenon has been discussed recently by Davis and Williamson (1979).

The second effect, a change in product distribution, is caused by secondary reactions. The rates of decomposition of higher α-olefins are comparable to those for paraffins, and thus, as these olefins are produced by paraffin cracking, they themselves decompose to lighter products. As conversion increases, the selectivity to lighter products (C_1–C_4) increases. Butadiene is not a primary product of paraffin cracking, but it and other diolefins are observed at moderate paraffin conversions. As conversion is pushed still further, butenes and propylene begin to decompose, yielding still more methane and ethylene, but also aromatics and heavier products. A schematic description of the effect of conversion on product yields for a gas oil feed is shown in Fig. 1. A more detailed description of higher olefin pyrolysis is given in Section III.

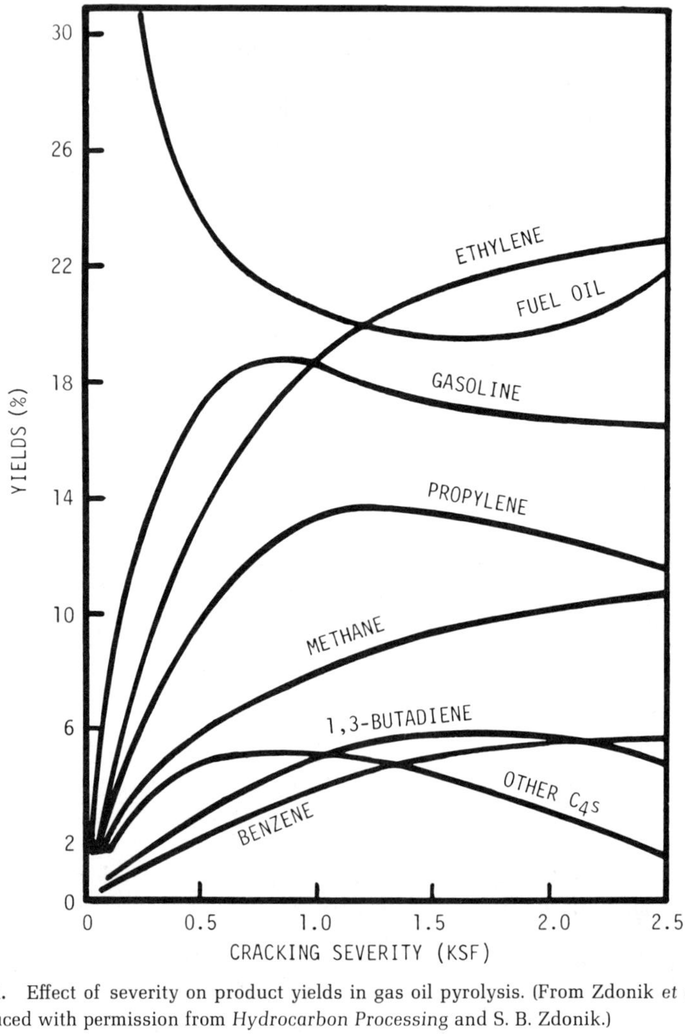

Fig. 1. Effect of severity on product yields in gas oil pyrolysis. (From Zdonik *et al.*, 1976. Reproduced with permission from *Hydrocarbon Processing* and S. B. Zdonik.)

D. Effect of Pressure

Pressure also affects both the rate and the selectivity of paraffin pyrolysis. Although the cracking reactions are nominally first order in paraffin, the first-order rate constant actually increases very gradually as pressure is raised. Typically, the rate constant may double as the pressure is increased from 1 to 50–100 atm (Fabuss *et al.*, 1964a).

The effect on product selectivity is much more interesting. As the pressure is raised, the rates of bimolecular reactions increase significantly faster than unimolecular processes. For example, the rate of hydrogen abstraction by radicals increases faster than radical decomposition. As a result the selectivity to paraffins increases. This first shows up as an increase in the relative yield of ethane and a corresponding decrease in the yields of ethylene and hydrogen as the competition between

$$C_2H_5\cdot + RH \longrightarrow C_2H_6 + R\cdot \tag{1}$$

and

$$C_2H_5\cdot \qquad\qquad C_2H_4 + H\cdot \tag{2}$$

is tipped more and more in favor of (1). As pressure is further increased, higher paraffins begin to appear, as larger alkyl radicals are stabilized before they can decompose. A complete study of the effect of pressure on hexadecane pyrolysis was conducted by Doue and Guiochon (1968). By looking at the effect of pressure on product distribution, they were able to deduce the average relative rates of decomposition and hydrogen transfer. At the highest pressures studied (about 150 atm), equal amounts of paraffins and olefins were made, indicating that each hexadecyl radical only decomposed once before being stabilized by hydrogen transfer (see Fabuss *et al.*, 1964a, Appendix, for a further analysis).

Other effects of pressure on product distribution were reported by Fabuss *et al.* (1962). In the early stages of hexadecane pyrolysis at 69 atm, they observed the formation of significant amounts of products heavier than C_{16}, presumably from alkylation or polymerization. At higher conversions, these heavy products decomposed. This is again most likely due to the favoring of bimolecular processes ($R\cdot$ + olefin → alkylated product) over unimolecular ones ($R\cdot$ → decomposition products) at higher pressures.

In general, these effects are largely unimportant at the pressure ranges of commercial interest. In the hydrocarbon partial pressure ranges of 0.5–2 atm, the main effect of pressure will be to increase the relative yields of ethane and propane, and perhaps to increase slightly the selectivity to aromatics and tars (Babash *et al.*, 1974).

E. Effect of Temperature

As was noted earlier, there have been very few low-conversion studies of heavy paraffins at temperatures much above 650°C; so any discussion of temperature effects is clouded by the influence of secondary reactions. Nonetheless, temperature is known to have significant effects on both rates and selectivities of paraffin pyrolysis.

Because thermal cracking reactions proceed with significant activation energies, temperature has an important effect on rate. Reported activation energies of the cracking of normal paraffins heavier than hexane are in the range of 55–65 kcal/mol (Fabuss *et al.*, 1964a). Isoparaffins generally crack more easily than straight-chain paraffins except if quaternary carbon atoms are present. The activation energy will depend greatly on how it is measured because of secondary reactions, with activation energies measured at constant conversion being 10–20 kcal/mol greater than those measured at constant residence time (Davis and Williamson, 1979).

The effect of temperature on product selectivity is more complex. Generally, an increase in temperature results in an increase in the selectivity to light products, especially to methane, ethylene, and hydrogen. Part of the increase in the latter two compounds occurs because reaction (2) has a higher activation energy than (1), so temperature has the opposite effect of pressure on the competition between (1) and (2). Decomposition of higher radicals by hydrogen elimination has a higher activation energy than C—C β scission by approximately 8–12 kcal/mol (Kondratiev, 1972). At high enough temperatures, this might become a factor. Also, as temperature is increased, one might expect an increase in the ratio of rates of radical decomposition to radical isomerization. This would make the product distribution look more like the simple Rice rather than the Rice–Kossiakoff, thus increasing the selectivity to ethylene (see Table I). Finally, the product selectivity at high conversions depends on secondary as well as primary reactions; thus the observed temperature effects will depend on the relative sensitivity to temperature changes of secondary versus primary processes.

Another interesting effect of temperature is its relationship to the chain length of the reaction. This is defined as the ratio of the total reaction rate to the rate of initiation. Typically, at 500–600°C, paraffins have pyrolysis chain lengths of several orders of magnitude. However, because the activation energy for initiation is significantly greater than that for the overall decomposition (75–80 versus 55–65 kcal/mol), as the temperature is raised, the chain length becomes shorter and shorter, and initiation and termination products make up a larger fraction of the total. At high enough temperatures, the chain length becomes unity, and only products resulting from initiation and termination processes are observed. For most heavy paraffins, this regime is not reached at temperatures of commercial interest.

III. PYROLYSIS OF PURE HEAVY OLEFINS

The cracking of heavy olefins is of interest because such compounds are among the primary products of higher paraffin pyrolysis, and their decomposition becomes significant as a secondary process in paraffin conversion at severities of commercial interest. Higher α-olefins decompose by similar paths to paraffins, the propagation chain consisting of radical decomposition and hydrogen-transfer steps. The products are similar to those observed in paraffin cracking—hydrogen, methane, and ethane, and the whole range of α-olefins lighter than the reactant, with ethylene and propylene predominating. However, there are some significant differences. First, the whole range of α,Ω-diolefins lighter than the reactant are also

TABLE II

Low-Conversion Product Yields from 1-Hexadecene and *n*-Hexadecane Pyrolysis[a,b]

	Yield (mol/100 mol cracked)	
Products	1-Hexadecene	*n*-Hexadecane
Space time (s)	9.2	20.9
Conversion (%)	3.1	4.9
H_2	<2	5.1
CH_4	28.2	48.8
C_2H_4	60.4	83.6
C_2H_6	18.6	48.1
C_3H_6	46.2	43.6
C_3H_8	3.7	4.4
i-C_4H_8	16.3	23.5
1,3-C_4H_6	8.7	—
C_5H_{10}	11.9	12.0
C_5H_8	1.1	—
C_6H_{12}	13.3	13.2
C_6H_{10}	5.1	0
C_7H_{14}	9.6	13.0
C_7H_{12}	3.0	0
C_8H_{16}	8.0	12.2
C_8H_{14}	2.7	0
C_9H_{18}	11.6	11.6
C_9H_{16}	1.9	—
$C_{10}H_{20}$	10.2	11.7
$C_{10}H_{18}$	1.3	0
$C_{11}H_{22} + C_{11}H_{20}$	4.9	10.3
$C_{12}H_{24} + C_{12}H_{22}$	8.0	9.4
$C_{13}H_{26} + C_{13}H_{24}$	24.6	8.2
$C_{14}H_{28} + C_{14}H_{26}$	3.5	8.9
$C_{15}H_{30} + C_{15}H_{28}$	1.5	2.8

[a] Rebick (1979, 1981).

[b] Experimental conditions: 500°C and 0.18 atm hydrocarbon partial pressure.

observed as primary products; such compounds are not primary products of paraffin cracking. Small yields of cyclic olefins are also observed at low conversions, and again these materials are not primary cracking products of paraffins. Finally, the distribution of olefin products is different from that for paraffin cracking, with larger than expected yields of propylene and of the olefin with 3 fewer carbon atoms than the reactant (Miller, 1963; Rumyanstev *et al.*, 1975; Giraud-Horvilleur and Blouri, 1977; Rebick, 1979). A comparison of the product yields from 1-hexadecene and *n*-hexadecane pyrolysis is shown in Table II.

This product distribution can be rationalized by assuming that the decomposition occurs by three parallel pathways—one molecular and the other two free radical (Rebick, 1979; Rumyantsev *et al.*, 1975). The molecular pathway is a retro-ene reaction that splits the olefin C_nH_{2n} into C_3H_6 and $C_{n-3}H_{2n-6}$. This pathway accounts for the peaks in the olefin product distribution. The two free-radical pathways both involve parent-radical decomposition by C—C β scission, but in one case the parent radical is created by hydrogen abstraction from the reactant by small radicals ($H\cdot$, $CH_3^{\cdot}$, $C_2H_5^{\cdot}$, $C_3H_5^{\cdot}$), whereas in the other, the small radicals add to the reactant olefin to make the parent radicals. This mechanism qualitatively accounts for all of the observed products and with some additional assumptions of radical stabilities, can provide a reasonable quantitative fit as well. With suitable modifications, the mechanism also describes the decomposition of internal olefins (Giraud-Horvilleur and Blouri, 1977).

Similar kinetic trends are observed for olefins as paraffins. Activation energies 5–10 kcal/mol lower than the corresponding paraffins have been reported (Rumyantsev *et al.*, 1975), and the reaction orders tend to be somewhat higher than unity. However, as with paraffins, the cracking rate increases slowly with increasing molecular weight. Less is known about the effect of pressure on olefin cracking.

IV. NAPHTHENES AND ALKYL AROMATICS

Besides paraffins, pyrolysis feedstocks contain naphthenes and aromatics. In virgin feedstocks, both compound types are alkylated, and side-chain decomposition dominates the pyrolysis mechanism.

A. Naphthenes

Very few studies have been made of the pyrolysis of pure alkylated naphthenes. The most extensive is that of Fabuss *et al.* (1964b). There have been some studies of unsubstituted cycloparaffins, most dealing with cyclohexane (Virk *et al.*, 1979; Levush *et al.*, 1969). In general, cycloparaffins are more refractory than paraffins, presumably because the initiation step, H· elimination, is slower. Because of this, one might assume that side-chain cracking would dominate the pyrolysis of alkylated naphthenes. Also, cycloparaffins tend to produce tars and coke more readily than paraffins, presumably because they can readily dehydrogenate to aromatics.

B. Alkyl Aromatics

The literature on alkyl aromatics pyrolysis is also sparse. It is presumed that side-chain cracking is the dominant process, and studies using the toluene-carrier technique have determined the activation energies for the initiation step in ethyl-, propyl-, and butylbenzene pyrolysis (Leigh and Szwarc, 1952; Crowne *et al.*, 1968; Barton and Stein, 1980). For those reactants, the side chain is believed to crack to form a benzyl and an alkyl radical. The overall pyrolysis can be described by a modified Rice mechanism, if the greater stability of the benzyl radical rather than primary or secondary alkyl radical is taken into account. A benzylic C—H bond is about 10 kcal weaker than secondary C—H (Kerr, 1966), so much of the hydrogen abstraction will be at the carbon α to the ring. β scission would then give styrene as the predominant aromatic product, and, indeed, this is what is observed in butyl benzene pyrolysis (Rebick, 1980). Alkyl aromatics tend to make coke and tars more readily than paraffins, presumably because of the tendency of benzyl radicals to combine or add to other aromatics to make heavier species, which ultimately condense to polynuclear aromatics (Badger and Spotswood, 1960; Smith, 1979). The pyrolysis of aromatics (particularly toluene) will be discussed more fully in Chapter 5 of this book.

V. PYROLYSIS OF MIXTURES

The previous three sections have dealt with pyrolysis of pure compounds. In order to translate this information into an understanding of how real feedstocks (naphthas, gas oils, crudes) crack, one needs to know how pure compounds interact with each other when they are cracking in admixture. Such information can be gained by studying the pyrolysis of simple mixtures of pure hydrocarbons. Unfortunately, few such studies have been carried out on mixtures of heavy ($>C_{10}$) hydrocarbons; however, much work has been done with lighter compounds, and the information gained there should be generalizable to heavier materials.

A. General Principles

If hydrocarbons cracked by simple unimolecular concerted processes, one might expect no interaction between species when they were cracked in admixture. However, because the predominant mechanism is a free-radical chain, the radical intermediates produced by one compound can interact with those produced by another, or with the reactant itself. Kinetics can be altered by the replacement of a rate-determining step, and selectivities can change because the distribution of intermediates is altered. Some general principles are as follows:

1. Compounds that crack by similar mechanisms have little effect on each other. The reason is straightforward—no new radical intermediates are introduced by either compound into the mechanism of the other.
2. If two compounds with significantly different cracking rates are cracked in admixture, the more refractory compound tends to inhibit the cracking of the more reactive compound, and the latter accelerates the cracking of the former. The relative magnitudes of these effects may not be equal, depending on the mechanism. In particular, if a compound is refractory because of a slow initiation step, its pyrolysis will be enhanced by the addition of a less refractory material without any significant reverse inhibition. On the other hand, if a compound is refractory because the radical intermediates produced in its cracking are stable, then it will have a significant inhibiting effect on the cracking of more reactive materials.

Examples of these effects are given in the following subsections.

B. Paraffin–Paraffin Mixtures

In general, if two paraffins are cracked in admixture, they behave as if they were cracked separately (Shah *et al.*, 1973). Both rates and selectivities are unchanged. The main exception to this is paraffin cracking in admixture with ethane. At moderate severities, the cracking of ethane is significantly accelerated by the addition of a higher paraffin, while little reverse inhibition is observed (Froment *et al.*, 1979). The net effect is a significant increase in ethylene yield over what would be observed in segregated cracking. At higher conversions, the olefins produced by the higher paraffin inhibit the ethane cracking, and the reverse effect is observed (Froment *et al.*, 1977, 1979). The net effect at high severities, then, is a decrease in ethylene yield. In general, in analyzing cracking patterns of naphthas and gas oils, one can neglect any interactions between paraffins in the feed.

C. Paraffin–Olefin Mixtures

In general, olefins inhibit the cracking of paraffins, and conversely, paraffins accelerate the cracking of olefins. The most dramatic effects are observed in the cracking of paraffins with propylene (Lavrovski *et al.*, 1971; Bakh *et al.*, 1974), but similar effects are observed with higher olefins (Shah *et al.*, 1973). In terms of real feed cracking, because olefins are not usually present in virgin feeds, the main effect this phenomenon has is in causing self-inhibition at high paraffin conversions (see Section II,C; see also Davis and Williamson, 1979).

D. Paraffin–Aromatics Mixtures

Few pure-compound studies have been made of such mixtures. At the low pressures and high temperatures characteristic of commercial pyrolysis, one would expect little selectivity change but perhaps some paraffin cracking rate decrease due to the introduction of stable benzyl radicals. Also, in mixtures of paraffins and unsubstituted aromatics (e.g., benzene, naphthalene), one might expect some increase in the rate of conversion of the aromatics because of the presence of radicals in the system from the paraffin. At present, there are no literature data to substantiate either of these hypotheses.

VI. EFFECTS OF ADDITIVES AND DILUENT GASES

A. Steam and Inert Gases

In much of the literature discussed in this and previous chapters, nitrogen or steam is used as a diluent in the cracking reactor to control hydrocarbon partial pressure. In commercial practice, steam dilution is used for this purpose and also to mitigate surface coking. It is thought to have no effect on the homogeneous cracking reaction; thus any changes due to the presence of steam are due to changes in surface effects (see Chapter 10, Albright and Tsai). However, Doue and Guiochon (1969) report that inert gas pressure can have an effect on homogeneous reaction selectivities. In particular, they show that a large excess of steam or argon reduces the selectivity to paraffins in the high-pressure pyrolysis of hexadecane. They attribute this to the ability of the inert gas molecules to activate intermediate radicals to allow them to decompose rather than abstract hydrogen. This effect should only be of concern at pressures much higher than those encountered in commerical practice. Thus, with the exception of surface effects and, of course, overall hydrocarbon partial pressure considerations, steam or inert gas dilution effects can be ignored.

B. Hydrogen

Considerable work has been done to investigate the effect on hydrocarbon pyrolysis of dilution with hydrogen under both high- and low-pressure conditions. Kunugi and co-workers (1967) showed for a variety of hydrocarbon feedstocks that replacement of steam by H_2 at typical steam-cracking conditions ($\sim$850°C, 1–3 s, H_2/hydrocarbon molar ratio of 5—7) resulted in an increase in the overall cracking rate as well as the yields of methane, ethylene, ethane, and higher saturates, and decreases in the yields of higher olefins and coke. Similar results have been reported by other researchers with both real feeds (Tsybulevskii *et al.*, 1971; Welther *et al.*, 1971) and pure compounds (Rybin and Yampol'skii, 1976; Amano and Uchiyama, 1963). These effects can be accounted for by several factors. The increase in saturates (methane, ethane, and propane) is most likely due to rapid stabilization of alkyl radicals by abstraction from molecular hydrogen ($R\cdot + H_2 \rightarrow RH + H\cdot$, where $R\cdot = CH_3\cdot$, $C_2H_5\cdot$, or $C_3H_7\cdot$). This will also decrease the net yield of hydrogen. The increase in ethylene and decrease in propylene and higher olefins can be explained by hydrogen-atom addition to the olefin, followed by more selective cracking of the resulting alkyl radical to ethylene. In particular, Amano and Uchiyama (1963) have shown that propylene cracks more selectively to ethylene and methane in the presence of hydrogen. A third effect of hydrogen dilution is to enhance the hydrodealkylation of alkyl aromatics. This should increase the overall yield of light products as well as the yield of benzene. The latter was reported by Kunugi but has not been observed by other researchers, most of whom state that H_2 reduces the yield of all aromatics (e.g., Tsybulevskii *et al.*, 1971). Finally, the increase in overall cracking rate has been attributed to the replacement of stable unreactive chain carriers such as allyl or benzyl radicals by very reactive hydrogen atoms.

Similar advantages are claimed for the pyrolysis of various hydrocarbon feeds in the presence of hydrogen at high pressures (10 to 70 atm) (Chahvekilian and Plichon, 1974; Shabtai *et al.*, 1979; Nowak *et al.*, 1975). This hydropyrolysis process, it is claimed, can crack heavy feeds with high ethylene yields and minimal coke and tar formation. Careful residence-time control appears to be necessary to prevent excessive hydrogenation at these pressures. In both low- and high-pressure reactors, hydrogen can also reduce metal oxides on the reactor surface, altering the heterogeneous chemistry. Such effects are discussed further in Chapter 10.

C. H_2S and Hydrogen Halides

Several researchers have investigated the effect of H_2S, HCl, and HBr on hydrocarbon pyrolysis (for a recent review, see Rebick, 1980). These additives have been shown to alter the rate of cracking of many hydrocarbons, and also to change the product selectivity. The result of adding an inorganic hydride HX (X = HS, Cl, or Br) is to replace a slow hydrogen-transfer step

$$R_1\cdot + R_2H \longrightarrow R_1H + R_2\cdot \tag{3}$$

with two faster reactions:

$$R_1\cdot + HX \longrightarrow R_1H + X\cdot \tag{4}$$

$$X\cdot + R_2H \longrightarrow HX + R_2\cdot \tag{5}$$

HX thus functions as a catalyst for the hydrogen-transfer step and accelerates the rate of cracking. This effect forms the basis for several patents (e.g., Porchey and Royer, 1974; Tischler and Wing, 1973). In some cases (e.g., ethane cracking), hydrogen transfer is not rate limiting, and HX either has no effect (e.g., HCl) or inhibits the reaction (H_2S, HBr). The latter is due to the participation of HS· or Br· in termination reactions (Scacchi *et al.*, 1970). For heavy paraffins and alkyl aromatics, H_2S accelerates the cracking because of hydrogen-transfer catalysis but also modifies the reaction selectivity (Rebick, 1980, 1981). When hexadecane is pyrolyzed in the presence of 0.1–0.3 atm of H_2S, the light gas yield (C_1, C_2) is greatly reduced, and paraffins heavier than propane appear in the products. This is analogous to the results reported by Doue and Guiochon at high hydrocarbon pressures, in the absence of any diluent, and is due to the stabilization of intermediate alkyl radicals by reaction (4). Changes in selectivity have also been reported for the cracking of 2-methyl-2-pentene in the presence of HBr and H_2S, and this chemistry forms the basis of a process for isoprene manufacture (Hutchings *et al.*, 1976).

It should be noted that in addition to the homogeneous effects just discussed, H_2S, HCl, and HBr can also modify the surface of a pyrolysis reactor, altering the heterogeneous chemistry (see Chapter 10 by Albright and Tsai).

VII. CONCLUSIONS

This chapter has summarized the large amount of work that has been done on the pyrolysis of pure heavy hydrocarbons. Our present understanding of the cracking of paraffins and olefins is good, although the applicability of the Rice–Kossiakoff mechanism at high temperatures

(>800°C) has yet to be determined. Less work has been done on naphthenes and aromatics, particularly at temperatures of commercial interest. Studies of pure compounds more typical of those found in heavy naphthas and gas oils are required. Finally, in order to make pure-compound data applicable to real feed cracking, more studies of compound mixtures are required. In particular, the extent of nonadditive interaction between paraffins and alkyl aromatics has to be established.

REFERENCES

Amano, A. and Uchiyama, M. (1963). *J. Phys. Chem.* **67,** 1242.

Babash, V. V., Mulchina, T. N., and Aerov, M. E. (1974). *Khim. Tekhnol. Topl. Masel* No. 10, 9–12.

Badger, G. M., and Spotswood, T. M. (1960). *J. Chem. Soc.* p. 4420.

Bakh, G., Novak, Z., Korochuk, S. I., and Kalinenko, R. A. (1974). *Kinet. Katal.* **15,** 1103–1109.

Barton, B. D., and Stein, S. E. (1980). *J. Phys. Chem.* **84,** 2141.

Chahvekilian, E., and Plichon, J. M. (1974). U.S. Patent 3,842,138.

Crowne, C. W. P., Grigolis, V. J., and Throssell, J. J. (1968). *Trans. Faraday Soc.* **65,** 1051–1058.

Davis, H. G., and Williamson, K. D. (1979). *In* "Thermal Hydrocarbon Chemistry" (A. G. Oblad, H. G. Davis, and R. T. Eddinger, eds), Advances in Chemistry Series, Vol. 183, pp. 41–66. Am. Chem. Soc., Washington, D.C.

Doue, F., and Guiochon, G. (1968). *J. Chim. Phys.* **64,** 395.

Doue, F., and Guiochon, G. (1969). *Can. J. Chem.* **47**, 3477.

Egloff, G., Thomas, C. L., and Linn, C. B. (1936). *Ind. Eng. Chem.* **28,** 1283.

Fabuss, B. M., Smith J. O., Lait, R. I., Borsanyi, A. S., and Satterfield, C. N. (1962). *Ind. Eng. Chem., Process Des. Dev.* **1,** 293.

Fabuss, B. M., Smith, J. O., and Satterfield, C. N. (1964a). *Adv. Pet. Chem. Refin.* **9,** 156–195.

Fabuss, B. M., Kafesjian, R., Smith, J. O., and Satterfield, C. N. (1964b). *Ind. Eng. Chem., Process Des. Dev.* **3,** 248.

Froment, G. F., Van de Steene, B. O., Vanden Berghe, P. J., and Goossens, A. G. (1977). *AIChEJ.* **23,** 93.

Froment, G. F., Van de Steene, B. O., and Sumedha, O. (1979). *Oil Gas J.* Apr. 16, **77**(16), 87–90.

Giraud-Horvilleur, F., and Blouri, B. (1977). *Inf. Chim.* **164,** 113.

Greco, M. (1978). *J. Chromatogr. Sci.* **16,** 36.

Hutchings, D. A., Frech, K. J., and Hoppstock, F. H. (1976). *In* "Industrial and Laboratory Pyrolyses" (L. F. Albright and B. L. Crynes, eds.), ACS Symposium Series, Vol. 32, pp. 178–196. Am. Chem. Soc., Washington, D.C.

Kerr, J. A. (1966). *Chem. Rev.* **66,** 465–500.

Kondratiev, V. N. (1972). "Rate Constants of Gas Phase Reactions," NSRDS-COM-72-10014, p. 177. U.S. Dep. Commer., Washington, D.C.

Kossiakoff, A., and Rice, F. O. (1943). *J. Am. Chem. Soc.* **65,** 590.

Kunugi, T., Tominaga, H., and Abiko, S. (1967). *Proc. World Pet. Congr., 7th, Mexico City* **5,** 239–245.

Laidler, K. J. (1965). "Chemical Kinetics," pp. 327–329, 390–393. McGraw-Hill, New York.

Lavrovski, K. P., Kalinenko, R. A., and Shevel'kova, L. V. (1971). *Proc. World Pet. Congr., 8th, Moscow* **4,** 357–366.

Leigh, C. H., and Szwarc, M. (1952). *J. Chem. Phys.* **20,** 403–411.

Levush, S. S., Abadzhev, S. S., and Shevchuk, V. U. (1969). *Neftekhimiya* **9,** 716.

Miller, D. B. (1963). *Ind. Eng. Chem., Prod. Res. Dev.* **2,** 220.

Murata, M. and Saito, S. (1974). *J. Chem. Eng. Jpn.* **7,** 389–391.

Murata, M., Saito, S., Amano, A., and Maeda, S. (1973). *J. Chem. Eng. Jpn.* **6,** 252–258.

Neuhaus, M., and Marek, L. F. (1932). *Ind. Eng. Chem.* **24,** 490.

Nowak, S., Keil, G., Gunschel, H., and Pechstein, G. (1975). *Proc. World Petr. Congr., 9th, Tokyo* **5,** 145–153.

Porchey, D. V., and Royer, D. J. (1974). U.S. Patent 3,803,260.

Rebick, C. (1979). *In* "Thermal Hydrocarbon Chemistry" (A. G. Oblad, H. G. Davis, and R. T. Eddinger, eds.), Advances in Chemistry Series, Vol. 183, pp. 1–19. Am. Chem. Soc., Washington, D.C.

Rebick, C. (1980). *In* "Frontiers of Free Radical Chemistry" (W. A. Pryor ed.), pp. 117–137. Academic Press, New York.

Rebick, C. (1981). *Ind. Eng. Chem., Fundam.* **20,** 54–59.

Rice, F. O. (1931). *J. Am. Chem. Soc.* **53,** 1959.

Rice, F. O. (1933). *J. Am. Chem. Soc.* **55,** 3035.

Rice, F. O., and Herzfeld, K. F. (1934). *J. Am. Chem. Soc.* **56,** 284.

Rumyantsev, A. N., Nametkin, N. S., Lavrovsky, K. P., Sanin, P. I., Masaev, I. A., Vinintsky, O. M., and Kurashava, E. K. (1975). *Proc. World Pet. Congr., 9th, Tokyo* **5,** 155.

Rybin, V. M., and Yampol'skii, Y. P. (1976). *Neftekhimiya* **16,** 729–736.

Scacchi, G., Dzierzynski, M., Martin, R., and Niclause, M. (1970). *Int. J. Chem. Kinet.* **2,** 115.

Shabtai, J., Ramakrishman, R., and Oblad, A. G. (1979). *In* "Thermal Hydrocarbon Chemistry" (A. G. Oblad, H. G. Davis, and R. T. Eddinger, eds.), Advances in Chemistry Series, Vol. 183, pp. 297–328. Am. Chem. Soc., Washington, D.C.

Shah, Y. T., Stuart, E. B., and Kunzru, D. (1973). *Ind. Eng. Chem., Process Des. Dev.* **12,** 344.

Smith, R. D. (1979). *J. Phys. Chem.* **83,** 1553.

Stinson, S. C. (1979). *Chem. Eng. News* **57**(22), 32–35.

Stubbs, F. J., Ingold, K. U., Spall, B. C., Danby, C. J., and Hinshelwood, C. (1952). *Proc. R. Soc. London, Ser. A* **214,** 20–35.

Tischler, L. G., and Wing, M. S. (1973). U.S. Patent 3,773,850.

Tsybulevskii, A. M., Danilushkina, K. P., and Zhuromskaya, T. A. (1971). *Neftekhimiya* **11,** 687–692.

Virk, P. S., Korosi, A., and Woebeke, H. N. (1979). *In* "Thermal Hydrocarbon Chemistry" (A. G. Oblad, H. G. Davis, and R. T. Eddinger, eds.), Advances in Chemistry Series, Vol. 183, pp. 67–76. Am. Chem. Soc., Washington, D.C.

Voge, H. H., and Good, G. M. (1949). *J. Am. Chem. Soc.* **71,** 593.

Welther, K., Illes, V., and Szepesy, L. (1971). *Ropa Uhlie* **13,** 121–131.

Wing, M. S. (1980). *Chemtech* **10**(1), 20–22.

Zdonik, S. B., Hayward, G. L., Fishtine, S. H., and Feduska, J. C. (1975). *Hydrocarbon Process.* **54**(12), 111–114.

Zdonik, S. B., Potter, W. S., and Hayward, G. L. (1976). *Hydrocarbon Process.* **55**(4), 161–166.

5

Thermal Reaction of Olefins and Diolefins, and Formation of Aromatics

TOMOYA SAKAI

I. INTRODUCTION

When a saturated hydrocarbon such as *n*-heptane is subjected to pyrolysis at 700–800°C, olefins such as ethylene, propylene, butenes, and 1,3-butadiene, and small amounts of methane and hydrogen are formed during the initial stages of the reaction. As the reaction proceeds, the selectivities of olefins decrease from their maxima; these decreased selectivities are more rapid for higher olefins, which are converted eventually to mono- and polynuclear aromatics, methane, and hydrogen. The C_5 and C_6 cycloolefins are intermediate compounds; the formation of C_6 cycloolefins is caused mainly by Diels–Alder reactions between butadiene and olefins, whereas C_5 cycloolefins are produced by the cycloaddition of the

Pyrolysis: Theory and Industrial Practice

ISBN 0-12-048880-9

allyl radical to olefins. In this chapter, kinetic features of the thermal reactions of olefins and diolefins are described, with emphasis on the formation of cycloolefins and aromatics. As used here, "thermal" means "noncatalytic reaction at high temperatures."

In Sections II–V, kinetic features of the thermal reactions of ethylene, propylene, butenes, and butadiene, respectively, are described. The overall rate constant and the initial and the successive reaction products from each substrate are indicated. In Sections VI and VII, the results of kinetic studies are summarized for the addition of butadiene to olefins and for the addition of the allyl radical to olefins. Thus the basic reactions for production of C_6 cyclic compounds are represented as follows:

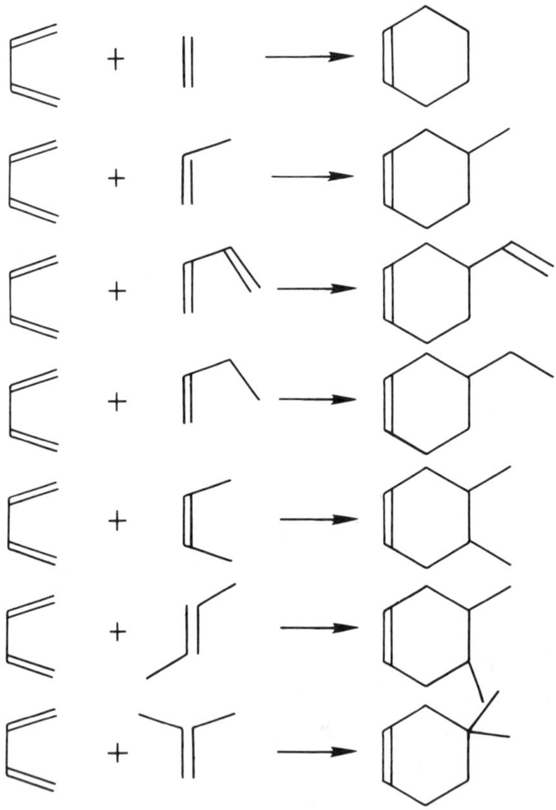

(1)
(2)
(3)
(4)
(5)
(6)
(7)

Analogously, the basic reactions for C_5 cyclic compound formation might be represented as follows, where the allyl radical plays a role somewhat similar to butadiene in the preceding reactions. In these cases, however, the efficiency of cyclization is not as high as that in the case of

Diels–Alder reactions of butadiene and olefins, and another addition reaction proceeds concurrently to produce the acyclic carbon skeleton.

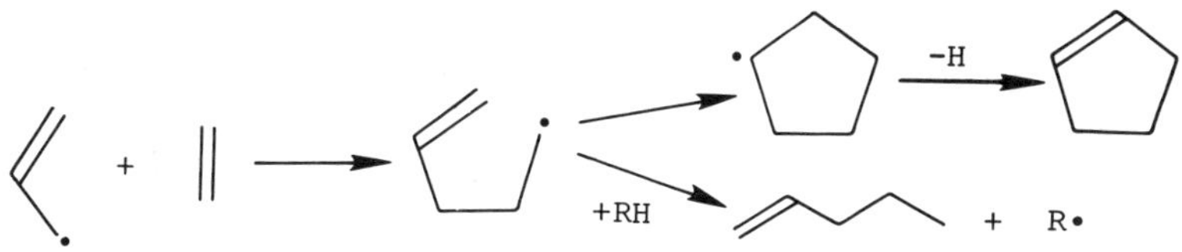

(8)

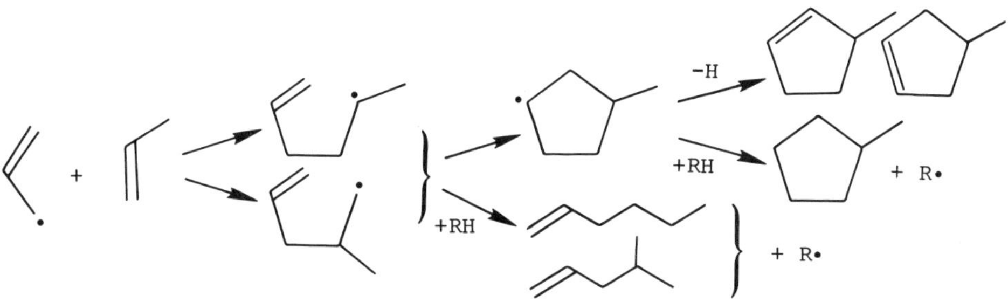

(9)

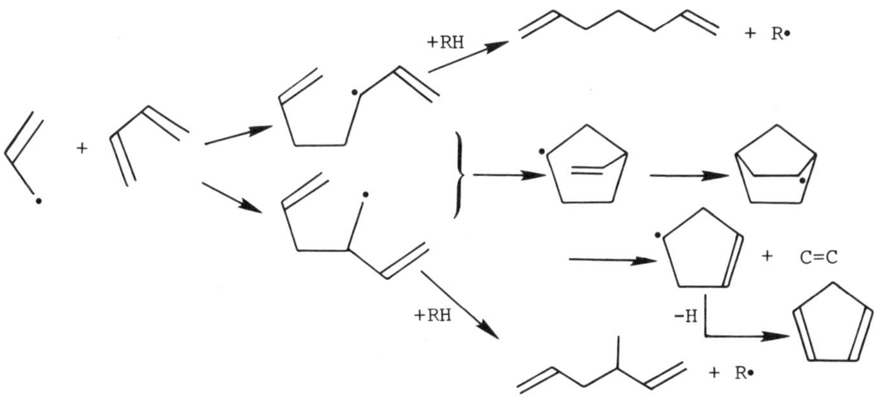

(10)

In Section VIII, kinetic results are described for reactions of 4-vinylcyclohexene, cyclohexene, and 4-methylcyclohexene. These compounds were predicted in Section VI to be typical substrates for the formation of aromatic compounds. In the final section, an example is given of the application of kinetic data obtained in the preceding sections relating to the formation of C_6 cyclic compounds for the actual pyrolysis data, and areas requiring additional information are also outlined.

Some comments should be made about the experimental conditions usually adopted to study these reactions in order to help provide a better understanding of several key factors in this research field. Feed hydrocarbons and their diluent gas should be free of oxygen. A tubular flow reactor constructed of quartz or Pyrex glass and heated electrically through a stainless-steel block or a dense fluidized phase of ceramic particles is often used. Reactant conversions should be as low as possible, provided that the initial products can be analyzed quantitatively. By adopting these experimental conditions, the influence of the specific surface area of the reactor on the reaction can be minimized. Confirmation of this conclusion can be made by inserting rods or chips of the same type of glass into the reactor. Significant differences always occur, however, whenever metallic surfaces such as steel or copper are provided in the reactor, as discussed in Section IX. The suggested reaction system is assumed to proceed as essentially a homogeneous gas-phase reaction.

II. THERMAL REACTIONS OF ETHYLENE

Only a relatively few investigations are available on the thermal reactions of ethylene. Dahlgren and Douglas (1958) reported that with ethylene at 480–582°C and at pressures of 1.2–1.83 kPa, the primary products were propylene, butenes, butadiene, and ethane. Addition of nitric oxide, hydrogen sulfide, and oxygen significantly affected the product distribution and the rate of ethylene disappearance. A radical-chain mechanism governs the reaction. Kunugi *et al.* (1969) studied the same reaction at temperatures of 700–850°C and at atmospheric pressure with and without nitrogen dilution, and they demonstrated that the chain mechanism was initiated by the C—C bond rupture of ethane, which is formed from ethylene in the reaction. The primary products were hydrogen, ethane, acetylene, propylene, 1-butene, and 1,3-butadiene. Secondary products were methane; cycloolefins such as cyclopentene, cyclopentadiene, and cyclohexene; and aromatics, mainly benzene and naphthalene. Production of hydrogen, acetylene, butadiene, and benzene are favored at higher temperatures, whereas production of ethane, propylene, 1-butene, and cycloolefins are favored at lower temperatures, in accordance with the thermal stability of these compounds. As the reaction proceeds further, 1-butene and butadiene formed in the initial stages are rapidly converted to secondary products. Cycloolefins, the intermediate compounds, increase during the initial stage of the reaction, pass through maxima, and eventually are converted to the final products, that is, aromatics. Hydrogen and methane, which are thermally stable, increase in amount as the reaction progresses.

Wheeler and Wood (1930) postulated the following scheme for the formation of aromatics in their early study on the pyrolysis of ethylene, which consisted of reactions (1)–(7) followed by dealkylation and/or dehydrogenation.

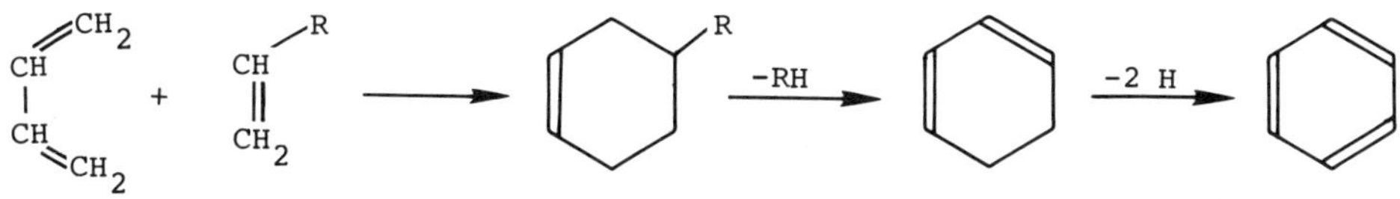

(11)

Sakai (1972) demonstrated, as indicated by Fig. 1, that the addition of a small amount of butadiene to ethylene increased the selectivity for the formation of cycloolefins and benzene to a greater degree than for 1-butene, propylene, ethane, methane, and hydrogen.

An alternative mechanism is the cyclization of 1,3,5-hexatriene formed through polymerization of ethylene followed by dehydrogenation, which was proposed by Kinney and Crowley (1954) in their article on the pyrolysis of ethylene. However, the latter mechanism is less likely for two reasons. First, the formation of hexatriene is not reported explicitly in other articles (Dahlgren and Douglas, 1958; Kunugi *et al.*, 1969; Wheeler and Wood, 1930). Second, during the thermal reaction of 1,5-hexadiene as described in

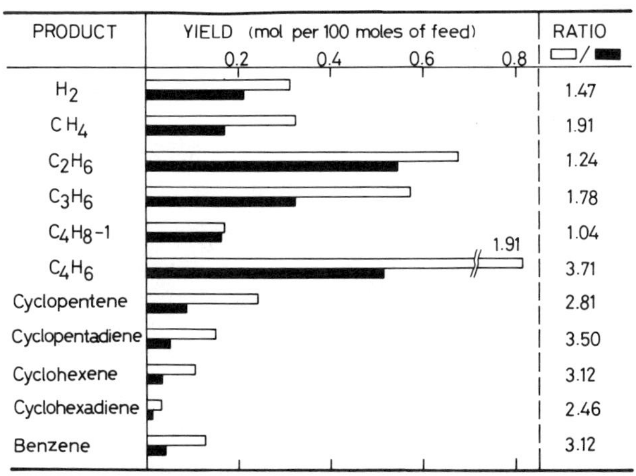

Fig. 1. Relative increase of cyclic product selectivity on addition of butadiene (C_4H_6) to thermal reaction of ethylene (753°C, 0.55 s). Unshaded bars, 2.9 mol % butadiene in feed; shaded bars, none. (Sakai, 1972).

Section VII, 1,3-cyclohexadiene was obviously produced from 1,3,5-hexatriene (which is a primary product from 1,5-hexadiene), but cyclohexene was never detected in the product stream. Both cyclohexene and cyclohexadiene were detected during the thermal reactions of ethylene. A plausible mechanism for the formation of aromatics hence is reaction (11), as supported by the following findings. A mixture of butadiene and ethylene reacted to produce 4-vinylcyclohexene and cyclohexene at 400–600°C (Rowley and Steiner, 1951). Smith and Gordon (1961) pyrolyzed cyclohexene at 425–535°C to obtain cyclohexadiene and benzene combined with concurrent dissociation products, butadiene and ethylene; more details on this will be presented later. It is worth noting that a concerted molecular process such as reactions (1–7) can proceed as fast as the radical-chain process, as evaluated on the products observed; Benson (1970) discussed such a possibility.

Ethylene is thermally the most stable compound among the substrates studied in this chapter. The order of the reaction is reported to be between first and second (Burk *et al.*, 1937; Kunugi *et al.*, 1969; Dahlgren and Douglas, 1958), and the experimental data of these workers can be represented by Eq. (12) for ethylene disappearance.

$$k = 10^{11.73} \exp(-207{,}600/RT)/\mathrm{cm}^{3/2}\,\mathrm{mol}^{-1/2}\,\mathrm{s}^{-1} \tag{12}$$

III. THERMAL REACTIONS OF PROPYLENE

Thermal reactions of propylene have been studied extensively in order to elucidate the kinetic parameters for the elementary reactions. The product analyses have been limited either to light hydrocarbons below C_4 or to the products formed in narrow ranges of temperatures and conversions. Lack of clear discrimination between the primary and secondary products and poor description of the cyclic compounds formed is caused by limitations of the analyses of the product gas. Seres and Ausloos (1973), and Simon and Back (1973) have both put emphasis on the molecular orbital calculation of the activated complex of the initiation reaction.

Laidler and Wojciechowski (1960) reported that the main products from propylene were ethylene, methane, and hydrogen, and that minor ones were ethane, propane, butenes, cyclopentadiene, cyclohexadiene, benzene, toluene, and 1,5-hexadiene. Their investigations in a static reactor system were at temperatures of 580–640°C and pressures of 5.34–53.4 kPa. They detected no allene, whereas Szwarc (1949) and Sakakibara (1964) did at higher temperatures. Kallend *et al.* (1967) performed a detailed analysis of the product gases obtained at 555–640°C and at pressures of 0.94–40.1 kPa. The main C_6 compounds formed were 1,3- and 1,4-hexadiene. Methylcyclohexene and cyclohexadienes were not detected. Kunugi *et al.* (1970a,b) investigated the thermal reaction of propylene at temperatures of 703–854°C and at atmospheric pressure both with and without nitrogen dilution. Change of product distributions were determined in detail as a function of propylene conversion. The primary products are thought to be ethylene, methane, hydrogen, butenes, butadiene, methylcyclopentene, hexadienes, acetylene, and ethane; and the secondary products are cyclopentene, cyclopentadiene, benzene, toluene, and condensed aromatics. All C_4 olefins were detected, and 1-butene was the major one. The distribution of butene isomers was nearly the same as that found by Kallend *et al.* (1967). Trace amounts of propane, allene, methylacetylene, cyclohexane, cyclohexadiene, 4-methylcyclohexene, xylenes, and styrene were detected by Kunugi *et al.* (1970a,b). Kallend *et al.* (1967), however, did not detect cyclohexadiene and methylcyclohexene formation.

Characteristic features of the thermal reactions of propylene can be summarized as follows: First, except for ethane and acetylene, selectivity of product formation is nearly independent of temperatures between 703 and 854°C. This conclusion is based on the findings of Kallend *et al.* (1967) and Hurd and Meinert (1930). The findings of Kunugi *et al.* (1970b), who evaluated the kinetics of the key elementary reactions, also support this conclusion. Second, selectivity of the formation of hydrogen, methane, benzene, and condensed aromatics increases, whereas that of butenes and butadiene decreases with conversion of propylene. Product selectivity of cycloolefins and toluene decreases at higher conversions after passing through their maxima. These facts indicate that dehydrogenation, demethylation, and cyclization predominate at higher conversion to produce hydrogen, methane, benzene, and condensed aromatics. The increase in selectivity of methane at higher conversions may be caused partly by the decomposition of 1-butene. The main products in pyrolysis of 1-butene at 640–760°C are methane and propylene (Sakai *et al.*, 1970).

Third, considerably larger amounts of C_5 cyclic compounds, such as methylcyclopentene, cyclopentene, and cyclopentadiene, are produced in reactions of propylene than those in the reactions of ethylene (Simon and Back, 1970a,b). Methylcyclopentene is produced from the initial stage of the reaction. Formation of these C_5 cyclic compounds may be attributed to the reactions (8–10). The allyl radical apparently exists in higher concentration in this reaction system than that in the ethylene case (Bryce and Ruzicka, 1960). Selectivity of ethylene shows little dependence on propylene conversion, which suggests an almost equal rate of ethylene production from propylene to that of ethylene consumption through reactions (1) and (8). Fourth, both ratios of toluene to benzene and of xylene to benzene obtained at temperatures from 703° to 854°C and at conversions up to 40% are about twice as large as those obtained in the thermal reaction of ethylene. On the contrary, the ratio of styrene to benzene is about one-third of that obtained in the ethylene case. Addition of butadiene in the thermal reaction of propylene increases the selectivity of cyclic compound formation, although the increment is smaller than that in the case of ethylene.

Laidler and Wojciechowski (1960), Kallend *et al.* (1967), and Kunugi *et al.* (1970a) reported the order of propylene disappearance as being 3/2, 7/5, and 3/2 to propylene concentration, respectively. Laidler and Wojciechowski (1960) reported a smaller *A* factor, $10^{13.34}/\mathrm{cm}^{3/2}\,\mathrm{mol}^{1/2}\,\mathrm{s}^{-1}$, and smaller activation energy, 237 $\mathrm{kJ\,mol}^{-1}$, than those reported here by Kunugi *et al.* (1970a):

$$k = 10^{15.06} \exp\left(-\frac{264{,}500}{RT}\right) \Big/ \mathrm{cm}^{3/2}\,\mathrm{mol}^{-1}\,\mathrm{s}^{-1} \qquad (13)$$

An induction period was observed by Kallend *et al.* (1967), Amano and Uchiyama (1963), and Kunugi *et al.* (1970a), but this induction period was shorter than that observed during ethylene pyrolysis at 703–854°C (Kunugi *et al.*, 1969).

IV. THERMAL REACTIONS OF BUTENES

Thermal reactions of 1-butene were investigated by Bryce and Kebarle (1958), Kunugi *et al.* (1967), and Shibatani and Kinoshita (1974); good agreement in results were obtained by these three investigators. Sakai *et al.* (1970) compared the thermal reactions of 1-butene and *cis*- and *trans*-2-butenes at 640–680°C, and their results are listed in Table I.

The main products from 1-butene are methane, propylene, butadiene, and ethylene; minor products are ethane, and *cis*- and *trans*-2-butenes. Almost the same product distribution was obtained at conversions in the range of 1.2 to 14.2%. At 700°C and higher temperatures and for 1-butene conversions above 40%, the selectivities of methane and ethylene increase and those of ethane, propylene, and 2-butenes decrease. Simultaneously, the formation of high molecular weight products becomes pronounced. Of interest, the formation of high-molecular-weight substances is less pronounced for *cis*- and *trans*-2-butene reactions even at high conversions as compared to the case of 1-butene. Possibly the larger amount of butadiene, ethylene, and propylene produced in the case of 1-butene pyrolysis as compared to pyrolyses of *cis*- and *trans*-2-butene accounts for the differences in high molecular weight products just noted.

TABLE I

Product Distributions in Thermal Reaction of Butenes[a]

	1-Butene				*cis*-2-Butene					*trans*-2-Butene			
Conversion (mol %)	5	15	30	40	10	20	30	40	50	10	20	30	40
Selectivity[b]													
CH_4	46	46	51	53	4	5	7	10	15	6	8	11	15
C_2H_6	5	5	4	4	—	—	—	—	—	—	—	—	—
C_2H_4	21	21	27	29	2	2.5	3	3.5	4	0.7	0.7	2.0	2.8
C_3H_6	46	46	44	42	4	5	7	10	15	6	8	11	15
1-C_4H_8	—	—	—	—	1	1	1	2	2	1	1	2	3
cis-2-C_4H_8	5	5	3	3	—	—	—	—	—	85	85	77	67
trans-2-C_4H_8	5	5	3	3	76	84	83	75	66	—	—	—	—
C_4H_6	30	30	30	30	5	7	9	13	15	6	8	11	15

[a] Sakai *et al.* (1970).

[b] Moles of product per 100 moles of converted butenes.

The overall rate of 1-butene disappearance is well represented by the following first-order rate equation:

$$k_{1\text{-}C_4H_8} = 10^{13.47} \exp\left(-\frac{256{,}600}{RT}\right)\Big/ s^{-1} \tag{14}$$

Isomerization between *cis*-2-butene and *trans*-2-butene is the single most important reaction in each case. Minor products in both cases were methane, propylene, and butadiene. Lesser amounts of ethylene, 1-butene, and hydrogen were formed in both cases. The first-order rate equations for *cis*-2-butene or *trans*-2-butene disappearance are as follows:

$$k_{cis\text{-}2\text{-}C_4H_8} = 10^{13.64} \exp\left(-\frac{236{,}501}{RT}\right)\Big/ s^{-1} \tag{15}$$

$$k_{trans\text{-}2\text{-}C_4H_8} = 10^{13.82} \exp\left(-\frac{239{,}400}{RT}\right)\Big/ s^{-1} \tag{16}$$

Marked differences in product selectivity and reaction rates between reactions of 1-butene and 2-butenes are attributed to the radical-chain mechanism of pyrolysis, which involves C—H and C—C bond rupture for the case of 1-butene, and the intramolecular reaction mechanism of isomerization for 2-butenes. The C—H and C—C bond rupture, however, also proceeds to a slight extent in the case of 2-butenes (Molera and Stubbs, 1952; Danby *et al.*, 1955). Shibatani and Kinoshita (1974) studied the thermal reaction of 2-butenes excluding isomerizations to each other. The radical-chain mechanism might govern the reaction of 2-butenes in this case. They found the first-order rate constant to be $k = 10^{14.8} \exp(-293{,}000/RT)/s^{-1}$; this value is either comparable with or 5 times smaller at 650°C than the k value of Eq. (14) for 1-butene.

The thermal reaction of isobutylene will not be considered here. It is, however, well known that the reaction proceeds by a radical-chain mechanism and that the main initial products from isobutylene are propylene and methane.

V. THERMAL REACTIONS OF 1,3-BUTADIENE

Dimerization of butadiene forms 4-vinylcyclohexene (VCH), as reported by Vaughan (1932). Kistiakowsky and Ransom (1939) reported the validity of the second-order kinetics of the reaction at high pressures, and described the rate constant as $k = 10^{9.96} \exp(-99{,}200/RT)/cm^3\ mol^{-1}\ s^{-1}$. They also obtained 4,4′-octahydrodiphenyl as a by-product. Rowley and Steiner (1951) reported the second-order rate constant of the reaction as $k = 10^{10.48} \exp(-115{,}100/RT)/\mathrm{cm^3\,mol^{-1}\,s^{-1}}$. Sakai *et al.* (1970) studied the thermal reaction of butadiene at higher temperatures of 550–750°C. VCH was the sole product below 600°C. As the reaction temperature was raised, cyclohexene, cyclohexadiene, benzene, butenes, propylene, ethylene, methane, and hydrogen were also formed. The second-order rate constant at 550–700°C is reported next; it is in reasonable agreement with information reported by Kistiakowsky and Ransom (1939), and Rowley and Steiner (1951):

$$k_{C_4H_6} = 10^{10.41} \exp\left(-\frac{103{,}800}{RT}\right)\Big/\mathrm{cm^3\,mol^{-1}\,s^{-1}} \qquad (17)$$

VI. THERMAL REACTIONS OF OLEFINS WITH 1,3-BUTADIENE

The rate constant reported by Rowley and Steiner (1951) for the thermal reaction of ethylene with butadiene agrees well with that obtained by Sakai *et al.* (1970) described later. Tarasenkova (1955) reported for thermal reaction between propylene and butadiene that the yield of toluene was twice as large as the combined yield of benzene and xylenes; for thermal reaction of 1-butene or 2-butene with butadiene, that ethylbenzene, or *o*-xylene, respectively, were produced.

Sakai *et al.* (1970) investigated the thermal reactions of some gaseous mixtures of ethylene, propylene, and butadiene at 510–670°C to obtain cyclohexene (HCH), 4-methylcyclohexene (MCH), and VCH as initial products of the reaction. The rates of formation of these products fitted well to the second-order kinetics as predicted here.

$$\frac{d(\mathrm{HCH})}{dt} = 2\,k_{\mathrm{HCH}}[C_2H_4][C_4H_6] \qquad (18)$$

$$\frac{d(\mathrm{MCH})}{dt} = 2\,k_{\mathrm{MCH}}[C_3H_6][C_4H_6] \qquad (19)$$

$$\frac{d(\mathrm{VCH})}{dt} = k_{\mathrm{VCH}}[C_4H_6]^2 \qquad (20)$$

Undoubtedly, the Diels–Alder reaction between olefins and butadiene, reactions (1)–(3), proceed in this system. A correction factor of 2 was, however, introduced for reactions (18) and (19) from considerations of the collision frequencies of the reactant molecules in the mixed gas. With k_{VCH} determined as in Eq. (17), then k_{HCH} and k_{MCH} can be determined

TABLE II

A Summary of Rate Data for Thermal Ractions of Olefins and Diolefins[a]

Reactant	Temperature (°C)	Product: Primary	Product: Secondary
C=C	700–850	C_4H_6, 1-C_4H_8, C_3H_6, C_2H_2, C_2H_6, H_2	CH_4, H_2, , 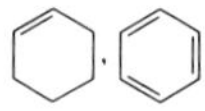
C=C—C	700–850	C_2H_4, C_4H_8, C_4H_6, , C_2H_2, CH_4, H_2	CH_4, H_2, 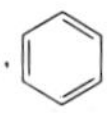, 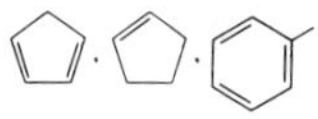, C_2H_6
C=C—C—C	640–680	CH_4, C_3H_6, C_4H_6, C_2H_6, 2-C_4H_8's	—
C—C=C—C (*cis*-)	640–680	*trans*-C_4H_8, C_4H_6, CH_4, C_3H_6	C_2H_4
C—C=C—C (*tr*-)	640–680	*cis*-C_4H_8, C_4H_6, CH_4, C_3H_6	C_2H_4
C=C—C=C + C=C—C=C	550–750	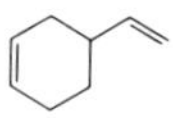	, C_2H_4, CH_4, C_3H_6, H_2
C=C—C=C + C=C	510–590		—
C=C—C=C + C=C—C	510–590	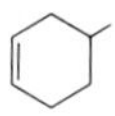	—
C=C—C=C + C=C—C—C	510–590	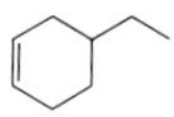	—
C=C—C=C + C—C=C—C (*cis*-)	510–590	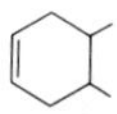	—
C=C—C=C + C—C=C—C (*tr*-)	510–590	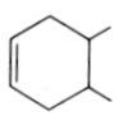	—
C=C—C=C + C=C(C)—C	510–590	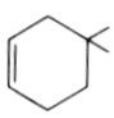	—
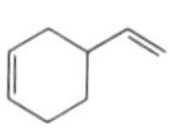	530–585	C_4H_6	—
		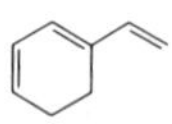	—
	585–665	C_2H_4, C_4H_6	C_3H_6

TABLE II (*continued*)

Reaction order	$\log_{10} k$ (cm^3, mol, s) 500	600	700	800	E (kJ mol^{-1})	$\log_{10} A$ (cm^3, mol, s)
1.5	—	—	0.59	1.63	207.6	11.73
1.5	—	—	0.86	2.19	264.6	15.06
1.0	—	−1.88	−0.30	—	256.6	13.47
1.0	—	−0.51	0.95	—	236.5	13.64
1.0	—	−0.50	0.97	—	239.4	13.82
2.0	3.40	4.20	4.84	5.36	103.8	10.41
2.0	2.54	3.47	—	—	120.6	10.68
2.0	2.00	2.96	—	—	124.3	10.40
2.0	0.77	2.48	—	—	220.6	15.67
2.0	0.50	2.27	—	—	229.0	15.97
2.0	2.14	3.14	—	—	128.1	10.80
2.0	1.11	2.59	—	—	191.7	14.06
1.0	−2.09	−0.12	—	—	253.7	15.05
1.0	−3.09	−1.52	—	—	202.6	10.60
1.0	−3.57	−1.44	—	—	275.0	15.01
1.0	−4.02	−2.25	—	—	229.8	11.50

TABLE II (*continued*)

Reactant	Temperature (°C)	Product	
		Primary	Secondary
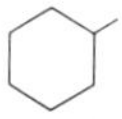	575–650	C_3H_6, C_4H_6	—
		CH_4, 	—
		C_2H_4, 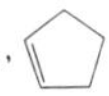	—
		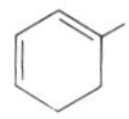	—

[a] Sakai *et al.* (1976).

from the experiment by use of these equations:

$$\frac{k_{\mathrm{HCH}}}{k_{\mathrm{VCH}}} = \frac{[C_4H_6]d[\mathrm{HCH}]}{2[C_2H_4]d[\mathrm{VCH}]} = \frac{[C_4H_6]\Delta[\mathrm{HCH}]}{2[C_2H_4]\Delta[\mathrm{VCH}]} \tag{21}$$

$$\frac{k_{\mathrm{MCH}}}{k_{\mathrm{VCH}}} = \frac{[C_4H_6]d[\mathrm{MCH}]}{2[C_3H_6]d[\mathrm{VCH}]} = \frac{[C_4H_6]\Delta[\mathrm{MCH}]}{2[C_3H_6]\Delta[\mathrm{VCH}]} \tag{22}$$

In a similar manner, they also obtained the second-order rate constants for reactions (4–7). The thermal reaction of 1-butene, *cis*- and *trans*-2-butene, or isobutylene with butadiene yielded 4-ethylcyclohexene, *cis*- and *trans*-4,5-dimethylcyclohexene, or 4,4-dimethylcyclohexene, respectively, accompanied by a large amount of VCH. Kinetic results are summarized in Table II; they are combined with those obtained from the thermal reactions of respective olefins and that of pure butadiene. The large A factors for the reaction of 1-butene and *cis*-2-butene with butadiene—that is, 15.67 and 15.97 in $\log_{10} A$ ($\mathrm{cm^3\ mol^{-1}\ s^{-1}}$)—should be reexamined in connection with their large activation energies (220.6 and 229.0 kJ $\mathrm{mol^{-1}}$, respectively).

The relative rates of formation of VCH, HCH, and MCH are about 10:4:1 at temperatures of 510–600°C, whereas those of 4-ethylcyclohexene, *cis*- and *trans*-4,5-dimethylcyclohexene, and 4,4-dimethylcyclohexene are less than 0.1 re that of MCH. The essential reactions for the formation of C_6 cyclic compounds can be predicted as reactions (1), (2), and (3). Thus, thermal reactions of VCH, HCH, and MCH will be discussed in Section VIII.

TABLE II (*continued*)

Reaction order	$\log_{10} k$ (cm^3, mol, s) 500	600	700	800	E (kJ mol^{-1})	$\log_{10} A$ (cm^3, mol, s)
1.0	−3.41	−1.31	—	—	271.7	14.94
1.0	−3.67	−1.69	—	—	255.8	13.61
1.0	−3.94	−1.92	—	—	261.2	13.70
1.0	−3.52	−1.92	—	—	207.4	10.49

VII. THERMAL REACTIONS OF OLEFINS WITH THE ALLYL RADICAL

Nohara and Sakai (1973) investigated the thermal reactions of 1,5-hexadiene at 500–620°C and with nitrogen dilution at atmospheric pressure. Considerable C_5 cyclic compounds were formed during the initial stage. The C_6 cyclic compounds, mostly cyclohexadiene and benzene, were formed as successive products. Table III lists the relative amounts of the primary products obtained by extrapolating the conversion of 1,5-hexadiene to zero.

Cleavage of the C—C bond of 1,5-hexadiene produces two allyl radicals and initiates the reaction (Lossing *et al.*, 1954; Ruzicka and Bryce, 1960; Akers and Throssell, 1967):

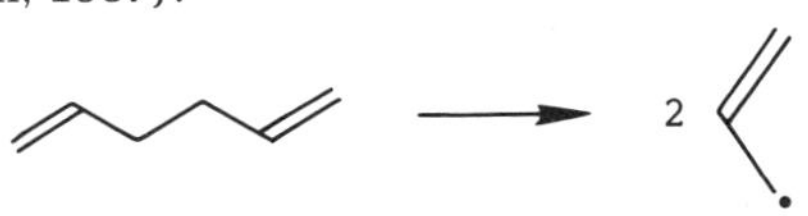

(23)

The C_5 cyclic compounds are formed as primary products by the reactions between the allyl radical and 1,5-hexadiene. Reaction (24) produces the cyclopentenyl radical and 1-butene. Another possible reaction produces cyclopentene and the 3-butenyl radical, or the cyclopentyl radical and butadiene.

(24)

TABLE III

Relative Amounts of Initial Products in Thermal Reaction of 1,5-Hexadiene[a]

Product	Molar ratio
Ethylene	1
Propylene	2.217
1-Butene	0.461
Butadiene	0.539
1-Pentene	0.199
Cyclopentene	0.223
Cyclopentadiene	0.355
1,3,5-Hexatriene	0.367
1,3-Cyclohexadiene	0
Benzene	0

[a] Nohara and Sakai (1973). Reprinted with permission from *Ind. Eng. Chem., Prod. Res. Dev.* Copyright 1973 by the American Chemical Society.

Benzene and 1,3-cyclohexadiene were apparently produced through the following sequence of reactions:

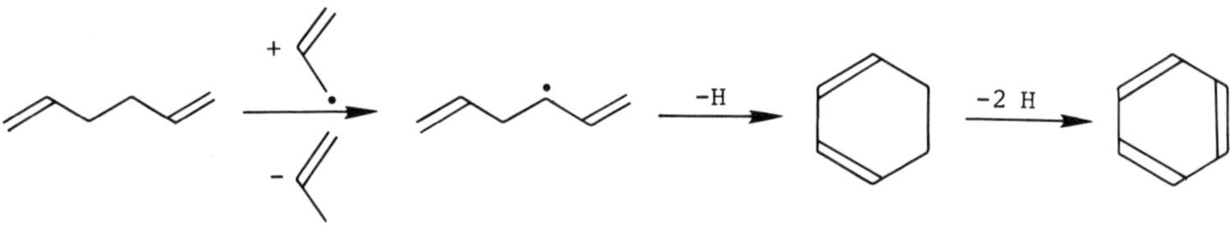

(25)

The existence of 1,5-hexadienyl radical is supported by the work of Ruzicka and Bryce (1960). Orchard and Thrush (1973) reported the thermal isomerization of 1,3,5-hexatriene to 1,3-cyclohexadiene at 400°C and the eventual formation of benzene at 550°C. It should be emphasized that no cyclohexene was found in this reaction, as was also reported in Section II.

Cycloaddition of the allyl radical to an olefinic bond is the basic reaction for formation of C_5 cyclic compounds. Sakai *et al.* (1976) reported on the addition of the allyl radical to ethylene by employing diallyl oxalate as a source of the allyl radical.

$$\begin{array}{l} COOCH_2\text{-}CH{=}CH_2 \\ | \\ COOCH_2\text{-}CH{=}CH_2 \end{array} \longrightarrow 2\ CH_2{=}CH\text{-}CH_2\cdot + 2CO_2 \qquad (26)$$

Nohara (1980) made a similar investigation of the reactions of the allyl radical with propylene, acetylene, methylacetylene, and butadiene. Reactions of the allyl radical with ethylene, propylene, and butadiene were considered to be reactions (8), (9), and (10), respectively, although 1-pentene formation in reactions (9) and (10) are not thoroughly understood. The reactions of the allyl radical with acetylene (Nohara and Sakai, 1980) and with methylacetylene were postulated as in reactions (27) and (28). The efficiencies of C_5 cyclic product formation were higher for the reaction of the allyl radical with acetylenes than for the reaction with olefins.

(27)

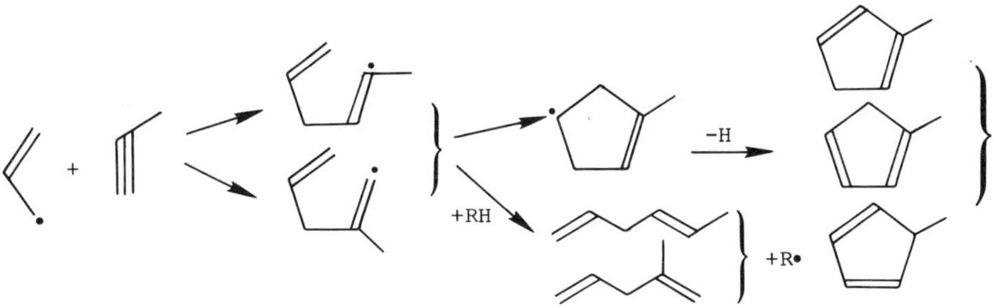

(28)

The ease with which these C_5 cyclic compounds are formed likely depends on how much the allyl radical has diene characteristics.

A summary of kinetic data for the presumed second-order reaction of olefins, diolefins, or acetylenes with the allyl radical is given in Table IV. Presumed second-order rate constants for the formation of C_5 cyclic compounds are large enough to compensate for the low concentrations of the allyl radical in the gas phase. For the reaction of allyl radicals with butadiene, the amount of cyclopentadiene produced was as large as the sum of benzene, cyclohexadiene, and VCH, as shown in Table V.

The rate of cyclopentadiene produced in the reaction of the allyl radical with acetylene is extraordinarily high as seen from Table IV. Moreover, the selectivity of the formation of cyclopentadiene is more than 90% and the formation of the parallel product, 1,4-pentadiene, is small. With methylacetylene, a similar or less selective formation of cyclic compounds results; however, it is much faster than the cyclic product formation by the reaction of allyl radical with monoolefins.

TABLE IV

Presumed Second-Order Rate Data for Each Product in Reactions of Olefins, Diolefins, and Acetylenes with the Allyl Radical[a]

Reaction	Cyclic/Acyclic	Product	$\log_{10} k_{500\,C}$ ($cm^3\ mol^{-1}\ s^{-1}$)	E ($kJ\ mol^{-1}$)	$\log_{10} A$ ($cm^3\ mol^{-1}\ s^{-1}$)
C=C—C· + C=C	1.5–2.5	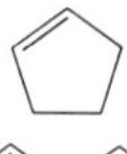	6.52	48.0	9.77
			6.38	69.8	11.1
C=C—C· + C=C—C	1.5–2.3		5.47	25.2	7.17
			5.48	36.9	7.97
		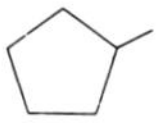	5.43	30.7	7.51
			5.10	76.9	10.3
			5.48	72.7	10.4
			5.22	(38.5)[b]	(7.82)[b]
			4.92	(23.5)[b]	(6.51)[b]
C=C—C· + C≡C	20		7.64	104	14.6
C=C—C· + C≡C—C	3–5	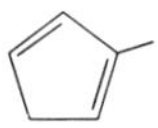	6.97	102	13.9
			6.89	107	14.1
			6.35	117	14.3
			6.61	100	13.4
			6.51	80.3	11.9
		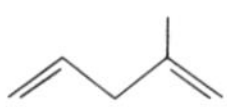	6.17	—	—
C=C—C· + C=C—C=C	1–2		7.54	46.8	10.8
			7.22	64.8	11.6
		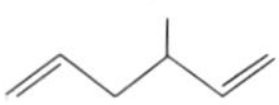	7.02	87.4	12.9
			6.87	61.0	11.0

[a] Nohara (1980).

[b] The figures in parentheses are relatively imprecise because of low concentrations of these two products.

ABLE V

oncentration of Cyclic Compounds Formed in Thermal Reaction of a Diallyl Oxalate–Butadiene Mixture[a,b]

Temperature (°C)	430		450		470		490		510	
Residence time (s)	3.6	7.7	3.8	7.9	3.5	7.1	3.3	7.2	3.5	7.4
DAO conversion (%)	0.7	1.2	1.4	2.9	3.3	6.0	8.0	15.4	15.2	27.0
Concentration of cyclic products (10^{-10} mol cm^{-3})										
	5.2	10.0	7.0	18.2	20.9	40.5	33.6	74.2	58.1	139
	0.8	0.9	0.6	1.4	1.9	4.6	4.9	12.9	13.5	44.3
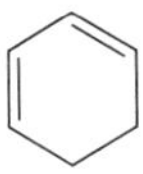	0.9	1.5	1.3	3.3	4.3	9.7	10.7	24.2	26.5	75.3
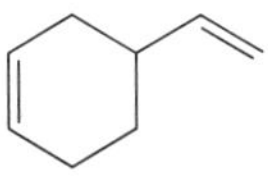	3.6	3.5	2.6	4.2	4.5	7.8	6.3	7.8	6.5	12.2

[a] Nohara (1980).
[b] Initial concentrations are 1.0 (diallyl oxalate) and 0.62 (butadiene) × 10^{-6} mol cm^{-3}.

The contribution of acetylenic compounds may be large for the production of cyclic products. Other routes may exist for formation of cyclic products from acetylenes in addition to the Diels–Alder reaction, however. The following route also is postulated for acetylene pyrolysis. When an acetylene–nitrogen mixture was heated at 580–740°C and at residence times of 0.09–0.30 s, vinylacetylene and benzene were the main products (Nohara and Sakai, 1980). These can be considered as being in the lower-temperature reaction region as designated by Back (1971). Apparent induction periods occurred with the formation of both vinylacetylene and benzene, such as reported by Minkoff *et al.* (1957) and Silcocks (1957); however, difficulty is experienced in discriminating the primary products of the reaction. The following scheme is postulated, which is somewhat analogous to the Diels–Alder reaction between butadiene and ethylene.

$$C{\equiv}C + C{\equiv}C \longrightarrow C{=}C{-}C{\equiv}C \quad (29)$$

$C{=}C{-}C{\equiv}C + C{\equiv}C \longrightarrow$  (30)

In addition, some trimolecular addition with acetylene might occur on the surface of the reactor. Thermal reactions of acetylene are known to be susceptible to the type of metal surfaces.

VIII. THERMAL REACTIONS OF CYCLOOLEFINS

In Section VI, VCH, HCH, and MCH were shown to be important intermediate compounds for the formation of aromatics during pyrolysis. Reactions (1), (2), and (3) proceed 5–100 times faster than reactions (4–7).

Considerable research has been conducted on the pyrolysis of HCH, but little on VCH and MCH. Smith and Gordon (1961), who pyrolyzed HCH at 425–535°C, concluded that three parallel reactions took part in this conversion, two unimolecular reactions and one free-radical reaction:

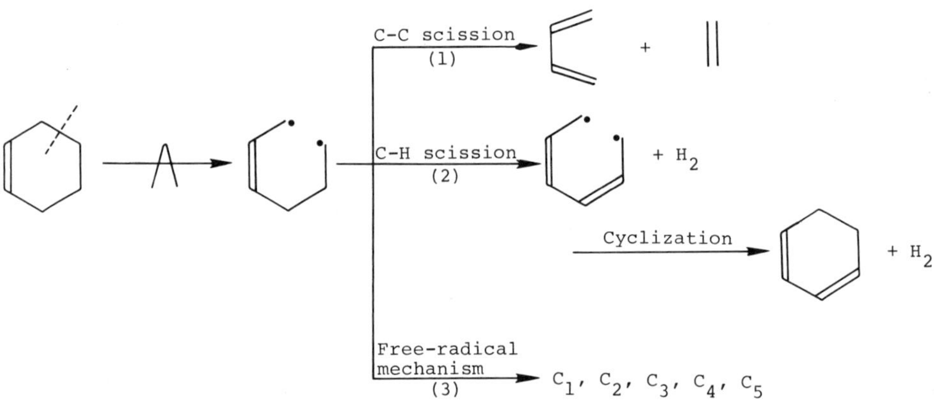

(31)

The overall reaction is reasonably well represented by first-order kinetics with the rate constant $k = 10^{15.89} \exp(-283{,}000/RT)/\mathrm{s}^{-1}$. The rate constants for the two unimolecular reactions are $k = 10^{17.15} \exp(-304{,}300/RT)/\mathrm{s}^{-1}$ for the reverse Diels–Alder reaction, and $k = 10^{16.28} \exp(-298{,}000/RT)/\mathrm{s}^{-1}$ for the cyclization reaction. They proposed the formation of the biradical intermediate during the first step of the reaction. This biradical then undergoes several scission and/or cyclization reactions. The bases of their arguments are as follows: first, the same activation energy is working for the reverse Diels–Alder reaction as for the cyclization reaction; second, the hexadienyl biradical is rather stable and easily converted to cyclohexadiene; third, the biradical intermediate can easily be converted to C_1–C_5 fragments through a free-radical mechanism. Salem and Wright (1969) performed a molecular orbital calculation to indicate the possibility of the biradical mechanism in the reaction of HCH. Basilevsky *et al.* (1977) stated that it is difficult to discriminate the concerted from the biradical mechanism in this reaction from such calculations.

Uchiyama *et al.* (1964) conducted the same reaction at 541–629°C, and found that ethylene and butadiene were obtained with yields of more than 96%. The first-order rate constant for the reverse Diels–Alder reaction was reported to be $k = 10^{15.18} \exp(-276{,}300/RT)/\mathrm{s}^{-1}$. They stated in addition that their kinetic equation fitted the data obtained by Smith and Gordon (1961). They however assumed a molecular activated complex rather than a biradical one, based on the comparison of activated entropies calculated for these two postulated complexes with that obtained from their experimental results. The calculation of activated entropies of the two postulated complexes were made using the bond-frequency data by Rowley and Steiner (1951).

Kuchler (1939) and Kraus *et al.* (1956) had reported still lower activation energies and *A* factors than those of Uchiyama *et al.* (1964), however. Tsang (1965) studied the reverse Diels–Alder reactions of HCH, MCH, and VCH in a single-pulse shock tube by use of the rate of elimination of hydrogen chloride from isopropyl chloride as the reference reaction. First-order rate constants were reported as $k_{\mathrm{HCH}} = 10^{15.02} \exp(-279{,}200/RT)/\mathrm{s}^{-1}$, $k_{\mathrm{MCH}} = 10^{15.13} \exp(-278{,}800/RT)/\mathrm{s}^{-1}$, and $k_{\mathrm{VCH}} = 10^{15.20} \exp(-259{,}500/RT)/\mathrm{s}^{-1}$.

Sakai *et al.* (1972) investigated thermal reactions of VCH, HCH, and MCH in a flow-type reactor with nitrogen dilution. Figures 2, 3, and 4 illustrate the product distributions for VCH, HCH, and MCH, respectively, against the reaction depth defined here as total moles of products formed per

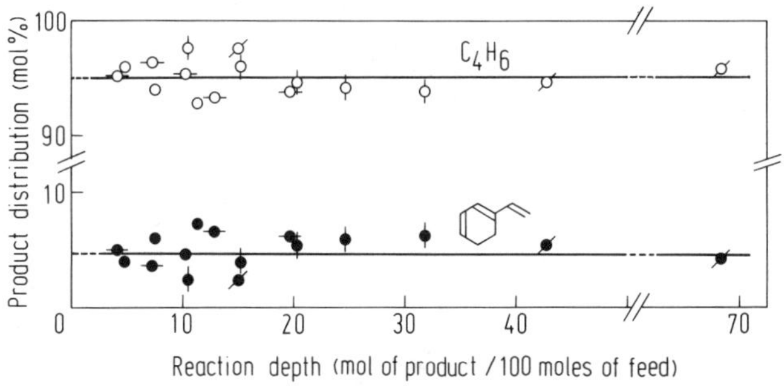

Fig. 2. Distribution of product on pyrolysis of VCH. Reaction temperatures shown: ○ ●, 530°C; -○- -●-, 550°C; ȯ ●̇, 570°C; ⌀ ●⁄, 585°C. (From Sakai *et al.*, 1972. Reprinted with permission from *Ind. Eng. Chem., Fundam.* Copyright 1972, ACS.)

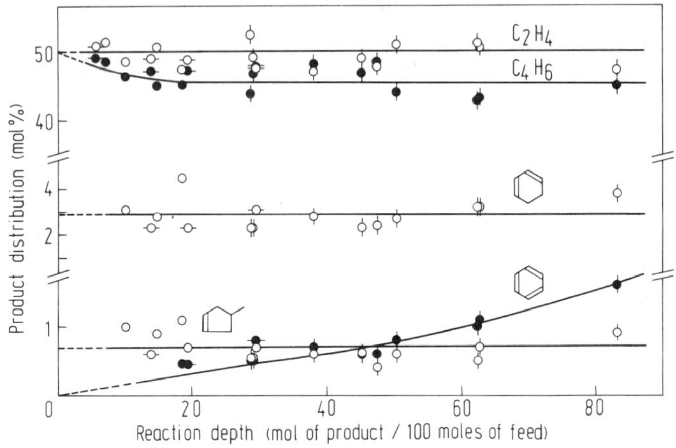

Fig. 3. Distribution of product on pyrolysis of HCH. Reaction temperatures shown: ○ ●, 585°C; -○- -●-, 625°C; ○ ●, 665°C. (From Sakai *et al.*, 1972. Reprinted with permission from *Ind. Eng. Chem., Fundam.* Copyright 1972, ACS.)

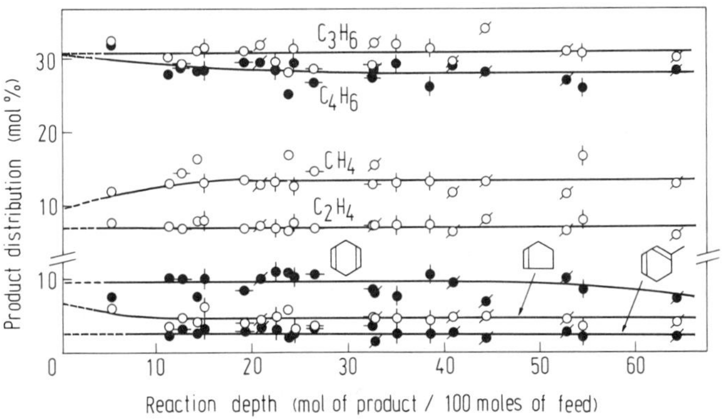

Fig. 4. Distribution of product on pyrolysis of MCH. Reaction temperatures shown: ○ ●, 575°C; -○- -●-, 600°C; ○ ●, 625°C; ○ ●, 650°C. (From Sakai *et al.*, 1972. Reprinted with permission from *Ind. Eng. Chem., Fundam.* Copyright 1972, ACS.)

100 moles of feed. In the reaction of VCH, butadiene and vinylcyclohexadiene were formed.

$$k_{VCH}(\text{butadiene}) = 10^{15.05}\exp(-253{,}700/RT)/\text{s}^{-1} \qquad (32)$$

$$k_{VCH}(\text{vinylcyclohexadiene}) = 10^{10.60}\exp(-181{,}700/RT)/\text{s}^{-1} \qquad (33)$$

HCH was converted to yield almost equimolar amounts of ethylene and butadiene plus some 1,3-cyclohexadiene, accompanied by a small amount of 4-methylcyclopentene. These were the initial products of the reaction.

$$k_{\mathrm{HCH}}(\text{ethylene}) = 10^{15.01}\exp(-275{,}000/RT)/\mathrm{s}^{-1} \tag{34}$$

$$k_{\mathrm{HCH}}(\text{cyclohexadiene}) = 10^{11.50}\exp(-229{,}800/RT)/\mathrm{s}^{-1} \tag{35}$$

The reaction of MCH was somewhat complicated. Other than 1-methyl-1,3-cyclohexadiene, three pairs of equimolar products resulted: propylene and butadiene, methane and 1,4-cyclohexadiene, and ethylene and cyclopentene.

$$k_{\mathrm{MCH}}(\text{propylene}) = 10^{14.94}\exp(-271{,}700/RT)/\mathrm{s}^{-1} \tag{36}$$

$$k_{\mathrm{MCH}}(\text{methane}) = 10^{13.61}\exp(-255{,}700/RT)/\mathrm{s}^{-1} \tag{37}$$

$$k_{\mathrm{MCH}}(\text{ethylene}) = 10^{13.70}\exp(-261{,}200/RT)/\mathrm{s}^{-1} \tag{38}$$

$$k_{\mathrm{MCH}}(\text{methylcyclohexadiene plus toluene}) = 10^{10.49}\exp(-207{,}400/RT)/\mathrm{s}^{-1} \tag{39}$$

In this study, two parallel reactions, namely the reverse Diels–Alder reaction and the dehydrogenation reaction, were demonstrated. These findings disagree with those reported by Smith and Gordon (1961) as reported in Eq. (31), however. Arrhenius plots for all ks of Eqs. (32–39) are shown in Figs. 5, 6, and 7. Sakai *et al.* (1972) postulated the mechanism for these reactions as shown in Eqs. (40), (41), and (42).

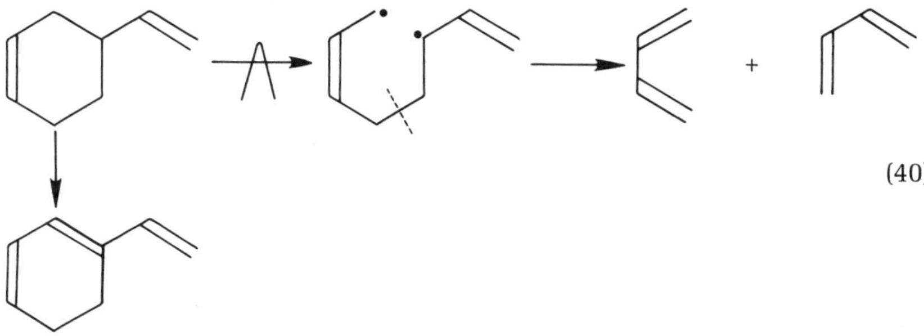

(40)

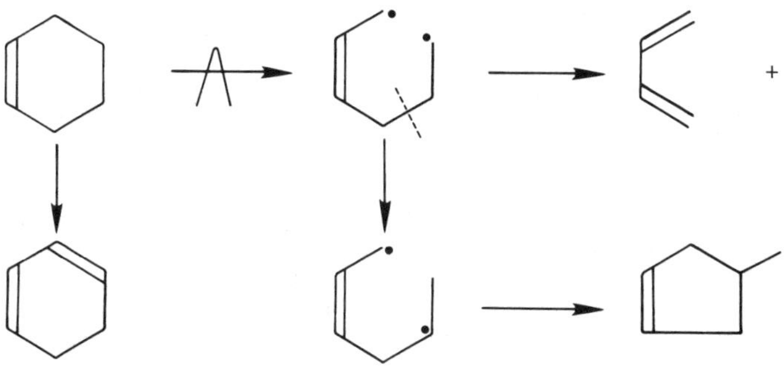

(41)

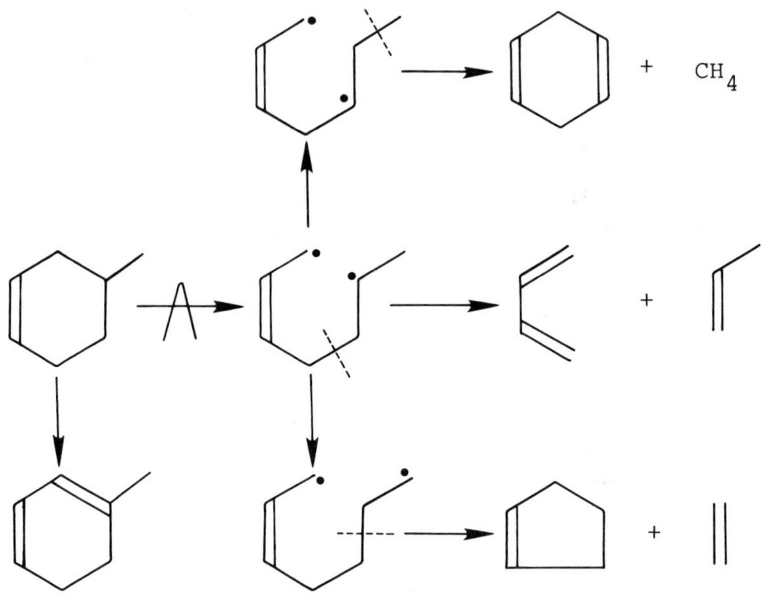

(42)

Because the activation energy is higher for the reverse Diels–Alder reaction than for the dehydrogenation reaction in all three cases, there is the possibility that the formation of aromatics via dehydrogenation can be

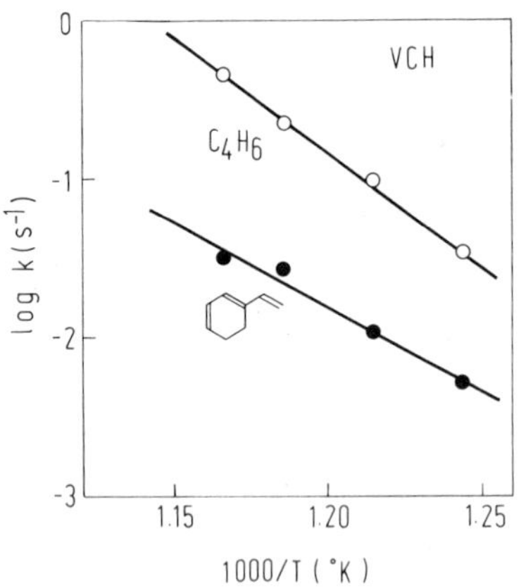

Fig. 5. Arrhenius plots (VCH). (From Sakai *et al.*, 1972. Reprinted with permission from *Ind. Eng. Chem., Fundam.* Copyright 1972, ACS.)

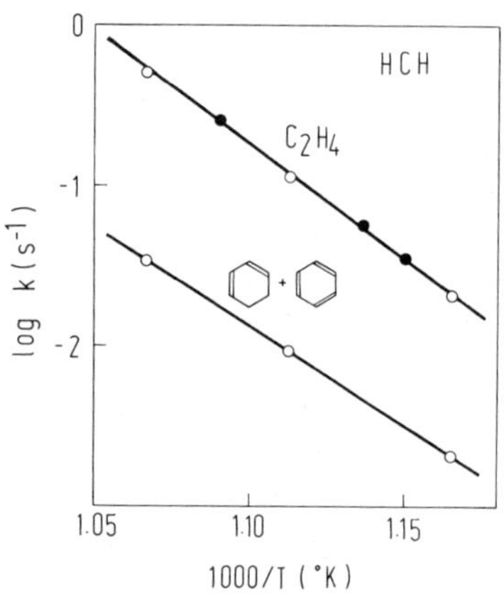

Fig. 6. Arrhenius plots (HCH) ○, from Sakai *et al.* (1972); ●, from Uchiyama *et al.* (1964). (From Sakai *et al.*, 1972. Reprinted with permission from *Ind. Eng. Chem.* Copyright 1972, ACS.)

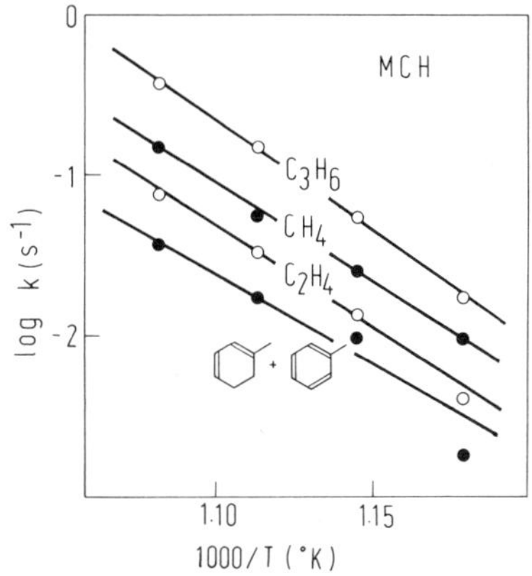

Fig. 7. Arrhenius plots (MCH). (From Sakai *et al.*, 1972. Reprinted with permission from *Ind. Eng. Chem., Fundam.* Copyright 1972. ACS.)

suppressed at high temperatures and at short residence times. Short residence times are needed to get similar conversion levels at high temperatures. These kinetic performances in a sense parallel the high-severity cracking technology developed for olefin production from naphtha with less coke formation.

Another interesting fact is that, in the case of MCH pyrolysis, cyclic products are formed not only through the dehydrogenation route but also through the reverse Diels–Alder reaction, which proceeds much faster than the former route. In this way, propylene should be considered as a significant substrate for the cyclic compound formation, although a mere comparison of rates of reactions (1) and (2) is unfavorable for propylene compared to ethylene.

IX. SUMMARY AND AREAS RECOMMENDED FOR FURTHER RESEARCH

Thermal reaction of olefins and diolefins were surveyed in this chapter in view of cyclic product formation. Diels–Alder reactions between olefins and butadiene—that is, reactions (1–7)—are basic methods for production of C_6 cyclic hydrocarbons. Kinetic data for various olefins are summarized in Table II.

The importance of the Diels–Alder reactions for aromatic compound formation during hydrocarbon pyrolysis was confirmed by use of the results for production of C_6 cyclic compounds in the thermal reactions of ethylene and propylene at 753°C reported by Kunugi *et al.* (1969, 1970a). In Figs. 8 and 9, both the rates of C_6 cyclic compounds formation observed in the experiments and the rates of combined HCH, MCH, and VCH

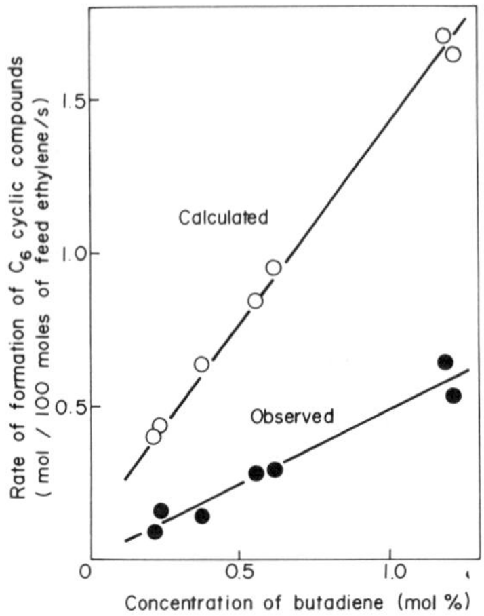

Fig. 8. Rate of formation of C_6 cyclic compounds in thermal reaction of ethylene at 753°C. From Sakai *et al.*, 1970.

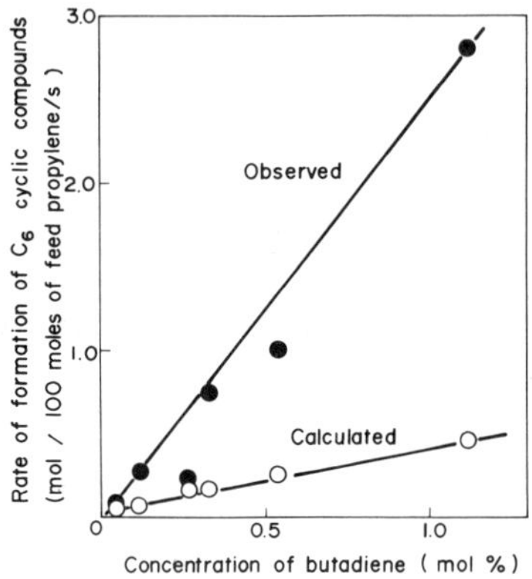

Fig. 9. Rate of formation of C_6 cyclic compounds in thermal reaction of propylene at 753°C. From Sakai *et al.*, 1970.

formations calculated from Eqs. (18), (19), and (20) have been plotted versus the butadiene concentration. The figures reveal the proportionality between the observed or calculated rates of C_6 cyclic compounds formation and the butadiene concentration.

A problem lies in the inconsistencies between the calculated and the observed rates for both cases. In the case of ethylene, the calculated rates are larger than the observed rates. Because the preceding calculation excluded the reverse Diels–Alder reaction of HCH, MCH, and VCH, the results in Fig. 8 support the premise that the cyclic C_6 products resulted from the Diels–Alder reactions of olefins and butadiene. In the case of propylene, however, the calculated rates are much smaller than the observed rates (Fig. 9). Existence of the allyl radical is likely to be related to the formation of C_6 cyclic compounds in the thermal reaction of propylene, although the reaction of the allyl radical with olefins yields mostly C_5 cyclic compounds as listed in Table IV. A further study is needed to understand the role of propylene more precisely (Sakai, 1972).

Another substance that is responsible for C_6 cyclic compound formation would be acetylene, especially at high temperatures where the concentration of acetylene would be relatively high. This was briefly commented on in Section VII.

Elucidation of the formation of condensed aromatics in thermal reactions of hydrocarbons will be a difficult field for basic research. Thermal reactions of benzene at 800–850°C do not give condensed aromatics but give biphenyl, and terphenyl, mostly *p*-terphenyl (Mead and Burk, 1935; Pease and Morton, 1933; Bauer and Aten, 1963; Hou and Palmer, 1965). An addition of a small amount of naphthalene to the reacting benzene results in the formation of white crystals upon quenching, which are mainly composed of biphenyl and phenylnaphthalene. However, when a small mol % of ethylbenzene is added to the reacting benzene, the reaction phase immediately changes color to brown, indicating that polycondensed aromatics are formed (Sakai *et al.*, 1971). These polycondensed aromatics are obviously caused by the presence of aliphatic fragments. Formation of tetralin isomers or octalin isomers from the reactions between benzene and butadiene, between styrene and ethylene, or between cyclohexane and butadiene is too slow to be adopted as a candidate reaction for the formation of polycondensed aromatics.

The surface effects are extraordinarily significant in the formation of aromatic structures from olefinic or acetylenic hydrocarbons. A copper surface retards the pyrolysis of *n*-heptane at 750°C and deposits little coke on the surface. On the other hand, a steel surface accelerates the reaction and deposits coke dramatically. Many studies have been done and are under way in this important area of research on surface effects.

REFERENCES

Akers, R. J., and Throssell, J. J. (1967). *Trans. Faraday Soc.* **63**, 124.
Amano, A., and Uchiyama, M. (1963). *J. Phys. Chem.* **67**, 1242.
Back, M. H. (1971). *Can. J. Chem.* **49**, 2199.
Basilevsky, M. V., Shamov, A. G., and Tikhomirov, V. A. (1977). *J. Am. Chem. Soc.* **99**, 1369.
Bauer, S. H., and Aten, C. F. (1963). *J. Chem. Phys.* **39**, 1253.
Benson, S. W. (1970). *In* "Refining Petroleum for Chemicals" (R. F. Gould, ed.), Advances in Chemistry Series, No. 97, pp. 1–19. Am. Chem. Soc., Washington, D.C.
Bryce, W. A., and Kebarle, P. (1958). *Trans. Faraday Soc.* **54**, 1660.
Bryce, W. A., and Ruzicka, D. J. (1960). *Can. J. Chem.* **38**, 835.
Burk, R. E., Baldwin, B. G., and Whitacre, C. H. (1937). *Ind. Eng. Chem.* **29**, 326.
Dahlgren, G., and Douglas, J. E. (1958). *J. Am. Chem. Soc.* **80**, 5108.
Danby, C. J., Spall, B. C., Stubbs, F. J., and Hinshelwood, C. (1955). *Proc. R. Soc. London, Ser. A* **228**, 448.
Hou, K. C., and Palmer, H. B. (1965). *J. Phys. Chem.* **69**, 863.
Hurd, C. D., and Meinert, R. N. (1930). *J. Am. Chem. Soc.* **52**, 4978.
Kallend, A. S., Purnell, J. H., and Shurlock, B. C. (1967). *Proc. R. Soc. London, Ser. A* **300**, 120.
Kinney, R. E., and Crowley, D. J. (1954). *Ind. Eng. Chem.* **46**, 258.
Kistiakowsky, G. B., and Ransom, W. W. (1939). *J. Chem. Phys.* **7**, 725.
Kraus, M., Varruska, M., and Bogart, V. (1956). *Chem. Listy* **50**, 553.
Küchler, L. (1939). *Trans. Faraday Soc.* **35**, 874.
Kunugi, T., Tominaga, H., Abiko, S., Uehara, K., and Ohno, T. (1967). *Kogyo Kagaku Zasshi*, **70**, 1477.
Kunugi, T., Sakai, T., Soma, K., and Sasaki, Y. (1969). *Ind. Eng. Chem., Fundam.* **8**, 374.

Kunugi, T., Sakai, T., Soma, K., and Sasaki, Y. (1970a). *Ind. Eng. Chem., Fundam.* **9**, 314.
Kunugi, T., Soma, K., and Sakai, T. (1970b). *Ind. Eng. Chem., Fund.* **9**, 319.
Laidler, K. J., and Wojciechowski, B. W. (1960). *Proc. R. Soc. London, Ser. A* **259**, 257.
Lossing, F. P., Ingold, K. U., and Henderson, I. H. S. (1954). *J. Chem. Phys.* **22**, 621.
Mead, F. C., and Burk, R. E. (1935). *Ind. Eng. Chem.* **27**, 299.
Minkoff, G. J., Newitt, D. M., and Rutledge, P. (1957). *J. Appl. Chem.* **7**, 406.
Molera, M. J., and Stubbs, F. J. (1952). *J. Chem. Soc.* p. 381.
Nohara, D. (1980). Ph.D. Thesis, Kyoto Univ., Kyoto.
Nohara, D., and Sakai, T. (1973). *Ind. Eng. Chem., Prod. Res. Dev.* **12**, 322.
Nohara, D., and Sakai, T. (1980). *J. Jpn. Pet. Inst.* **23**, 133.
Orchard, S. W., and Thrush, B. A. (1973). *J. C. S. Chem. Commun.* p. 14.
Pease, R. N., and Morton, J. M. (1933). *J. Am. Chem. Soc.* **55**, 3190.
Rowley, D., and Steiner, H. (1951). *Discuss. Faraday Soc.* **10**, 198.
Ruzicka, D. J., and Bryce, W. A. (1960). *Can. J. Chem.* **38**, 827.
Sakai, T. (1972). *Kagaku No Ryoiki* **26**, 10.
Sakai, T., Soma, K., Sasaki, Y., Tominaga, H., and Kunugi, T. (1970). *In* "Refining Petroleum for Chemicals" (R. F. Gould, ed.), Advances in Chemistry Series, No. 97, pp. 68–91. Am. Chem. Soc., Washington, D.C.
Sakai, T., Wada, S., and Kunugi, T. (1971). *Ind. Eng. Chem., Process Des. Dev.* **10**, 305.
Sakai, T., Nakatani, T., Takahashi, N., and Kunugi, T. (1972). *Ind. Eng. Chem., Fundam.* **11**, 529.
Sakai, T., Nohara, D., and Kunugi, T. (1976). *In* "Industrial and Laboratory Pyrolyses" (L. F. Albright and B. L. Crynes, eds.), ACS Symposium Series, No. 32, pp. 152–177. Am. Chem. Soc., Washington, D.C.
Sakakibara, Y. (1964). *Bull. Chem. Soc. Jpn.* **37**, 1262.
Salem, L., and Wright, J. S. (1969). *J. Am. Chem. Soc.* **91**, 5947.
Seres, L., and Ausloos, P. (1973). *In* "Mech. Hydrocarbon React., Symp. 1973" (F. Marta and D. Kallo, eds.), pp. 539–550. Elsevier, Amsterdam.
Shibatani, H., and Kinoshita, H. (1974). *Sekiyu Gakkai Shi* **17**, 226.
Silcocks, C. G. (1957). *Proc. R. Soc. London, Ser. A* **242**, 411.
Simon, M., and Back, M. H. (1970a). *Can. J. Chem.* **48**, 317.
Simon, M., and Back, M. H. (1970b). *Can. J. Chem.* **48**, 3313.
Simon, M., and Back, M. H. (1973). *Can. J. Chem.* **51**, 2934.
Smith, S. R., and Gordon, A. S. (1961). *J. Phys. Chem.* **68**, 1124.
Szwarc, M. (1949). *J. Chem. Phys.* **17**, 284.
Tarasenkova, E. M. (1955). *Tr. Leningr. Inzh.-Ekon. Inst.* **9**, 123.
Tsang, W. (1965). *J. Chem. Phys.* **42**, 1805.
Uchiyama, M., Tomioka, T., and Amano, A. (1964). *J. Phys. Chem.* **68**, 1878.
Vaughan, W. E. (1932). *J. Am. Chem. Soc.* **54**, 3863.
Wheeler, R. V., and Wood, W. L. (1930). *J. Chem. Soc.* p. 1819.

6

Pyrolysis: Hydrogenolysis of Toluene

MUNEYOSHI YAMADA AND AKIRA AMANO

I. INTRODUCTION

The mechanism for thermal breakdown of toluene, producing mainly benzene, methane, hydrogen, and biphenyl, has been known for over a century, since Berthelot (1867). The reaction has recently been restudied by Kershaw (1978) under conditions at which the so-called *supercritical gas extraction of coal* is usually conducted, resulting in essentially the same conclusions. Thus, when toluene is subjected to temperatures above 350°C, fragmental hydrocarbons in its product are invariably accompanied

Pyrolysis: Theory and Industrial Practice

ISBN 0-12-048880-9

by hydrogen and higher molecular moieties. This statement is further supported by Braekman-Danheux *et al.* (1977). In their study on very-low-pressure pyrolysis of toluene, they reported the formation of cyclohexanes, implying in turn the formation of hydrogen-lean components in the product.

A very fundamental feature of aromatic pyrolysis can be deduced from the preceding and numerous other observations. The pyrolysis in its literal meaning can occur only at the cost of a part of the reactant, which at the same time acts as a hydrogen donor upon its condensation. Conversely, as has been pointed out earlier by Hofmann and Lang (1929), the formation of the higher molecular moieties is minimized when aromatic pyrolysis is undertaken in the presence of a substantial amount of hydrogen or any suitable hydrogen donor; this technique is referred to as hydrogenolysis.

Hydrogenolysis has thrown considerable light on the chemistry of aromatic pyrolysis, and has also been adapted to the commercial production of benzene from various aromatics. Hydrogenolysis is currently under serious investigation relative to coal liquefaction.

In what follows, the background experiments and theories that are thought to be most important in understanding homogeneous degradation of aromatic hydrocarbons will be described.

II. THERMOCHEMISTRY

A. General Scope

Valuable information as to overall pyrolysis and elementary reactions can be obtained from equilibrium studies. Provided that a balanced equation is established either experimentally or theoretically, the equilibrium constant (K) can be evaluated at standard temperature (298°K) according to the following equation:

$$-RT \ln K = \Delta H^\circ - T\,\Delta S^\circ \tag{1}$$

where ΔH° and ΔS° are standard enthalpy and entropy changes associated with the reaction. Thus, to calculate K(298°K), only the values of standard heats of formation (ΔH_f°) and standard entropies (S°) are required. The values of $K(T)$ at any other temperature (T) can be calculated with sufficient precision, if heat capacities, $C_p^\circ(T)$, of the species involved are known over a range of temperatures, by using the following approximate relations:

$$\begin{aligned} \Delta H^\circ(T) &= \Delta H^\circ + \Delta C_p^\circ(T_m)\cdot(T-298) \\ \Delta S^\circ(T) &= \Delta S^\circ + 2.303\,\Delta C_p^\circ(T_m)\cdot\log(T/298) \end{aligned} \tag{2}$$

where T_m is the mean temperature between 298 and T.

For relatively simple molecules, radicals, and ions, the values of C_p°, ΔH_f°, and S° are either known from experiments or estimated by conventional procedures of statistical mechanics. However, such is not the case for more complex species that are likely to be involved in aromatic pyrolysis. To this end, the empirical group additivity method proposed by Benson and Buss (1958) is a powerful predictive tool for thermochemical properties, and has recently been elaborated on by Stein *et al.* (1977) and by Shaw *et al.* (1977) to include group values for polycyclic aromatic hydrocarbons.

B. Hydrogenolysis

Hydrogenolytic demethylation of polymethylbenzenes is exothermic by about 12 kcal/mol for each methyl group being removed. Therefore, at sufficiently high temperatures where the hydrogenolysis occurs at significant rates, the equilibrium is less favorable. Such a state of affairs is illustrated in Fig. 1 for the hydrogenolysis of toluene. A large excess of hydrogen is required for the reaction to go to completion.

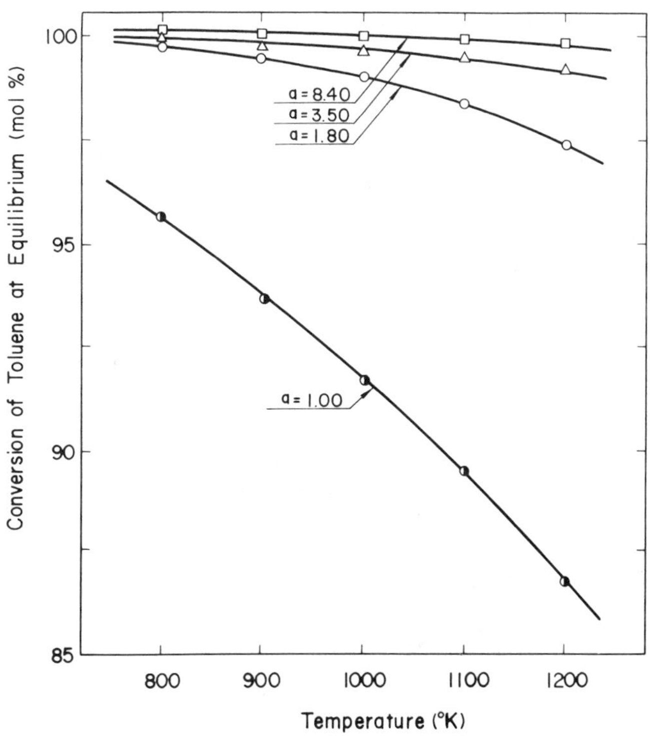

Fig. 1. Equilibrium conversion of toluene at various ratios (a) of hydrogen to toluene. (From Matsui *et al.*, 1959.)

The exothermicity of the hydrogenolysis of nonaromatic molecules is still higher. Thus when a feed consisting of 70 wt% toluene and 30 wt% nonaromatics is completely hydrogenolyzed in an adiabatic reactor with 3.0 mol ratio of pure hydrogen to the feed, temperature rise is as much as 250°C. Therefore, in any commercial unit, careful control of the reactor temperature is of prime importance.

Because reactions in this class are the main theme of this chapter, thermochemical consequence of the hydrogenolysis will be discussed in more detail later.

C. Hydrogasification

The thermochemistry of polycyclic aromatic molecules is also of considerable interest in view of the recent growing emphasis on the hydrogasification of coal and char.

If the aromatic nucleus is to decompose, the delocalization energy must first be overcome. Summarized in Table I are the values of the delocalization energy estimated by the method proposed by Dewar (1969) for a few selected aromatic hydrocarbons. Because this energy is large, about 40 kcal/mol, the initial destabilization of the aromatic nucleus should invariably be the rate-determining step of the overall hydrogasification. Among those listed in Table I, benzene is the least reactive whereas anthracene is the most reactive.

The destabilized nucleus then undergoes hydrogenation and hydrogenolysis, and also pyrolysis if hydrogen is in short supply, giving off a variety of products in the ratios compatible with thermochemical predictions. Illustrated in Fig. 2 are the most probable products obtained from anthracene. Less probable products are indicated in parentheses.

TABLE I

Delocalization Energy of Aromatic Hydrocarbons[a]

Benzene	1.155
Naphthalene	0.904
Phenanthrene	0.899
Chrysene	0.833
Pyrene	0.755
Anthracene	0.632

[a] Figures are nondimensionalized by the Hückel factor 2β.

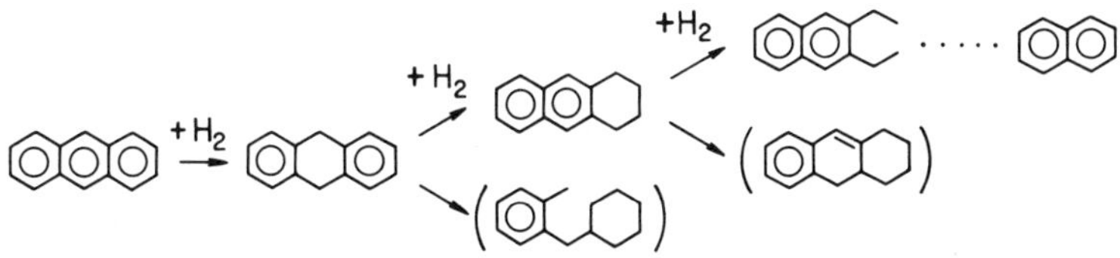

Fig. 2. Hydrogasification sequence of anthracene. Less probable products are in parentheses.

Because experimental results are sparse and mechanistic theories are even more incomplete, the reactions in this class will not be discussed further.

III. KINETICS OF HYDROGENOLYSIS

A. Experiments

A considerable number of kinetic investigations were reported during the decade 1956–65 on the gas-phase thermal hydrogenolytic demethylation of toluene. A static reactor was used by Matsui *et al.* (1959); and conventional flow reactors were adopted by Silsby and Sawyer (1956), Betts *et al.* (1957), Tsuchiya *et al.* (1959), Masamune *et al.* (1960), Morii *et al.* (1963), and Burr *et al.* (1964). These studies covered a rather wide range of reaction conditions, from 580 to 900°C in temperature and from 0.02 to 100 atm in pressure.

Toluene, when mixed with a substantial excess of hydrogen, reacts amazingly cleanly as follows:

$$\text{toluene} + H_2 \longrightarrow \text{benzene} + CH_4$$

Reproduced in Figs. 3 and 4 are the typical results of a now classical study reported by Matsui *et al.* (1959). The results of all the investigations cited previously are in excellent agreement. The observed rates can be described, within the deviations specified, by the following equation:

$$-\frac{d[\text{toluene}]}{dt} = 10^{(11 \pm 1)-(55 \pm 5)/\theta}[\text{toluene}] \cdot [H_2]^{1/2} \text{ mol/1 s} \tag{3}$$

where $\theta = 4.575\, T$ kcal/mol.

Hydrogenolyses of alkyl derivatives of benzene, naphthalene, and phenol have also been reported frequently. However, because the results are in line with those of toluene, they will not be further considered in this chapter.

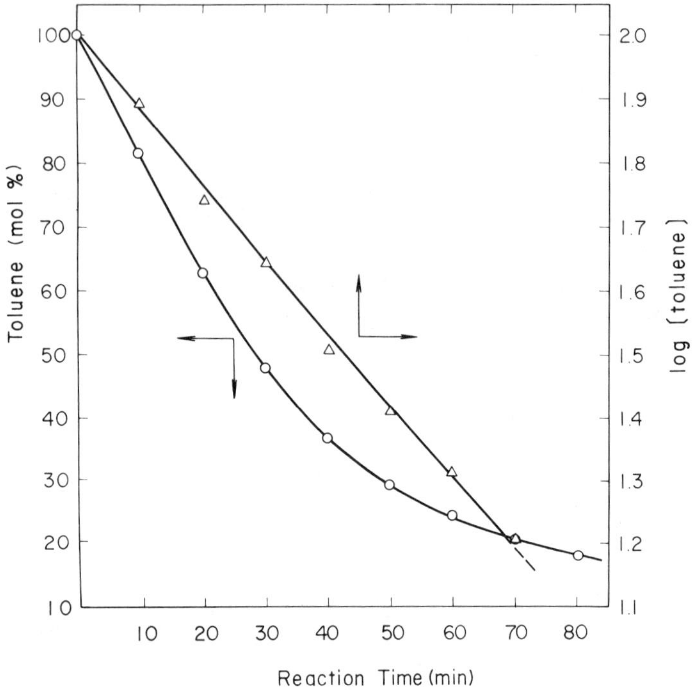

Fig. 3. Rate of toluene hydrogenolysis. Initial pressures: toluene, 50 torr; hydrogen, 599 torr. Temperature, 853°K. ○, relation between toluene concentration and reaction time; △, linear relation between the logarithm of toluene concentration and reaction time. (From Matsui *et al.*, 1959.)

The results for *p*-xylene are somewhat different from the results of other aromatic compounds mentioned already. Szwarc (1976) indicated the formation of *p*-cyclophane, an interesting monomer in polymer chemistry, and acetone-soluble linear oligomers, when *p*-xylene is heated to more than 800°C in a flow reactor and quenched by toluene. The reaction presumably takes place *via* the following sequence:

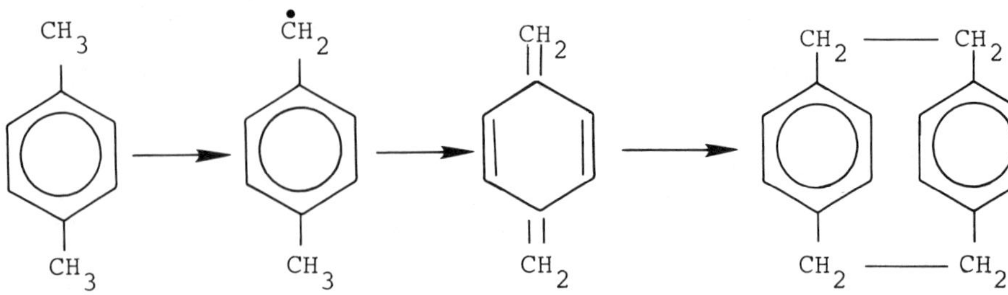

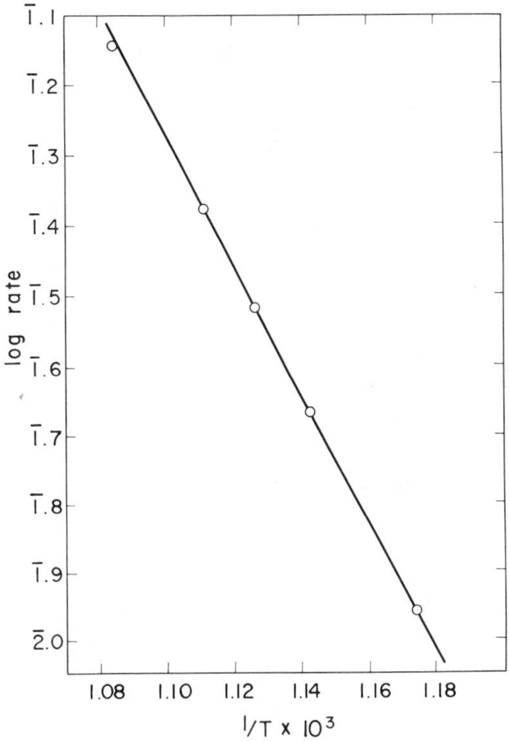

Fig. 4. Arrhenius plots of toluene hydrogenolysis. (From Matsui *et al.*, 1959.)

B. Mechanism

The following chain mechanism has been proposed by Amano *et al.* (1965) to explain toluene hydrogenolysis described in the preceding section (Section III,A).

$$\text{toluene} \underset{2}{\overset{1}{\rightleftarrows}} \text{benzyl} + \text{H}$$

$$\text{H} + \text{toluene} \underset{4}{\overset{3}{\rightleftarrows}} \text{H}_2 + \text{benzyl}$$

$$\text{H} + \text{toluene} \xrightarrow{5} \text{CH}_3 + \text{benzene}$$

$$\text{CH}_3 + \text{H}_2 \xrightarrow{6} \text{CH}_4 + \text{H}$$

$$\text{CH}_3 + \text{toluene} \underset{8}{\overset{7}{\rightleftarrows}} \text{CH}_4 + \text{benzyl}$$

The chain is initiated by reaction (1) and terminated by reaction (2), and is propagated by the cycle comprising reactions (5) and (6). Reactions (3), (4),

(7), and (8) are the metathetical processes. With a few trivial approximations, the mechanism leads to the following rate equation:

$$-\frac{d[\text{toluene}]}{dt} = k_5\left(\frac{k_1 k_4}{k_2 k_3}\right)^{1/2} [\text{toluene}][H_2]^{1/2} \tag{4}$$

where ks are the rate constants as indicated earlier. All the values of ks except k_5 are known with high precision. Thus the comparison between Eqs. (3) and (4) can place the values of the A factor and activation energy of reaction (5) to be approximately 10^8 1/mol s and 5 kcal/mol, respectively.

Because reaction (5) is the key reaction for both pyrolysis and hydrogenolysis, the remainder of this chapter will be devoted to both mechanistic and energetic features of reaction (5).

As was pointed out later by Benson and Shaw (1967), the replacement of reactions (1) and (2) in the above chain mechanism by $H_2 \rightleftarrows 2\ H$ still results in essentially the same form as Eq. (4).

C. Resonance-Stabilized Radicals

The kinetic consequence of adding hydrogen to the reaction mixture is explicitly shown by reaction (4). In the absence of this latter reaction, the mechanism simply reduces to the one proposed by Szwarc (1974) for the toluene pyrolysis, and the long-lived benzyl radical becomes the source of higher molecular moieties leading eventually to carbonaceous materials. Therefore, resonance-stabilization energy (E_{RSE}) of conjugated radicals is closely related to the complexity in aromatic pyrolysis.

At present, the only E_{RSE} value that is known with reasonable accuracy is the one for benzyl radical. The value of 13 ± 1.5 kcal/mol has been accepted by Golden and Benson (1969). According to Stein and Golden (1977), however, the values for other alternate conjugated radicals can be estimated by the following hypothetical relation:

$$E_{RSE}(\text{kcal/mol}) = 16.75 \ln[SC(R)] - 20.19 \ln[SC(RH)] \tag{5}$$

where $SC(R)$ and $SC(RH)$ stand for numbers of the stable Kékule isomers in the radical and in the parent aromatic molecule, respectively. It can be shown from Eq. (5) that the kinetic complexity in pyrolysis tends to increase as the number of condensed rings increases. The estimated values will be referred to in connection with coal liquefaction.

The allyl radical in alkene pyrolysis is expected to play the same role as the benzyl radical in aromatic pyrolyses. This was experimentally demonstrated by Bryce and Ruzicka (1960).

IV. THEORIES OF HYDROGENOLYSIS

A. Molecular Orbitals

The empirical molecular orbital theory known as the extended Hückel method was developed by Hoffmann (1963) and has proven useful for various conformational problems of hydrocarbons; it has been applied by Tominaga *et al.* (1970) for reaction (5) mentioned earlier (Section III,B) and also for the following reaction:

$$H + \text{toluene} \xrightarrow{9} CH_4 + \text{phenyl}$$

The latter reaction is the one proposed by Silsby and Sawyer (1956) as an alternative to reaction (5).

Perspective views of the two transition states, corresponding to reactions (5) and (9), are illustrated in Figs. 5 and 6, respectively. In these figures, H-16 stands for the attacking hydrogen atom, C-1 the phenyl carbon atom adjacent to the methyl group, and C-7 the methyl carbon atom being removed.

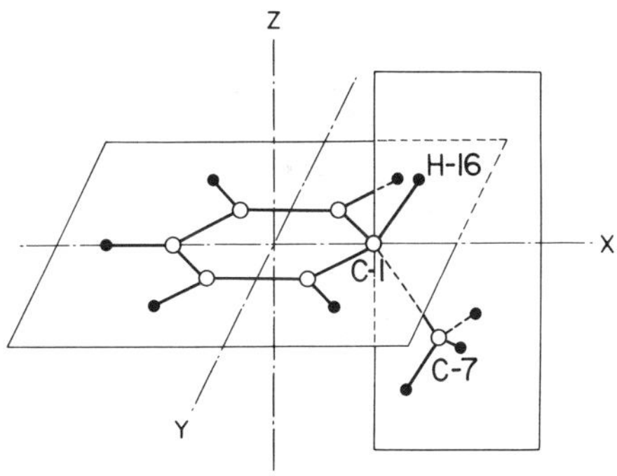

Fig. 5. Perspective view of the transition state of reaction (5). (From Tominaga *et al.*, 1970.)

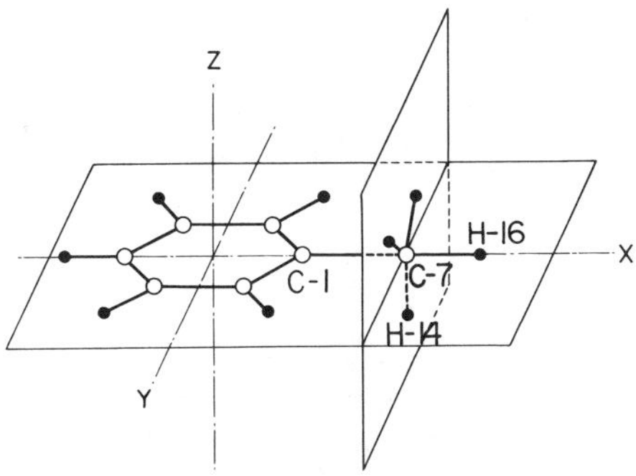

Fig. 6. Perspective view of the transition state of reaction (9). (From Tominaga *et al.*, 1970.)

TABLE II

Changes of Total Electron Energy with Vertical Approach of Attacking Hydrogen Atom: Reaction (5)[a]

Species number	Bond length (Å)		Bond angle (deg)		Conformation	Total electron energy (eV)	Energy level of half-filled orbital (eV)
	C-1–H-16	C-1–C-7	<H-16–C-1–x	<C-7–C-1–x			
I	5.0	1.51	90	0	Toluene	−645.79	—
II	3.0	1.51	90	0	Toluene	−645.75	—
III	2.0	1.52	80	25	Toluene	−645.53	—
IV	1.5	1.53	70	35	Toluene	−645.67	−10.750
V	1.5	1.53	70	35	Cyclohexadienyl	−646.04	−10.432
	1.22	1.53	60	60	Cyclohexadienyl	−646.81	—
	1.22	1.68	60	60	Cyclohexadienyl	−647.50	—
VI	1.11	1.54	55	55	Cyclohexadienyl	−647.21	−10.439
	1.11	1.68	60	60	Cyclohexadienyl	−647.62	—
VII	1.10	2.00	25	80	Benzene and methyl	−647.24	—
VIII	1.09	3.00	0	90	Benzene and methyl	−649.28	—
IX	1.09	5.00	0	90	Benzene and methyl	−649.42	—

[a] From Tominaga *et al.*, 1970.

TABLE III

Changes of Total Electron Energy with Horizontal Approach of Attacking Hydrogen Atom: Reaction (9)[a]

Species number	Bond length (Å) C-7–H-16	Bond length (Å) C-1–C-7	Bond angle <H-14–C-7–C-1	Conformation Phenyl	Conformation Methyl	Total electron energy (eV)
I′	5.0	1.51	109°28′		Tolueme	−645.79
II′	3.0	1.51	109°28′		Toluene	−645.71
III′	2.0	1.51	100°		Toluene	−644.32
IV′	1.5	1.60	95°	Benzene	Shallow pyramid	−643.18
V′	1.3	1.66	90°	Benzene	Planar	−643.55
	1.3	1.81	90°	Benzene	Planar	−644.11
	1.3	1.90	90°	Benzene	Planar	−644.33
VI′	1.2	1.81	90°	Benzene	Planar	−644.19
	1.3	1.66	90°	Benzene	Planar	−643.66
VII′	1.15	2.0	85°	Benzene	Shallow pyramid	−644.71
VIII′	1.09	3.0	70°32′	Benzene	Methane	−648.23
IX	1.09	5.0	70°32′	Benzene	Methane	−648.44

[a] From Tominaga *et al.*, 1970.

The secular determinant of the 37th order to describe the C_7H_9 radical has been solved for 12 transient species for each reaction, with their conformational parameters being chosen as in Tables II and III. The results are reproduced in Tables II and III for reactions (5) and (9), respectively. Illustrated in Fig. 7 are the changes in the total electron energies for the two different angles of approach of the attacking hydrogen atom. The calculation indicates that reaction (5) is much preferred to reaction (9). Thus, *atomic cracking* of toluene by the hydrogen atom most probably takes place via the formation of methylcyclohexadienyl radical.

B. Elementary Reaction: Toluene + H

The toluene + H reaction was studied by Mulcahy *et* al. (1966) at 300°K and 663°K at a pressure of 0.6 torr, and later more thoroughly by Amano *et al.* (1972) at temperatures randing from 300° to 780°K and a pressure of 6 torr. The product pattern changed drastically with the variation of temperature. Figure 8, based on the findings of Amano *et al.* (1972), indicates that ring hydrogenation predominates at room temperature, but benzene formation becomes significant at temperatures above about 600°K.

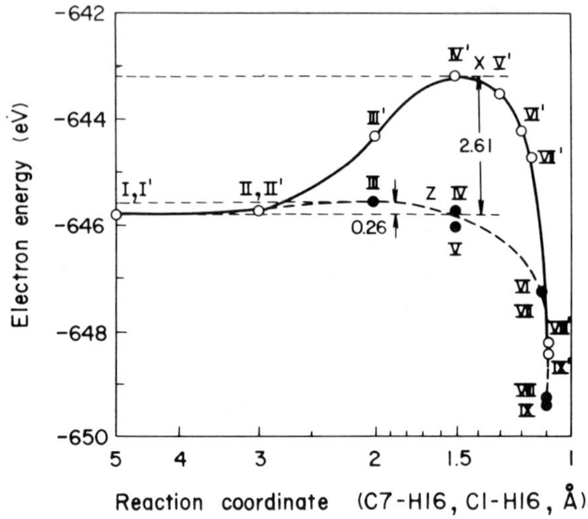

Fig. 7. Changes of total electron energy associated with reaction (5), ○—○; and reaction (9), ●---●. (From Tominaga *et al.*, 1970.)

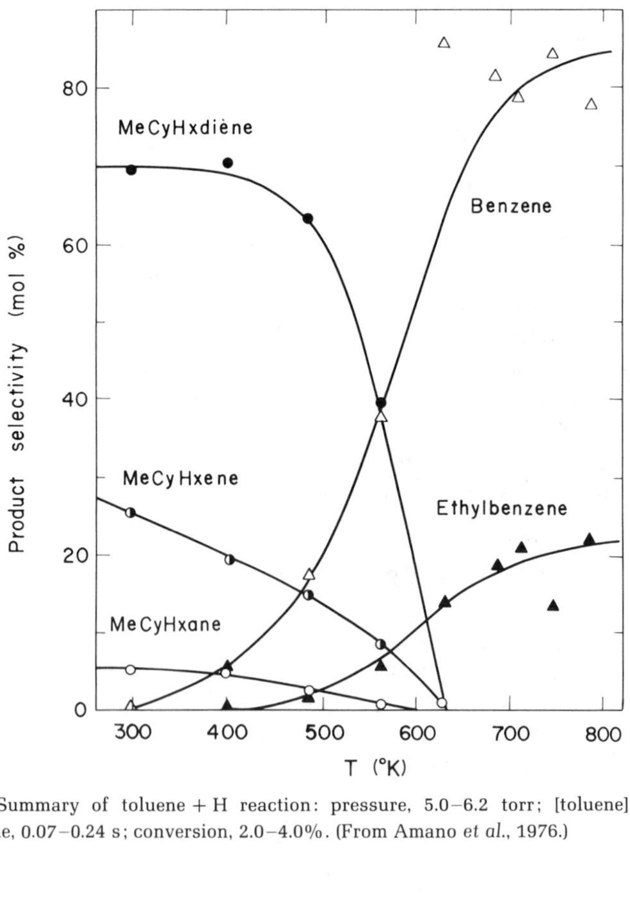

Fig. 8. Summary of toluene + H reaction: pressure, 5.0–6.2 torr; [toluene]/[N_2], 0.01; reaction time, 0.07–0.24 s; conversion, 2.0–4.0%. (From Amano *et al.*, 1976.)

C. Application of Unimolecular Reaction Theory

Based on the following discussions, intermediate steps for reaction (5) are suggested in Fig. 9. A double dagger indicates an activated complex, and an asterisk indicates an energized radical. Addition of hydrogen to the ring carbon of toluene leads to the formation of energized methylcyclohexadienyl radical via the association complex. The energized radical undergoes unimolecular decomposition to form D via decomposition complex with the rate proportional to kE. Alternatively, the energized radical may stabilize on collision with bath molecules, giving rise to the formation of hydrogenated products S. A potential energy profile of the scheme in Fig. 9, illustrated in Fig. 10, can be estimated based on the thermochemistry of the species involved.

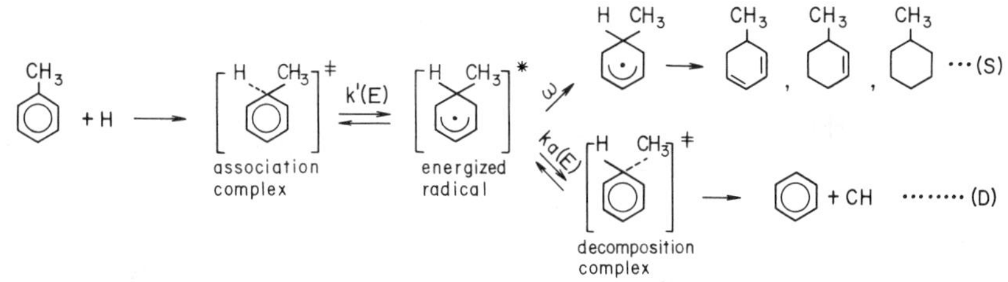

Fig. 9. Detailed reaction scheme of toluene hydrogenolysis. (D), Decomposition products; (S), stabilization products. See text for further details.

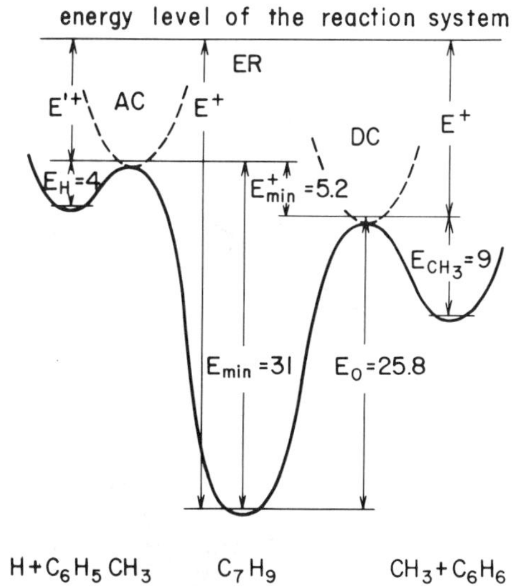

Fig. 10. Potential energy profile of the system: Toluene + H → Benzene + CH_3. (From Amano *et al.*, 1976.)

The experimentally determined average decomposition-rate constant, $\langle k_a \rangle$, is defined by:

$$\langle k_a \rangle = \omega(\mathrm{D/S}) \tag{6}$$

where ω is the collision frequency between the energized radical and bath molecules.

The system can be looked on as a unimolecular decomposition of methylcyclohexadienyl radical chemically activated by 31 kcal/mol as illustrated in Fig. 10. A quantum-statistical treatment known as RRKM theory, developed essentially by Marcus (1952), has been proven most realistic and successful for the types of reactions presently being considered. According to the theory, $\langle k_a \rangle$ can also be expressed by the following equation:

$$\langle k_\mathrm{a} \rangle = \frac{\int_{E_0}^{\infty} \{k_\mathrm{a}(E)/[k_\mathrm{a}(E) + \omega]\} f(E)\, dE}{\int_{E_0}^{\infty} \{\omega/[k_\mathrm{a}(E) + \omega]\} f(E)\, dE} \tag{7}$$

where, now, the decomposition rate constant, $k_\mathrm{a}(E)$, is a function of energy described by the energy-distribution function, $f(E)$.

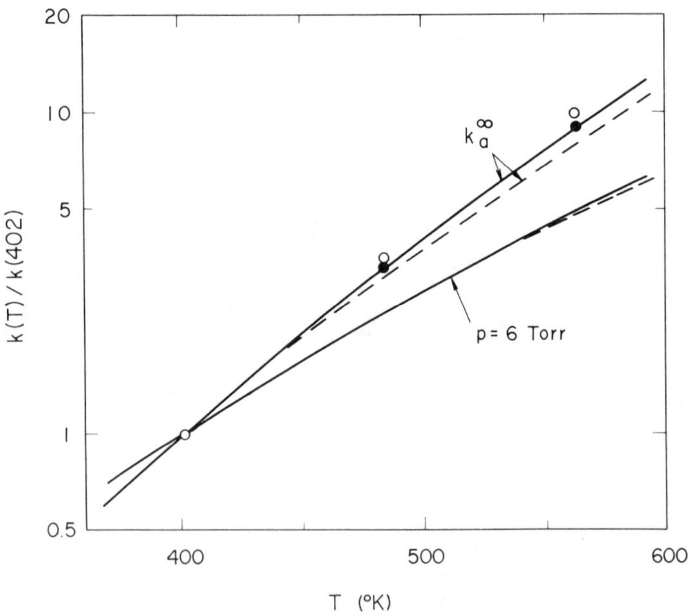

Fig. 11. Comparison of theoretical and experimental relative rate constants. Both hard-sphere (○) and gas-kinetic (●) collision diameters are tested. ——, Normal energized radical; --- modified energized radical. (From Amano *et al.*, 1976.)

Computation of the theoretical $\langle k_a \rangle$ has been performed by Amano *et al.* (1976) following the standard procedures recommended by Robinson and Holbrook (1972). In the computation, the proper assignment of mechanical properties to both the complex and energized radical, and also the accurate heat of formation values of the species involved are all essential. The results are compared with those from experiment in Fig. 11, based on the values at 402°K. Agreement between the theory and the experiment can be obtained by freezing the degree of freedom of the complex corresponding to the reaction coordinate indicated in Fig. 7, and the fact may be taken as support for the mechanism proposed.

V. ADDITIONAL COMMENTS

Aromatic pyrolysis has so far been discussed strictly in terms of 6-membered cyclic intermediates. Inasmuch as the rapid interconversion between benzyl and tropylium has been argued frequently, one should also consider the possible participation of an intermediate with 7 carbon atoms in the ring.

The superdelocalizability, a dynamic reactivity index proposed by Fukui *et al.* (1954, 1957), has been tested with the present system. This theory predicts well the experimental results of the demethylation of many polymethylbenzenes as observed by Amano *et al.* (1970), and also the results for the ring-hydrogen exchange of toluene as observed by Sato and Amano (1973). Such agreement between this theory and experimental results is only rational if the aromatic ring retains its structural identity during the reaction.

McLafferty and Winkler (1974), in their study of collisional activation mass spectroscopy, have also demonstrated that the interconversion between benzyl and tropylium requires a prohibitively high energy.

ACKNOWLEDGMENTS

This article is dedicated to the late Professor Hiroshi Tokuhisa, whose words, "pyrolysis in its ideal technology produces hydrogen and carbon" remain in the authors' memory.

REFERENCES

Amano, A., Tominaga, H., and Tokuhisa, H. (1965). *Bull. Jpn. Pet. Inst.* **7**, 59.

Amano, A., Uchiyama, M., Sato, Y., Tominaga, H., Arai, H., and Kunugi, T. (1970). *Bull. Chem. Soc. Jpn.* **43**, 3653.

Amano, A., Horie, O., and Hanh, N. H. (1972). *Chem. Lett.* p. 917.

Amano, A., Horie, O., and Hanh, N. H. (1976). *Int. J. Chem. Kinet.* **8**, 321.
Benson, S. W., and Buss, J. H. (1958). *J. Chem. Phys.* **29**, 546.
Benson, S. W., and Shaw, R. (1967). *J. Chem. Phys.* **47**, 4052.
Berthelot, M. (1867). *Ann. Chem. Pharm.* **142**, 254.
Betts, W. D., Popper, F., and Silsby, R. I. (1957). *J. Appl. Chem.* **7**, 497.
Braekman-Danheux, C., Delaunois, C., and Quyen, N. C. (1977). *Fuel Process. Technol.* **1**, 57.
Bryce, W. A., and Ruzicka, D. J. (1960). *Can. J. Chem.* **38**, 827, 835.
Burr, J. G., Meyer, R. A., and Strong, J. D. (1964). *J. Am. Chem. Soc.* **86**, 3846.
Dewar, M. J. S. (1969). "Molecular Orbital Theory of Organic Chemistry." McGraw-Hill, New York.
Fukui, K., Yonezawa, T., and Nagata, C. (1954). *Bull. Chem. Soc. Jpn.* **27**, 423.
Fukui, K., Yonezawa, T., and Nagata, C. (1957). *J. Chem. Phys.* **27**, 1247.
Golden, D. M., and Benson, S. W. (1969). *Chem. Rev.* **69**, 125.
Hoffmann, R. (1963). *J. Chem. Phys.* **39**, 1397.
Hoffmann, F., and Lang, K. (1929). *Brennst.-Chem.* **10**, 203.
Kershaw, J. R. (1978). *S. Afr. J. Chem.* **31**, 15.
McLafferty, F. W., and Winkler, J. (1974). *J. Am. Chem. Soc.* **96**, 5182.
Marcus, R. A. (1952). *J. Chem. Phys.* **20**, 359.
Masamune, S., Amano, A., and Tokuhisa, H. (1960). *Technol. Rep. Tohoku Univ.* **25**, 27.
Matsui, H., Amano, A., and Tokuhisa, H. (1959). *Bull. Jpn. Pet. Inst.* **1**, 67.
Morii, H., Hashimoto, A., and Tominaga, H. (1963). *World Pet.* **34**(6), 131.
Mulcahy, M. F. R., Harrisson, R. J., and Wilmshurst, J. R. (1966). *Aust. J. Chem.* **19**, 1431.
Robinson, P. J., and Holbrook, K. A. (1972). "Unimolecular Reactions." Wiley (Interscience), New York.
Sato, Y., and Amano, A. (1973). *Bull. Chem. Soc. Jpn.* **46**, 2646.
Shaw, R., Golden, D. M., and Benson, S. W. (1977). *J. Phys. Chem.* **81**, 1716.
Silsby, R. I., and Sawyer, E. W. (1956). *J. Appl. Chem.* **6**, 347.
Stein, S. E., and Golden, D. M. (1977). *J. Org. Chem.* **42**, 839.
Stein, S. E., Golden, D. M., and Benson, S. W. (1977). *J. Phys. Chem.* **81**, 314.
Szwarc, M. (1947). *Nature (London)* **160**, 403.
Szwarc, M. (1976). *Polym. Eng. Sci.* **16**, 473.
Tominaga, H., Arai, H., Kunugi, T., Amano, A., Uchiyama, M., and Sato, Y. (1970). *Bull. Chem. Soc. Jpn.* **43**, 3658.
Tsuchiya, A., Hashimoto, A., Tominaga, H., and Masamune, S. (1959). *Bull. Jpn. Pet. Inst.* **1**, 73.

7

Mathematical Modeling of Hydrocarbon Pyrolysis Reactions

MARIO E. DENTE and ELISEO M. RANZI

I. PURPOSE OF THE PYROLYSIS SIMULATION MODELS

Reliable mathematical models for the simulation of hydrocarbon pyrolysis are generally accepted as being a formidable tool for several purposes.

1. The economics associated with the different feedstocks and the different effluents of the steam-cracking units are large and continuously changing, necessitating careful choice of feeds and conditions.
2. Increasing flexibilities of the reactors are claimed both for the effluents distribution and for the feedstock utilization.
3. For some pyrolysis units, certain fractions should be recycled.
4. The runtime of the pyrolysis coils should be as high as is possibly compatible with the optimal yields of the process.

Pyrolysis: Theory and Industrial Practice

ISBN 0-12-048880-9

5. New cracking reactors and technologies are being developed, requiring reliable prediction of the effluents distribution.
6. Existing units are sometimes revamped.

The mathematical models can help solve these problems and optimize operations; in addition, costly and time-consuming experimental programs can often be minimized. There is now a general trend toward the development and the use of models over a wider range of application as compared to several pioneering works mainly related to the cracking of light gases. It seems appropriate to remark that the increasing completeness and complexity of the models proposed recently is related to the evolution of the computation facilities that permit handling of much larger numerical problems.

Mathematical modeling of a process is, in general, based on at least three approaches. The first is based on a phenomenological description of the process system, which should be as accurate and complete as possible; for instance, the identification of the complete set of chemical species and of their overall reaction mechanisms (i.e., reactions network). The second feature is based on the translation into mathematical terms (expressions, equations, boundary conditions, etc.) of the chemical and physical steps, including assignment of the parameters values for heat, mass, and momentum balances and also complete kinetic parameters. Qualitative and, if possible, quantitative assessment should be included of any simplifications made. The third item is concerned with the numerical methods to solve the set of equations and conditions: for instance, general purpose or specifically conceived programs are developed to handle large system of stiff differential equations.

The main objective of a mathematical model is of course to simulate accurately the system; namely, to predict the process behavior both inside and outside the range of experimental data. Because of the large number of chemical species involved, the number and type of reactions, and uncertainties of the mechanisms, the first two aspects of the modeling activity (i.e., setup of the reactions network and quantitative assessment of the associated kinetic scheme) are critical in establishing the degree of simplification that can be accepted.

Other limitations are related to the inadequacy of details and reliability of considerable available experimental data: for instance, inadequate description of the liquid feedstocks (such as naphthas and gas oils) and effluents components; wall or catalytic effects and/or coke formation that interferes with the gas-phase reactions (mainly in the small-diameter bench-scale reactors), as demonstrated by Dunkleman and Albright (1976a,b); sometimes incomplete or incorrect analyses of minor but significant components (such as acetylenics) in the gaseous effluents; incomplete or unavailable information on the gas temperatures and residence-time distributions (specifically in the bench-scale, laminar-flow reactors). Such faults in the quality of the experimental information may lead to misleading or contradictory interpretations of data and mechanisms.

Finally, the availability of efficient and adequate numerical algorithms to handle the numerous equations is perhaps the most critical reason necessitating restriction of the reaction network even at the possible expense of oversimplifications.

The foregoing considerations, and furthermore the personal preferences of the people working in the pyrolysis field toward a specific modeling philosophy, provide motivation for different types of models. In some respects, mathematical modeling of pyrolysis reactions is more like a developing art than a completely scientific approach.

II. STATUS OF THE ART: TYPES OF PYROLYSIS MODELS

To be suitably flexible, an overall model for the simulation of pyrolysis reactors should, of course, include the heat, momentum, mass-balances differential equations, physicochemical properties (specific heats, enthalpies, viscosities, thermal conductivities, etc.), firing-box patterns, and sometimes fluid dynamic characterization (as in the advanced cracking reactors). As is usual for chemical reactor problems, however the keystone should be the reaction modeling (i.e., kinetic scheme, reaction rates expressions, and related kinetic constants). A convenient classification of the model types can be based on the degree of sophistication. With some arbitrariness, at least three major types can be considered in order of increasing complexity.

A. Empirical Modeling

The main feature here is the absence of even the simplest reaction hypothesis; of course the integration of the fundamental differential equations as a function of tube length or residence time is absent. Only the overall material and thermal balances around the unit are possible. This empirical modeling has in the past been probably the most common method employed by olefins producers. In general, a pilot cracking unit was used, and empirical or semitheoretical scale-up rules were adopted for transferring the base information to the commercial units. In the simplest approach, a single feedstock is evaluated for standard conditions in the test unit and the yields of the effluents are measured; then the scale-up (for the estimation of the operative conditions and product distribution) is made to the commercial unit.

In a more sophisticated approach, several standard feedstocks have been tested to cover a range of compositions of future available feedstocks. The behavior of a new feedstock is estimated on the basis of suitable interpolation or additivity rules; a single naphtha or gas oil can, for example, be estimated as a properly weighted mixture of previously studied naphthas or gas oils.

A further improvement, at least for the naphthas cracking (where the PIONA, ASTM range, and specific gravity are usually known), is to single out from the basic information obtained from the test unit, the specific behaviors of *n*-paraffins, of isoparaffins, of naphthenes, and of aromatics fractions; in this manner data can be synthesized for new naphthas.

In the brief discussion of this chapter, modeling techniques strictly related to the local narrow-range regression models, such as are often used for control purposes (usually directly developed on the commercial units), have been ignored. Yet these highly empirical models, which can be developed quickly and are simple to use, are highly regarded by numerous olefins producers. At least three different reasons can explain this fact. First, olefin producers can readily use their experiences and limited data to interpolate and perform scale-ups; second, this approach seems to them more closely related to the practice of the operating plant people; finally, a group or team inside the company specifically devoted to the development of more sophisticated models cannot always be readily organized.

The following are the main disadvantages of empirical modeling:

1. The development of each model requires expensive experimental work (with the accompanying analytical problems) to produce the necessary extensive experimental data.
2. Once a model has been set up, its validity is strictly limited to a specific range of feedstocks and to a specific furnace. Accurate predictions are possible over only a limited range of operating conditions, and extrapolations are of doubtful value.
3. These models are not to be trusted for design or revamping decisions.
4. Additivity rules for feedstocks may result in inaccurate predictions, hence the evaluation of new feedstocks cannot be made reliably.
5. New mixtures (e.g., with partial recycling of some effluents) or new feedstocks necessitate additional experimental investigations.

B. Molecular (or Stoichiometric) Modeling

Much effort has been devoted in recent years to molecular or stoichiometric modeling. Although thermal cracking involves mainly free-radical reactions, only the apparent global molecular reactions are accounted for in this approach. Some models adopt partial mechanistic considerations to the correct molecular kinetic scheme, and a large set of high-quality experimental data are always necessary to estimate the kinetic parameters through nonlinear optimization programs.

Recently Sundaram and Froment (1977a) proposed for ethane and propane pyrolysis an example of discrimination of the best kinetic model between rival molecular schemes. These molecular kinetic schemes are generally drastically simplified in order to reduce the number of rate equations to only a few global reactions. Some authors prefer to distinguish for this type of model between more sophisticated stoichiometric models and simpler semikinetic models (Ross and Shu, 1979).

This molecular approach has been more or less successfully applied to the pyrolysis of ethane, propane, butanes, and their mixtures (sometimes also mixtures with ethylene and propylene). Without any pretension to give exhaustive examples, the following have developed models in this category: Myers and Watson (1946), Lichtenstien (1964), Shah (1967), Crynes and Albright (1969), van Damme *et al.* (1975), Ross and Shu (1977), Sundaram and Froment (1977b), Williamson and Davis (1979). Global kinetic correlations and product distributions for pure higher hydrocarbon pyrolysis are reported, for example, by Zdonik *et al.* (1967), Davis and Farrell (1973), Murata and Saito (1975), and Arai *et al.* (1977).

The pyrolysis of complex mixtures such as naphthas and gas oils is approached mainly by ultrasimplified semi-kinetic models, made up of a single overall reaction in which the feed is considered as a single pseudocomponent. The extent of hydrocarbon decomposition along the reactor is evaluated as a function of at least an appropriate severity index (several are proposed) and feedstock-characterization parameters. The products distribution is correlated with the extent of decomposition, using global stoichiometry depending on the feedstock (as determined from extensive experimental work). Examples of such models for the naphthas are reported by de Blieck and Goossens (1971), Fernadez-Baujin and Solomon (1976), Illes *et al.* (1976), Shu *et al.* (1978); for gas oils, Green *et al.* (1975) and Lohr and Dittmann (1977) have developed models. Temperature and pressure profiles along the reactors are determined by coupling the enthalpy and momentum equation with the appropriate material balances. Empirical coking-rates' correlations are also given.

A further discrimination factor among the several models depends on the ability to describe correctly the reaction behavior. With the exception of Sundaram and Froment (1980), the mathematical description of the cracking furnace is based on lumped one-dimensional models, considering gradients only in the axial direction. Adequate description of pressure losses and temperature profiles along the coil and in the metal skin is an advantage of this type of modeling as compared to the empirical one. The prediction of the effluents composition is adequate only when the feed composition and the overall cracking conditions are in the data base range. As compared to the more mechanistic approach, the obvious advantages include a simpler formulation, shorter development time, and substantial savings of computer-storage and computing time. Yet the intrinsic empirical nature of the overall molecular scheme constitutes the major disadvantage, in terms of extrapolating both operative conditions and feedstocks. The extent to which these molecular models are reliable at conditions outside those of the original data base is controlled by their ability to reflect the correct kinetics.

Sometimes these models seem to suggest a fundamental approach. Such a conclusion is misleading because apparent frequency factors and activation energies are by no means real constants, and only exceptionally do they have a real physical meaning because they change substantially (van Damme *et al.*, 1975) with changes of the reactant concentrations, temperature, and pressure. The limitations of this type of model are evident when different kinetic parameters are proposed for the same reactions depending on whether a gas mixture or pure hydrocarbon is employed. As an example, the paradoxical behavior of decreased ethane conversions when in mixture is not easy to explain. Global reactions are unable to represent the true mechanistic effect of the overall H abstraction reactions of the ethyl radical in the case of pure ethane cracking. A more mechanistic approach is needed to explain some apparent conflicting interpretations regarding the ethylene selectivity in the case of ethane–propane co-cracking (see, e.g., Ross and Shu, 1977; Hatch and Matar, 1978; Froment *et al.*, 1979). Yet, Sundaram and Froment (1978a) apparently recognize the shortcomings of the global molecular models by proposing a radical reaction scheme for gas cracking.

C. Mechanistic Modeling

"There is now sufficient understanding and both kinetic and thermochemical data available to describe the behavior of even the most complex pyrolysis in terms of a finite number of elementary step reactions . . . . On this basis, it might be expected that hydrocarbon pyrolyses are ripe candidates for quantitative modeling." This forecast was made by Benson (1970) and his final suggestion was ". . . the inclusion of too many starting reactants can impose intolerable burdens on even large computers. Under the circumstances, the scheme must be simplified by a combination of methods which include careful adjustment of 'averaged' kinetic parameters with available data. Many of these can be done reasonably well by analyzing product ratios."

The advances made in the experimental techniques to obtain accurate and detailed thermochemical kinetic data and the increased availability of pyrolysis data for both single components and mixtures over a wide range of conditions permit development of well-balanced mechanistic models. On the basis of the experience made in the development stage of this type of modeling by Dente *et al.* (1970), an essential philosophy, also suggested by Allara and Edelson (1975), has developed. In establishing the reactions network, account is taken of the most relevant species and reactions. Simplifications are made only in the second stage in order to reduce the complexity judiciously. To avoid an initial enormous number of species and reactions, some partitioning of the hypothetical global network has to be done. The simplification procedures adopted will be presented later.

Because the same model is operating on several components, its validation is performed simultaneously on a set of experimental data related to different hydrocarbons (from different sources and different operating conditions). Using structural analogies among similar reactions, it is possible to cross-check and to "average" the kinetic parameters related to single reaction classes, as previously suggested by Benson. Hence it is not surprising that the "averaged" kinetic parameters (for example for primary, secondary, tertiary hydrogen abstractions, or radical additions on double bonds) estimated independently by Allara and Edelson (1975), Kerr and Parsonage (1976), and Dente *et al.* (1979a) are substantially coincident and, moreover, coherent with the thermochemical kinetic provisions.

To handle these complex kinetic schemes at least two major difficulties have to be overcome, depending on the largeness and stiffness of the differential equations system. In addition to the necessity of efficient numerical algorithms, it is also compulsory to have a chemical compiler. This compiler provides in the most suitable way the matrices of stoichiometric and kinetic coefficients necessary for the builiding of the mathematical model (Dente *et al.*, 1970; Allara and Edelson, 1975; Côme *et al.*, 1979).

Even the simplest cases (e.g., pure ethane or propane pyrolysis) require an extensive kinetic scheme in order to represent the real system behavior. Otherwise kinetic constants for the elementary reactions cannot be extrapolated; and this constitutes loss of one of the main advantages of this type of modeling. In fact the mechanistic models, conceived as very specific "ad hoc" models, suffer the same limitations and contradictions as the molecular ones. Consequently, rather wide discrepancies that are often found in the literature about elementary reaction parameters can be at least partially explained on the basis of the previous consideration. To mention only a few examples, interpretative effort was spent to develop more or less sophisticated mechanistic models by Snow *et al.* (1959), Davis and Williamson (1959), Blakemore and Corcoran (1969), Kunugi *et al.* (1969), Dunkleman and Albright (1976a), and Goossens *et al.* (1978a,b). The main advantages of well-balanced and adequately extensive mechanistic models are as follows:

1. There is a capacity for wide extrapolation and considerable flexibility (in terms of components, mixtures, and operating conditions). After characterization of the types of reactions and their intrinsic kinetic parameters and by proper use of structural or mechanistic analogies, *a priori* modeling of reactions, kinetic constants, and product distribution can be developed for previously untested hydrocarbons. For example, starting from a well-tested mechanistic model of light paraffins and naphthas (Ranzi *et al.*, 1977), the SPYRO model was quickly extended to atmospheric gas oils with low cost (Goossens *et al.*, 1978a,b).
2. Only with this type of model can advantage be taken simultaneously of theoretical kinetic background and of all available experimental data (bench scale, pilot, and commercial).
3. Once the model has been developed, it is no longer necessary to employ an expensive pilot unit.
4. Simulations of numerous modifications are readily produced, such as recycling of low-market-value products to the cracking unit and new cracking technologies.

Such a model, however, requires a more expensive and time-consuming initial development stage. After development though, the marginal cost for new improvements tends to be small. With the modern computers, memory allocation and computing time cannot really be considered a limitation for a judiciously simplified mechanistic model.

In the following section, attention will be focused on the mechanistic approach, particularly on the experience acquired with the so-called SPYRO model.

III. MECHANISM CONSIDERATIONS AND MAIN REACTION CLASSES

The main reason to classify the elementary reactions for a mechanistic model is the analogies for reactions of the same class. Such a classification is illustrated in Table I. Most reactions involve radicals, but some purely molecular reactions play a significant role.

TABLE I

Main Reaction Classes

1. *Chain-initiation reactions*
   unimolecular:
   $R—R' \rightarrow R\cdot + R'\cdot$ (example: $C_2H_6 \rightarrow 2\ CH_3^{\cdot}$)
   or bimolecular (R'H in an unsaturated hydrocarbon):
   $RH + R'H \rightarrow R\cdot + R'H\cdot$ (example: $C_2H_6 + C_2H_4 \rightarrow 2\ C_2H_5^{\cdot}$)
2. *Hydrogen-abstraction (metathetical) reactions*
   $R\cdot + R'H \rightarrow RH + R'\cdot$ (example: $CH_3^{\cdot} + C_2H_6 \rightarrow CH_4 + C_2H_5^{\cdot}$)
3. *Radical-decomposition reactions*
   $R\cdot \rightarrow RH + R'\cdot$ (example: $n\text{-}C_3H_7^{\cdot} \rightarrow C_2H_4 + CH_3^{\cdot}$)
4. *Radical-addition reactions to unsaturated molecules*
   $R\cdot + R'H \rightarrow R''\cdot$ (example: $CH_3^{\cdot} + C_2H_4 \rightarrow n\text{-}C_3H_7$)
5. *Chain-termination reactions*
   by recombination of radicals:
   $R\cdot + R'\cdot \rightarrow R—R'$ (example: $CH_3^{\cdot} + C_2H_5^{\cdot} \rightarrow C_3H_8$)
   by disproportionation of radicals:
   $R\cdot + R'\cdot \rightarrow RH + R''H$ (example: $C_2H_3^{\cdot} + C_2H_5^{\cdot} \rightarrow 2\ C_2H_4$)
6. *Purely molecular reactions*
   $RH + R'H \rightarrow R''H + R'''H$
   e.g., "four-center" concerted reactions (example: $C_2H_6 \rightarrow C_2H_4 + H_2$)
   "six-center" concerted reactions (example: pentene-1 $\rightarrow C_2H_4 + C_3H_6$
   or hexadienes 1-3 $\rightarrow$ hexadienes 2-4 isomerization)
   Diels–Alder type dissociations (cyclohexene $\rightarrow C_2H_4 + C_4H_6$)
7. *Radical-isomerization reactions*
   $R'\cdot \rightarrow R''\cdot$ (example: $CH_3CH_2CH_2CH_2^{\cdot} \rightarrow CH_3CH_2\dot{C}HCH_3$)

Benson (1970) observed that the exclusion of molecular reactions that occur simultaneously with radical reactions has been responsible for some misleading conclusions of rate parameters, particularly for olefins and diolefins. The structural (or mechanistic) analogies are essential to extend the knowledge acquired and tested for a few specific elementary reactions to similar reactions. Taking advantage of them, the really independent kinetic constants can be reduced to a realistic number.

In fact, once the values for the frequency factor and the activation energy for the reference reaction have been obtained experimentally and found to be in agreement with the theoretical predictions of thermochemical kinetics, analogous reactions can be predicted (after proper corrections are made for symmetrical variations and bonds positions) with little or no experimental investigation.

The analogy rules rely partially on theoretical and mainly on intuitional bases in regard to the identity of the reacting portions of the species, and are related to similar species and reactions. Actually the thermokinetics theory can give support to those rules by assuming that the transition-state complexes (at least for the hydrocarbon pyrolysis reactions) are subject to additivity rules of group properties for both molecules and radicals.

Because this assumption implies only short-range forces (i.e., in the order of magnitude of the bond lengths), it is mainly an intuitive hypothesis. However, it is also essential to indicate that the globally available experimental information on simple reactions is not in contradiction (within the limits of confidence on these data) with these similarity rules; and experience with some extended mechanistic models actually seems to confirm this approach.

Perhaps the most widely adopted analogy concerns the H abstraction reactions; the rate constants for a single abstracted H atom are assumed to depend only on the abstracting radical and on the type of H atom (i.e., primary, secondary, or tertiary paraffinic; primary or secondary allylic, etc.). For the same similarity rule, the frequency factor is assumed for any H abstraction rate constant to depend only on the abstracting radical and to be independent of the type of abstracted H atom. A third simplification is the assumption that the differences in activation energies among the different types of H atoms is independent of the abstracting radical: this last assumption may be slightly less correct from a strictly theoretical point of view, because small deviations from the rule should arise for highly resonant radicals (such as allylic or cyclopentadienylic radical).

Another similarity rule generally accepted concerns the addition of radicals to double bonds (and also triple, dienic, aromatic, etc., bonds); the rate constants depend only on the radical and on the bond position (α, β) and type (double, triple, etc.).

Further rules are related to the radical–radical recombination constants (exclusion has to be made for the recombination that give rise to molecules with less than 6 or 7 atoms, because they are usually third-body reactions for energetic reasons); for them the activation energy is nil and the frequency factors can be proposed in terms of the proper collision rates and the probability of effective orientation of the colliding bodies. Furthermore, recombination constants of dissimilar radicals can be considered the geometric means of the constants for identical radicals. Consequently, the initiation reactions (reverse of the previous ones) are also easily estimated by thermodynamic considerations.

General rules are also used for the isomerization of radicals, both through internal H abstraction (via ring formations in the transition state) and through internal addition on unsaturated bonds and subsequent cyclic radicals formation. The frequency factors for these reactions are deduced from the universal expression $ekT/h \exp(\Delta S^{\neq}/R)$ (e, bases of the natural logarithms; k, Boltzmann constant; h, Planck constant; T, absolute temperature; $\Delta S\neq$, entropy variation in the transition state; R, gases constant); $\Delta S\neq$ accounts for the free or hindered rotation lost in forming the ring, for the symmetry, and for the eventual spin degeneracy (Benson, 1976).

Activation energies for these isomerization reactions are deduced from the general ones, taking into account the nature of the radical abstracting or adding a hydrogen atom, the type of the abstracted hydrogen atom, the position and the ype of the unsaturated bond, and the eventual strain for the ring formation. Other analogies are needed for molecular decomposition, isomerization, molecule–molecule four-center or six-center reactions (e.g., direct or reverse Diels–Alder, ENE reaction, etc.) of olefins, diolefins, cycloolefins, cyclodiolefins.

All the previous and related similarity rules provide a guideline for predicting and testing rate constants for elementary reactions. Simplifications have to be introduced in a mechanistic model, however, in order to make it realistic and practically manageable, as will be discussed in the next paragraph. First a realistic kinetic scheme must cover the largest possible but not an unlimited range of conditions.

In the usual range of temperatures, pressures, and compositions, some obvious simplifications can be adopted, at the expense of a modest loss of accuracy. All simplifications should be carefully evaluated by testing the more complete structure versus the simplified one; the deviations are compared with some appropriate tolerance limits and on these bases the "judicious" simplifications are accepted (or rejected). The most immediate and fruitful simplifications concern the fate of the relatively heavier radicals as outlined next.

1. The H abstraction rate constants decrease with larger radicals (because of fewer collisions and a decreasing orientation factor); this reduction is almost inversely proportional to the molecular weight of the radical, as supported by the experimental evidence for pure components (except in the low-temperature, high-pressure range). No significant amounts of C_4^+ paraffins or naphthenes are formed. Similar behavior occurs with unsaturated radicals. Hence for a range of operating conditions greater than that of commercial interest, the H abstraction reactions are becoming progressively weaker sinks as the molecular weight of the radicals increases.

2. The addition of radicals on unsaturated bonds rapidly decreases with increased molecular weight of the radicals, for the same reasons previously discussed.

3. Larger radicals have, however, an increased number of routes and rates relative to internal isomerization and decomposition that yield smaller radicals and unsaturated molecules. Exceptions are unsaturated resonant

radicals (such as cyclopentadienyl, isoprenyl, and benzyl) that have no significant isomerization or decomposition alternatives.

4. Chain-termination reactions are of decreasing importance as the number of C atoms on the radicals increases because of decreased oriented collision frequencies and lower concentrations. Resonance-stabilized unsaturated radicals, as previously discussed, enter into mainly termination reactions; indeed their H abstraction or their addition on to unsaturated molecules is penalized by an increase of activation energy related to the resonance energy.

These considerations permit drastic simplifications in the reaction scheme. In fact very stabilized radicals with more than four C atoms undergo essentially comparable isomerizations and decompositions; the most probable final decomposition products (smaller radicals and unsaturated molecules) can thus be substituted for each in the scheme. The distribution of the decomposition products of these radicals is only a weak function of temperature and is practically independent of the composition of the reacting system or the pressure.

An example of this simplification is reported in Fig. 1, for the radical 1-hexene-6-yl formed for example by H abstractions from cyclohexane. Using this simplification, all the reactions (such as H abstraction or radical additions to unsaturated hydrocarbons) can be significantly reduced in complexity by combining isomerization and decomposition reactions as a sequence to produce the final products. A qualitative example of this approach is given in Fig. 2 for H abstraction reactions involving *n*-octane. The numerical elaboration to provide these simplifications (to be inserted into a global kinetic scheme) should be conveniently developed as an auxiliary program. Because of the weak dependence of the product composition in relation to temperature, the assumption that there is no change results in little error.

Other simplifications concern the intermediate molecular species that are quickly isomerized. For example, 1,3-hexadienes are quickly isomerized (through a concerted six-center reaction) to 2,4-hexadienes, and vice versa; hence the two isomers are in a quasi-equilibrium state and can be considered as a single equivalent component in subsequent reactions. A similar assumption occurs for homologues of methylhexadienes. As another example, cyclohexadienes are quickly transformed to benzene and H_2 (essentially by a molecular path): actually it is difficult to find them as significant pyrolysis products (except in carefully designed tests).

Based on the foregoing assumptions and simplifying rules, a quite "complete" and wide-range mechanistic model has been developed. The "complete" model has now been verified and the global rules adopted (more

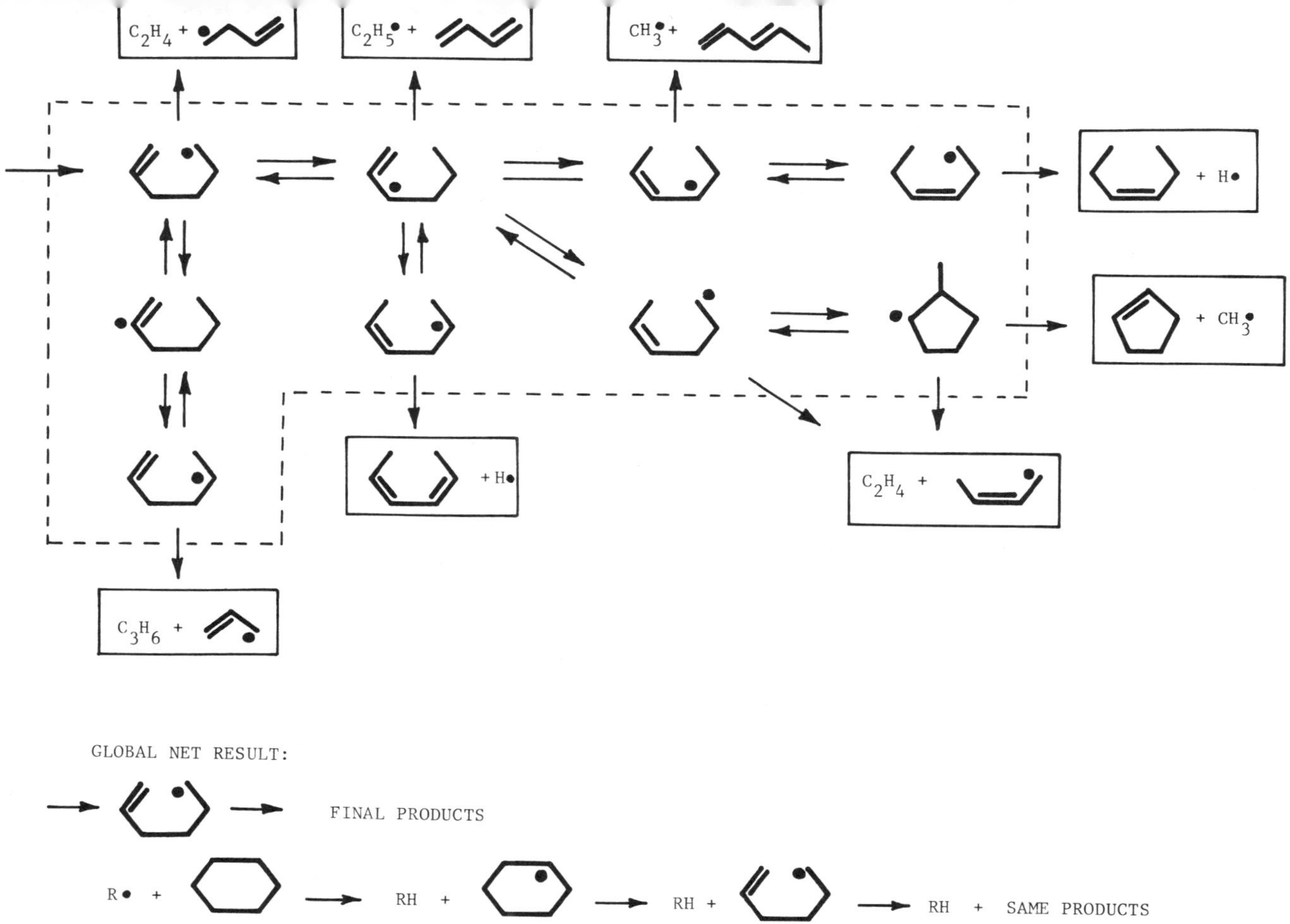

Fig. 1. Simplified reaction path for 1-hexene-6-yl radical.

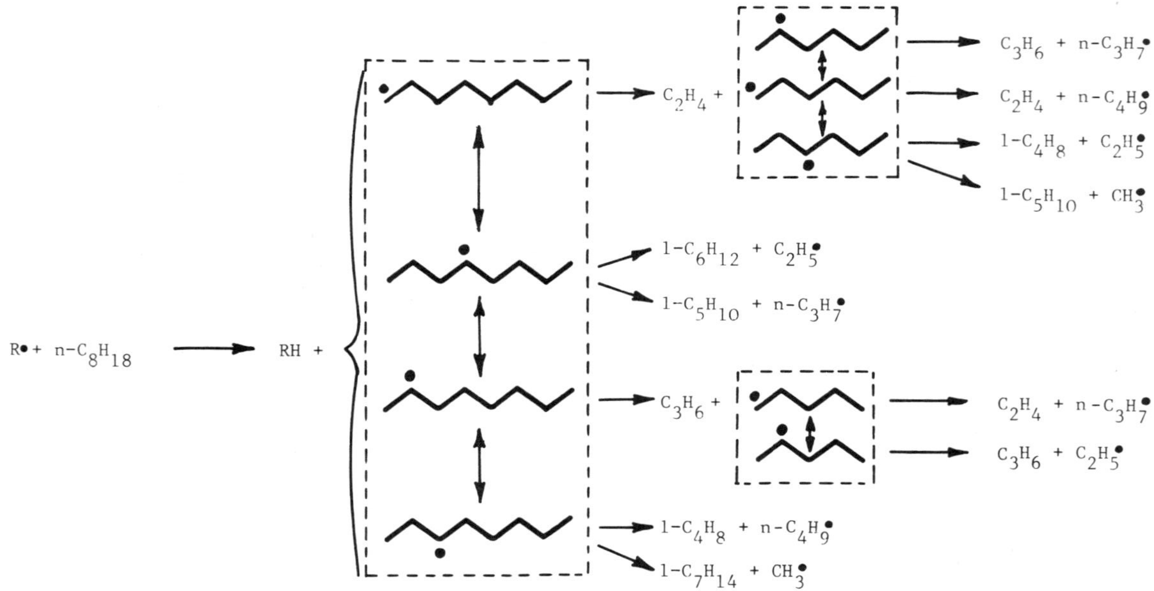

Fig. 2. Hydrogen abstraction reactions on *n*-octane.

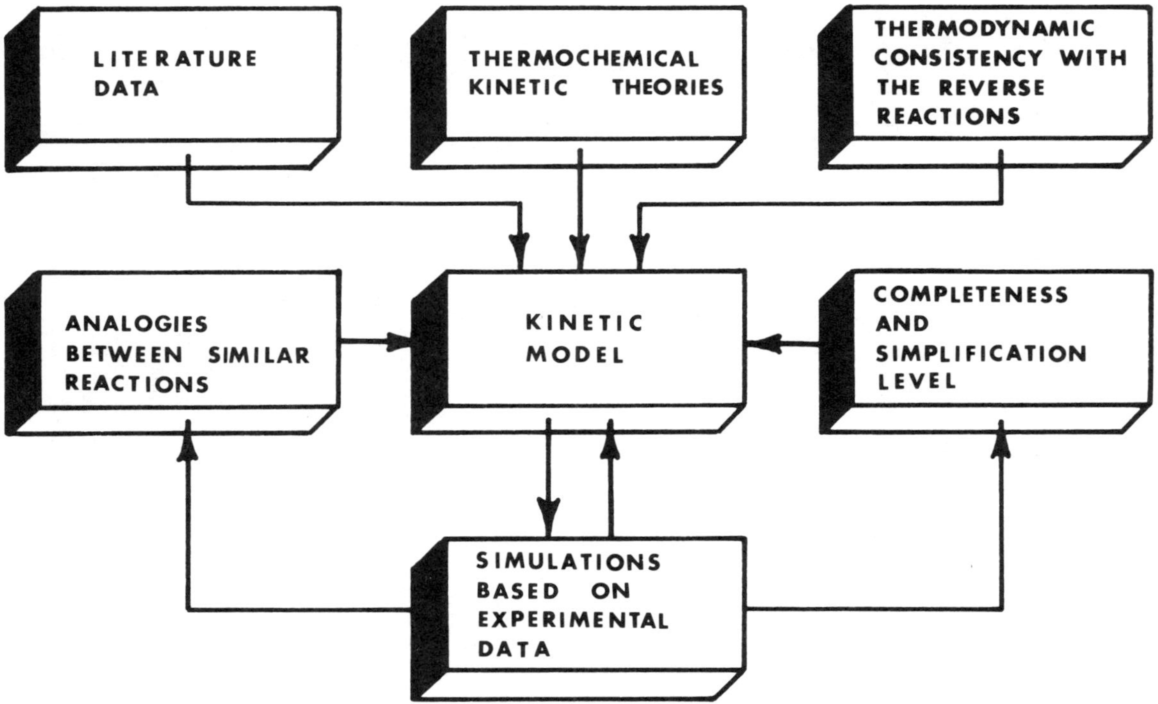

Fig. 3. Block diagram of SPYRO model development.

than the single kinetic parameter) are checked simultaneously using selected experimental information. Obviously adjustments of hypotheses, insertion of previously disregarded reactions, and tuning of the basic set of kinetic parameters had to be made. A schematic representation is given in Fig. 3 of the information sources, criteria and philosophy, finalization, and validation of the SPYRO model.

Elaboration of data obtained using very-small-diameter laboratory bench-scale reactors is complicated because of surface effects and particularly because of tars and carbon formation on the tube. Because the product composition is partially affected by surface reactors, experimental data have to be properly evaluated relative to these effects. The catalytic activity of the wall, the presence or absence of sulfur or other poisons, the runtime, the pretreatment of the inner surface, and other related aspects that are of considerable importance in small reactors (Dunkleman and Albright, 1976a,b), can usually be disregarded in industrial reactors.

IV. MATHEMATICAL MODEL OF THE PYROLYSIS REACTOR

The thermal cracking process can be mathematically described using the fundamental balance equations (specifically, the material balances for each molecular and radicalic component, the energy—tube side and firebox side—and the momentum balances). The solution of this well-known set of nonlinear ordinary differential equations is complicated by two specific problems.

1. Not all the *initial values* for the differential equations are known (typically, industrial data are given for the reactor *inlet* composition and temperature and the *outlet* pressure and temperature); an iterative solution method is hence required.
2. The "stiffness" of the material balance equations results from the wide range of kinetic-parameters values and the fact that the (low) radical concentrations are the result of very fast formation and disappearance reactions (close to steady state); extremely small integration (time) steps and hence prohibitively long computer time would be required if usual numerical integration methods were applied.

As elsewhere reported (Dente *et al.*, 1979a), to overcome the first of these complications and to save computing time, the solution procedure adopted into the SPYRO program is divided into two phases. In the first phase, simplified kinetics and normal numerical integration methods are employed; the pressure profile is established with good accuracy, and the approximate temperature and composition profiles are determined. In the second phase, rigorous kinetics are used, together with a specific solution technique for the material balances to overcome the "stiffness" problem.

Employing the gas temperature profiles calculated in the first phase, it is possible to separate partially—within each integration step—the enthalpy balance from the material balances (which are most expensive in computing time). In fact, the kinetic constants are evaluated from the average temperature within each step; this average temperature is estimated by the actual step-inlet temperature plus one-half of the temperature increment of the step as determined from the previously determined temperature profile. An updated step-outlet temperature is calculated for each step from the enthalpy balance, after solution of the material balances. If the updated average temperature inside the step differs from the preestimated one by more than an accepted deviation, the computation procedure is repeated. Usually two iterations are sufficient. In addition, the average step temperature for the evaluation of the kinetic constants is tentatively corrected for the nonuniformity of the radial temperature profile (boundary film cracking) because of the prevailing high heat flux. This correction becomes significant for those reactions having high activation energies, and for reactors having small diameters, low Reynolds numbers, and high heat fluxes. It is estimated from

$$T_{\mathrm{eff}} = T_{\mathrm{b}} + \frac{R}{E_{\mathrm{a}}} T_{\mathrm{b}}^2 \ln \left[1 + \frac{4}{\mathrm{Nu}} \left(\frac{\exp \sigma - (1 + \sigma)}{\sigma} \right) \right]$$

where E_{a} = activation energy, $\sigma = (E_{\mathrm{a}}/RT_{\mathrm{b}}^2)(T_{\mathrm{f}} - T_{\mathrm{b}})$, T_{f} = maximum film temperature (°K), T_{b} = bulk temperature (°K), Nu = Nusselt number, and R = gas constant (kcal/°K, kmol).

The SPYRO program overcomes the stiffness of the material balance equations by using a particular semiimplicit integration method. This method, when applied to the radical species, yields a *numerical* "simplification," which is often equivalent to the so-called continuously varying steady-state assumption (CVSS).

The convergence of this method was tested by increasing drastically the number of integration steps and the number of iterations (20 instead of 2) for each step. The difference between the two "extremes" was in general less than 5% for the radical concentrations and negligible for the molecular species.

Various authors have questioned the CVSS approach. It is, however, generally agreed that most radicals do not (Sundaram and Froment, 1978b; Edelson and Allara, 1973) attain a steady state along the coil, as found in the CVSS approach. However, Edelson and Allara report a "surprising" agreement between calculated values obtained by integration with Gear's method (Gear, 1971) and Snow's procedure. Snow (1966) assumed "pseudo-steady state" and used a coupled solution, for each integration stage, of the differential balance equations for molecules and algebraic equations for radicals. After theoretical and practical testing of the CVSS assumption of radicals at conventional cracking conditions, it is difficult to agree with the rejection of the CVSS principle. The main problem in applying the CVSS approach is to have a suitable method to solve the algebraic nonlinear equations for radical concentrations.

A further remark is needed on radical concentrations at the inlet, when using the non-steady-state approach. For industrial temperatures and contact times, it is not correct to assume zero (radical) concentrations at the reactor inlet. Except, however, for the first few integration steps, radical concentration profiles are remarkably independent of the concentration values assumed at the inlet.

Fouling of the tubes with coke directly influences furnace performance. First, higher pressure drops occur at the same coil outlet temperature, increasing the tube-skin temperatures. Second, coke causes changes of the temperature profile along the coil, modification of the effluent composition (with losses in olefins selectivity), and reduction in firebox efficiency. Increases of tube-skin temperature up to the allowable maximum limits the runtime of the furnace. With heavy feedstocks, this limitation may arise from transfer-line exchanger fouling. The periodic decoking required decreases considerably the overall capacity and affects optimal operation (Ranzi *et al.*, 1980).

To predict these effects, the overall model has to be supplied with a realistic coking model. Several qualitative and quantitative simplified coking models have been presented based on both commercial data and experience; see, for example, Atkins (1962), de Blieck and Goossens (1971), and Fernandex-Baujin and Solomon (1976).

The coking model, of course, has to be coupled with the equations describing the effects both on the circumferential tube-skin temperature patterns and temperature variations in the coke. The knowledge of this last variation is essential to evaluate the nonuniform growing of the coke deposit inside the tube. This nonuniformity (strictly connected to the inequality of the heat fluxes) has been reported by Atkins (1962) and by Goossens *et al.* (1978a,b). Only those mechanisms that consider the kinetics of the most important polymerization reactions (followed by dehydrogenation reactions) on the surface explain the experimental evidence. Mass-transfer controlled mechanisms do not explain the experimental data and are hence rejected.

The SPYRO model provides a mechanistic approach to the rates of carbon deposition at commercial conditions. Briefly, the coke deposit is continuously being covered by polymeric material that is gradually transformed into coke. Steam is very inactive as a decoking agent at most industrial conditions. Carbon deposition is controlled by the concentration of unsaturated radicals or molecules at the surface of the polymer layer. Typical coke precursors are vinylic or phenylic-like radicals. Unsaturated compounds, particularly acetylenics, may combine directly with the growing polymeric surface through Diels–Alder and other concerted reactions. Considering the life of the radicals, the effective radicals for coke formation are those produced by gaseous reactions at or near the growing coke surface.

Particularly for gas oil cracking, high-boiling-point compounds are also absorbed (dissolved) in the polymeric layer. Through subsequent H abstraction reactions involving attacking gas-phase radicals, they are incorporated permanently in the growing cross-linked polymeric network. The coating effect of the polymeric layer precludes the assumption of increased surface activity by porosity. In fact, the porosity of the deposit interiors must be considered to result from the subsequent (and slower) transformation of the reticulated polymer in coke, through dehydrogenations, changes in morphological characteristics, and finally, decreasing of the specific volume with the unavoidable cleavage and fracturing.

The catalytic effects of the metal surfaces (or metal particles included in this polymer-coated structure) and their inhibition or modification due to poisons (like sulfur compounds) or surface pretreatments, also play essential roles during the first stage of coke formation, as indicated by Brown and Albright (1976), by Albright and McConnell (1979), and by Albright *et al.* (1979). Some very interesting morphological aspects of the initial deposits are presented, often showing surprising similitude with the nascent polymeric materials produced by heterogeneous catalysis. In summary, only mechanistic models seem able to correlate experimental results on the coking rate after the initial catalytic stage.

Some discussion is also needed on the formation of carbon oxides. After a uniform layer of coke, covered by the continuously evolving polymeric film, is formed, the coke itself does not contribute to significant formation of carbon oxides. The main source of carbon oxides (CO, CO_2) involves the hydroxyl radical ($HO\cdot$) in gas-phase reactions. The hydroxyl radical is formed by H abstraction reactions between radicals in the gaseous phase and steam. The $HO\cdot$ radical (among other reactions) interacts with the unsaturated hydrocarbons by addition on unsaturated bonds, followed by the internal rearrangement (H internal abstraction through a 4-membered ring), and subsequent decomposition and dehydrogenation of the intermediate aldehyde:

$$HO\cdot + C_2H_4 \rightleftharpoons HO{-}CH_2{-}CH_2\cdot \longrightarrow \cdot O{-}CH_2{-}CH_3 \longrightarrow CH_3\cdot + (CH_2O)$$
$$= CH_3\cdot + CO + H_2$$

Furthermore, carbon monoxide is oxidized by a hydroxyl radical to CO_2 (and carbon dioxide is reduced to CO mainly by the H radical):

$$HO\cdot + CO \longrightarrow CO_1 + H\cdot$$

During decoking, however, the polymer layer (stabilized during hydrocarbons pyrolysis) reacts, forming initially a highly porous deposit. The metal particles incorporated in the coke become catalytically active.

V. PYROLYSIS OF GASES

The pyrolysis of essentially pure components or simple mixtures is of commercial interest for the light paraffins (ethane, propane, and butanes). (The study of heavier components has been directed mainly to the understanding of the reaction mechanisms.) In fact in some locations, the relative abundance of light hydrocarbons favors their use for olefins production.

Simple characterizations of feedstocks and of product compositions were logical starting points for pyrolysis modeling. Hence many models at different sophistication levels have been proposed for ethane, propane, butanes, and, more recently, their mixtures. Some of these are in reasonable agreement with the experimental data as already mentioned. Difficulties have been experienced in the modeling of gas mixtures co-cracking, particularly involving ethane, versus the pure-additivity rule; contradictory interpretations of the experimental results sometimes occur. Similar problems are also noted for pyrolysis of "pure" gases, when considering the prediction problem of the formation and interactions of minor products. Actually, in the range of high severity and decomposition, the description of a "pure component" pyrolysis becomes almost equivalent to decomposition and interactions of a complex mixture. It is mainly for these reasons that the extrapolability of most molecular and mechanistic models for gas cracking is of limited value. To overcome these limitations, rather extensive and complete mechanistic models are needed.

In Table II, several relevant reactions in the cracking of ethane, propane, and their mixtures are presented: these are a part of the overall kinetic

TABLE II

Main Reactions in Ethane and Propane Cracking (Reduced Scheme)[a,b,c,d]

	$\log_{10} A$[e]	E[f]
1. *Chain-initiation reactions*		
$C_2H_6 \rightarrow 2\ CH_3^{\cdot}$	17.04	89.0
$C_3H_8 \rightarrow CH_3^{\cdot} + C_2H_5^{\cdot}$	16.85	86.0
$1\text{-}C_4H_8 \rightarrow CH_3^{\cdot} + a\text{-}C_3H_5^{\cdot}$	15.78	4.0
$CyC_5H_6 \rightarrow H^{\cdot} + CyC_5H_5^{\cdot}$	15.40	73.5
$C_6H_{10} \rightarrow CH_3^{\cdot} + H^{\cdot} + CyC_5H_6$	16.20	72.0
Diallyl $\rightarrow 2\ a\text{-}C_3H_5^{\cdot}$	14.78	59.0
$Me\text{-}CyC_5H_7 \rightarrow CH_3^{\cdot} + H^{\cdot} + CyC_5H_6$	16.00	72.0
$Me\text{-}CyC_5H_5 \rightarrow [H^{\cdot} + MeCyC_5H_4^{\cdot}] \rightarrow [H^{\cdot} + CyC_6C_6H_7^{\cdot}] \rightarrow$		
$2\ H^{\cdot} + C_6H_6$	15.48	73.5
$C_2H_4 + CyC_5H_6 \leftrightarrow$ [Norbornene] $\rightarrow [MeCyC_6H_7] \rightarrow$		
$H^{\cdot} + CH_3^{\cdot} + C_6H_6$	8.70	30.0
$C_3H_6 + CyC_5H_6 \leftrightarrow$ [Me-Norbornene] $\rightarrow [Di\text{-}Me\text{-}CyC_6H_6] \rightarrow$		
$2\ CH_3^{\cdot} + C_6H_6$	8.70	30.0
$n\text{-}C_4H_{10} \rightarrow 2\text{-}C_2H_5^{\cdot}$	16.70	83.0
2. *H Abstraction reactions*		
$R + H_2 \rightarrow RH + H^{\cdot}$		
$R + CH_4 \rightarrow RH + CH_3^{\cdot}$		
$R + C_2H_4 \rightarrow RH + C_2H_3^{\cdot}$		
$R + C_2H_6 \rightarrow RH + C_2H_5^{\cdot}$		
$R + C_3H_6 \rightarrow RH + a\text{-}C_3H_5^{\cdot}$		
$R + C_3H_6 \rightarrow RH + v\text{-}C_3H_5^{\cdot}$		
$R + C_3H_8 \rightarrow RH + 1\text{-}C_3H_7^{\cdot}$		
$R = H^{\cdot}$	11.06	8.0
$R = CH_3^{\cdot}$	9.18	11.9
$R = C_2H_3^{\cdot}$	9.10	7.8
$R = C_2H_5^{\cdot}$	9.08	14.0
$R = a\text{-}C_3H_5^{\cdot}$ (allylic)	9.60	23.5
$R = v\text{-}C_3H_5^{\cdot}$ (vinylic)	9.10	7.8
$R = 2\text{-}C_3H_7^{\cdot};\ 2\ C_4H_9^{\cdot}$	8.90	15.5
$R = a\text{-}C_4H_7^{\cdot}$	9.60	24.0
$R = 4\text{-}C_4H_7^{\cdot}$ ()	8.90	12.2
$R = iso\text{-}C_4H_7^{\cdot}$	9.60	23.0
$R = 1\text{-}C_4H_9^{\cdot};\ iso\text{-}C_4H_9^{\cdot}$	8.90	14.0
$R = CyC_5H_5^{\cdot}$	9.60	32.5
$R + C_3H_8 \rightarrow RH + 2\text{-}C_3H_7^{\cdot}$		
$R + 1\text{-}C_4H_8 \rightarrow RH + a\text{-}C_4H_7^{\cdot}$		
$R + 1\text{-}C_4H_8 \rightarrow RH + 4\text{-}C_4H_7^{\cdot}$		
$R + 2\text{-}C_4H_8 \rightarrow RH + a\text{-}C_4H_7^{\cdot}$		
$R + iso\text{-}C_4H_8 \rightarrow RH + iso\text{-}C_4H_7^{\cdot}$		
$R + C_5H_8 \rightarrow RH + [C_5H_7^{\cdot}] \rightarrow RH + H^{\cdot} + CyC_5H_6$		
$R + CyC_5H_8 \rightarrow RH + [CyC_5H_7^{\cdot}] \rightarrow RH + H^{\cdot} + CyC_5H_6$		
$R + CyC_5H_6 \rightarrow RH + CyC_5H_5^{\cdot}$		
$R + G_6H_{10} \rightarrow RH + CH_3^{\cdot} + CyC_5H_6$		
$R + MeCyC_5H_7 \rightarrow RH + CH_3^{\cdot} + CyC_5H_6$		
$R + MeCyC_5H_7 \rightarrow RH + H^{\cdot} + MeCyC_5H_5$		
$R + MeCyC_5H_5 \rightarrow RH + [Me\text{-}CyC_5H_4^{\cdot}] \rightarrow RH + H^{\cdot} + C_6H_6$		
$R + C_7H_8 \rightarrow RH + [C_7H_7^{\cdot}]$		
$R + C_8H_{10} \rightarrow RH + [C_8H_9^{\cdot}]$		
$R + n\text{-}C_4H_{10} \rightarrow RH + 1\text{-}C_4H_9^{\cdot}$		
$R + n\text{-}C_4H_{10} \rightarrow RH + 2\text{-}C_4H_9^{\cdot}$		
$R + CyC_6H_{10} \rightarrow RH + [CyC_6H_9^{\cdot}] \rightarrow RH + H^{\cdot} + H_2 + C_6H_6$		

TABLE II *(continued)*

3. *Radical-addition reactions*		
$H^{\cdot} + C_2H_2 + T \rightarrow C_2H_3^{\cdot} + T$	9.00	−10.0
$H^{\cdot} + C_2H_4 \rightarrow C_2H_5^{\cdot}$	11.0	2.0
$H^{\cdot} + C_3H_4^{==} \rightarrow a\text{-}C_3H_5^{\cdot}$	10.54	2.0
$H^{\cdot} + C_3H_4^{\equiv} \rightarrow v\text{-}C_3H_5^{\cdot}$	10.54	2.0
$H^{\cdot} + C_3H_6 \rightarrow 1\text{-}C_3H_7^{\cdot}$	10.64	2.0
$H^{\cdot} + C_3H_6 \rightarrow 2\text{-}C_3H_7^{\cdot}$	10.64	2.0
$H^{\cdot} + C_4H_6 \rightarrow a\text{-}C_4H_7^{\cdot}$	11.08	1.9
$H^{\cdot} + C_4H_6 \rightarrow 4\text{-}C_4H_7^{\cdot}$	10.61	2.0
$H^{\cdot} + 1\text{-}C_4H_8 \rightarrow 1\text{-}C_4H_9^{\cdot}$	10.40	2.0
$H^{\cdot} + 1\text{-}C_4H_8 \rightarrow 2\text{-}C_4H_9^{\cdot}$	10.88	2.0
$H^{\cdot} + 2\text{-}C_4H_8 \rightarrow 2\text{-}C_4H_9^{\cdot}$	11.00	2.5
$H^{\cdot} + \text{iso-}C_4H_8 \rightarrow \text{iso-}C_4H_9^{\cdot}$	11.00	2.0
$H^{\cdot} + C_7H_8 \rightarrow C_6H_6 + CH_3^{\cdot}$	10.70	5.0
$H^{\cdot} + C_8H_{10} \rightarrow C_7H_8 + CH_3^{\cdot}$	10.90	5.0
$H^{\cdot} + C_8H_8 \rightarrow C_6H_6 + C_2H_3^{\cdot}$	11.48	12.0
$CH_3^{\cdot} + C_3H_6 \rightarrow 1\text{-}C_3H_7^{\cdot}$	8.48	7.6
$CH_3^{\cdot} + C_3H_6 \rightarrow 2\text{-}C_4H_9^{\cdot}$	8.60	7.6
$CH_3^{\cdot} + C_3H_6 \rightarrow \text{iso-}C_4H_9^{\cdot}$	8.30	7.6
$C_2H_3^{\cdot} + C_2H_4 \rightarrow 4\text{-}C_4H_7^{\cdot}$	8.55	6.0
$C_2H_3^{\cdot} + C_3H_6 \rightarrow [C_5H_9^{\cdot}\ \text{isomers}] \rightarrow$ 50% $\{CH_3^{\cdot} + C_4H_6$; 50% $\{H^{\cdot} + C_5H_8$	8.20	6.0
$C_2H_3^{\cdot} + C_4H_6 \rightarrow [C_6H_9^{\cdot}] \rightarrow$ 75% $\{CyC_5H_6 + CH_3^{\cdot}$; 25% $\{[CyC_6H_8 + H^{\cdot}] \rightarrow H^{\cdot} + H_2 + C_6H_6$	8.64	5.0
$C_2H_3^{\cdot} + C_6H_6 \rightarrow C_8H_8 + H^{\cdot}$	8.40	6.0
$C_2H_3^{\cdot} + C_8H_8 \rightarrow C_9^{+}$	8.70	5.0
$C_2H_3^{\cdot} + CyC_5H_6 \rightarrow [CyC_7H_9^{\cdot}] \rightarrow C_7H_8 + H^{\cdot}$	8.60	5.0
$C_2H_5^{\cdot} + C_2H_4 \rightarrow 1\text{-}C_4H_9^{\cdot}$	8.30	7.6
$a\text{-}C_3H_5^{\cdot} + C_2H_2 \rightarrow CyC_5H_6 + H^{\cdot}$	8.30	13.0
$a\text{-}C_3H_5^{\cdot} + C_2H_4 \leftrightarrow [C_5H_9^{\cdot}] \rightarrow C_5H_8 + H^{\cdot}$	7.82	21.5
$a\text{-}C_3H_5^{\cdot} + C_2H_4 \leftrightarrow [CyC_5H_9^{\cdot}] \rightarrow CyC_5H_8 + H^{\cdot}$	7.08	12.5
$a\text{-}C_3H_5^{\cdot} + C_3H_6 \leftrightarrow [\text{Me-}CyC_5H_8^{\cdot}\ \text{isomers}] \rightarrow$ $\{4\text{-}C_4H_7^{\cdot} + C_2H_4$	7.88	21.5
$C_2H_5^{\cdot} + C_4H_6$	7.78	21.5
$CH_3^{\cdot} + CyC_5H_8$	7.60	21.5
$H^{\cdot} + C_6H_{10}$	7.88	21.5
$a\text{-}C_3H_5^{\cdot} + C_3H_6 \leftrightarrow [\text{Me-}CyC_5H_8^{\cdot}\ \text{isomers}] \rightarrow$		
$\{\text{Me-}CyC_5H_7(2) + H^{\cdot}$	6.57	12.5
$[\text{Me}CyC_5H_7(3) + H^{\cdot}] \rightarrow C_6H_{10} + H^{\cdot}$	6.57	12.5
$a\text{-}C_3H_5^{\cdot} + C_4H_6 \leftrightarrow [C_7H_{11}^{\cdot}\ \text{isomers}] \rightarrow$		
$\{C_2H_5^{\cdot} + CyC_5H_6$	6.01	5.0
$[CH_3^{\cdot} + CyC_6H_8] \rightarrow CH_3^{\cdot} + H_2 + C_6H_6$	6.38	5.0
$a\text{-}C_3H_5^{\cdot} + CyC_5H_6 \leftrightarrow [C_8H_{11}^{\cdot}] \rightarrow CH_3^{\cdot} + C_7H_8$	8.65	14.5
$a\text{-}C_3H_5^{\cdot} + C_6H_6 \rightarrow C_9^{+}$	8.78	25.0
$a\text{-}C_3H_5^{\cdot} + C_8H_8 \rightarrow C_9^{+}$	7.08	12.0
$a\text{-}C_4H_7^{\cdot} + C_2H_4 \leftrightarrow [C_6H_{11}^{\cdot}] \leftrightarrow [\text{Me}CyC_5H_8] \rightarrow CH_3^{\cdot} + CyC_5H_8$	8.48	13.0
4. *Radical-decomposition reactions*		
$C_2H_3^{\cdot} + T \rightarrow C_2H_2 + H^{\cdot} + T$	11.90	29.7
$C_2H_5^{\cdot} \rightarrow C_2H_4 + H^{\cdot}$	13.90	41.0
$a\text{-}C_3H_5^{\cdot} \rightarrow C_3H_4^{==} + H^{\cdot}$	12.95	59.0
$v\text{-}C_3H_5^{\cdot} \rightarrow C_3H_4^{\equiv} + H^{\cdot}$	12.15	37.5
$v\text{-}C_3H_5^{\cdot} \rightarrow C_2H_2 + CH_3^{\cdot}$	12.20	37.5
$1\text{-}C_3H_7^{\cdot} \rightarrow C_2H_4 + CH_3^{\cdot}$	13.70	34.0

TABLE II (*continued*)

$1\text{-}C_3H_7^{\bullet} \rightarrow C_3H_6 + H^{\bullet}$	14.30	39.0
$2\text{-}C_3H_7^{\bullet} \rightarrow C_3H_6 + H^{\bullet}$	14.30	43.0
$4\text{-}C_4H_7^{\bullet} \rightarrow C_2H_4 + C_2H_3^{\bullet}$	14.30	39.5
$4\text{-}C_4H_7^{\bullet} \rightarrow C_4H_6 + H^{\bullet}$	13.00	34.0
$a\text{-}C_4H_7^{\bullet} \rightarrow C_4H_6 + H^{\bullet}$	14.00	49.0
$\text{iso-}C_4H_7^{\bullet} \rightarrow C_3H_4^{==} + CH_3^{\bullet}$	14.08	55.0
$1\text{-}C_4H_9^{\bullet} \rightarrow 1\text{-}C_4H_8 + H^{\bullet}$	14.11	39.0
$1\text{-}C_4H_9^{\bullet} \rightarrow C_2H_4 + C_2H_5^{\bullet}$	14.18	31.0
$2\text{-}C_4H_9^{\bullet} \rightarrow 1\text{-}C_4H_8 + H^{\bullet}$	13.70	41.0
$2\text{-}C_4H_9^{\bullet} \rightarrow 2\text{-}C_4H_8 + H^{\bullet}$	14.00	41.0
$2\text{-}C_4H_9^{\bullet} \rightarrow C_3H_6 + CH_3^{\bullet}$	14.00	34.0
$\text{iso-}C_4H_9^{\bullet} \rightarrow \text{iso-}C_4H_8 + H^{\bullet}$	14.00	38.0
$\text{iso-}C_4H_9^{\bullet} \rightarrow C_3H_6 + CH_3^{\bullet}$	14.30	34.0
5. *Radical-isomerization reactions*		
$1\text{-}C_3H_7^{\bullet} \rightarrow 2\text{-}C_3H_7^{\bullet}$	12.40	34.0
$2\text{-}C_3H_7^{\bullet} \rightarrow 1\text{-}C_3H_7^{\bullet}$	13.00	38.0
$a\text{-}C_4H_7^{\bullet} \rightarrow 4\text{-}C_4H_7^{\bullet}$	13.44	49.0
$4\text{-}C_4H_7^{\bullet} \rightarrow a\text{-}C_4H_7^{\bullet}$	12.70	35.0
6. *Chain-termination reactions*		
$2\ CH_3^{\bullet} \rightarrow C_2H_6$	10.34	0.
$CH_3^{\bullet} + C_2H_5^{\bullet} \rightarrow C_3H_8$	10.00	0.
$CH_3^{\bullet} + a\text{-}C_3H_5^{\bullet} \rightarrow 1\text{-}C_4H_8$	9.85	0.
$2\ a\text{-}C_3H_5 \rightarrow$ Diallyl	8.70	0.
$C_2H_5^{\bullet} + C_2H_5^{\bullet} \rightarrow n\text{-}C_4H_{10}$	9.30	0.
7. *Molecular reactions*		
$C_2H_6 \rightarrow C_2H_4 + H_2$	13.90	71.0
$C_2H_4 + H_2 \rightarrow C_2H_6$	8.90	38.0
$C_3H_4^{==} \rightarrow C_3H_4^{\equiv}$	13.50	63.0
$C_3H_4^{\equiv} \rightarrow C_3H_4^{==}$	13.20	61.5
$C_3H_8 \rightarrow C_3H_6 + H_2$	13.70	70.0
$C_3H_6 + H_2 \rightarrow C_3H_8$	8.90	38.0
$1\text{-}C_4H_8 \rightarrow 2\text{-}C_4H_8$	12.35	60.0
$2\text{-}C_4H_8 \rightarrow 1\text{-}C_4H_8$	12.29	62.0
$CyC_5H_8 \rightarrow CyC_5H_6 + H_2$	13.30	58.0
Diallyl $\rightarrow [H_2 + C_6H_8] \rightarrow [H_2 + CyC_6H_8] \rightarrow 2\ H_2 + C_6H_6$	13.78	59.0
$Me\text{-}CyC_5H_7 \rightarrow Me\text{-}CyC_5H_5 + H_2$	13.68	64.0
$Me\text{-}CyC_5H_7 \rightarrow C_6H_{10}$	13.86	64.0
$CyC_5H_6 + C_2H_2 \leftrightarrow$ [Norbornadiene] $\rightarrow C_7H_8$	8.00	21.5
$C_4H_6 + C_2H_2 \leftrightarrow [CyC_6H_8] \rightarrow C_6H_6 + H_2$	8.18	22.5
$C_4H_6 + C_3H_4^{\equiv} \leftrightarrow [MeCyC_6H_7] \rightarrow C_7H_8 + H_2$	7.87	22.5
$CyC_5H_6 + C_3H_4^{\equiv} \leftrightarrow$ [Me-Norbornadiene] $\rightarrow C_8H_{10}$	8.00	21.5
$C_2H_4 + C_4H_6 \rightarrow CyC_6H_{10}$	7.48	26.5
$CyC_6H_{10} \rightarrow .9(C_2H_4 + C_4H_6) + .1(2\ H_2 + C_6H_6)$	15.04	66.7

[a] Species within brackets are unstable intermediate (reaction-transition) compounds.

[b] For conciseness, the fates of $\text{iso-}C_4H_7^{\bullet}$, $CyC_5H_5^{\bullet}$, $C_7H_7^{\bullet}$, and $C_8H_9^{\bullet}$ are not reported; except for $\text{iso-}C_4H_7^{\bullet}$, they are mainly involved in C_9^+ formation.

[c] For simplicity, the kinetic parameters for the primary H abstraction reactions are only reported for propane and the most effective radicals. (These values are similar to those for ethane.)

[d] T, third body.

[e] For unimolecular reactions, $A = (s^{-1})$; for bimolecular reactions, $A = (L\ mol^{-1}\ s^{-1})$.

[f] $E =$ (kcal/mol).

TABLE III

Ethane Cracking

Commercial data		
Feed (wt%)		
C_2H_4	0.75	
C_2H_6	99.25	
Operating conditions		
Residence time at 650°C (s)	0.79	
Hydrocarbon partial pressure out (atm)	1.24	
Coil outlet temperature (°C)	838	
	Yields (wt%)	
	Exp.	SPYRO
H_2	4.06	3.93
CH_4	5.0	4.90
C_2H_2	0.37	0.37
C_2H_4	50.05	49.66
C_2H_6	34.42	34.80
C_3H_4 C_3H_6	1.30	1.65
C_3H_8	0.17	0.20
C_4H_6	1.32	1.28
C_4H_8 C_4H_{10}	1.02	0.65
C_5^+	2.25	2.5
C_5^{2-}	0.22	0.06
$C_5^{\pm}$	0.06	0.07
$CyC_5^- + CyC_5^{2-}$	0.47	0.51
$C_6H_6 + C_6$ n.a.[a]	1.13	1.05
$C_7H_8 + C_7$ n.a.	0.17	0.21
Other C_5^+	0.20	0.60
C in $(CO + CO_2)$	0.04	0.05

[a] n.a., non-aromatics.

scheme adopted in the SPYRO model for the simulation of hydrocarbon cracking from ethane to gas oils. Comparisons are reported in Tables III and IV between commercial data and predictions given by this model for both ethane and propane feedstocks. The good agreement for the relatively minor components, whenever this information was available, confirms the correctness of the mechanistic structure for the C_5^+ prediction. In an

TABLE IV

Propane Cracking

	Test					
	1		2		3	
Feed (wt%)[a]						
C_2H_6	0.15					
C_3H_6	4.34					
C_3H_8	86.40					
n-C_4H_{10}	5.72					
iso-C_4H_{10}	2.51					
C_4H_8	0.85					
Operating conditions						
Residence time at $T > 650°C$ (s)	0.80		0.80		0.80	
Hydrocarbon partial pressure out (atm)	1.94		1.93		1.91	
Coil outlet temperature (°C)	847		833		824	
	Yields (wt%)					
	Exp.	SPYRO	Exp.	SPYRO	Exp.	SPYRO
H_2	1.23	1.16	1.3	1.1	1.2	1.0
CH_4	24.4	24.1	22.2	21.9	20.9	20.3
C_2H_2	0.2	0.4	0.1	0.3	0.1	0.2
C_2H_4	31.3	31.6	29.1	29.3	27.5	27.7
C_2H_6	4.9	4.8	4.8	4.7	4.6	4.6
C_3H_4	0.12	0.23	0.1	0.2	0.1	0.2
C_3H_6	16.1	16.3	18.0	18.4	19.1	19.3
C_3H_8	9.8	9.3	13.8	13.7	17.0	17.2
C_4H_6	1.9	2.34	1.7	2.1	1.4	1.8
C_4H_8	1.5	1.2	1.8	1.5	2.0	1.6
C_4H_{10}	0.25	0.32	0.5	0.4	0.7	0.7
Other	8.30	8.25	6.6	6.4	5.4	5.4
Benzene	3.1	2.9	2.2	2.3	1.5	1.9
Toluene	0.8	0.75	0.7	0.55	0.4	0.5
Xylene + ethylbenzene	0.14	0.18	0.18	0.14	0.05	0.12
C in ($CO + CO_2$)	0.08	0.04	—	0.03	—	0.03

[a] Feed composition for all three tests was identical.

example reported in Fig. 4, the reaction network is shown for cyclopentadiene, which is an important precursor of the aromatic fraction.

Several contributions found in the literature provide information useful for cross-checking with other models. Table V compares data coming from

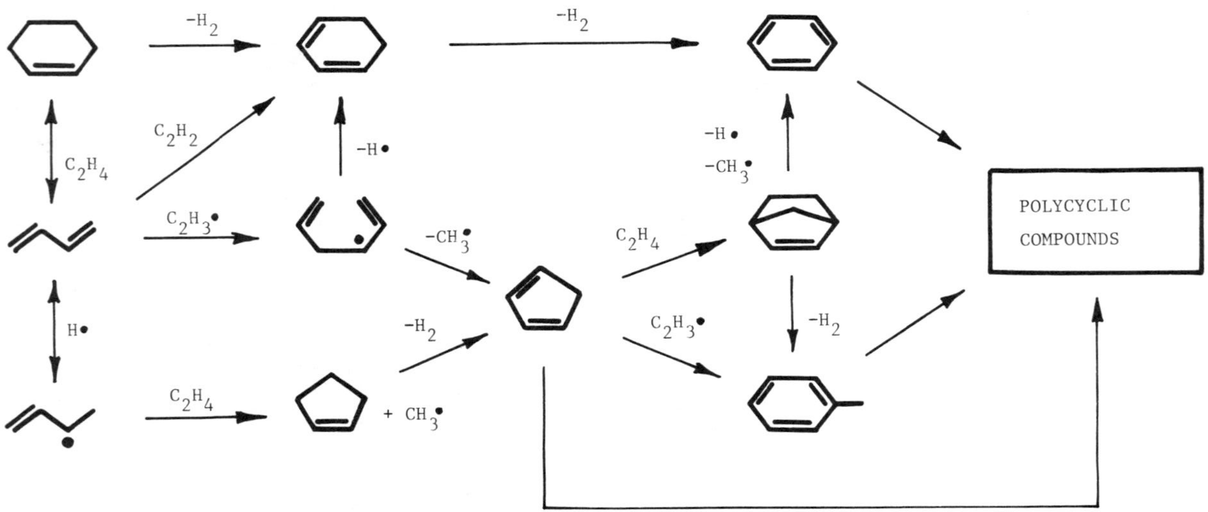

Fig. 4. Ethane cracking—formation of liquid compounds (C_5^+).

Table V

Gases Cracking: Comparison of Models

	Test 1	Test 2	Test 3
Feed (wt%)			
Methane	1.08	0.01	—
Ethylene	0.02	—	—
Ethane	97.20	1.54	—
Propylene	1.35	0.70	—
Propane	0.35	97.12	1.57
n-Butane	—	0.63	95.33
Iso butane	—	—	3.10
Operating conditions			
Residence time above 650°C (s)	1.39	1.35	1.07
Hydrocarbon partial pressure out (atm)	1.30	1.35	1.51
Outlet temperature (°C)	829/830/829[a]	834/834/834	793/793/795

Yields (wt% on feed)	Test 1			Test 2			Test 3		
	Observed	Predicted	SPYRO	Observed	Predicted	SPYRO	Observed	Predicted	SPYRO
Hydrogen	3.51	3.61	3.52	1.50	1.45	1.30	0.63	0.61	0.74
Methane	6.43	6.10	6.75	23.74	23.35	24.60	18.94	19.05	18.98
Acetylene	0.19	0.28	0.26	0.50	0.63	0.40	0.19	0.26	0.14
Ethylene	46.59	46.32	46.13	33.49	33.36	33.33	27.02	26.18	27.30
Ethane	37.50	37.53	37.41	4.94	5.02	5.46	6.92	7.02	6.04
Methylacetylene and Propadiene	1.70	1.50	1.60	0.20	0.54	0.18	0.30	0.34	0.13
Propylene				15.05	14.86	14.85	21.96	23.64	21.36
Propane	0.09	0.18	0.16	9.72	9.71	9.72	0.91	1.41	1.03
Butadiene	1.25	1.30	1.23	1.92	2.64	2.21	2.19	2.35	2.28
Butenes	0.32	0.20	0.26	0.90	1.04	0.85	2.54	2.87	2.92
Butanes	0.29	0.21	0.37	0.11	0.29	0.05	11.18	11.41	12.30
C_5^+	2.13	2.77	2.31	7.93	7.11	7.05	7.22	4.86	6.78
Total	100.00	100.00	100.00	100.00	100.00	100.00	100.00	100.00	100.00

[a] Temperature for the three models are given as follows: Observed/Predicted/SPYRO.

commercial units (ethane, propane, mixtures), as reported by Ross and Shu (1977), both with predictions given by the same authors (on the basis of an accurate *ad hoc* stoichiometric model) and with those given by the general mechanistic model. (It should be also mentioned that in the reported article only overall indices for the actual operating conditions were given,

TABLE VI

Propane and Mixture Cracking: Comparison of Models

Test 1	
Feed	C_3H_8
Outlet temperature (°C)	850
Outlet pressure (atm)	3
Steam/hydrocarbon ratio (kg/kg)	0.4
Contact time (s)	0.88

	Yields (wt%)	
	SPYRO	Molecular
CH_4	22.5	22.1
C_2H_4	30.8	32.7
C_3H_6	16.6	16.5
C_4^+	8.8	9.6
Propane conversion (%)	85.8	86.5

Test 2	
Feed (wt%)	
CH_4	0.29
C_2H_4	0.02
C_2H_6	69.31
C_3H_6	0.46
C_3H_8	28.96
n-C_4H_{10}	0.96
Outlet temperature (°C)	849
Hydrocarbon partial pressure out (atm)	1.32
Contact time (s)	0.85

	Yields (wt%)			
	Exp.	SPYRO	Molecular	"Ad hoc" molecular
CH_4	11.9	11.9	12.6	11.2
C_2H_4	46.2	46.2	46.0	45.5
C_2H_6	27.4	27.8	21.0	27.4
C_3H_6	4.1	3.8	6.9	4.0
C_3H_8	1.7	2.1	4.4	1.8
C_4^+	5.0	5.0	5.8	5.3

and the match had to be made on this basis.) Both models indicate excellent agreement; minor deviations appear to be in the range of experimental errors. The mechanistic model has the advantage in terms of flexibility if one considers that its kinetic parameters were not rearranged to fit those data. The stoichiometric models can be reliable over limited ranges of operating conditions, but the more fundamental models can be extrapolated with more confidence.

In Table VI, a comparison is reported between the SPYRO results and another accurate molecular model (Sundaram and Froment, 1977a). The agreement is good in the original range of conditions, but outside it large discrepancies occur. Sundaram and Froment (1978a) later presented a mechanistic model for the gases cracking that is claimed to cover all industrial conditions. This scheme involves 133 reactions, 15 molecular species, and 11 radical species. Some of the kinetic constants were found in the literature and some were estimated from experimental data of a pilot plant. The agreement shown in that article in the case of ethane, propane, *n*-butane, and isobutane is excellent. To compare further with a more general mechanistic program, Figs. 5 and 6 depict the experimental and simulated yields of the major effluents graphically reported by

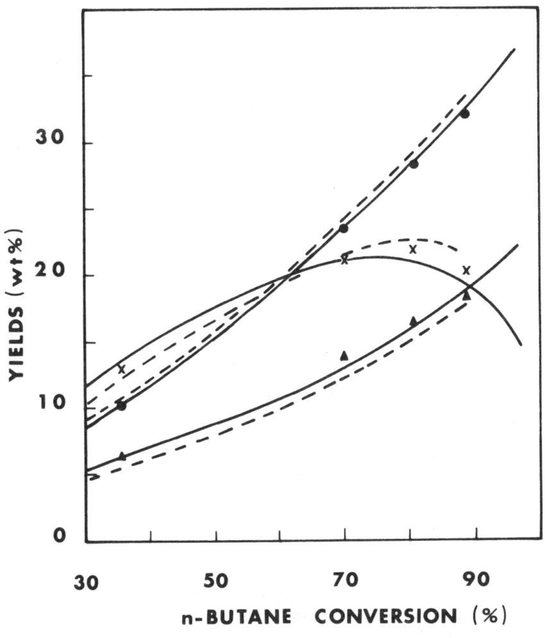

Fig. 5. Experimental and simulated yields of major effluents for *n*-butane cracking. Experimental: ▲, CH_4; ●, C_2H_4; ×, C_3H_6. Simulations: -----, Sundaram and Froment (1978a); ——, SPYRO.

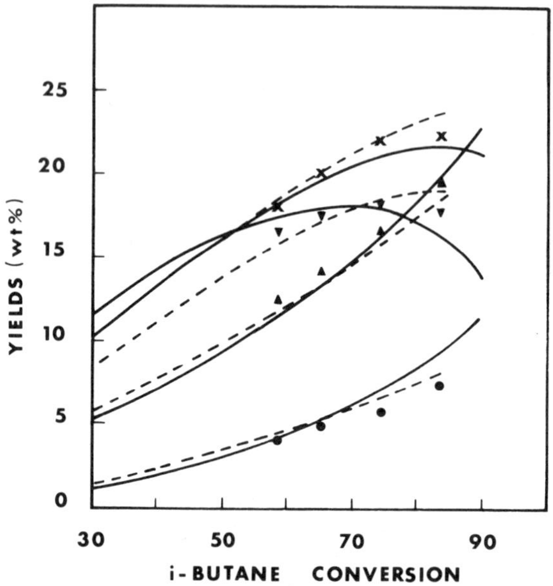

Fig. 6. Experimental and simulated yields of major effluents for isobutane cracking. Experimental: ▲, CH_4; ●, C_2H_4; ×, C_3H_6; ▼, iso-C_4H_8. Simulations: -----, Sundaram and Froment (1978a); ——, SPYRO.

Sundaram and Froment for *n*-butane and isobutane cracking and the predictions of the SPYRO model. The agreement is satisfactory within the range of experimental uncertainties.

A particularly severe testing procedure of a model is to use mixtures of light paraffins, particularly ethane–propane mixtures. Table VI, already discussed, is an example of a comparison between molecular and mechanistic models using commercial data. Figure 7 compares results for a mixture containing 75% (w/w) propane. Good agreement is noted between the SPYRO model and the experimental results of Froment *et al.* (1976). Slightly poorer agreement is probable for C_5^+ species, perhaps because of less accurate analyses of these compounds.

Models with less mechanistic approaches or models developed only for pure hydrocarbons often fail for paraffin mixtures. For instance, the good but nonextensive kinetic model proposed by Dunkleman and Albright (1976a) for ethane and propane was reported by Dunkleman and Albright (1976b) to be unsatisfactory for the co-cracking simulation. In fact, a slight decrease of propane conversion was predicted in the presence of large amounts of ethane; experimental evidence indicated the reverse to be true. Although surface reactions may have been partially responsible in their small-diameter reactor for this contradiction, the main explanation remains

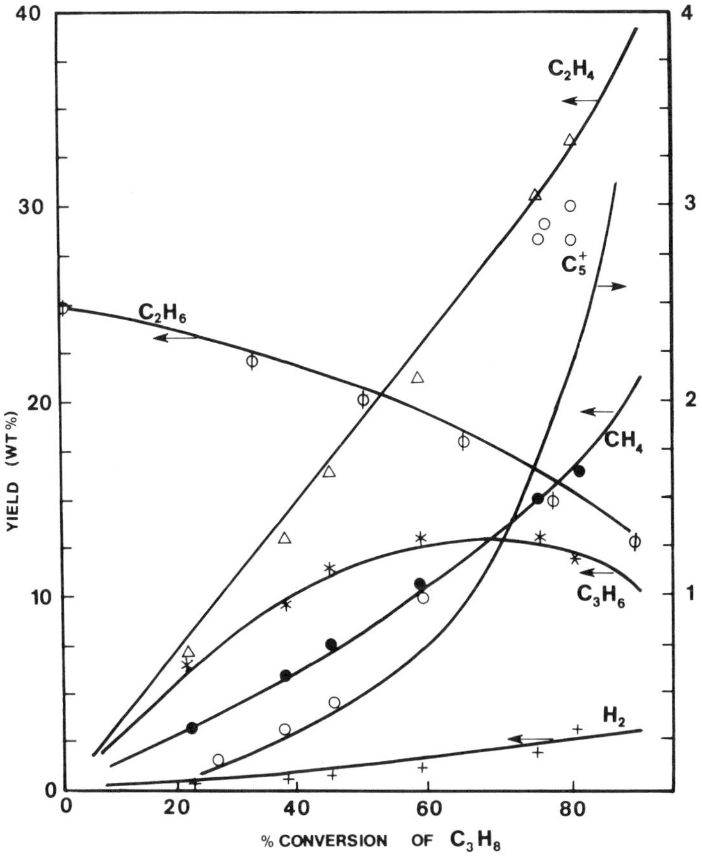

Fig. 7. Comparison of experimental and SPYRO results of cracking an ethane–propane mixture containing 75% (w/w) propane. Experimental: +, H_2; ●, CH_4; △, C_2H_4; ɸ, C_2H_6; *, C_3H_6; ○, C_5^+. Simulation, ——.

in the limitations of the homogeneous gas-phase kinetic scheme they adopted. In fact, Fig. 8 shows the good agreement between the predictions of the SPYRO model versus the experimental data for C_2H_6–C_3H_8 mixtures presented in that article.

Although numerous experimental data and models have been reported, at most only a few of these models have been used in industrial applications. Sufficient space is not available here to recognize all of them. Even those that have questionable features or are quite incomplete have often made some contribution to the overall field of pyrolysis knowledge. In this context, numerous comparisons have been made to indicate how a wide-range kinetic scheme such as the SPYRO model can help to unify theoretical and experimental information. Particularly, then, mention should be made

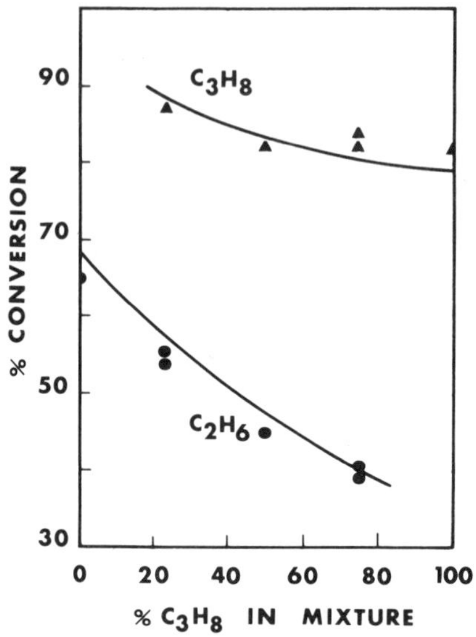

Fig. 8. Conversion of ethane and propane in mixtures. Outlet temperature, 792°C; outlet pressure, 1 atm; residence time, 0.9 s

of the comparison presented by Dente *et al.* (1979b) with the results of the following investigators: Davis and Williamson (1959) on minor components from ethane; Kunugi *et al.* (1969) on ethylene; Kunugi *et al.* (1970) on propylene; Haraguchi *et al.* (1976) on mixtures of hydrogen and propane.

Dente *et al.* (1979b) discuss the main routes for the formation and disappearance of several components, particularly for ethylene and propylene pyrolysis. However, the very complex and extensive reactions network indicates that many reaction paths contribute to a single product, with the relative importance of the different paths sometimes varying significantly with the reaction time, the reacting system, and the operating conditions.

VI. PYROLYSIS OF NAPHTHAS AND GAS OILS

Detailed mechanistic modeling of the pyrolyses of complex mixtures such as virgin naphthas and gas oils would require detailed composition of the feedstock. Such modeling would also include several hundred species and several thousand elementary chemical reactions. It would be impractical because of unacceptable computing times and computer memory allocations. Hence several simplifications are necessary for modeling of these complex systems.

Fortunately, several simplifications are applicable. Because of the relative abundance of particular isomers, a large fraction of the naphtha (or gas oil) is represented by a relatively small number of individual hydrocarbons. By comparing different naphthas and gas oils, and in spite of the very different origin of these mixtures (in terms of age as well as in terms of temperature and pressure condition of formation), a surprising regularity of the composition can be observed. For three different naphthas, the internal distributions of the branched isomeric C_8H_{18} compounds are compared in Table VII. By means of eight isomers (three monomethylheptanes, ethylhexane and four dimethylhexanes having a tertiary C structure), the whole fraction can be described with fair accuracy. Slightly branched paraffins (with only a few methyl groups in the branch) are predominant in isoparaffinic fractions. For example, for branched octanes the relative predominances are as follows: monomethylheptanes > dimethylhexanes > trimethylpentanes; moreover, the monomethylheptanes prevail over monoethylhexanes.

Considering the same number of side chains, the quaternary structures are less abundant than tertiary structures (2,2- and 3,3-dimethylhexanes are present in small quantities). Similar considerations also occur for heavier isomeric hydrocarbons, and for naphthenic and aromatic fractions.

Despite the lack of a theoretical explanation of these and similar regularities in naphtha and gas oil composition, there are at least two relevant

TABLE VII

Relative Amounts of Branched Isomers C_8H_{18} (wt%) Commercial Source

Isomers	Ponca	Occidental	Texas
2-methylheptane	46.3	36.9	42.1
3-methylheptane	15.4	28.5	23.4
4-methylheptane	10.3	10.2	9.3
2,3-dimethylhexane	3.6	5.4	6.3
2,4-dimethylhexane	3.1	5.5	4.2
2,5-dimethylhexane	3.1	5.7	4.0
3,4-dimethylhexane	6.7	2.6	3.7
2,2-dimethylhexane	0.5	—	0.3
3,3-dimethylhexane	1.5	1.7	0.4
2,3,4-trimethylpentane	0.3	—	1.1
2,2,3-trimethylpentane	0.2	—	—
2,3,3-trimethylpentane	0.3	—	0.6
3-ethylhexane	4.6	3.5	3.1
2-methyl-3-ethylpentane	3.1	—	1.5
3-methyl-3-ethylpentane	1.0	—	—

TABLE VIII

Initial Product Distribution from Pyrolysis of Branched C_8H_{18} Paraffins[a]

	2-methyl-heptane	3-methyl-heptane	4-methyl-heptane	3-ethyl-hexane	2,3-dimethyl-hexane	2,4-dimethyl-hexane	2,5-dimethyl-hexane	3,4-dimethyl-hexane
Methyl	0.194	0.192	0.163	0.330	0.381	0.220	0.215	0.233
Ethyl	0.097	0.246	0.485	0.487	0.226	0.145	0.025	0.386
1-Propyl	0.170	0.173	0.320	0.161	0.153	0.006	—	—
2-Propyl	0.273	0.016	0.020	—	0.240	0.293	0.490	—
1-Butyl	0.141	0.158	—	0.009	—	—	—	—
2-Butyl	—	0.215	—	0.012	—	0.209	—	0.382
Isobutyl	0.124	—	0.012	—	—	0.128	0.269	—
Ethylene	0.244	0.220	0.147	0.207	0.070	0.105	0.199	0.140
Propylene	0.308	0.219	0.666	0.176	0.466	0.211	0.261	—
1-Butene	0.116	0.240	—	0.276	—	0.058	0.012	0.114
2-Butene	—	0.100	—	—	0.170	0.088	—	0.408
Isobutene	0.141	—	—	—	—	0.209	0.269	—
1-Pentene	0.148	0.003	0.089	0.040	0.077	0.003	—	—
2-Pentene	—	—	0.159	0.149	0.114	—	—	—
2-Methyl-1-butene	—	0.125	—	—	—	0.202	—	—
3-Methyl-1-butene	0.096	—	—	—	0.027	—	0.291	—
2-Methyl-2-butene	—	—	—	—	0.104	—	—	—
1-Hexene	—	0.041	—	—	—	—	—	—
2-Hexene	—	0.066	—	0.149	—	—	—	—
3-Hexene	—	—	—	0.117	—	—	—	—

2-Methyl-1-pentene	—	—	0.206	—	—	—	—	—
3-Methyl-1-pentene	—	0.066	—	—	—	—	—	0.040
4-Methyl-1-pentene	0.050	—	—	—	—	0.037	—	—
3-Methyl-2-pentene	—	—	—	—	—	—	—	0.205
4-Methyl-2-pentene	—	—	—	—	—	0.062	—	—
2-Ethyl-1-butene	—	—	—	0.109	—	—	—	—
2,3-Dimethyl-1-butene	—	—	—	—	0.113	—	—	—
1-Heptene	0.042	—	—	—	—	—	—	—
2-Heptene	0.088	0.046	—	—	—	—	—	—
3-Heptene	—	0.030	0.073	—	—	—	—	—
2-Methyl-1-hexene	0.050	0.058	—	—	—	—	—	—
3-Methyl-1-hexene	—	—	0.073	—	0.022	—	—	—
4-Methyl-1-hexene	—	0.026	—	—	—	0.037	—	—
5-Methyl-1-hexene	—	—	—	—	—	—	0.109	—
2-Methyl-2-hexene	—	—	—	—	0.048	—	—	—
3-Methyl-2-hexene	—	—	—	—	0.104	—	—	—
4-Methyl-2-hexene	—	—	—	—	—	0.057	—	0.044
5-Methyl-2-hexene	—	—	—	—	—	0.041	0.095	—
2-Methyl-3-hexene	—	—	—	—	0.019	0.028	—	—
4-Methyl-3-hexene	—	—	—	—	—	—	—	0.094
2-Ethyl-1-pentene	—	—	—	0.100	—	—	—	—
3-Ethyl-1-pentene	—	—	—	0.027	—	—	—	—
2,3-Dimethyl-1-pentene	—	—	—	—	—	—	—	0.094
2,4-Dimethyl-1-pentene	—	—	—	—	—	0.033	—	—
3,4-Dimethyl-1-pentene	—	—	—	—	0.019	—	—	—

[a] Stoichiometric coefficient of H abstraction reactions.

simplifications. First, the approximate amount of individual hydrocarbons can possibly be predicted using very simple probabilistic rules. For example, a well-defined mixture of individual hydrocarbons (e.g., the branched C_8H_{18} isomers) can be considered as a single "equivalent" component. In Table VIII, the initial product distribution (in terms of stoichiometric coefficients of the H abstraction reactions) from the decomposition of the eight most relevant branched C_8H_{18} paraffins is reported. These results are evaluated employing the modified Kossiakoff and Rice (1943) theory for free-radical reactions at 700°C. The initial product distributions for the "equivalent component," mixture of branched octanes, can be directly derived from Table VIII with simple additivity rules. Despite the large differences in the initial product distribution, it is possible to justify theoretically and to verify experimentally (Illes *et al.*, 1974) the reliability of the method.

Referring again to Table VIII, it seems relevant to indicate the production of at least 7 branched C_6H_{12} olefins and 15 C_7H_{14} olefins; to avoid extreme expansion of kinetic schemes, intermediate products are grouped. Fortunately the regularities in the feedstock composition are still approximated in the pyrolysis products; and for these components, similar and even more drastic simplifications are applicable.

The net result of these and similar considerations is in one respect to describe (in terms of detailed composition) the naphtha or gas oil feedstocks by very simple and usually available physical properties (PIONA content, ASTM distillation range, H:C ratio, specific gravity, and sulfur content), and in another respect reduce significantly the total number of components to be individually considered. For particularly gas oils, the "equivalent components" concept is severely extended as a compromise between an increased number of components and hydrocarbons versus a somewhat poor characterization (Dente *et al.*, 1979c). Such simplifications have been adopted in the SPYRO program, but the overall kinetic scheme nevertheless involves about 2000 reactions accounting for about 100 radicals and molecular species (or "equivalent species"). In Figures 9 and 10 two typical scatter diagrams for ethylene and C_5^+ fraction are reported for the cracking of gases, naphthas, and gas oils. A total of 44 experimental points (29 feeds, 17 sources) were used, and they are truly independent data.

Recently Ross and Shu (1979) presented detailed results obtained with a generalized semikinetic model for naphtha pyrolysis. The naphtha feed is treated as a single "pseudocomponent," and feed decomposition is assumed to follow first-order kinetics. An overall inhibition effect is accounted for by reducing the zero-conversion rate constant as the conversion increases.

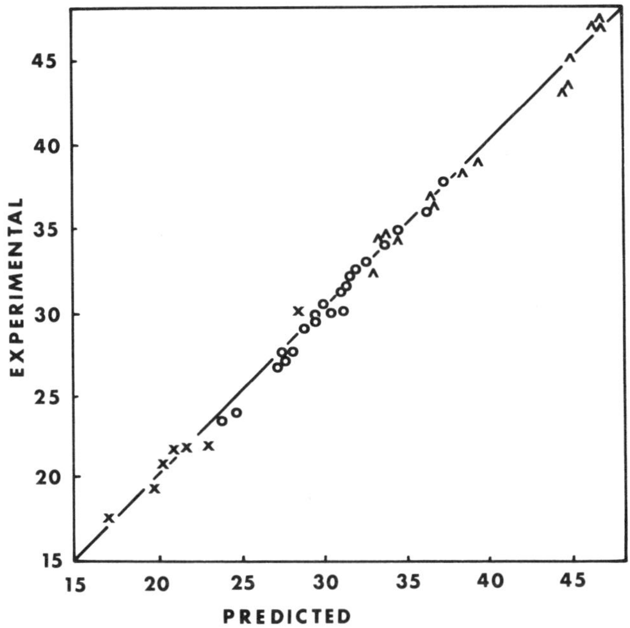

Fig. 9. Typical scatter diagram showing the ethylene fraction for the cracking of gases (Λ), naphthas (○), and gas oils (×).

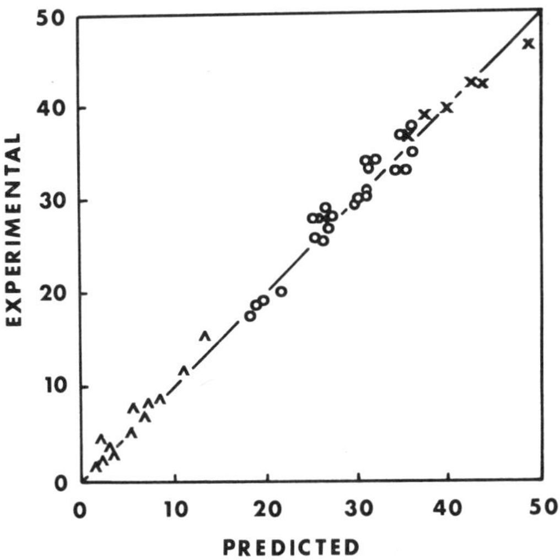

Fig. 10. Typical scatter diagram showing the C_5^+ fraction for the cracking of gases (Λ), naphthas (○), and gas oils (×).

Pyrolysis yields are evaluated primarily on the basis of a cracking severity index (CSI), defined as:

$$CSI = -\ln(1 - X)$$

where X is the feed conversion.

A further yield correction is based on the following two "selectivity indices": σ, residence time above incipient cracking; KAPP, kinetic average hydrocarbon partial pressure (weighted with respect to cracking severity).

A comparison between semikinetic (or stoichiometric) and mechanistic models is shown in Table IX; the predictions of the SPYRO model are compared with the predicted and experimental results reported by Ross

TABLE IX

Naphtha Cracking: Comparison of Models

Feed description	Light naphtha	Heavy naphtha
API 60°F	79.4	67.4
ASTM D 86 (°F)		
IBP	100	102
50%	149	210
95%	220	310
Paraffin content (wt%)	86	67
Hydrogen content (wt%)	16.0	15.2
Operating conditions		
CSI	2.66	2.81
Contact time (s)	0.70	0.53
KAPP (psia)	15.8	14.8

		Predicted			Predicted	
Yields (wt%)	Exp.	Ross and Shu	SPYRO	Exp.	Ross and Shu	SPYRO
H_2	0.73	0.93	0.77	0.35	0.85	0.81
CH_4	16.11	16.27	16.58	14.80	14.90	15.71
C_2H_2	0.37	0.37	0.36	0.50	0.40	0.49
C_2H_4	28.71	28.04	28.46	26.60	26.09	26.80
C_2H_6	4.87	4.83	5.32	4.09	4.07	4.00
C_3H_4	0.48	0.87	0.72	1.07	0.81	0.76
C_3H_6	16.37	16.25	16.40	13.84	14.43	14.29
C_3H_8	0.47	0.54	0.61	0.16	0.46	0.45
C_4H_6	4.20	4.27	4.27	4.65	4.46	4.49
C_4H_8	5.19	4.54	5.15	3.78	4.03	3.99
C_4H_{10}	0.48	0.34	0.41	0.40	0.43	0.26
C_5^+	22.02	22.85	20.96	29.16	29.08	27.96
Total	100.00	100.00	100.00	100.00	100.00	100.00

and Shu (1979) for two naphthas in a commercial reactor. Both the models agree fairly well with the experimental results, but presumably still better results could be obtained using the SPYRO program with improved knowledge of the operating conditions and coil geometry. Actually SPYRO simulations were performed assuming in the two cases respectively 1550° and 1587°F as the coil outlet temperature in order to match the reported "integrated" operating conditions. For the same criterion, coil geometry, outlet pressure, and steam dilution were also selected.

A further application of the SPYRO program for a full-range naphtha, chosen for its wide range of residence time, is reported in Table X. SPYRO predictions are in this case also in excellent agreement with the experimental results of conventional and millisecond furnaces reported by Leftin *et al.* (1976). Table XI reports still another comparison of SPYRO predictions with the experimental results of an advanced cracking reactor as reported by Kearns *et al.* (1979). Despite the large extrapolation to very high process temperatures (more than 1200°C) and the relatively poor indication of operating conditions and feedstock characterization, agreement between observed and simulated results is impressive. Only small

TABLE X

Wide-Range Naphtha Cracking: Yields for Two Different Furnace Types [a]

	Yield by furnace type (wt%)			
	Conventional		Millisecond	
Product	Exp.	SPYRO	Exp.	SPYRO
H_2	17.6	0.9	16.7	1.2
CH_4		17.6		16.8
C_2H_2	0.6	0.7	1.0	1.8
C_2H_4	28.2	27.0	31.0	31.0
C_2H_6	3.9	4.4	3.3	2.9
C_3H_4	0.6	1.0	1.3	1.7
C_3H_6	12.5	12.6	12.7	11.2
C_3H_8	0.5	0.4	0.2	0.2
C_4H_6	3.6	4.1	4.9	4.9
C_4H_8	2.7	2.8	2.8	1.9
C_4H_{10}	0.2	0.1	0.5	0.1
C_5^+	29.6	28.4	25.6	26.4

[a] Commercial data: Specific gravity, 0.70. ASTM, IBP, 37°C; 50%, 82°C; FBP, 196°C, molecular weight, 93 H:C atomic ratio, 2.16. paraffins, 73.5% vol. naphthenes, 21.2% vol. aromatics, 5.3% vol.

TABLE XI

Advanced Cracking Reactor

	Light naphtha		Light gas oil		Vacuum gas oil	
Sp. gravity (assumed)	0.65		0.83		0.88	
Pressure (atm)	5		5		5	
Temperature (°C)	1250–950		1220–970		1200–970	
Contact time (s)	0.16		0.15		0.16	
	Yields (wt%)					
	Exp.	Pred.	Exp.	Pred.	Exp.	Pred.
H_2	{17.04	2.27	{11.84	1.59	{12.03	1.50
CH_4		12.20		9.86		9.58
C_2H_2	4.43	4.37	4.56	4.86	4.48	4.61
C_2H_4	44.50	43.50	34.09	32.92	29.05	28.15
C_2H_6	3.20	4.72	1.86	2.40	1.62	2.39
C_3H_4	1.83	1.90	1.86	1.34	2.17	1.12
C_3H_6	11.70	11.13	7.15	6.84	5.80	5.42
C_3H_8	0.38	0.27	0.16	0.13	0.07	0.11
C_4H_4–C_4H_6	3.69	5.21	3.67	4.64	3.92	3.78
C_4H_8–C_4H_{10}	1.68	1.64	0.65	0.73	0.37	0.58
C_5	1.55	1.66	0.96	1.24	1.18	0.93
C_6^+	10.00	11.42	33.20	33.46	40.55	41.81

deviations, mainly in ethane and butadiene, suggest that, in this range of temperatures and pressures, other reactions or small modifications of kinetic parameters are needed.

Recycling to the cracking furnace of certain by-products (and particularly C_4 compounds) seems to be of increasing interest. With models other than the mechanistic ones, it is hard to account for the interactions that arise. Comparison of SPYRO predictions with experimental pilot plant data of C_4 and naphtha co-cracking demonstrates the effectiveness of a mechanistic program in simulating also these difficult situations, as reported by Goossens *et al.* (1978b).

VII. CONCLUSIONS

In this chapter the philosophy and application of pyrolysis models have been briefly presented, without any pretension of being exhaustive in a rapidly growing field. Attention has been focused mainly on the mechanistic approach, with indications on how to make it both extensive and realistic through the proper analogies and simplification levels. The SPYRO model was used as a term of comparison, because of its relative completeness and flexibility. Less extensive mechanistic, molecular, or semikinetic models show reliable predictions over a more limited range of conditions. Future trends in cracking technology and the necessity of using more widely differentiated feedstocks and mixtures, under the impact of the hydrocarbons shortage, will necessitate more complete and flexible fundamental models.

REFERENCES

Albright, L. F., and McConnell, C. F. (1979). *In* "Thermal Hydrocarbon Chemistry" (A. G. Oblad, H. G. Davis, and R. T. Eddinger, eds.), Advances in Chemistry Series, No. 183, Chap. 12. Am. Chem. Soc., Washington, D.C.

Albright, L. F., McConnell, C. F., and Welther, K. (1979). *In* "Thermal Hydrocarbon Chemistry" (A. G. Oblad, H. G. Davis, and R. T. Eddinger, eds.), Advances in Chemistry Series, No. 183, Chap. 10. Am. Chem. Soc., Washington, D.C.

Allara, D. L., and Edelson, D. (1975). *Int. J. Chem. Kinet.* **7**, 479.

Arai, Y., Murata, M., Tanaka, S., and Saito, S. (1977). *J. Chem. Eng. Jpn.* **10**, 303.

Atkins, G. T. (1962). *Petro/Chem Eng.* **34**(4), 20.

Benson, S. W. (1970). *In* "Refining Petroleum for Chemicals," Advances in Chemistry Series, No. 97, Chap. 1. Am. Chem. Soc., Washington, D.C.

Benson, S. W. (1976). "Thermochemical Kinetics," 2nd ed. Wiley, New York.

Blakemore, J. E., and Corcoran, W. H. (1969). *Ind. Eng. Chem., Process Des. Dev.* **8**, 206.

Brown, S. M., and Albright, L. F. (1976). *In* "Industrial and Laboratory Pyrolyses" (L. F. Albright and B. L. Crynes, eds.), ACS Symposium Series, No. 32, Chap. 17. Am. Chem. Soc., Washington, D.C.

Côme, G. M., Azai, P., Alraw, D., and Niclause, M. (1979). *AIChE Annu. Meet., 72nd, San Francisco, Calif.* Pap. 21-A.

Crynes, B. L., and Albright, L. F. (1969). *Ind. Eng. Chem., Process Des. Dev.* **8**, 25.

Davis, H. G., and Farrell, T. J. (1973). *Ind. Eng. Chem., Process Des. Dev.* **12**, 171.

Davis, H. G., and Williamson, K. D. (1959). *Proc. World Pet. Congr., 5th.*

de Blieck, J., and Goossens, A. G. (1971). *Hydrocarbon Process.* **50**(3), 76.

Dente, M. E., Ranzi, E. M., Antolini, G., and Losco, F. (1970). *Event Eur. Fed. Chem. Eng., 97th, Florence, Italy* Pap. II-4.

Dente, M. E., Ranzi, E. M., and Goossens, A. G. (1979a). *CACE '79, Symp. Comput. Appl. Chem. Eng., 12th, Montreux, Switz.* Pap. 2A-1.

Dente, M. E., Ranzi, E. M., Barendregt, S., and Goossens, A. G. (1979b). *AIChE Annu. Meet., 72nd, San Francisco, Calif.* Pap. 38-b.

Dente, M. E., Ranzi, E. M., Pierucci, S. J., Biardi, G. E., Goossens, A. G., and Losco, F. (1979c). *ICP, Ind. Chim. Petrolifera* **7**(2), 67.

Dunkleman, J. J., and Albright, L. F. (1976a). *In* "Industrial and Laboratory Pyrolyses" (L. F. Albright and B. L. Crynes, eds.), ACS Symposium Series, No. 32, Chap. 14. Am. Chem. Soc., Washington, D.C.

Dunkleman, J. J., and Albright, L. F. (1976b). *In* "Industrial and Laboratory Pyrolyses" (L. F. Albright and B. L. Crynes, eds.), ACS Symposium Series, No. 32, Chap. 15. Am. Chem. Soc., Washington, D.C.

Edelson, D., and Allara, D. L. (1973). *AIChE J.* **19**, 638.

Fernandez-Baujin, J. M., and Solomon, S. M. (1976). *In* "Industrial and Laboratory Pyrolyses" (L. F. Albright and B. L. Crynes, eds.), ACS Symposium Series, No. 32, Chap. 20. Am. Chem. Soc., Washington, D.C.

Froment, G. F., van de Steene, B. O., van Damme, P. S., Narayanan, S., and Goossens, A. G. (1976). *Ind. Eng. Chem., Process Des. Dev.* **15**, 495.

Froment, G. F., van de Steene, B. O., and Sumedha, O. (1979). *Oil Gas J.* Apr., 16, **77**, 87.

Gear, C. W. (1971). "Numerical Initial Value Problems in Ordinary Differential Equations," Chap. 11. Prentice-Hall, Englewood Cliffs, New Jersey.

Goossens, A. G., Dente, M. E., and Ranzi, E. M. (1978a). *Hydrocarbon Process.* **57**(9), 227.

Goossens, A. G., Dente, M. E., and Ranzi, E. M. (1978b). *Oil Gas J.* Sept. 4, **76**, 89.

Green, E. J., Zdonik, S. B., and Halle, L. P. (1975). *Hydrocarbon Process.* **54**(9), 164.

Haraguchi, T., Nakashio, F., and Sakai, W. (1976). *In* "Industrial and Laboratory Pyrolyses" (L. F. Albright and B. L. Crynes, eds.), ACS Symposium Series, No. 32, Chap. 7. Am. Chem. Soc., Washington, D.C.

Hatch, L. F., and Matar, S. (1978). *Hydrocarbon Process.* **57**(1), 135.

Illes, V., Pleszkats, I., and Szepsy, L. (1974). *Acta Chim. Acad. Sci Hung.* **80**, 267.

Illes, V., Szalai, O., and Czermely, Z. (1976). *In* "Industrial and Laboratory Pyrolyses" (L. F. Albright and B. L. Crynes, eds.), ACS Symposium Series, No. 32, Chap. 24. Am. Chem. Soc., Washington, D.C.

Kearns, J. D., Milks, D., and Kamm, G. R. (1979). *In* "Thermal Hydrocarbon Chemistry" (A. G. Oblad, H. G. Davis, and R. T. Eddinger, eds.), Advances in Chemistry Series, No. 183, Chap. 7. Am. Chem. Soc., Washington, D.C.

Kerr, J. A., and Parsonage, M. J. (1976). "Evaluated Data on Gas Phase Hydrogen Transfer Reactions on Methyl Radicals." Butterworth, London.

Kossiakoff, A., and Rice, F. O. (1943). *J. Am. Chem. Soc.* **65**, 580.

Kunugi, T., Sakai, T., Soma, K., and Sasaki, Y. (1969). *Ind. Eng. Chem., Fundam.* **8**, 374.

Kunugi, T., Soma, K., and Sakai, T. (1970). *Ind. Eng. Chem., Fundam.* **9**, 319.

Leftin, H. P., Newsome, D. S., Wolff, T. J., and Yarze, J. C. (1976). *In* "Industrial and Laboratory Pyrolyses" (L. F. Albright and B. L. Crynes, eds.), ACS Symposium Series, No. 32, Chap. 21. Am. Chem. Soc., Washington, D.C.

Lichtenstein, I. (1964). *Chem. Eng. Prog.* **60**(12), 64.

Lohr, B., and Dittmann, H. (1977). *Oil Gas J.* July 4, **75**, 53.

Murata, M., and Saito, S. (1975). *J. Chem. Eng. Jpn.* **8**, 39.

Myers, P. S., and Watson, K. M. (1946). *Natl. Pet. News* **38**(18), R-388.

Ranzi, E. M., Dente, M. E., Losco, F., Biardi, G. E., and Pierucci, S. J. (1977). *Proc. Symp. Comput. Chem. Eng., 5th, High Tatras, CSSR* p. 505.

Ranzi, E. M., Dente, M. E., Pierucci, S. J., Goossens, A. G., and Barendregt, S. (1980). *Proc. Chem Plant '80, Symp. Use Comput. Chem. Eng., 13th, Heviz, Hung.* p. 864.

Ross, L. L., and Shu, W. R. (1977). *Oil Gas J.* Oct. 17, **75**, 58.

Ross, L. L., and Shu, W. R. (1979). *In* "Thermal Hydrocarbon Chemistry" (A. G. Oblad, H. G. Davis, and R. T. Eddinger, eds.), Advances in Chemistry Series, No. 183, Chap. 8. Am. Chem. Soc., Washington, D.C.

Shah, M. J. (1967). *Ind. Eng. Chem.* **59**, 70.

Shu, W. R., Ross, L. L., and Pang, K. H. (1978). *AIChE Natl. Meet., 85th, Philadelphia, Pa.* Pap. 27d.

Snow, R. H. (1966). *J. Phys. Chem.* **70**, 2780.

Snow, R. H., Peck, R. E., and von Fredersdorff, C. G. (1959). *AIChE J.* **5**, 304.

Sundaram, K. M., and Froment, G. F. (1977a). *Chem. Eng. Sci.* **32**, 601.
Sundaram, K. M., and Froment, G. F. (1977b). *Chem. Eng. Sci.* **32**, 609.
Sundaram, K. M., and Froment, G. F. (1978a). *Ind. Eng. Chem., Fundam.* **17**, 174.
Sundaram, K. M., and Froment, G. F. (1978b). *Int. J. Chem. Kinet.* **10**, 1189.
Sundaram, K. M., and Froment, G. F. (1980). *Chem. Eng. Sci.* **35**, 364.
van Damme, P. S., Narayanan, S., and Froment, G. F. (1975). *AIChE J.* **21**, 1065.
Williamson, K. D., and Davis, H. G. (1979). *In* "Thermal Hydrocarbon Chemistry" (A. G. Oblad, H. G. Davis, and R. T. Eddinger, eds.), Advances in Chemistry Series, No. 183, Chap. 4. Am. Chem. Soc., Washington, D.C.
Zdonik, S. B., Green, E. J., and Halle, L. P. (1967). *Oil Gas J.* Aug. 21, **65**, 86.

8

Nonpetroleum Feedstocks

OTIS C. DERMER and BILLY L. CRYNES

I. ONCE AND FUTURE SOURCES OF LIGHT OLEFINS AND BTX

Ethylene was made, both in laboratory and for industrial use, from fermentation-derived ethanol before it was manufactured from petroleum and natural gas liquids, and it is well known that for many years the main commercial source of benzene, toluene, and xylenes (BTX) was the products of coal pyrolysis. It may be less familiar that ethylene was produced commercially from coke-oven gas as early as 1920 (Asinger, 1968) and for some years thereafter. Unconventional feedstocks for making gaseous olefins and BTX are reviewed in this chapter, not because they are of current industrial importance (most are not, at least not in the United States), but because some of them will surely become so in future decades.

Three primary types of nonpetroleum feedstocks require consideration: fossil carbonaceous materials other than petroleum and natural gas, biomass (mostly cellulose), and scrap plastics of hydrocarbon type. The fossil materials include tar sands, natural asphalts (but these are scarce and pyrolysis for chemicals is unstudied), oil shales, and of course coal, lignite, and peat. Biomass is represented by municipal and agricultural

Pyrolysis: Theory and Industrial Practice

ISBN 0-12-048880-9

wastes and potentially by crops grown for the purpose, as in tree plantations. Among plastics, only scrap polyethylene, polypropylene, and rubber are available in volumes large enough to be of any interest.

Each of these classes of materials can be pyrolyzed by various means to gases, liquids, and solids (char, coke). For some, complete gasification of the carbon contained is thus possible, especially by flash heating, but this is not a direct or useful route to olefins and aromatics. The gases and liquids can themselves be repyrolyzed, with or without preliminary fractionation, hydrogenation, and so on; this will be discussed.

Although the coking of coal gives BTX and smaller amounts of gaseous olefins, as just noted, coal and coal-derived liquids are too deficient in hydrogen to serve as a major source of such olefins unless hydrogenation is used at some point. The problem with cellulosic materials, $(C_6H_{10}O_5)_n$, is related but different; again hydrogenation is required, to effect deoxygenation and to improve the yield of hydrocarbons.

Ordinarily, pyrolysis to produce light olefins—from ethane, other light alkanes, naphtha, or even whole petroleum—does not involve catalysis. Although catalytic effects of reactor walls are virtually inevitable, they cause undesirable secondary reactions, such as carbon formation, rather than dehydrogenation to olefins. The use of steam in such cracking to minimize coking and other problems is of course standard practice. Other additives such as H_2S and HBr may serve as homogeneous catalysts, but they probably retard rather than promote dehydrogenation (Niclause *et al.*, 1976). Formation of aromatics in pyrolysis of light alkanes is also due to uncatalyzed reactions; such formation by reforming gasoline fractions, involving dehydrogenation-cyclizations, is conducted catalytically but is not simple pyrolysis and is not considered here.

II. COAL (INCLUDING LIGNITE AND PEAT) AND COAL TAR

The industrial carbonization of coal is well over 100 years old, and the aromatics derived from coal tar in Germany and England were the basis of the early synthetic organic chemicals industry. Carbonization for fuel gas production, now obsolete, gives good yields of BTX because the high temperature used, up to 1000°C, favors aromatization, whereas low-temperature (450–650°C) carbonization to make coke for use as domestic fuel does not. By far the principal use of coke, however, is in iron metallurgy, all over the world, and the gaseous olefins formed are insignificant by-products. Even the production of benzene from such high-temperature coking (1000°C)—because yields of BTX are only about 20 pounds per ton of coal (West, 1970)—amounted to only 10% of world production in 1972 (French, 1957), or 6% (Rao and Skeist, 1976) to 10% (Squires, 1976) of

American production in 1976. Brown coals and peats containing 22–36% oxygen and sulfur coked at 800–850°C gave similarly low yields (0.76–4.5%) of aromatics, mostly benzene (Kashirskii, 1974).

BTX recovery is mostly from coke-oven gases, which are cooled, variously purified, and passed through scrubbers containing petroleum or creosote oils, whence crude BTX is stripped with steam. Refining is necessary to remove sulfur compounds and aliphatic hydrocarbons that are in the BTX boiling range. The ratio of BTX product obtained from coal light oil is about 66/13/5. Even greater proportions of toluene and xylenes than of benzene are currently produced from petroleum; even in 1966 only 4% of the American toluene was derived from coal (Cier, 1969; McNeil, 1969).

The necessity to minimize air and water pollution and to conserve energy has led to improvements in the coking of coal. One of the most extensive is the "Clean Coke" process, in which part of the coal is pyrolyzed and the rest hydrogenated, the liquids from these steps are combined and processed, and the heavy residues are used as a binder for briquetting the coke (Schowalter and Boodman, 1974). This United States Steel process gives 38% coke and 20% chemicals, instead of 73% coke and 2% chemicals (Purcell, 1978).

The gases from high-temperature carbonization of coal contain 3–4% ethylene and about 1% propylene. Low in ethylene as this is, it has served as a commercial source of the olefin under special circumstances (Asinger, 1968, pp. 39–40; Kniel *et al.*, 1980). Oil absorption is useless as a technique for extracting these light hydrocarbons at concentrations so low; adsorption on activated carbon can serve (Engelhardt, 1938), but this obviously is not a significant source of ethylene. Flash pyrolysis of coal particles, preferably with rapid quenching, has been variously conducted and shown to give better yields of ethylene (Mainhood and Pitt, 1966; Granger and Ladner, 1970; Squires, 1976; Banerjee *et al.*, 1977), but this too appears to be of only academic interest.

Because this book excludes pyrolysis in which some reagent besides the cracking stock is used, various recent studies of coal hydropyrolysis are not cited.

The cracking of low-temperature coal tar to produce gaseous olefins has been repeatedly investigated, as reviewed by Rao (1979). Heavy oil fractions, being too aromatic, gave poor yields, at best 10% ethylene and 8% propylene at 700°C and 10 mm Hg (Kishida and Chuto, 1952). Slightly higher yields of these olefins (22% total) were obtained by cracking in steam with a contact time of about 0.8 s at 780°C; the optimum temperature for ethylene production was somewhat higher than that (Marizewska and Musierowicz, 1960). At shorter contact time (0.1 s), a higher optimum temperature for olefin production (880°C) was found for a Siberian coal tar;

under these conditions, gas (61% ethylene and 32% propylene) was produced in 7.9% yield (Apter and Kalinichenko, 1962). In the next study, Hight-Splint coal (Kentucky) was carbonized at 485°C in a steam-fluidized bed and the vapors cracked on similarly fluidized coke at 600–800°C, mostly without intermediate condensation. Olefin yields were highest at 600°C, but still poor, and not defined for gaseous products. The 600°C temperature was also optimum for total aromatics production, which at a tar vapor/steam ratio of 8/100 amounted to 4.3 gallons per ton of coal (Naugle *et al.*, 1965). It has been shown that yields of light olefins are better when the tar feedstock is free of tar acids (Imai *et al.*, 1957; Rao *et al.*, 1968); thus neutral oil gave 22% ethylene, whereas oil containing acids gave only 17% (Imai *et al.*, 1957). The lower-boiling (210°C) fraction of neutral oil is also preferable, the higher fractions producing more aromatics (Rao *et al.*, 1968). The maximum yield of ethylene (19.6%) was observed at 700°C and moderate feed rate; that for propylene (11.5%) was at 550°C with contact time about four times as long.

Tars from fixed-bed and fluid-bed carbonization of Australian coal at 500°C, and vapors from the former, were cracked at 500–1000°C on a fixed bed of coke. The best yield of ethylene (22%) was obtained at 800°C with a contact time of 1 s (Hesp *et al.*, 1969; Hesp and Waters, 1970).

The tar oil from carbonization of lignite can also be cracked to produce the gaseous olefins, best at 700–800°C (higher temperatures as usual cause more coking and more aromatization), with a contact time of 0.05 s (Tolubinskii *et al.*, 1962). In a Hoechst coker, which uses preheated coke balls as reaction sites, another lignite tar was converted at 630°C to 55% gases and only 3% coke; about half the gases were olefins (Krekeler, 1955). When a middle-tar fraction from brown coal (boiling range 81–324°C) was hydrogenated and then steam-pyrolyzed at 700–850°C, it gave 9–14.9% C_2–C_4 olefins (Kubicka *et al.*, 1962). Vapors from lignite carbonization were best cracked without condensation to avoid forming nonvolatile polymeric residues on revaporization. Texas lignite vapor was better than that from North Dakota lignite for producing olefins and BTX. The vapor-cracking unit was held at 600–800°C while the lignite was gradually heated up to 800°C. With this regime, Texas lignite carbonized up to 550°C and the vapors pyrolyzed at 800°C for about 1.5 s gave 2% ethylene, 0.39% propylene, and about 0.6% BTX on the dried lignite charged; as would be expected, the yield of propylene was better in pyrolysis at 600°C (Montgomery *et al.*, 1959). Similar work, but with steam cracking, likewise gave small yields of ethylene and BTX (Montgomery and Decker, 1964b). A cannel coal is even better for this purpose, giving 8.5–12% ethylene as well as 1.5–2.5% benzene (Montgomery and Decker, 1964a), but the economics are not attractive.

A pitch (45% boiling above 300°C) from the low-temperature carbonization of Texas lignite was subjected to delayed coking in a small reactor at 650–790°C. This gave up to 30 wt% gases, in which the ethylene/ethane ratio varied from about 1.25 at 650°C to about 3 at 790°C. Contact time had to be at least 0.7 s but beyond that it had little effect (Rice *et al.*, 1970).

III. COAL-DERIVED LIQUIDS

Coal liquefaction processes under development are potentially large sources of petrochemicals, including ethylene and BTX. A conceptual SRC-II plant processing 3×10^4 tons of coal per day could yield enough ethane and propane to produce 1.1 billion pounds per year of ethylene, the amount from one of the conventional ethylene plants in operation today (Harris *et al.*, 1980). Much of the information in this study is based on data obtained in a 1 ton/day pilot plant, as well as a 50 ton/day unit at Fort Lewis, Washington. Pyrolysis yields were obtained by both experiment and computer generation. Besides an equiweight mixture of ethane and propane that could be pyrolyzed to produce ethylene and propylene, the SRC-II yields naphtha (30–205°C) that would serve as a potential source of BTX and ethylene. The light fuel oil from the SRC-II process (205–280°C) would supply BTX, phenol, cresols, and naphthalene. The heavy fuel oil ($\geq$280°C) would be another feedstock for making BTX. This standard 3×10^4 ton/day SRC processing unit could yield approximately 340 million gallons per year of BTX. Table I compares the pyrolysis yields from two SRC liquids, a Kuwait light gas oil, and a Kuwait full-range naphtha. Caution should be used in assessing these data because the yields from the

TABLE I

Light Olefin Yields from Cracking of Heavy Coal Liquids and Petroleum Liquids[a,b]

	Yield (wt% of charge)[b]			
	Hydrotreated SRC-II naphtha	Hydrotreated SRC-II light fuel oil	Kuwait full-range naphtha	Hydrotreated Kuwait light gas oil
Ethylene	23.7	4.3	34.9	26.1
Propylene	8.0	4.9	12.3	15.0
1,3-Butadiene	2.1	1.2	3.6	4.4

[a] Harris *et al.* (1980).
[b] Typical commercial cracker conditions assumed.

SRC-II liquids are computed and not experimental data. Nevertheless, they suggest that the hydrotreated SRC-II light fuel oil is totally unsatisfactory for the production of olefins because it yields only 4 wt% ethylene and 5 wt% propylene. The yield from the SRC-II naphtha is significantly better and comparable to that obtained with the Kuwait light gas oil. However, the hydrotreated SRC-II naphtha, which has high naphthenic and aromatic contents, is an exceptionally good feedstock for the production of BTX, yielding from 50 to 80% more BTX than petroleum naphtha does. One must recognize that the SRC process under development is not directed toward the production of petrochemicals; it is a fuel-producing system. Nevertheless, it has the potential to be a significant supplier of petrochemicals, so much so that the process could have a significant effect on the market.

In more definitive recent studies, a comparison has been made of the thermal response of four different coal-derived liquids (Krishnamurthy and Shah, 1979; Krishnamurthy *et al.*, 1979, 1980). These studies were conducted in transfer-line reactors of 1 or 1.6 cm diameter and 76 or 90 cm length. The coal liquid mixed with an inert alumina powder was forced through the reactor with nitrogen pressure. Liquid feed rates were controlled at 120 ml/h, alumina feed rate was 60 g/h, reaction temperature 650–780°C, and residence time 0.08–0.13 s. Feedstocks utilized are described in Table II. Also shown for comparison is a liquid from another study (Korosi *et al.*, 1976). All four feedstocks contained approximately the same percentage of carbon, but the unhydrotreated SRC-II recycle solvent was higher in nitrogen, as expected. Comparisons of the ethylene and

TABLE II

Characteristics of Coal Liquid Feedstocks[a]

	Hydrotreated Western Kentucky COED		Kentucky SRC-II recycle solvent	Hydrotreated Kentucky Synthoil[b]
	Light oil	Heavy oil		
Specific gravity (ASTM D287)	0.818	0.906	0.98	0.895
Boiling range (°C)	13–500	100–510	115–497	250–380
Carbon (wt%)	84.7	85.5	87.2	87.1
Hydrogen (wt%)	12.5	11.6	9.0	12.8
H/C ratio	1.77	1.63	1.24	1.76
Saturates (wt%)	71	52	—	84

[a] Krishnamurthy and Shah (1979) and Krishnamurthy *et al.* (1979, 1980).
[b] From Korosi *et al.* (1976).

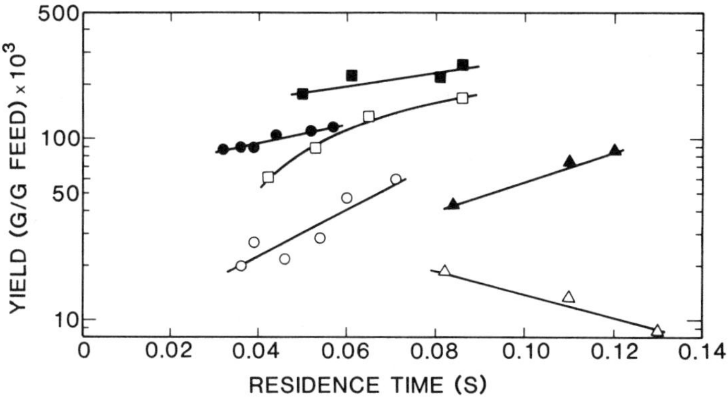

Fig. 1. Variation of ethylene yield from various feedstocks. ■, COED light oil (LO), 780°C; ●, COED heavy oil (HO), 750°C; □, COED LO, 740°C; ○, COED HO, 690°C; ▲, SRC-II, 730°C; △, SRC-II, 650°C. [From Krishnamurthy and Shah (1979) and Krishnamurthy *et al.* (1979).]

propylene yields from the three feedstocks at various temperatures and residence times are shown in Figs. 1 and 2. The yield of ethylene from the SRC-II recycle solvent is lower than from the COED oils. Because the SRC-II was not hydrotreated, it presumably contained a lower percentage of saturates than did the other oils, and thus produced less ethylene. A

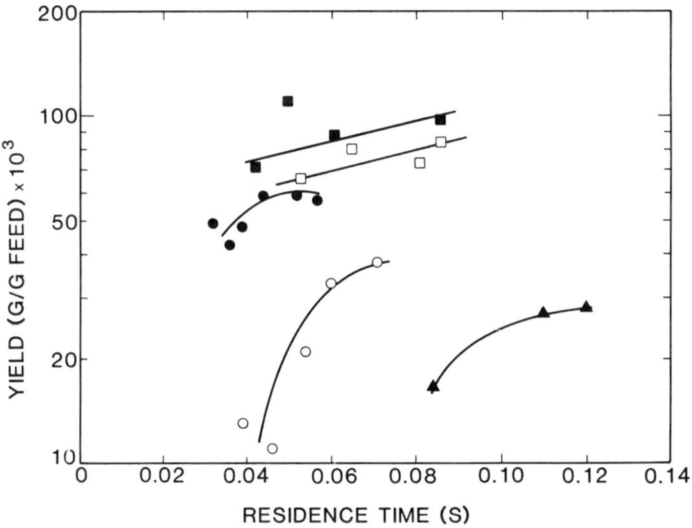

Fig. 2. Variation of propylene yield from various feedstocks. ■, COED LO, 780°C; □, COED LO, 740°C; ●, COED HO, 750°C; ○, COED HO, 690°C; ▲, SRC-II, 730°C. [From Krishnamurthy and Shah (1979) and Krishnamurthy *et al.* (1979).]

similar observation can be made about propylene yields in Fig. 2. This study further revealed that high molecular weight aromatics in some of the pyrolysates arose by polymerization and cyclization reactions. The H/C atom ratios of the liquid pyrolysates decreased with increasing cracking severity, that is, with increasing temperature and residence times. The preferential cracking of saturates over unsaturates emphasizes the importance of hydrotreatment of coal-derived liquids before thermal cracking.

A study of the steam cracking of two H–Coal liquids and one COED liquid resulted in the product yields shown in Table III (Peters, 1975). Unfortunately, neither the feedstock properties nor reaction conditions were stated. Clearly the H–Coal naphtha produced much more light olefins and benzene than other feedstocks, whereas the heavier H–Coal and COED mid-distillates were alike except that the one from H–Coal gave more benzene and the one from COED more fuel oil. The high fuel oil yields from these stocks again show the need for a preliminary hydrotreating step if petrochemical production from such coal liquids is desired.

TABLE III

Product Distribution in Steam Pyrolysis of Various Coal Liquids Feedstocks (in wt %)[a,b]

	Type of feedstock		
Product	H–Coal naphtha (IBP to 177°C)	H–Coal mid-distillate (177–343°C)	COED mid-distillate (177–343°C)
Hydrogen	0.8	0.7	0.9
Methane	16.3	12.1	11.8
Ethylene	23.4	14.4	13.9
Propylene	8.5	5.9	6.5
1,3-Butadiene	3.0	1.5	1.3
Benzene[c]	23.6	18.5	9.9
Fuel oil	24.4	46.9	55.8

[a] Peters (1975).
[b] Calculated yields assuming recycle gas extinction and hydrodealkylation of toluene and C_3 aromatics.
[c] Includes toluene and xylene equivalents.

Decahydronaphthalene (Decalin), a hydrogenated Synthoil coal liquid, and a petroleum-derived feedstock were steam cracked in a tubular reactor under conditions simulating those of a commercial reactor (Korosi *et al.*, 1976). These conditions were: outlet temperature, 875°C; outlet pressure, 27 psig; contact time, 0.3 s; and steam/hydrocarbon weight ratio, 1. Feed-

TABLE IV

Characteristics and Pyrolysis Products of Various Hydrocarbon Liquids[a]

	Liquid type		
Characteristics	Hydrogenated Synthoil liquid	Light gas oil Kuwait desulfurized	Decahydro-naphthalene
Boiling range (°C)	184–348	193–389	187–196
Carbon (wt%)	87.14	86.16	86.88
Hydrogen (wt%)	12.76	13.70	13.12
H/C ratio	1.74	1.91	1.81
Hydrocarbon group types (wt%)			
Straight-chain paraffins	0.0	23.5	0.0
Branched-chain paraffins	1.0	33.4	0.0
Alicyclics	83.1	23.0	100.0
Product yields (wt%), (partial list)			
Hydrogen	1.03	0.60	1.05
Methane	11.60	12.00	10.35
Ethylene	18.83	22.66	19.36
Ethane	2.40	2.96	2.86
Propylene	7.83	13.02	6.57
1,3-Butadiene	4.68	4.82	4.65
Butenes	1.63	3.12	1.41
Benzene	12.15	6.71	17.97
Toluene	6.76	3.51	10.34
C_8 Aromatics	3.43	2.11	4.88
Fuel oil (b. > 205°C)	22.46	20.51	11.66

[a] Korosi *et al.* (1967)

stock properties, shown in the first part of Table IV, show that except for the boiling range, Decalin is a good model for the coal liquid. The latter part of Table IV also shows the similarity of the two in pyrolysis, the main difference being that Decalin gives about 50% more BTX and only half as much residual fuel oil. Both are much better sources of BTX than the light gas oil, and not much inferior to this oil in producing light olefins. Thus the hydrotreated coal liquid is a promising feedstock for manufacturing these petrochemicals; probably more severe hydrogenation, to convert more aromatics to alicyclics, would have made it even better.

Pyrolysis at 700°C and near atmospheric pressure of a liquid obtained from the zinc chloride-promoted hydrocracking of a Utah coal gave 24 wt% conversion to ethylene and propylene and 51 wt% to BTX (Qader, 1976). Feedstock properties and product yields are summarized in Table V, but few other details are available.

TABLE V

Product of Pyrolysis of a Hydrotreated Oil Made from Utah Coal[a]

Product distribution (wt%)	
Hydrogen	0.5
Methane	8.0
Acetylene	0.2
Ethylene	15.0
Propylene	9.0
Butanes	4.3
BTX	50.7
Naphthalenes, etc.	12.3
BTX analysis (vol%)	
Benzene	51
Toluene	20
Ethylbenzene	8
Xylenes	21

[a] Qader (1976).

In Polish work, various light fractions, some hydrogenated, of coal extracts steam-pyrolyzed at 700–800°C gave up to 17% ethylene and 5% propylene, along with BTX in the liquid products (Taniewski and Krajewski, 1978).

The vapor-phase pyrolysis of hydrotreated coal extracts (laboratory preparations) was compared to that of a highly paraffinic North Sea gas condensate and three model compounds (National Coal Board, 1979). Table VI presents yields of ethylene and BTX from these five feedstocks. The cracking was conducted at 860–1000°C and 1–2 s contact time in nitrogen at essentially 1 atm pressure. The increased naphthene content of the treated coal liquid and Decalin compared to Tetralin led to markedly improved BTX yields, whereas the highly paraffinic feedstock gave high

TABLE VI

Ethylene and BTX Yields from Pyrolysis of Various Feedstock (wt%)[a]

	Hydrogenated coal extracts	Paraffinic North Sea gas condensate	Decalin	Tetralin	Undecane
Ethylene	≤20	26.4	20	2.7	37.4
BTX	≤20	12:5	24.8	4.0	6.8

[a] National Coal Board (1979).

ethylene yields. This demonstrated once more the need to transfer significant quantities of hydrogen into a coal liquid preceding thermal cracking operations. In spite of the difference in conditions, the yields of ethylene and BTX from Decalin agree reasonably with those in Table IV.

In an expanded study of the foregoing project, eight different coal-derived feedstocks were pyrolyzed and the results compared with those obtained from the pyrolysis of a natural gas condensate and four model compounds: undecane, Tetralin, Decalin, and mesitylene (Bernhardt *et al.*, 1980). These studies were conducted at essentially atmospheric pressure in a tubular reactor of 30 mm internal diameter and 100 mm heated zone. Over this length, the reactor was controlled within 20°C of the maximum reactor temperature. This system was tested for surface effects by packing with copper and stainless-steel material, and no effects were noted. Also, an eightfold reduction in the feedstock concentration at a constant vapor-residence time revealed no significant effect. The replacement of diluent nitrogen by a 70% steam plus 30% nitrogen mixture also had no observable effect on the yields of BTX and ethylene. The feedstock properties are shown in Table VII and product yields in Table VIII.

The Linby coal was carbonized in a prereactor, and the vapors were then passed into the pyrolysis system. The pyrolysis of material called "unhydrogenated coal-derived feedstocks" gave only low BTX and ethylene yields. This would be from the bituminous coal vapors, anthracene oil, and the anthracene oil extract.

In conformity with many other studies, the BTX and ethylene yields increased with increasing extent of hydrogenation. About 27% BTX and 24% ethylene were obtained from the highly hydrogenated coal extracts, By comparison, the paraffinic North Sea gas condensate gave only 16% BTX but 30% ethylene.

The model compounds were pyrolyzed separately and in mixtures. The mixture product yields were within about 2% of the values determined from the constituent yields as assessed from pyrolysis of the individual components. Not surprisingly, the distribution of product yields from these pure compounds was strongly related to the compound type. Maximum ethylene was obtained from the paraffin, undecane (40 wt%). The minimum ethylene yield was obtained from the aromatic, mesitylene (approximately 2 wt%); the maximum BTX was obtained from the naphthene, Decalin (24 wt%); and minimum BTX yield from the hydroaromatic, Tetralin (3 wt%).

This detailed study confirms several points that have been made in other investigations:

1. The low yields of BTX and ethylene from unhydrogenated feedstocks result from the stability of the condensed aromatic structures.

TABLE VII

Analyses of Pyrolysis Feedstocks[a]

Solid feedstock	Proximate analysis (wt%)			Ultimate analysis (wt%, dry, ash-free)					
	As delivered	Ash, as delivered	V.M., dry, ash-free	C	H	O	N	S	H/C Ratio
Linby coal	8.3	6.0	39.1	82.4	5.3	9.0	1.95	1.00	0.77
Anthracene oil extract of Linby coal	0.3	0.3	50.5	88.7	4.8	3.8	1.90	0.55	0.65

Liquid feedstocks	Distillation range (°C)	Hydrocarbon analysis (wt%)									
		Paraffins	Naphthalene	Other aromatics + hydroaromatics	Unidentified						
Anthracene oil	573–673	N.D.[b]	N.D.	N.D.	N.D.	90.6	6.9	1.9	0.6	0.55	0.91
Partially hydrogenated anthracene oil	IBP–503	1.7	20.7	64.4	13.2	88.2	10.7	0.5	0.4	<0.1	1.46
Hydrogenated coal extracts	443–523	0	28.5	71.5	0	89.4	10.1	0.7	0.25	<0.1	1.36
	523–573	0	23.7	76.2	0.1	89.6	9.4	N.D.	N.D.	N.D.	1.26
More highly hydrogenated coal extracts	443–523	0	95.5	2.2	2.3	86.4	13.2	0.5	<0.1	<0.1	1.83
	473–573	0	90.7	6.9	2.4	86.9	13.0	N.D.	N.D.	N.D.	1.78
Natural gas condensate	IBP–623	56.0	11.9	14.0	18.1	85.1	14.4	<0.2	<0.1	<0.1	2.03

[a] Bernhardt *et al.* (1980).
[b] N.D., Not determined.

TABLE VIII

Cracking Yields from Coal-Derived Fluids (wt%)[a]

	Cracking temperature, 1273°K				Cracking temperature, 1133°K				Cracking temperature, 1158°K
	VRT[b] = 2.0 s			VRT = 1.0 s	VRT = 0.4 s				
Feedstock	Linby coal[c]	Anthracene oil extract of Linby coal[c]	Anthracene oil	Partially hydrogenated anthracene oil	Hydrogenated coal extract	Hydrogenated coal extract	More highly hydrogenated coal extract	More highly hydrogenated coal extract	Natural gas condensate
Product									
Benzene	3.1	1.8	2.1	6.6	3.5	2.5	17.1	13.5	11.9
Toluene	0.1	0	0	2.7	1.9	1.3	5.3	5.1	2.5
Xylenes	0	0	0	1.3	1.7	1.1	3.0	2.8	1.7
Total BTX	3.2	1.8	2.1	10.6	7.1	4.9	25.4	21.4	16.1
C_2H_4	0.3	0.1	0.09	10.8	9.1	7.5	23.1	21.5	30.3
C_3–C_4 unsaturates	0	0	0	1.8	3.4	2.7	7.9	10.6	11.2

[a] Bernhardt *et al.* (1980).
[b] VRT, vapor-residence time.
[c] Carbonized at 600°C.

2. As the H/C atom ratios are increased by hydrotreatment, the corresponding naphthene contents increase, which in turn results in significantly greater yields of BTX and ethylene. The polynuclear aromatic hydrocarbons simply must be hydrogenated to the naphthenic structures.

3. The most important variables are cracking temperature and vapor-residence time with respect to yields from a given group.

4. The gas condensate reflects the desirability of highly paraffinic feedstocks for producing significant quantities of ethylene.

5. The molecular structure of the species being pyrolyzed is of paramount importance, as demonstrated by the model compound studies.

6. The hydroaromatics are preferentially dehydrogenated to aromatics; naphthenes crack primarily to BTX and ethylene; and larger olefinics yield small olefinic products such as ethylene and propylene.

IV. OIL SHALE AND SHALE OIL

Like pyrolysis of other carbonaceous materials, that of oil shale, whether mined or *in situ*, produces liquids, gases, and solid residue. The liquid, shale oil, is much richer in sulfur, oxygen, and nitrogen compounds than petroleum is. However, the extraction from it of useful individual petrochemicals, such as phenols and pyridine, suggested some time ago (Thorne, 1956), is only a remote possibility. The gas from oil shale pyrolysis alone or in steam is like coke-oven gas in containing ethylene and benzene, but is even leaner in both. Yields of ethylene from oil shale have been 1.0% (Sredin, 1960), 1.4% (Schachter, 1979), 1.9% (Brantley *et al.*, 1951), 1.5–2.5% (Kyll and Stepanov, 1959), and 2.5% (Sohns *et al.*, n.d.). Corresponding yields of benzene were 0.8% (Sohns *et al.*, n.d.) and 1.6% (Brantley *et al.*, 1951); although these are greater than the yield of benzene in conventional carbonization of coal (about 0.7%), oil shale pyrolysis is not a promising source of BTX.

Whereas the upgrading of shale oil to motor fuels is enormously complicated by the nitrogen compounds present—in that they deactivate refinery catalysts—these nitrogen compounds do not interfere with pyrolysis of the oil to produce petrochemicals. The first systematic study of pyrolysis of whole shale oils was by Shultz *et al.* (1955). Characteristics of the feedstocks are given in Table IX. In the range 640–750°C, at pressure, with contact time of 1 to 4.5 s, gaseous products ranged from 33 to 53 wt%, liquids (largely aromatic) 42–66%, and coke 1–11%. Both yields and compositions of products could be correlated by means of an empirical pyrolysis severity function $S = t\theta^{0.06}$ (t = temperature in degrees F, θ = residence time in seconds); a part of this is shown in Fig. 3. As usual, the more severe

TABLE IX

Characteristics of Whole Shale Oil Freedstocks[a]

Characteristics	Bureau of Mines or Union Oil Co. shale oil	TOSCO II shale oil	TOSCO II shale oil	NTU simulated-*in situ* shale oil
Specific gravity (at 60°F)	0.934–0.935		0.934	0.905
Gravity (°API)	19.8	22.0	19.8	24.3
Boiling range (C)	115 or 185 (initial) to 478 or 466 (final)		82% at 75–540, rest higher	92% at 125–540, rest higher
Carbon (wt%)	83.2 or 84.7		84.7	83.9
Hydrogen (wt%)	11.3		10.7	11.6
C/H Ratio	7.4–7.5		7.9	7.2

[a] Shultz, *et al.* (1955), Griswold *et al.* (1976), and Fritzler *et al.* (1977).

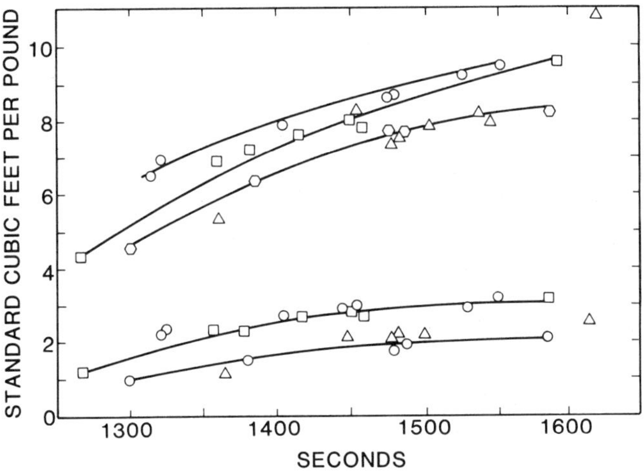

Fig. 3. Variation in yields with severity factor for whole shale oils. ○, NTU (gravity, 24.3° API); △, TOSCO II (gravity, 19.8° API); □, TOSCO II (gravity, 22.0° API); ⬡, United States Bureau of Mines (gravity, 20.0° API). Upper curves, total gas; lower curves, ethylene. [From Shultz *et al.* (1955), Griswold *et al.* (1976), and Fritzler *et al.* (1977).]

the pyrolysis, the greater the yield of gas and pitch; ethylene production was maximum, about 2.1 scf/pound feed, at 2.5 s and 800°C. Both this and the relatively high conversion to pitch (about 30%) are similar to the processes occurring in the pyrolysis of Bunker C fuel oil, but the latter gives somewhat better ethylene yields. Similar propylene and 1,3-butadiene yields from heavy petroleum oils and crude unfiltered shale oil were observed on cracking in steam with coke-laden inert solids under like conditions (Wynne, 1977; McKinney *et al.*, 1978). Rapid cracking in a special reactor at 700–800°C gave more gas (62–67%) and more liquids from petroleum-derived feedstocks than from shale oil (52–55%) (Bezmozgin *et al.*, 1961).

The most detailed study of steam pyrolysis of whole shale oils was directed at maximizing light olefin production (Griswold *et al.*, 1976; Fritzler *et al.*, 1977). Feedstocks were mainly TOSCO II shale oil and a slightly lighter NTU simulated-*in-situ* oil, as described in Table IX. Operating variables were: temperatures, 730–890°C; residence time (in tubular reactor containing porcelain balls), 0.3–1.3 s; and steam/oil ratio, 0.4–1.6. Increased steam/oil ratios decreased coke formation but affected gas yields little. Total gaseous olefins yield (28–36 wt%) varied little with a pyrolysis severity factor, but as Fig. 3 shows, total gas yield increased materially, and ethylene yields increased because of the degradation of propylene and other compounds, and reached 20–22% (Fritzler *et al.*, 1977) or 24% (Griswold *et al.*, 1976).

As a variant on pyrolyzing shale oils, several groups have cracked the vapors from retorted oil shale, without condensing them, to produce either ethylene or benzene (Dow Chemical Co., 1959; Kyll *et al.*, 1960). It does not appear that this technique much alters the process.

The extensive use of petroleum naphthas as pyrolysis feedstocks for manufacturing light olefins and BTX has suggested similar use of shale oil naphthas, especially in the Soviet Union. The lightest fractions were found to give best yields of olefins and least coking (Aarna and Lillie, 1959). As would be expected, steam cracking of such cuts also reduces coking (Aarna and Lillie, 1960). Rather more recent studies on pyrolysis of shale gasoline at 700°C with space velocity 0.8 found 20% gas, 65% light oil from which high-grade benzene could be produced, 15% tar, and less than 1% coke (Semenov, 1970); or with heavier fractions, 35–37% gas (mostly methane and ethylene), 59–62% liquids, and 3% coke (Semenov *et al.*, 1968). At lower temperature (550°C), cracking over manganese oxide supported on alumina produced about 40% gas, which was 40% olefins as well as an aromatics-rich light oil (Abarenkova and Tarasenkova, 1961). The best yield of such olefins (52%) was obtained by fluidized-bed cracking at 695°C of shale oil fractions boiling up to 200°C, with steam injection and short residence times (Elenurm *et al.*, 1967). The most recent study, and the only American one on unhydrogenated shale oil fractions, used combined distillates (at 1 atm and in vacuum) comprising about 70% of the original shale oils (Smith *et al.*, 1978) (for characteristics see Table X). Steam cracking (water/oil ratios 0.26/1.12) was conducted in a tubular packed reactor at 760–910°C and 0.4–1.2 s residence time. The effects of changes of variables on yields was as usual: steam/oil ratio had no effect, whereas increased temperature and/or increased residence time caused a maximum in ethylene production, a decrease in other olefins, and an increase in hydrogen, methane, and coke formation. Maximum ethylene yield was 22–29% at 830–910°C (different values for different shale oils); maximum total olefin yields (33–38%) were at somewhat lower temperatures.

Because increasing the hydrogen/carbon ratio of a feedstock usually improves the yield of olefins in pyrolysis, a few studies of hydrogenated shale oil fractions have been conducted. Analysis of such fractions has shown that the lightest ones (b. <200°C) are 52% alkanes, 36% cycloalkanes, and 12% aromatics (Vogh *et al.*, 1977); even heavier fractions contain 39–40% alkanes, more than in petroleum or coal liquid fractions of comparable boiling range. This suggests that hydrogenated shale oil fractions should be at least as good as petroleum naphthas as feedstocks for producing ethylene, a suggestion borne out by experiment (Griswold *et al.*, 1979, 1980). Fractions of two different shale oils were hydrogenated at three levels of severity (for characteristics see Table X), and the products pyrolyzed in steam at 700–875°C and 0.4–1.4 s residence time, in the same way that shale oil vacuum distillates were. As before, a little steam (40 wt% of oil fed) was as good as much more, and the controlling variables were temperature and time. The product yields could not be well correlated with

TABLE X

Characteristics of Fractions of Shale Oil Used as Feedstocks[a]

Feedstock	Gravity (°API)	Boiling range (°C)	Carbon (wt%)	Hydrogen (wt%)	C/H Ratio
TOSCO II distillate	27.7	95% at 73–540, rest higher	84.7	11.8	
Simulated-*in situ*	29.5	95% at 115–540, rest higher	84.4	12.2	
TOSCO II distillate					
Mildly hydrogenated	30.4		84.7	12.4	6.8
Moderately hydrogenated	33.8		86.0	12.9	6.7
Severely hydrogenated	37.2		88.0[b]	13.9[b]	6.3
Simulated-*in situ* distillate					
Mildly hydrogenated	31.1		84.2	12.0	7.0
Moderately hydrogenated	34.3		85.5	13.1	6.5
Severely hydrogenated	38.4		86.3[b]	15.0[b]	5.7

[a] Smith *et al.* (1978) and Griswold *et al.* (1979, 1980).
[b] As in original; total exceeds 100%.

a severity factor. However, except from one severely hydrogenated oil, which gasified unduly, the yield of ethylene increased with degree of hydrogenation of the feed. Moderately hydrogenated oils gave up to 35% ethylene and 55% total gaseous olefins, at about 820°C and 0.4 s residence time. These yields were better than those from comparable petroleum fractions (Griswold *et al.*, 1979, 1980). However, naphtha from a Paraho shale oil, and a light gas oil thence, with or without preliminary hydrogenation on roughly similar steam cracking but with longer (2.6–4.7 s) contact time, gave only 13–18% ethylene or 18–32% total gaseous olefins (Rudershausen and Thompson, 1978). In a continuation and extension of this study (Rudershausen and Thompson, 1979), the Paraho oil was fractionated to yield 4% naphtha (b. 117–205°C), 25.6% light gas oil (b. 218–336°C), and 35.9% heavy gas oil (b. 350–492°C). Each was lightly hydrotreated and then steam-cracked at about 1 atm and 744–802°C for 0.3 s. The hydrogenated naphtha gave the highest yield of gaseous olefins, 40% (21% to ethylene); the light gas oil, hydrogenated or not, gave about 16% ethylene and 35% olefins. The heavy gas oil was more improved as a feedstock by hydrotreating, but even so was inferior, producing 15% ethylene and 30% olefins, and coking badly. There was no measurement of BTX production, but this was obviously considerable. Both studies of Paraho oil, unlike those just cited (Smith *et al.*, 1978; Griswold *et al.*, 1979, 1980), led to the conclusion that shale oil is not as good a source of feedstocks for manufacturing olefins as petroleum fractions are.

The considerable literature on production of a crude oil by mining and processing tar sands, especially the extensive deposits in Athabaska, is not reviewed here. Pyrolysis of whole sands or of extracted asphaltenes and malthenes thence gave C_1–C_7 hydrocarbons along with some oxygen and sulfur compounds (Jha *et al.*, 1978), and that of some fractions gave *n*-alkanes and 1-alkenes up to C_{34} (Ritchie *et al.*, 1979); that is, the distillates are relatively paraffinic. The oils derived by coking the tarry material obtained from tar sands also contain aliphatic compounds. Thus they or fractions thereof doubtless could serve as feedstocks for light olefins manufacture in the same way as crude petroleum and its distillates do; the higher content of sulfur and nitrogen compounds in the sand-derived oils would not be a major problem. This deserves study.

V. POLYOLEFINS

Pyrolysis of polyolefins may be applied to either virgin or scrap polymers, and to either synthetic or natural ones. An example of pyrolyzing synthetic virgin polymers is the steam cracking of Fischer–Tropsch alkanes to make ethylene; this is easy chemically but out of the question economically (Kniel *et al.*, 1980; Dressel, 1977), as might be surmised. Much attention has been devoted to pyrolysis of polyethylene, but the bulk of the work has been concerned with decomposition (degradation) at temperatures too low to produce much ethylene, or else in a hydrogen atmosphere that produced chiefly alkanes. However, cracking polyethylene in hydrogen over a hydrogenation catalyst has given 75% yields of a gas that was 70% gaseous olefins (Yamamoto, 1978). Pyrolysis at 500°C in SO_2 plus a little air gave (in milliliters of gas) CH_4(60), C_2H_6(24), C_3H_8(8), C_4H_{10}(2), C_2H_4(46), C_3H_6(27), C_4H_8(11), CO_2(29), and CO(11) (Kominami and Tamura, 1973). Decomposition in a fluidized bed in nitrogen at 840°C converted polyethylene to ethylene in 44% yield, and polypropylene at 740°C to propylene (35%); at these higher temperatures some aromatization occurred (Menzel, 1974). Another Japanese study of pyrolysis of polyethylene and polypropylene gave up to 80% of gas containing 60–80% olefins, and 10–40% of olefinic oils boiling up to 200°C (Inomata *et al.*, 1976).

A study of the effects of residence time and temperature on the fixed-bed pyrolysis (in steam) of polyolefins gave the results shown in Table XI (Sawaguchi *et al.*, 1977). The yield from polyisobutylene was particularly high, 64%. More severe treatment of polyethylene and polypropylene in a fluidized sand bed gave the results shown in Table XII (Kaminsky and Sinn, 1978).

In comparison, polystyrene gives higher yields of monomer than any other polymer mentioned here, especially under reduced pressure, but the process is not used commercially.

A waste obviously available in substantial quantity over the world is scrapped rubber tires. Their pyrolysis gave more oil, and less residue and gas, at 500°C than at 900°C (Beckman and Laman, 1970), but no reference was made to olefin and BTX production. The higher temperature is to be preferred for BTX formation, which is almost seven times that from coal

TABLE XI

Products of Fixed-Bed Pyrolysis of Polyolefins[a]

Polyolefin	Residence time (s)	Temperature (°C)	Dilution ratio (steam/polymer)	Yield (%)
Polyethylene	3.2–3.4	700	2.5–2.7	C_2H_4(32)
Polypropylene	2.7	600	3.1	C_3H_6(26)
Polyisobutylene	3.6	600	2.1	iso-C_4H_8(64)

[a] Sawaguchi *et al.* (1977).

TABLE XII

Pyrolysis Products (Yields, %) from Polyethylene (PE) and Polypropylene (PP)[a]

	Laboratory apparatus		Miniplant	
Product	PE, 830°C	PP, 740°C	PE, 810°C	PP, 760°C
H_2	1.0	0.5	1.19	0.74
CH_4	23.7	17.0	18.8	29.3
C_2H_6	3.8	4.4	6.16	3.60
C_2H_4	27.6	13.9	16.92	13.62
C_3H_6	3.8	13.7	7.16	2.11
Other aliphatics	8.1	27.0	4.87	3.30
C_6H_6	16.2	10.4	21.75	18.79
Toluene	4.0	5.7	3.85	6.45
C_8 aromatics	0.6	1.8	0.17	0.40
Other aromatics	8.7	2.4	6.83	9.27
Other products	3.5	3.3	11.06	10.10

[a] Kaminsky and Sinn (1978).

pyrolysis (Karwat, 1978). Application of the TOSCO II process, with heat supplied by ceramic balls, to pyrolyzing shredded tires at 200–500°C gave carbon black of good quality and value, about 10 wt% of gas, mostly CO_2, CH_4, C_2H_6, C_3H_6, and C_4H_8, and three-fourths barrel of a heavy oil, about like No. 6 fuel oil, per ton of scrap (Schulman and White, 1978).

VI. BIOMASS

In the pyrolysis of wood to make charcoal, methanol (hence the old name wood alcohol) and acetic acid were at one time recovered as by-products, but their purification was difficult, and the process has long been obsolete. However, the utilization of waste cellulosic materials, or even wood grown for the purpose, to produce synthetic fuels has had considerable attention lately. The gases and liquids obtained by pyrolysis of such materials normally contain insignificant proportions of olefins and BTX. The one exception among the latter products is a patented pyrolysis of cellulosic wastes on a moving bed of molten lead at 760°C, which gave 14% benzene and toluene (Fassell, 1977).

Municipal solid waste or dried sewage sludge flash-pyrolyzed at 870°C by the Occidental process gave about 10% oil, 55–70% gases, and the rest char; but the gases were only 8–19% ethylene (Cone, 1978). At lower temperatures the process yields more oil, and this itself rapidly cracked at 750–850°C gave 4–11% ethylene (Lawler, 1976; Huffman and Liberick, 1978). Other examples: fluidized-bed pyrolysis of municipal waste gave gas that contained only 7% ethylene (Bailie and Richmond, 1978); cattle manure pyrolyzed with steam and a little air gave synthesis gas but only 1–3% ethylene by weight (Huffman *et al.*, 1977, 1978); and oak sawdust similarly treated gave 4% ethylene (Beck and Wang, 1980). Although it has been claimed (Antal, 1979a,b) that rapid gasification of biomass in steam can give "commercially interesting yields of ethylene, propylene, and methane," details are lacking, and the prospects of such a process are certainly poor. A fundamental problem in converting whole biomass to hydrocarbons is the highly oxygenated nature of carbohydrates; as already pointed out, more than 50% of cellulose is oxygen. Thus pyrolysis at best yields liquids that are themselves oxygen-rich and poor feedstocks from which to generate lower olefins.

Obviously, severe hydrogenation somewhere along the line would increase hydrocarbon yields. Cellulose is not easy to hydrogenolyze, and such a reaction does not proceed at all without a catalyst. Heating a 20–30% slurry of cellulose in mineral oil containing 0.2% nickel at 460°C and about 100 atm for 10–50 min converted only 75% of the carbon and produced only 45% of a hydrocarbon oil, some of it derived from the mineral oil (Kaufman *et al.*, 1974; Gupta *et al.*, 1976). Iodine catalyst also produces hydrocarbons in moderate yields (Geerards *et al.*, 1958; Grover *et al.*, 1972), but is totally out of the question commercially in spite of the fact that its use for hydrogenating coal tar has been patented (Friedman and Eddinger, 1969). The United States Bureau of Mines process of liquefying biomass with carbon monoxide coal tar has also been patented (Friedman and Eddinger, 1969). The Bureau of Mines process of liquefying biomass with carbon monoxide and water (Appell *et al.*, 1975) forms oils containing somewhat less oxygen (8–20%) than those from pyrolysis, but it is doubtful that they would be good for steam cracking to make ethylene. Apparently there has been no study of hydrogenated pyrolysis oils or hydrogenated Bureau of Mines oils as ethylene feedstocks, probably because the economic prospect is so unfavorable.

Pyrolysis of plant or animal products rich in hydrocarbon chains or rings (e.g., fats, fatty oils, and terpenes) must in principle give more gaseous olefins than cellulose does. This is ordinarily absurd economically, but the pyrolysis of sesquiterpene-rich latex of *Euphorbia lathyris*, a desert plant, is regarded by Calvin (1979) as a viable route to petrochemicals (Anonymous, 1979). Industrial laboratory research has shown that light olefins and aromatics can indeed be formed by cracking plant oils and latexes (Sherwood, 1980; Haag, 1980), but the conversion is only of laboratory interest at present.

REFERENCES

Aarna, A. Y., and Lillie, Y. E. (1959). *Tr. Tallin. Politekh. Inst., Ser. A* No. 165, 65–74, 75–82 [*C.A.* **56,** 7573].

Aarna, A. Y., and Lillie, Y. E. (1960). *Khim. Tekhnol. Goryuch. Slantsev Prod. Ikh Pererab.* No. 9, 147–154 [*C.A.* **56,** 15722].

Abarenkova, E. A., and Tarasenkova, E. M. (1961). *Tr. Leningr. Inzh.-Ekon. Inst., Khim. Khim. Proizvoid.* No. 36, 65–75 [*C.A.* **56,** 2632].

Anonymous (1979). *Chem. Week* **125**(24), 53–54.

Antal, M. J. (1979a). *Am. Chem. Soc., 177th Meet., Abstr. Pap., Honolulu, Hawaii* PETR-11.

Antal, M. J. (1979b). *Am. Chem. Soc., Div. Pet. Chem., Prepr.* **24,** 445–455.

Appell, H. R., Fu, Y. C., Illig, E. G., Steffgen, F. W., and Miller, R. D. (1975). *U.S. Bur. Mines, Rep. Invest.* No. 8013.

Apter, D. M., and Kalinichenko, F. I. (1962). *Izv. Sib. Otd. Akad. Nauk SSSR* No. 2, 120–122 [*C.A.* **57,** 6230].

Asinger, F. (1968). "Mono-Olefins: Chemistry and Technology" (B. J. Hassard, transl.). Pergamon, Oxford.

Bailie, R., and Richmond, C. A. (1978). *In* "Solid Wastes and Residues: Conversion by Advanced Thermal Processes" (J. L. Jones and S. B. Radding, eds.). ACS Symposium Series, No. 76, pp. 21–43. Am. Chem. Soc., Washington, D.C.

Banerjee, N. N., Ghosh, B., and Nair, C. S. B. (1977). *Fuel* **56,** 192–194.

Beck, S. R., and Wang, M. J. (1980). *Ind. Eng. Chem., Process Des. Dev.* **19,** 312–317.

Beckman, J. A., and Laman, J. R. (1970). *Proc. Natl. Ind./Solid Wastes Manage. Conf., Houston, Tex.* pp. 388–392.

Bernhardt, R. S., Ladner, W. R., Newman, J. O. H., and Sag, P. W. (1980). *Am. Chem. Soc., Div. Fuel Chem., Prepr.* **25,**(1), 188.

Bezmozgin, E. S., Nemchenko, A. G., and Yudkevich, Y. D. (1961). *Tr. Vses. Nauchno-Issled. Inst. Pererab. Ispol'z. Topl.* No. 10, 49–58 [*C.A.* **57,** 3704].

Brantley, F. E., Cox, R. J., Sohns, H. W., Barnet, W. I., and Murphy, W. I. R. (1951). *Ind. Eng. Chem.* **44,** 2641–2647.

Calvin, M. (1979). *Science* **206,** 436–437.

Cier, H. E. (1969). *In* "Kirk–Othmer Encyclopedia of Chemical Technology" (A. Standen, H. F. Mark, J. J. McKetta, Jr., and D. F. Othmer, eds.), 2nd ed., Vol. 20, pp. 532–534. Wiley, New York.

Cone, E. J. (1978). *In* "Solid Wastes and Residues: Conversion by Advanced Thermal Processes" (J. L. Jones and S. B. Radding, eds.). ACS Symposium Series, No. 76, pp. 287–306.

Dow Chemical Co. (1959). Br. 816,537 [*C.A.* **53,** 22873].

Dressel, G. M. (1977). *In* "Synthetic Fuels Processing. Comparative Economics" (A. H. Pelovsky, ed.), pp. 319–331. Dekker, New York.

Elenurm, A., Laus, T., and Gubergrits, M. (1967). *Eesti NSV Tead. Akad. Toim., Keem., Geol.* **16**(2), 88–96 [*C.A.* **67,** 92510].

Engelhardt, A. (1938). *Brennst.-Chem.* **19,** 297–304.

Fassell, W. M. (1977). U.S. Patent 4,010,098 (Barber-Colman Co.).

French, K. H. V. (1957). *In* "Benzene and Its Industrial Derivatives" (E. G. Hancock, ed.), pp. 26–54. Wiley, New York.

Friedman, L. D., and Eddinger, R. T. (1969). U.S. Patent 3,453,202 (FMC Corp.).

Fritzler, E. A., Yesavage, V. F., and Dickson, P. F. (1977). *Proc. Pac. Chem. Eng. Conf., 2nd*, Denver, Colorado, pp. 542–549 [*C.A.* **89**, 200086].

Geerards, J. J. T. M., Kemkes, J. F., van Krevelen, D. W., Meulenbeld, F., and Waterman, H. I. (1958). *Erdoel Kohle* **11,** 11–12.

Granger, A. F., and Ladner, W. R. (1970). *Fuel* **49,** 17–25.

Griswold, C. F., Yesavage, V. F., and Dickson, P. F. (1976). *Am. Chem. Soc., Div. Fuel Chem., Prepr.* **21**(6), 207–218.

Griswold, C. F., Ballut, A., Kavianian, H. R., Dickson, P. F., and Yesavage, V. F. (1979). *Chem. Eng. Prog.* **75**(9), 78–80.

Griswold, C. F., Ballut, A., Kavianian, H. R., Dickson, P. F., and Yesavage, V. F. (1980). *Energy Commun.* **6,** 153–178.

Grover, R. R., Barbour, J. F., and Freed, V. H. (1972). *AiChE Symp. Ser.* **68,** 28–34.

Gupta, C. V., Kranich, W. L., and Weiss, A. H. (1976). *Ind. Eng. Chem., Process Des. Dev.* **15,** 256–260.

Haag, W. O. (1980). *Chem. Eng. News* **58**(37), 43.

Harris, G. A., Sinnett, C. E., and Swift, H. E. (1980). *Hydrocarbon Process.* **59**(5), 152–126.

Hesp, W. R., and Waters, P. L. (1970). *Ind. Eng. Chem., Prod. Res. Dev.* **9,** 194–202.

Hesp, W. R., Hamor, R., and Waters, P. L. (1969). *CSIRO, Div. Miner. Chem., Invest. Rep.* No. 77.

Huffman, G. L., and Liberick, W. W. (1978). *In* "Solid Wastes and Residues: Conversion by Advanced Thermal Processes" (J. L. Jones and S. B. Radding, eds.), ACS Symposium Series, No. 76, pp. 349–355.

Huffman, W. J., Halligan, J. E., and Peterson, R. L. (1977). **CONF-770142-2** [*Energy Res. Abstr.* **4,** 22049].

Huffman, W. J., Halligan, J. E., and Peterson, R. L. (1978). **PB-280189** [*Energy Res. Abstr.* **4,** 18258].

Imai, K., Suzuki, M., and Nomura, S. (1957). *Nenryo Kyokai-Shi* **36**, 526–534 [*C.A.* **52**, 1591].

Inomata, O., Ando, H., Abe, T., Matsuzawa, S., Shimizu, Y., and Miyazaki, S. (1976). *Nenryo Kyokai-Shi* **55,** 331–342 [*C.A.* **86,** 60196].

Jha, K. N., Montgomery, D. S., and Strausz, O. P. (1978). *Oil Sand Oil Shale Chem., Proc. Symp., 1977* pp. 33–54.

Kaminsky, W., and Sinn, H. (1978). *Erdoel Kohle, Erdgas, Petrochem. Brennst.-Chem.* **31,** 199.

Karwat, E. (1978). *Erdoel Kohle, Erdgas Petrochem. Brennst.-Chem.* **31,** 568–571.

Kashirskii, V. G. (1974). *Khim. Tverd. Topl.* No. 5, 65–68 [*C.A.* **82,** 75386].

Kaufman, J. A., Gupta, C. V., Statkowski, T. S., and Weiss, A. H. (1974). *Chem.-Ing.-Tech.* **46,** 609.

Kishida, M., and Chuto, M. (1952). *Koru Taru* **4,** 398–402; [*C.A.* **47,** 6120].

Kniel, L., Winter, O., and Tsai, C.-H. (1980). *In* "Kirk–Othmer Encyclopedia of Chemical Technology" (M. Grayson and D. Eckroth, eds.), 3rd ed., Vol. 9, pp. 400–413. Wiley, New York.

Kominami, N., and Tamura, N. (1973). Jpn. Patent 73, 33,721 (Asahi Chem. Ind. Co., Ltd.) [*C.A.* **81,** 4345].

Korosi, A., Woebcke, H. N., and Virk, P. S. (1976). *Am. Chem. Soc., Div. Fuel Chem., Prepr.* **21**(6), 190–197.

Krekeler, H. (1955). *Proc. World Pet. Congr., 4th*, Rome, Italy, **3,** 381–390.

Krishnamurthy, S., and Shah, Y. T. (1979). *Ind. Eng. Chem., Process Des. Dev.* **18,** 466–474.

Krishnamurthy, S., Shah, Y. T., and Stiegel, G. J. (1979). *Ind. Eng. Chem., Process Des. Dev* **18,** 254–261.

Krishnamurthy, S., Shah, Y. T., and Stiegel, G. J. (1980). *Fuel* **59,** 738–746.

Kubicka, R., Kvapil, Z., and Sykora, M. (1962). *Chem. Prum.* **12,** 598–601 [*C.A.* **58,** 5420].

Kyll, A. T., and Stepanov, I. I. (1959). *Goryuch. Slantsy, Khim. Tekhnol.* No. 3, 176–183 [*C.A.* **55,** 944].

Kyll, A. T., Laus, T., and Elenurm, A. A. (1960), *Eesti NSV Tead. Akad. Toim., Fuus.-Mat. Tehnikatead. Seer.* **9**(2), 105–112 [*C.A.* **55,** 13821].

Lawler, S. P. (1976). *In* "North Eastern Regional Antipollution Conference" (S. M. Barnett, D. Sussman, and C. J. Wilson, eds.), pp. 307–314. Technomic, Westport, Connecticut.

McKinney, J. D., Sebulsky, R. T., and Wynne, F. E., Jr. (1978). U.S. Patent 4,080,285 (Gulf Res. Dev. Co.).

McNeil, D. (1969). *In* "Kirk–Othmer Encyclopedia of Chemical Technology" (A. Standen, H. F. Marks, J. J. McKetta, Jr., and D. F. Othmer, eds.), 2nd ed., Vol. 19, pp. 653–682. Wiley, New York.

Mainhood, J. S. R., and Pitt, G. J. (1966). Br. Patent 1,027,331 (Coal Ind. Patents, Ltd.) [*C.A.* **65,** 5269].

Marizewska, K., and Musierowicz, J. (1960). *Przem. Chem.* **39,** 97–104 [*C.A.* **55,** 20406].

Menzel, J. (1974). *Chem.-Ing.-Tech.* **46,** 607.

Montgomery, R. D., and Decker, D. L. (1964a). *Hydrocarbon Process. Pet. Refin.* **43**(12), 101–105.

Montgomery, R. S., and Decker, D. L. (1964b). *Am. Chem. Soc., Div. Fuel Chem., Prepr.* **8**(3), 25–35.

Montgomery, R. S., Decker, D. L., and Mackey, J. C. (1959). *Ind. Eng. Chem.* **51,** 1293–1296.

National Coal Board (1979). *Comm. Eur. Communities* [*Rep.*] *EUR* **EUR-6453** [*C.A.* **93,** 50187].

Naugle, B. W., Ortuglio, C., Mafrica, L., and Wolfson, D. E. (1965). *U.S. Bur. Mines, Rep. Invest.* No. 6625.

Niclause, M., Baronnet, F., Scacchi, G., Muller, J., and Jezequel, J. Y. (1976). *In* "Industrial and Laboratory Pyrolyses" (L. F. Albright and B. L. Crynes, eds.), ACS Symposium Series, No. 32, pp. 17–34. Am. Chem. Soc., Washington, D.C.

Peters, B. C. (1975). "Chemicals From Coal," Q. Tech. Prog. Rep., FE-1534-T-5, Dow Chemical Co. ERDA, Oak Ridge, Tennessee.

Purcell, W. P. (1978). *In* "Kirk–Othmer Encyclopedia of Chemical Technology" (M. Grayson and D. Eckroth, eds.), 3rd ed., Vol. 3, pp. 751–756. Wiley, New York.

Qader, S. A. (1976). *AIChE Natl. Meet. 82nd, Atlantic City, N.J.*

Rao, K. V., and Skeist, I. (1976). *Oil Gas J.* **74**(5), 90–93.

Rao, M. J. (1979). *Erdoel Kohle, Erdgas, Petrochem. Brennst.-Chem.* **32**(3), 127–131.

Rao, M. J., Ramacharyulu, M., and Krishna, M. G. (1968). *Brennst.-Chem.* **49,** 6–10.

Rice, R. L., Fortney, D. R., and Berber, J. S. (1970). *Ind. Eng. Chem., Prod. Res. Dev.* **9,** 30–33.

Ritchie, R. G. S., Roche, R. S., and Steedman, W. (1979). *Chem. Ind. (London)* pp. 25–26.

Rudershausen, C. G., and Thompson, J. B. (1978). *Am. Chem. Soc., Div. Pet. Chem., Prepr.* **23**(1), 241–252.

Rudershausen, C. G., and Thompson, J. B. (1979). *In* "Thermal Hydrocarbon Chemistry" (A. G. Oblad, H. G. Davis, and R. T. Eddinger, eds.), Advances in Chemistry Series, No. 183, pp. 91–105. Am. Chem. Soc., Washington, D.C.

Sawaguchi, T., Kuroki, T., and Ikemura, T. (1977). *Bull. Jpn. Pet. Inst.* **19,** 124–130 [*C.A.* **88,** 105970].

Schachter, Y. (1979). *Isr. J. Technol.* **17**(1), 51–57.

Schowalter, K. A., and Boodman, N. S. (1974). *Chem. Eng. Prog.* **70**(6), 76–82.

Schulman, B. L., and White, P. A. (1978). *In* "Solid Wastes and Residues: Conversion by Advanced Thermal Processes (J. L. Jones and S. B. Radding, eds.) ACS Symposium Series 76, pp. 274–285. Am. Chem. Soc., Washington, D.C.

Semenov, S. S. (1970). *Razrab. Ispol'z. Zapasov Goryuch. Slantsev* pp. 520–523, 557 [*C.A.* **76,** 143129].

Semenov, S. S., Kobyl'skaya, M. V., Pyshkina, N. I., and Solov'ev, Y. A. (1968). *Tr. Nauchno-Issled. Inst. Dobyche Pererab. Slantsev,* 203–211 [*C.A.* **71,** 103753].

Sherwood, M. (1980). *Chem. Ind. (London)* pp. 759–760.

Shultz, E. B., Jr., Guyer, J. J., and Linden, H. R. (1955). *Ind. Eng. Chem.* **47,** 2479–2482.

Smith, P. D., Dickson, P. F., and Yesavage, V. F. (1978). *Am. Chem. Soc., Div. Pet. Chem., Prepr.* **23**(2), 756–764.

Sohns, H. W., Jukkola, E. D., and Murphy, W. I. R. (n.d.). *U.S. Bur. Mines, Rep. Invest.* No. 5522.

Squires, A. M. (1976). *Science* **191,** 689–694, 699.

Sredin, V. V. (1960). *Khim. Tekhnol. Goryuch. Slantsev Prod. Ikh Pererab. Sbornik*, pp. 51–74 [*C.A.* **56,** 8990].

Taniewski, M., and Krajewski, J. (1978). *Przem. Chem.* **57,** 121–123 [*C.A.* **88,** 194259; *Eng. Index* p. 012778 (1978)].

Thorne, H. M. (1956). *Pet. Refin.* **35**(7), 155–160.

Tolubinskii, V. I., Shkuratov, I. Y., Govorova, R. P., and Klimenko, Y. G. (1962). *Zb. Pr., Inst. Teploenerg., Akad. Nauk Ukr. RSR* No. 25, 3–8 [*C.A.* **58,** 6617].

Vogh, J. W., Holmes, J. A., Sturm, G. P., Jr., Woodward, P. W., and Dooley, J. E. (1977). *U.S. Dep. Energy, Rep.* **BERC/RI-77/17**.

West, H. J. (1970). *Ind. Eng. Chem.* **62,** 28–33.

Wynne, F. E., Jr. (1977). U.S. Patent 4,057,490 (Gulf Res. Dev. Co.).

Yamamoto, M. (1978). *Nippon Kagaku Kaishi* pp. 1547–1551 [*C.A.* **90,** 24045].

9

Fundamental Aspects of the Formation and Gasification of Coke

DAVID L. TRIMM

I. INTRODUCTION

The high-temperature pyrolysis of hydrocarbons is a process of considerable theoretical and industrial importance. The major reactions occur in the gas phase and are free radical in nature. The formation of carbonaceous material is a minor reaction that is of major importance in that it leads to process inefficiencies. Coke formation in industrial systems is dealt with elsewhere in this book: this chapter is concerned with fundamental aspects of the formation and gasification of coke in pyrolysis systems. It will be seen that understanding coke formation requires some appreciation

Pyrolysis: Theory and Industrial Practice

ISBN 0-12-048880-9

of the fundamentals of pyrolysis. Important aspects are considered here and full details are given in other chapters.

The formation of carbonaceous material under pyrolysis conditions is a complex process that involves different types of tars and carbons. The overall deposit (coke) may contain high-boiling-point polycyclic aromatic compounds (tars) and ordered or disordered carbon originating in the gas phase (gas-phase carbon) or on a surface (surface carbon). Formation of surface carbon may be promoted by the nature of the material in the surface (catalytic carbon).

Tars generally consist of high-boiling polycyclic aromatics that can be liquids at high temperatures and solids at low temperatures (La Cava, 1977). Some constituents tentatively identified by mass spectrometry are listed in Table 1 and typical formulas are given in Fig. 1. One important comment

TABLE I

Molecules Tentatively Identified as Components of a Pyrolysis Tar Using Mass Spectrometry

m/e	Possible compound	Number of rings	*m/e*	Possible compound	Number of rings
128	Naphthalene	2	302	Dibenzopyrene	6
142	Methylnaphthalene	2	326	Benzanthanthrene	7
152	Diphenylene	2	328	Dibenzochrysene	6
154	Diphenyl	2	340	Dibenzonaphthanthrene	7
166	Fluorene	2	350	Benzocoronene	8
168	Methyldiphenyl	2	352	Tribenzpyrene	7
178	Phenanthrene, anthracene	3	354	Dihydrodibenzopentacene	7
180	Methylfluorene	2	364	Methylbenzcoronene	8
182	C_2-Diphenyl	2	366	Methyltribenzpyrene	7
192	Methylphenanthrene	3	374	Dibenzoperopyrene	9
194	C_2-Fluorene	2	376	Dibenzanthranthrene	8
202	Pyrene	4	378	Tribenzochrysene	7
204	Dihydropyrene	4	390	Methyldibenzanthranthrene	8
216	Methylpyrene	4	400	Dibenzocoronene	9
218	Methyldihydropyrene	4	402	Tetrabenzopyrene	8
226	Benzofluoranthrene	4/5	404	C_2-Dibenzanthanthrene	8
228	Chrysene	4	426	Tribenzanthranthrene	9
230	*p*-Terphenyl	3	428	Octacene	8
240	Naphthanthrene	5	440	Methyltribenzanthranthrene	9
252	Benzopyrene	5	442	Methyloctacene	8
254	Methylnaphthanthrene	5	450	Tribenzocoronene	10
266	Naphthoperinaphthene	5	452	Pentabenzopyrene	9
276	Anthanthrene	6	478	Nonacene	9
278	Benzochrysene	5	528	Hexabenzotriphenylene	10
290	Benzonaphthanthrene	6	578	Undecacene	11
300	Coronene	7			

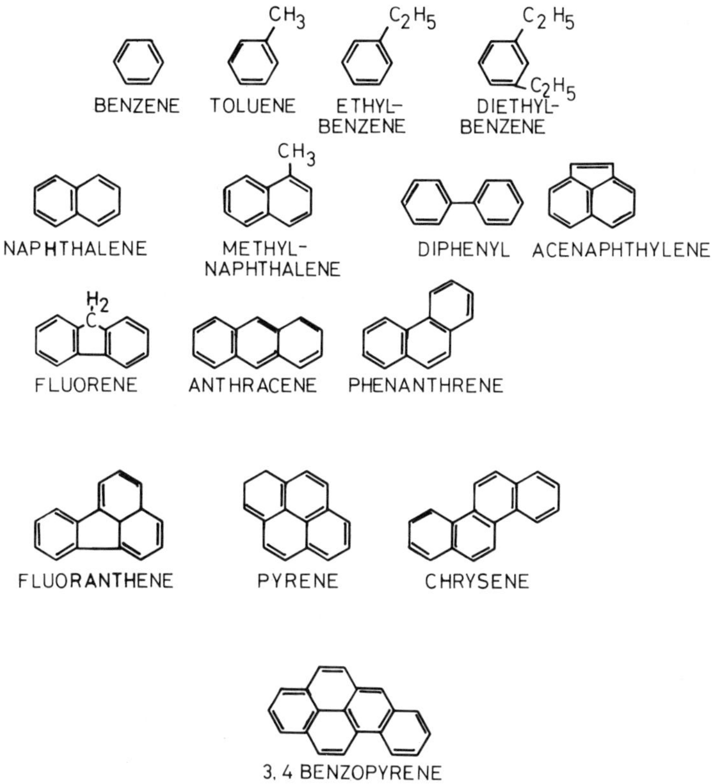

Fig. 1. Typical aromatics from hydrocarbon pyrolysis.

should be made. These materials are typical of those produced from pyrolysis of any hydrocarbon. *They contain carcinogenic compounds and should be handled with great care.*

Carbons can present a number of different structures ranging from nearly amorphous to highly crystalline. Gas-phase carbon (soot) is produced in spherical particles formed by small crystallites, which tend to link together in chainlike structures. This carbon has a structure apparently independent of the parent hydrocarbon (Conroy *et al.*, 1957). Surface carbons, on the other hand, have bigger crystallites, lower interlayer spacings, and higher densities. They are obtained as films with well-ordered crystallites deposited on solid substrates.

In heterogeneous pyrolysis, the surface process appears to be the controlling factor in determining the crystallinity of the carbon. On some metallic substrates, highly crystalline deposits are formed at low temperatures. Three-dimensional order, together with considerable crystallite size, have been observed in many of these deposits. Similar degrees of order could only be obtained on other surfaces if the temperature were substantially raised, and these substrates are "catalytic" as compared to "inert" surfaces, such as silica or porcelain (Banarjee *et al.*, 1961). Nickel, cobalt, and iron are highly effective substrates for catalytic carbon formation and have been much investigated, as will be discussed later.

Catalytic carbon deposits can be characterized according to their appearance and microstructure into three categories: laminar graphite, nonoriented carbon, and fibrous carbon. Laminar graphite has a structure close to that of ideal graphite, oriented with its basal plane almost parallel to the deposition surface (Presland and Walker, 1969; Robertson, 1969). Films of this carbon have been described as featureless, except for a polygonal network of ridges that covers the whole surface. The ridges are believed to have been formed by compressive stresses set up in the film on cooling (Presland and Walker, 1969; Derbyshire and Trimm, 1975). Laminar films can occur with low hydrocarbon pressure and high substrate temperature (Banarjee *et al.*, 1961; Derbyshire and Trimm, 1975).

Nonoriented carbon, also described as "polycrystalline carbon," is formed of small crystallites lacking orientation with the substrate. It is produced under conditions favoring high supersaturation of carbon in the metal (Derbyshire and Trimm, 1975). Laminar graphite is usually also present in the deposits, and these two types apparently are interchangeable.

Carbon fibers are produced on metals. They are usually long and thin (up to 7 μm long and 0.1 μm wide) and have a hollow core and a metal particle at the tip (Baker *et al.*, 1972a; Rostrup-Nielsen and Trimm, 1977; Trimm, 1977); the particle size correlates well with the fiber diameter (Baker *et al.*, 1973; Rostrup-Nielsen, 1972). Metal particles are also present along the fiber length (Baird *et al.*, 1972; Baker *et al.*, 1975b). The carbon

TABLE II

Typical Properties of Pyrolytic Carbons[a]

Type of carbon	Typical temperature of formation (°C)	d spacing (Å)	L_c[b] (Å)	L_a[b] (Å)	C/H Atomic ratio
Gas phase	1000	3.61–3.70	12.2	42	8
Surface	1000	3.46–3.54	28	51	>80
Catalytic (laminar)	600	3.37	200	400	20
Graphite	—	3.35	∞	∞	∞

[a] (Bernardo, 1977)
[b] L_c and L_a are characteristic dimensions of the crystal in the a and c axes.

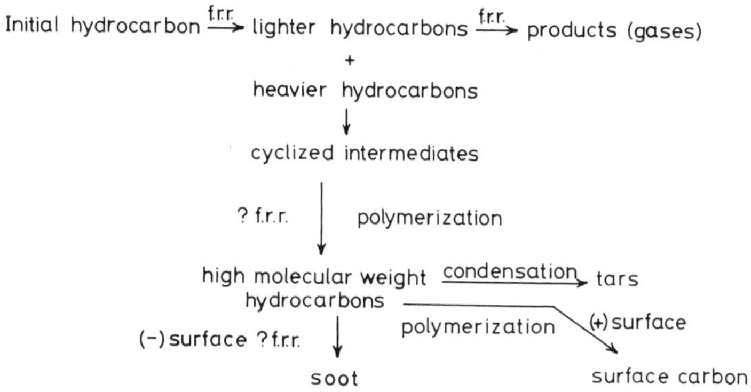

Fig. 2. Flow diagram for the production of coke. Any hydrocarbon can yield catalytic carbon in the presence of some materials. f.r.r., Free-radical reactions.

is mainly turbostratic with the basal planes preferentially oriented parallel to the fiber axis (R. Baird *et al.*, 1972; T. Baird *et al.*, 1971).

Some properties of the various carbons are presented in Table II.

Even from this brief description of the different compounds present in coke, it is obvious that the mechanisms of formation of the materials may be very different. This must be discussed in detail, but the general scheme for coke formation, as given in Fig. 2, is useful.

II. THE FORMATION OF GAS-PHASE COKE

Coke formation in the gas phase has received much attention over the years, and several excellent articles are available (Palmer and Cullis, 1975; Brooks and Taylor, 1968; Fitzer *et al.*, 1971; Trimm and Turner, 1981; Trimm, 1981). Some of the more important concepts are discussed in this chapter, but a more complete account of gas-phase reactions leading to carbon and coke can be found in these articles.

Free-radical reactions involved in pyrolysis are complex, as shown by a typical reaction sequence proposed for the pyrolysis of propane (Table III). There are two important factors to note about this sequence in the present context. The first of these is that the chain reaction involved in pyrolysis is very complex and there are many places at which significant interference with the chain might be possible. This could be as a result of chain-transfer reactions,

$$RH + H\cdot = R\cdot + H_2 \tag{1}$$

TABLE III

Chemical Reactions Proposed to Explain Propane Pyrolysis[a]

Initiation		
C_3H_8	$\longrightarrow$	$CH_3\cdot + C_2H_5\cdot$
Propagation reactions		
$CH_3\cdot + C_3H_8$	$\longrightarrow$	$C_3H_7\cdot + CH_4$
$C_2H_5\cdot + C_3H_8$	$\longrightarrow$	$C_3H_7\cdot + C_2H_6$
$C_2H_5\cdot$	$\longrightarrow$	$H\cdot + C_2H_4$
$CH_3\cdot + H_2$	$\longrightarrow$	$H\cdot + CH_4$
$H\cdot + CH_4$	$\longrightarrow$	$CH_3\cdot + H_2$
$C_2H_5\cdot + H_2$	$\longrightarrow$	$H\cdot + C_2H_6$
$C_2H_5\cdot + C_2H_4$	$\longrightarrow$	$CH_3\cdot + C_3H_6$
$H\cdot + C_2H_4$	$\longrightarrow$	$C_2H_5\cdot$
C_2H_6	$\longrightarrow$	$2\ CH_3\cdot$
$CH_3\cdot + C_2H_6$	$\longrightarrow$	$C_2H_5\cdot + CH_4$
$CH_3\cdot + C_2H_4$	$\longrightarrow$	$C_3H_7\cdot$
$C_3H_7\cdot + C_2H_6$	$\longrightarrow$	$C_2H_5\cdot + C_3H_8$
$H\cdot + C_2H_6$	$\longrightarrow$	$C_2H_5\cdot + H_2$
$H\cdot + C_3H_6$	$\longrightarrow$	$C_3H_7\cdot$
$C_3H_7\cdot$	$\longrightarrow$	$CH_3\cdot + C_2H_4$
$C_3H_7\cdot$	$\longrightarrow$	$H\cdot + C_3H_6$
$C_3H_7\cdot + H_2$	$\longrightarrow$	$H\cdot + C_3H_8$
$C_3H_7\cdot + CH_4$	$\longrightarrow$	$CH_3\cdot + C_3H_8$
$H\cdot + C_3H_8$	$\longrightarrow$	$C_3H_7\cdot + H_2$
$CH_3\cdot + C_3H_6$	$\longrightarrow$	$C_4H_9\cdot$
$C_2H_5\cdot + C_2H_4$	$\longrightarrow$	$C_4H_9\cdot$
$C_4H_9\cdot$	$\longrightarrow$	$H\cdot + C_4H_8$
$C_4H_9\cdot + H_2$	$\longrightarrow$	$H\cdot + C_4H_{10}$
$C_4H_9\cdot$	$\longrightarrow$	$CH_3\cdot + C_3H_6$
$C_4H_9\cdot$	$\longrightarrow$	$C_2H_5\cdot + C_2H_4$
$C_2H_5\cdot + C_4H_{10}$	$\longrightarrow$	$C_4H_9\cdot + C_2H_6$
C_4H_{10}	$\longrightarrow$	$CH_3\cdot + C_3H_7\cdot$
C_4H_{10}	$\longrightarrow$	$2\ C_2H_5\cdot$
$H\cdot + C_4H_{10}$	$\longrightarrow$	$C_4H_9\cdot + H_2$
$CH_3\cdot + C_4H_{10}$	$\longrightarrow$	$C_4H_9\cdot + CH_4$
Termination reactions		
$2\ H\cdot$	$\longrightarrow$	H_2
$H\cdot + CH_3\cdot$	$\longrightarrow$	CH_4
$H\cdot + C_2H_5\cdot$	$\longrightarrow$	C_2H_6
$H\cdot + C_3H_7\cdot$	$\longrightarrow$	C_3H_8
$2\ CH_3\cdot$	$\longrightarrow$	C_2H_6
$CH_3\cdot + C_2H_5\cdot$	$\longrightarrow$	C_3H_8
$2\ C_2H_5\cdot$	$\longrightarrow$	$C_2H_4 + C_2H_6$
$2\ C_2H_5\cdot$	$\longrightarrow$	C_4H_{10}
$CH_3\cdot + C_3H_7\cdot$	$\longrightarrow$	$CH_4 + C_3H_6$

[a] (Trimm and Turner, 1981)

or as a result of chain termination, for example, in the presence of a third body,

$$2\,R\cdot + M = R{-}R + M \tag{2}$$

The importance of these possibilities will be brought out later.

The second factor to note is that, as complex as the reaction scheme is, C_4H_{10} is the highest-molecular-weight product suggested to be formed. In fact this is common practice, with authors tending to consider either the production of gases or the production of tars and carbons, and not to consider both together. This is somewhat surprising because there is some evidence that the nature of tars produced is independent of the nature of the original hydrocarbon (Conroy *et al.*, 1957). Certainly it is true that the chemical natures (but not the amounts) of tar components produced from different hydrocarbons are very similar (La Cava, 1977).

Tars and carbons involve high molecular weight compounds normally containing condensed aromatic nuclei (Table I). Because it is possible to obtain approximately the same products from low or high molecular weight feedstocks, it is obvious that some increase in molecular size must occur during pyrolysis. It is rewarding to consider briefly how this could occur, using the formation and subsequent reactions of aromatic molecules as a pertinent example.

Studies of the synthesis of aromatic during the thermal reactions of paraffins and olefins led to the suggestion that their production could be related to the concentration of olefins in the reactor (Thompson and Brooks, 1973). The easiest way to envisage the reaction is in terms of a Diels–Alder type of condensation involving butadiene (Fitzer *et al.*, 1971).

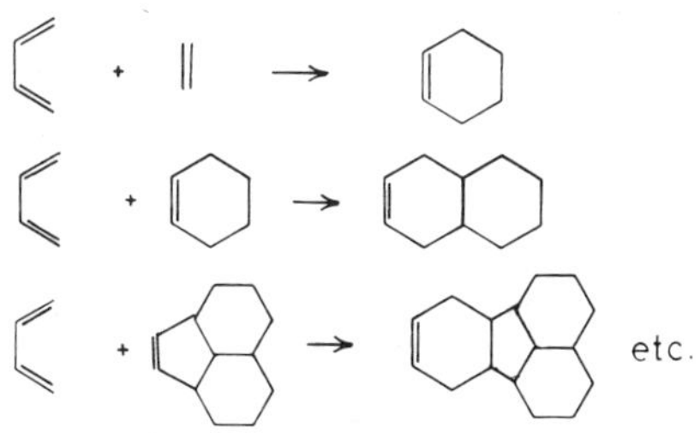

The scheme shown here is probably too simplistic a picture to account for the total production of aromatics, particularly because the reactions as written, involve molecular rather than radical species. Nonetheless, addition/polymerization reactions of this type can be expected to be involved in the formation of aromatics.

The subsequent reactions of aromatics may be complex. Degradation of the ring and polymerization to form polyphenyls are possible (Fitzer *et al.*,

1971). This latter reaction is favored by unsubstituted aromatics, and hydrocarbons with an anthracene configuration are most reactive (Badger, 1965). Alternatively, side-chain substituents on aromatic molecules may be involved in reaction; alkyl substitution increases reactivity in direct proportion to the number and length of the alkyl groups, the reaction products often being linked through these groups (Fitzer *et al.*, 1971).

The kinds of reactions that may occur can be illustrated in terms of polycylic aromatics that have been identified among the tarry products of pyrolysis (La Cava, 1977). These are listed in Table I and some typical formulas are given Fig. 1.

Thompson and Brooks (1973) and Virk *et al.* (1974) assumed that some polycyclic aromatics are formed via reactions of the type shown below.

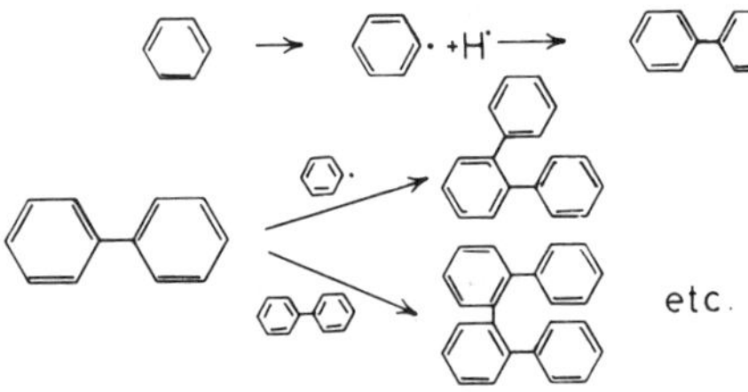

Alkyl-substituted aromatic compounds can also be expected to undergo free-radical reactions leading to ring closure.

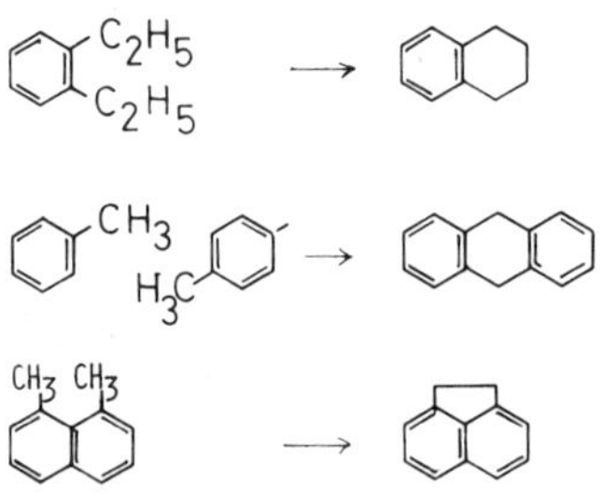

On the other hand, Diels–Alder type condensations lead to an increase in molecular size.

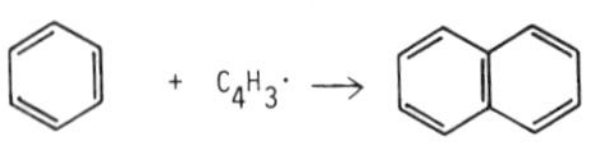

What is obvious from the preceding is that a sequence of free-radical reactions can take place that can lead to a wide variety of high molecular weight products. The size of the molecules so produced can be expected to be large enough that their vapor pressure is not high, even at high temperatures (La Cava, 1977). As a result, they may exist in a pyrolysis reactor as liquid drops.

We are now in a position to begin to turn our attention to the formation of coke in the gas phase. From the preceding it is obvious that this coke consists in part of tar materials and in part of carbon.

Several alternative models have been proposed for carbon formation in the gas phase (Palmer and Cullis, 1975; Brooks and Taylor, 1968; Fitzer *et al.*, 1971).

1. *The atomic C theory* suggests that monoatomic carbon may be a significant species in nucleation.
2. *The C_2 theory* suggests that solid carbon results from the polymerization of C_2 species.
3. *The C_3 theory* supports the idea that solid carbon in flames arises by condensation of carbon vapor, the major component of which is believed to be C_3.
4. *The acetylene theory*, first advanced by Berthelot in 1866, in essence is that carbon particles are formed from acetylene by simultaneous polymerization and dehydrogenation.
5. *Hydrocarbon polymerization theories* assume that fuels undergo a polymerization process that ends in carbon. Some assume that liquid polymers or polymers of high molecular weight are formed as intermediates. Some suspicion exists that polycyclic and polynuclear aromatics could be intermediates in the path toward carbon.
6. *The surface-decomposition theory*, of which Tesner (1973) is the principal advocate, suggests that carbon is formed through a surface chemical reaction of the original hydrocarbon on a carbon surface.
7. *The condensation theory* has been revived recently by Lahaye *et al.* (1974) in a model for formation of carbon black in a tubular reactor. According to their model, the initial hydrocarbon is transformed by a gas-phase reaction into macromolecules. The partial pressure of the macromolecules increases with reaction time until the supersaturation is high enough to induce condensation of the macromolecules into droplets. The formation of the liquid nuclei eliminates the supersaturation, and the formation of additional liquid nuclei becomes impossible. The macromolecules that continue to be formed maintain the nuclei growth. The liquid droplets are further pyrolyzed into a solid material.

The condensation theory has many attractions, not the least being the proven presence of high molecular weight polycyclic aromatic compounds of low vapor pressure at the temperatures concerned. There is some evidence, however, that condensation is important neither in the nucleation of gas-phase soot particles nor in their growth. Direct evidence against the formation of nuclei by condensation has been obtained by Graham *et al.* (1975). Formation of soot aerosols from benzene and other aromatics was studied in a shock tube equipped with instrumentation capable of measuring the concentration of aromatics, the amount of drops or particles (by light scattering), and the adsorption at 3.39 μm produced by the formation of solid carbon. The appearance of liquids as aerosols was found to be simultaneous with the production of solid carbon and, as a result, at least some of the carbon nuclei must originate from intermediates other than condensed aromatics.

The importance of condensation during particle growth has been examined by La Cava and Trimm (1978), who modeled mathematically results obtained by Murthy and Edge (1963) on the growth of carbon particles. The results show that a model based on a chemical reaction on the surface (Tesner, 1973) predicts much better experimental data than a model based on condensation.

The condensation theory has many attractions and it explains most experimental observations in soot formation and the effect of operation variables on the particles formed. Nonetheless, the evidence just reported shows that, at least in some cases, serious objections may be raised. The main quantitative support for the condensation theory comes from the work of Lahaye *et al.* (1974). If the theory is unacceptable, a correct theory must lead to similar mathematical models, and this would be true of the chemical nucleation mechanism proposed by Tesner (1973).

The intermediates that, through the yet unknown mechanism, finally form the carbon particles or deposits, are still the object of controversy. Experiments with carbon formation in flames tend to support the idea that acetylene is the intermediate that polymerizes, forming polyacetylenes that grow until particles form (Palmer and Cullis, 1975). Research on the pyrolysis and thermal hydrogenation of hydrocarbons supports the concept that aromatics and polycyclic aromatics are involved (Fitzer *et al.*, 1971; Palmer and Cullis, 1975; Virk *et al.*, 1974). This has been supported by a recent article dealing with soot formation from toluene. Wang *et al.* (1981) have used advanced technology to study the kinetics of the process, and they find that the aromaticity of the hydrocarbons is the most important factor in determining their importance as intermediates in the formation of soot. Thermochemical calculations indicated that $C_{96}H_{24}$ appears to be an important precursor, but no direct proof of this was obtained. However, the problems with this and every other study remain the same. Possible intermediates are very similar in chemical structure and difficult to identify, particularly because it is highly probable that gas-phase carbon is produced by several different mechanisms. As a result, the formation of this type of carbon is difficult to understand in detail and may be expected to offer opportunities for study for many years to come.

III. THE FORMATION OF COKE ON SURFACES: INTRODUCTION

With respect to the deposition of gaseous carbon on surfaces, very little has been reported that could give a quantitative description of the phenomena, and less to explain the mechanism. The treatment of Tesner (1973) of the deposition of carbon over particles assumes a surface reaction of the parent hydrocarbon, but neglects any effect of the homogeneous pyrolytic reactions. Other models, reviewed by Bokros (1969), correlate the conditions of deposition over particles in a fluid bed with the physical properties of the carbons, but do not try to explain the mechanism of deposition.

In some ways the formation of coke on surfaces is less complex than in the gas phase although, at first sight, this would seem not to be the case. In addition to the factors that can affect gas-phase coking, a noncatalytic surface can be expected to (*a*) affect mass and heat transfer in the system, (*b*) provide nucleation and growth sites, and (*c*) affect coking as a result of local effects induced by physical properties such as porosity.

A catalytic surface, in addition, may be expected to (*a*) affect the nature of reactants, intermediates, and products; (*b*) offer nucleation and growth sites on a catalyst; (*c*) influence the structure and compositions of coke; and (*d*) become involved in secondary reactions.

The role of surfaces in coking is perhaps best considered initially in connection with the formation of catalytic carbon: other aspects of coking can be discussed later.

IV. THE KINETICS AND MECHANISM OF CATALYTIC COKE FORMATION

Table IV indicates publications that recently considered deposition of carbon from hydrocarbons on catalytic metals (Bernardo, 1977). Some important conclusions on the kinetics and the mechanism of carbon formation can be drawn from this literature survey:

1. Long periods of steady-state deposition have been observed in a large number of different systems.
2. The carbon deposits contain metal particles, often at the tips of carbon fibers.

The mechanism of filament growth is discussed next, but is believed to involve deposition of carbon onto a metal, diffusion through metal, and precipitation underneath the metal. This results in the growth of a fiber

TABLE IV

Some Studies of Catalytic Carbon Formation[a]

Reference		Parent gas	Catalyst	Temperature (°K)	Comments
Cunningham and Gwathmey	(1957)	C_2H_4	Ni	720	Effect of crystal face
Karu and Beer	(1966)	CH_4	Ni	1175–1375	Graphite growth
Lafitau and Jacque	(1968)	CH_4	Ni	1270	C diffusion controls
Escoubes *et al.*	(1967)	CH_4, C_2H_2, C_2H_4, C_2H_6	Ni	460–580	C diffusion controls
Tamai *et al.*	(1968)	CH_4, C_2H_4, C_2H_6	Fe, Ni	1140–1300	Kinetic study
Tamai *et al.*	(1969)	CH_4, C_2H_4, C_2H_6	Fe, Ni	1170–1270	Effect of hydrogen mechanism
Presland and Walker	(1969)	C_2H_2	Pt, Ni	1270	Mechanism proposed
Robertson	(1969)	CH_4	Fe, Co, Ni	920–1020	Growth of graphite
Robertson	(1970)	CH_4	Fe, Co, Ni	920–1020	Flake and polycrystalline carbon; metal found in carbon
Tesner *et al.*	(1970)	C_2H_2	Ni–Cr	720–970	Carbon fibers; kinetics: metal in carbon
Moayeri	(1970)	CH_4	Ni, Co	1070–1370	Growth of graphite; Ni(111), Ni(110)
Blair and Presland	(1971)	C_2H_4	Ni	1070–1470	Effect of pressure and temperature on crystallinity
Lobo and Trimm	(1971)	C_2H_2	Ni	670–970	Complex temperature dependency; gas phase above 870°K
Baird *et al.*	(1971)	CH_4, C_3H_8	Ni	970	Structure of carbon fibers
Tomita	(1972)	Benzene	Ni	870–1240	Flake and sooty carbon; metal in carbon
Baker *et al.*	(1972b)	$CH_4/CO/CO_2$	Fe	725–1225	Mechanisms for amorphous C and graphite formation
Baker *et al.*	(1972a)	C_2H_2	Ni	870–1300	Mechanism for fibrous C formation; C diffusion in Ni
Rostrup-Nielson	(1972)	CH_4, CO	Ni	720–970	Equilibrium studies; correlation with crystallite size
Robertson	(1972)	CH_4	Fe, Co, Ni	920–1020	Factors affecting flake or polycrystalline C formation

Reference		Parent gas	Catalyst	Temperature (°K)	Comments
Lobo *et al.*	(1973)	C_2H_2, C_2H_4, C_3H_6, C_4H_6, C_4H_8	Fe, Co, Ni, Pd, Pt, Cu, Ag, Au	570–1070	Pd, Pt, Cu, Ag, Au, inactive; mechanism: C diffusion in Ni, Ni crystallite transported
Lobo and Trimm	(1973)	C_2H_2, C_2H_4, C_3H_6, C_4H_6, C_4H_8, C_3H_8, CH_4, C_2H_6	Ni	670–870	Autocatalytic deposition from olefins, very slow with paraffins
Baker *et al.*	(1973)	C_2H_2	Fe, Co, Cr	785–1270	Mechanism of fiber growth: C diffusion in metal
Fryer and Paal	(1973)	C_6H_{12}	Pt	630	C fibers of graphitic nature
Nishiyama and Tamai	(1974)	Benzene	Ni/Cu	850–1170	Diffusion of C in metal
Baird *et al.*	(1974)	CH_4, C_3H_8, C_3H_6, C_4H_6	Fe, Ni	720–970	Mechanism: surface diffusion of metal on carbon
Renaud and Bonnetain	(1974)	CH_4, CO	Ni	720–970	Correlation between particle size and excess free energy of formation of carbon
Derbyshire and Trimm	(1975)	CH_4, C_2H_4, C_2H_2, C_4H_6, benzene	Fe, Co, Ni	770–1370	Conditions of formation of laminar, nonoriented, and fibrous carbon and metal carbides
Toei *et al.*	(1975)	n-C_4H_{10}	Cr_2O_3/γ-Al_2O_3	820	Dual active sites: dehydrogenation on chromia coking on both chromia and alumina
Baker *et al.*	(1975b)	C_2H_2	Co	800–1150	Metal found in carbon fibers
Baker *et al.*	(1975a)	C_2H_4, C_6H_6 1,3-C_4H_6 C_3H_6	Co	1000	Ratio of area filament wall/area of filament, apparently correlates with the heat of formation of the hydrocarbon
Baker and Waite	(1975)	C_2H_2, C_2H_4	Pt–Fe	690–1175	Filaments extended from particles; mechanism for extrusion growth proposed
Nishiyama and Tamai	(1976b)	C_6H_6	Ni/Cu	840–970	Effect of hydrogen: removal of chemisorbed species and of surface carbon
Wentrcek *et al.*	(1976)	CO	Ni	300–723	Surface carbon from CO dissociation precursor to CH_4 formation
Keep *et al.*	(1977)	C_3H_8	Ni	970	Carbon filaments formed from intermediate unsaturated hydrocarbons

TABLE IV (*continued*)

Reference		Parent gas	Catalyst	Temperature (°K)	Comments
Rostrup-Nielsen and Trimm	(1977)	—	Ni	—	Review of the mechanisms of carbon formation on nickel
McCarty and Madix	(1977)	C_2H_4	Ni(110)	600	Dicarbon fragments formed on nickel surface
Trimm	(1977)	Various	Various	Various	Review
Buyanov *et al.*	(1977)	—	Fe–Cr	—	C formation via carbides
Walkov *et al.*	(1977)	C_6H_6, C_6H_{14}, C_6H_{12}	Pt, Pt–Sn	—	Effect of Sn on coke formation
McCarty *et al.*	(1977)	C_2H_2	Ni	—	Different forms of C
Dalmon and Martin	(1978)	CO, CO_2	Ni	—	Different forms of C
Ducros *et al.*	(1978)	C_2H_4, C_2H_2	Re	780	Coverage versus C
Hamilton and Blakeley	(1978)	—	Pt	Various	Bulk and surface C
Martin and Hudson	(1978)	—	Pt	Various	Surface diffusion C
Matsumoto and Bennett	(1978)	CO	Fe	580	C in Fisher–Tropsch
Chesnokov *et al.*	(1979)	—	Fe	—	Carbide intermediates
Low and Bell	(1979)	CO	Ru	300–400	Methanation intermediate
Sachtler *et al.*	(1979)	CO	Ru, Co	—	Methanation intermediate
McCarty and Wise	(1979)	—	Ni	550	Four types of surface C
Oya and Otani	(1979)	—	22 metals	2500–3250	Catalysis of graphitization
Moeller and Bartholomew	(1979)	CO	Ni, Ni molybdate	723	Kinetics: relative resistance to poisoning
Bonzel and Krebs	(1979)	CO	Fe	—	Carbides and intermediates
Nishiyama and Tamai	(1980)	C_2H_4, C_3H_6	TiC	—	Wall effects on soot
Brown *et al.*	(1980)	CO, CH_4	Fe, steel	893–1123	Morphology
Baird *et al.*	(1980)	C_4H_8	Fe, Ni	623–648	Kinetics graphite formation
Bertolini *et al.*	(1981)	CO	Ni single crystal	300–600	Effect carbides
Gardner and Bartholomew	(1981)	CO	Ni	473–573	Kinetics deposition and removal
Kneale and Ross	(1981)	C_2H_6	Ni	—	C as intermediate
De Deken *et al.*	(1981)	—	Ni	—	Morphology C
Bennett and Price	(1981)	Naphtha	Hk40	1193–1203	Steam cracking
Shohachi *et al.*	(1981)	Benzene	Fe	83–1473	Graphite

[a] Adapted from Bernardo, 1977.

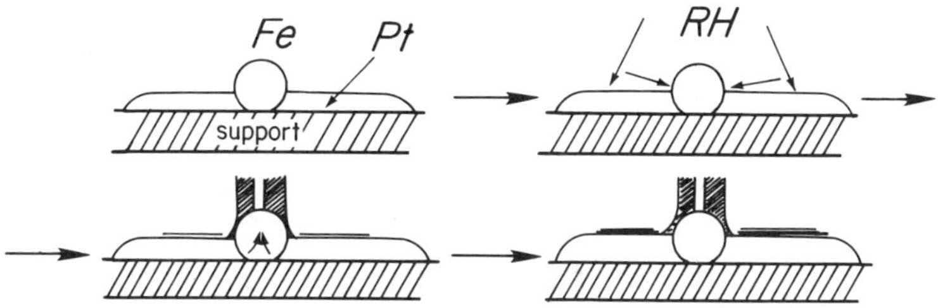

Fig. 3. Stages in the growth of carbon filaments for iron and platinum particles on a support. Fe, Iron; Pt, platinum; RH, hydrocarbon.

with a metal particle at the tip of the column. In the case of platinum–iron alloys a different mode of filament growth has been observed (Baker and Waite, 1975), involving extrusion of the filament from a catalyst particle that remains in contact with the support (see Fig. 3).

3. The dependence on temperature of the rate of deposition is complex, and a maximum between 500° and 650°C has been observed by many researchers. Zero-order kinetics and apparent activation energies independent of the parent hydrocarbon (but dependent on the metal) were observed at lower temperatures.
4. As the temperature increases, the surface reaction becomes the rate-determining step in some systems. The negative apparent activation energies observed over this temperature zone have been ascribed to either desorption effects or the formation of surface carbon films that deactivate the catalyst.
5. At even higher temperatures the coke-forming reaction has been identified as the gas-phase pyrolysis of the hydrocarbon.
6. Hydrogen can have a complex effect on coke formation. Catalytic carbon formation is usually accelerated by the presence of hydrogen.
7. The ease of formation of catalytic carbon is found to decrease in the order: acetylenes > olefins > paraffins.

Rostrup-Nielsen (1979) has reported coking tendencies for several hydrocarbons on nickel.

The mechanism for formation of catalytic carbon on nickel has been considered in some detail (Trimm, 1977); a similar mechanism probably operates for carbon formation on other metals. Hydrocarbons in the gas phase adsorb on a metal or undergo a series of reactions to form an intermediate that adsorbs on a metal. Adsorbed species undergo reaction to produce carbon atoms/groups of atoms. These then either dissolve in the metal or migrate across the surface to a nucleation/growth center. Where carbon dissolves in the metal, it migrates to a grain boundary and precipitates out, leading to separation of the nickel particle at the head of a growing column of carbon. Where carbon stays on the surface it forms an encapsulating layer and prevents further deposition of carbon on or in the catalyst.

This mechanism explains many of the observations just reported. Unless encapsulation occurs, the growth of carbon will continue, because the catalytic surface is always lifted at the tip of a growing column of carbon. Long periods of steady-state deposition can, then, be expected. At low temperatures, the diffusion of carbon through nickel is rate determining but, at higher temperatures, the surface reaction controls the overall rate (Trimm, 1977). Hence the maximum in the rate of carbon formation observed as the temperature is increased can be correlated with interaction between these processes.

Sometimes the heat liberated during the decomposition of the hydrocarbon establishes a temperature gradient across the metal particle (Baker *et al.*, 1975b). This probably is not the general case because the decomposition of some hydrocarbons is endothermic (Rostrup-Nielsen and Trimm, 1977). Nevertheless, temperature gradients may assist carbon growth.

Dissolution–reprecipitation of carbon in metals has also been suggested to be responsible for the formation of laminar graphite catalyzed by metals (Derbyshire and Trimm, 1975). In this case there is some question as to the relative importance of carbide intermediates in the system.

Given this general mechanism of formation of catalytic carbon, it is possible now to examine some detailed questions concerning the formation of coke on surfaces. This will include some aspects of catalytic carbon formation.

V. THE FORMATION OF COKE ON ANY SURFACE

Gas-phase coke has been discussed already, and it is obvious that one role of a surface is to collect coke deposits. Possibilities include (*a*) condensation of tar materials, (*b*) collection of gas-phase carbons, and (*c*) concentration of tars and carbons in an environment where further reactions can take place (e.g., dehydrogenation, ordering).

Any of the foregoing processes will lead to deposition of coke on a surface, to encapsulation of the surface, and, as a result, to a change of properties of the original surface to those of a carbon surface. The exact nature of the deposit obviously depends on the relative importance of tars and carbons in the gas phase, which, in turn, depends on the S/V ratio in the reactor, the reaction conditions, and the feed (La Cava, 1977).

It is important to realize, however, that the nature of the gas-phase species also depends on the nature of the surface. To take the most obvious example, if paraffins are passed over nickel at high temperatures, then olefins are produced. As a result, the feed may be paraffinic while the reactants inside the reactor are olefinic.

There is very good evidence that such effects are important during pyrolysis. Coke formation downstream of a reactor has been shown to be affected by the material of the reactor because this affects the amount and nature of coke-forming intermediates in the gas phase (Trimm, 1977; Slotboom and Penninger, 1974a,b).

The composition of the gas phase also affects coke formation in other ways. The importance of catalytic carbon formation obviously depends on the free metal surface, and this may be reduced by encapsulation. If the gas phase contains a gas such as steam or hydrogen, encapsulating carbon may be removed and the net rate of coke formation may increase.

The chemistry of the surface is very important because this will control the rate of production of carbonaceous intermediates, the rate of dissolution of carbon, and the rate of surface migration and encapsulation. In addition, however, the structure of the surface is important. On a macroscopic scale (Bernardo, 1977; Bernardo and Trimm, 1976; Cooper and Trimm, 1980), the roughness and crystallite orientation of a surface have a considerable effect on coke formation. On a microscopic scale, the same is true, and this can be very important for prevention of carbon formation.

The concept on which this is based is very simple: a given chemical reaction occurring on a surface needs an ensemble of sites on which molecules can be adsorbed in positions favoring that reaction (Dowden, 1978). In these terms, then, is it possible to arrange an ensemble of sites such that a desired reaction (say, steam reforming) can occur, while an undesired reaction (coke formation) cannot? In fact, the answer appears to be yes. By careful control of the amount of sulfur added to a feedstream (and, as a result, of the amount of sulfur adsorbed on a nickel surface), Rostrup-Nielsen and Tottrup (1979) were able to steam reform with no carbon formation. This control of ensemble size is very interesting and offers much hope for the future.

The chemistry of the surface may also influence the further reactions of tars and carbons from the gas phase. Although catalytic carbon may not be produced, dehydrogenation or ordering of deposited coke may well be favored by the material of the surface.

The relative importance of catalytic carbon also depends on the dissolution in, the migration through, and the precipitation from the metal. It should be emphasized that the composition of the surface may be different from that of the bulk and, as a result, different species must be considered.

Thus, for example, a steel containing 24.5% Cr, 20.5% Ni, 1.8% Mn, and 52.5% Fe was used in a pyrolysis reactor and, before use, was either preoxidized or prereduced (Holmen, 1981). The preoxidized steel surface examined by Auger spectroscopy had a composition similar to the bulk. The prereduced steel surface contained essentially only Mn and Cr, the content of Ni and Fe being less than 0.3 wt%. The effect of the surface will obviously be very different: the effect of the bulk will be much the same.

Although data can be limited, migration of carbon through metals (or the growth of carbon behind metals) depends first on dissolution in the metal. Any additive that inhibits dissolution—alumina (Long *et al.*, 1975) or silica (Karcher and Glaude, 1971: Brown *et al.*, 1976)—will limit catalytic carbon formation. Second, carbon must diffuse through the metal, which obviously depends on the individual metal. Finally, carbon must precipitate from the metal at a suitable location, which may well be a grain boundary (Trimm, 1977). If it is possible to limit the number of grain boundaries in the metal, or to supply an alternative "sink" for carbon—for example, by forming a metal carbide (Pelissier, 1976)—then catalytic carbon formation will be limited. Additives inhibiting dissolution and precipitation of carbon from metals have been developed (Thuiller, 1976); they reduce significantly the formation of carbon.

VI. SUMMARY: COKE FORMATION

Coke formation in a pyrolysis system is obviously very complex and it is convenient at this point in the discussion to summarize the overall picture.

The main reactions that occur in a pyrolysis system are free-radical chain reactions in the gas phase. Coke formation is a minor reaction that is important because the coke accumulates in the system.

A reactant entering a pyrolysis reactor may undergo a series of free-radical reactions leading to gaseous, liquid, or solid products. These reactions occur in the gas phase.

The reactants and products of reaction may be affected by surfaces in the following ways.

Noncatalytic surfaces may (*a*) act to collect condensed tar materials, (*b*) act to collect soot, (*c*) affect heat and mass distribution in the system, and (*d*) concentrate tars or soots, thereby allowing further noncatalytic reactions to occur.

Catalytic surfaces may, in addition, (*a*) promote the formation of carbon via a dissolution–precipitation mechanism; (*b*) alter the nature of the gases present in the reactor by catalyzing gas-phase reactions; (*c*) alter, in turn, the nature and amount of tars or carbon formed in the gas phase or on a downstream surface; (*d*) become coated by inactive layers either deliberately (alumina, silica) or accidentally (encapsulating coke).

The mechanisms of these processes are discussed in the text, but one question remains. If a catalytic surface becomes covered in encapsulating coke (as indeed it does), then the limits of any catalytic action would be expected to be defined by the time needed to encapsulate the surface. Why, then, do catalytic effects play such an important role in surface reactions during pyrolysis?

There are, in fact, two reasons for this: the first is obvious, based on Fig. 4 (La Cava, 1977). During the pyrolysis of hydrocarbons there is a distinctive induction period for the production of tars and carbon. For a flow reactor this may be translated in terms of the fact that the early part of the reactor never becomes covered in carbon or tar. As mentioned before, however, this tube material will influence downstream coking. Catalytic effects, under these circumstances, will continue unless the early part of the tube is coated.

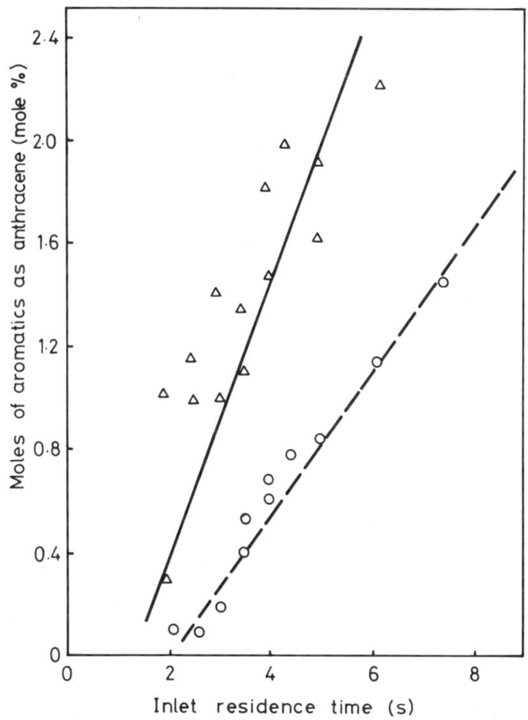

Fig. 4. The formation of tars (Δ—— Δ) and carbon (○---○) during pyrolysis. Butane pyrolysis at 800°C in a quartz reactor. Carbon uptake measured continuously using a nickel foil. Tar measured by condensation at the outlet.

Second, the process of carburization of the metal wall can be expected to continue, no matter whether coke is deposited catalytically or noncatalytically. Carburization, and the subsequent disruption of the tube, will be affected by the metals present and will appear to be a "catalytic" effect. Perhaps, as discussed before, solid–solid reactions and dissolution–precipitation–crystallization concepts are better means to describe what actually occurs.

VII. GASIFICATION OF COKE

The gasification of coke is a subject that deserves a book in its own right. There are four basic reactions involved, although it is common practice to include the water–gas shift reaction in any considerations of gasification.

$$C + O_2 = CO_2 \tag{3}$$
$$C + CO_2 = 2\ CO \tag{4}$$
$$C + 2\ H_2 = CH_4 \tag{5}$$
$$C + H_2O = CO + H_2 \tag{6}$$
$$CO + H_2O = CO_2 + H_2 \tag{7}$$

The reaction of coke with oxygen will not be considered here, except insofar as results pertinent to gasification with other reagents will be discussed. Excellent articles on the reaction between coke and oxygen have appeared recently (Chemical Society, 1978; Laurendeau, 1978), and the interested reader is referred to these.

Before considering individual systems, three general points can be made. The first concerns the importance of metallic impurities. There is no doubt that metal catalysts accelerate gasification markedly and, in this connection, the formation and gasification of coke are related. Where carbon formation leads to metal particles such as at the tip of carbon fibers (for example), gasification will be catalyzed. Noncatalytic gasification will be important when encapsulating carbons or tars are present. Second, the oxidation state of the metallic impurities is important (Baker and Sherwood, 1980). McKee (1974) found that the gasification of graphite by hydrogen was catalyzed by iron, cobalt, and nickel, was slightly catalyzed by manganese, chromium, vanadium, and molybdenum, but was not catalyzed by copper, silver, zinc, cadmium, or palladium. Gasification by water gave generally the same reactivity if hydrogen was present, but if no hydrogen was added, the metal could oxidize and the catalytic activity of all additives was small or zero.

Finally, it is important to recognize that various components of tars and carbons may have different reactivities. This may result from their structure or from their chemical composition. Thus, for example, a polycyclic aromatic compound would be expected to react with steam more easily than carbon, but if the carbon is porous and the tar is in the form of a solidified drop, the effective reactivity will be much less because the the surface area is less.

Various forms of carbon also show a range of reactivities with, as a general rule, reactivity decreasing as the degree of order increases. This finding has been extended in one important direction in articles from McCarty *et al.* (1977) and McCarty and Wise (1979). The basis of the studies was to investigate the nature of intermediates in the conversion of carbon monoxide to methane. Four types of carbon were observed on nickel catalysts. Chemisorbed carbon atoms were reactive to hydrogen at 550°K, and methane production was fast. Bulk nickel carbide, amorphous carbon, and crystalline carbon were all much less reactive, and significant hydrogenation did not occur till higher temperatures were reached (McCarty and Wise, 1979). Any discussion of gasification must recognize that differences in reactivity may be important and that what may be observed overall could well be a composite effect involving several types of carbon. For example, the importance of gasification has been suggested to be small in many systems (Trimm, 1977) as a result of measurements of kinetics of gasification of stable carbons. The identification of a very reactive form of carbon by McCarty *et al.* (1977) and McCarty and Wise (1979) opens this finding to considerable question.

Carbon gasification has been studied for many years and has been the subject of several articles (Chemical Society, 1978; Laurendeau, 1978; B.P. Trading Co., 1964; Walker *et al.*, 1968). Most earlier studies were connected with oxygen gasification, but gasification by steam, carbon dioxide, and hydrogen is also of considerable interest. Steam and carbon dioxide modify the porosity of carbons, thereby producing activated carbons of different adsorptive behavior (Walker *et al.*, 1959). Steam is often added partially as a diluent and partially to gasify deposited coke (Trimm, 1977; Rostrup-Nielsen, 1975; Catalyst Handbook, 1970). Hydrogen and carbon dioxide, produced in these systems, can have the same effect, as in steam reforming and steam cracking (Zdonik *et al.*, 1970). Thus in some industrial reactors, coke formation is a balance between coke production and coke gasification. As a result, there is a great deal of interest in gasification catalyzed by metals (Bernardo and Trimm, 1976), because catalysts and reactors are generally fabricated from metallic material.

In addition, however, recent years have seen an enormous increase in interest in the gasification of coals by oxygen, steam, or carbon dioxide, in connection with the program to produce synthetic crude oil. Although some reference will be made to the role of alkalis or alkaline earths as gasifying catalysts, this will not be primarily with reference to coal or coal char. This is partially because of the extent of the literature and partially because the reactivities of different coals make coal gasification studies complex and relevant mainly only to individual systems.

VIII. THE CATALYZED GASIFICATION OF CARBON: NICKEL

Most of the earlier studies of carbon gasification were catalyzed reactions, because the carbons used always contained a certain amount of catalytic impurities. The possibility of producing high-purity carbons to which a controlled amount of catalyst could be added has enabled more detailed studies of catalyzed gasification by Walker *et al.* (1968). Two general mechanisms of metal catalysis of the gasification were advanced: oxygen transfer and electron transfer.

The electron-transfer mechanism supposes that the unpaired electrons in the d-band of a transition metal will pair up with π electrons of the graphite in contact with the metal. As a result, the number of mobile π electrons in the graphite decreases. This, in turn, weakens the average strength of the bonds between peripheral carbon atoms and carbon–oxygen complexes that are intermediates during gasification. Liberation of these to the gas phase will hence be facilitated and gasification enhanced.

This mechanism does not explain the observed experimental features of the carbon gasification catalyzed by iron (Walker *et al.*, 1965). The oxygen-transfer mechanism, originally advanced by Newman *et al.* (1931), assumes that the catalyst cycles between two oxidation states. Walker *et al.* (1968) extended this theory and proposed the following equation as representing the mechanism for the case of the C–CO_2 reaction:

$$aCO_2 + bFe = Fe_bO_a + aCO \tag{8}$$
$$Fe_bO_a + C = bFe + aCO \tag{9}$$

This mechanism implies catalyst–oxygen interaction and excludes interaction between carbon and catalyst. It explains successfully observed features of the carbon dioxide gasification, including the deactivation (by oxidation) of the catalyst (Walker *et al.*, 1968). Experimental evidence was presented to support the view that the metal acts as a dissociation center, producing active species that diffuse across the metal to react at the graphite–metal interface.

It is fairly obvious that this latter mechanism cannot be important, for example, in the case of gasification by hydrogen. In fact, both mechanisms have been suggested to operate, with the exact importance of each depending on the particular system.

One of the most comprehensive studies of gasification has been made for nickel (Trimm, 1977; Bernardo, 1977), which is an important catalyst and is a major component of most high-temperature alloys. It is useful to discuss other systems in the light of findings for the nickel-catalyzed reaction.

A. Gasification by Carbon Dioxide

One obvious question with this system is whether the presence of carbon dioxide may alter the nature of the catalyst,

$$CO_2 + Ni \rightleftharpoons CO + NiO \tag{10}$$

and hence the reaction kinetics. Under steam reforming/steam cracking conditions, sufficient carbon monoxide is produced to maintain the equilibrium to the left (Rostrup-Nielsen, 1975), although sufficient NiO was probably present during the gasification of nickel-doped carbons (Marsh and Adair, 1975). The amount of oxide present under any given conditions can easily be checked from thermodynamic data.

The kinetics of the gasification reaction are reported in Table V. It would seem that mass-transfer effects are important above about 675°C, and that the low-temperature kinetics can be explained in terms of a model that bears many similarities to that proposed for the uncatalyzed gasification (Rao and Taplan, 1972).

TABLE V

Gasification of Carbon Deposited on Nickel Alumina Catalysts[a]

Gas 0	Temperature (°C)	Order wrt. gas	E_a (kJ mol^{-1})
H_2O	<600	0.6	164
H_2O	600–725	~0.8	61
H_2O	>725	1.0	0
H_2	<550	Fractional	265
H_2	550–650	1	180
H_2	>600	2	−82
CO_2	<625	0.5	160
	>625	~1.0	70
CO_2	<625	0.2	140
	>625	~1.0	65

[a] (Bernardo, 1977)

Carbon dioxide adsorbs and dissociates in the metal, giving atomic oxygen (oxygen-transfer mechanism). Eventually the oxygen will form a metal–oxygen–carbon linkage and carbon monoxide will desorb to the gas phase:

$$CO_2 \rightleftharpoons CO + (O)_M \tag{11}$$

$$(O)_M + C \longrightarrow (CO)_M \tag{12}$$

$$(CO)_M \rightleftharpoons CO \tag{13}$$

If the forward reaction in step (13) is assumed to be rate controlling, then the overall kinetic expression obtained would be similar to that of the uncatalyzed reaction:

$$\text{Rate} = \frac{kP_{CO_2}}{k_1 + k_2P_{CO_2}} \tag{14}$$

The derivation assumes that the amount of carbon monoxide in the gas phase does not affect the overall rate via the reverse reaction in step (13). The trend observed in the order (Table V) can now be understood, as the increasing importance of the term P_{CO_2} in the rate expression will eventually bring about a zero-order reaction with respect to carbon dioxide.

The presence of metal aggregates in the coke influences both the reactivity for gasification and the activation energies (Walker *et al.*, 1968). The efficiency of the metal action depends on its concentration and state of dispersion. In the detailed study, nickel was finely dispersed in a carbon deposit of high internal surface area at a concentration of about 1.5% (w/w) (Trimm, 1977). The rates observed were much higher than those reported by Marsh and Adair (1975), probably because of the distribution of the metal in the carbon.

In fact, comparisons of the observed rates with those predicted on the basis of carbon diffusion through metal show a striking similarity (Figueiredo and Trimm, 1975; Figueiredo, 1981). Hence gasification may involve the reverse process to carbon formation, with carbon diffusing through a metal particle to a base surface where gasification occurs. There seems to be good evidence that such a sequence is sometimes important.

B. Gasification by Hydrogen

The main features of the carbon dioxide gasification were also observed in this system, suggesting that the mechanism previously described may well be operative in the gasification of the carbon deposits, irrespective of the gasifying gas. The kinetics observed, however, will obviously depend on the nature of the rate-controlling steps for particular conditions of gasification.

The most interesting feature observed in the gasification was a decrease of rate with increasing temperature above 650°C ($E_a \simeq -82$ kJ mol^{-1}). (Table V). Orders with respect to hydrogen close to 2 were observed in this zone. As the temperature decreased, the apparent activation energy increased sharply to 180 kJ mol^{-1} and then to 265 kJ mol^{-1}. At the same time, the orders decreased to one and fractional numbers, respectively. Detailed kinetics are reported in Table V.

In the lower-temperature zone, the kinetics observed support the view that the surface reaction is the rate-controlling step. First, the apparent activation energy is similar to that reported by Nishiyama and Tamai (1976a) at conditions where one of the surface reactions was stated to be rate controlling. Second, the orders showed the same "saturation effect" as was observed with carbon dioxide, decreasing from one to near zero as the partial pressure of hydrogen increased. This effect can be understood on the basis of a reaction scheme where hydrogen adsorbs and dissociates on the metal, giving atomic hydrogen that combines with carbon supplied from the bulk:

$$H_2 \rightleftharpoons 2\,(H)_M \tag{15}$$

$$2\,(H)_M + C \longrightarrow (CH_2)_M \cdots \longrightarrow (CH_4)_M \tag{16}$$

$$(CH_4)_M \longrightarrow CH_4 \tag{17}$$

If, in step (15), the simultaneous combination of 2 hydrogen atoms with carbon is assumed to be the rate-controlling step, a rate expression can be derived using the Langmuir isotherm:

$$\text{Rate} \simeq k'(\theta)_M^2 = \frac{k' b_H P_H}{(1 + (b_H P_H)^{1/2} + c_x)^2} \tag{18}$$

where c_x stands for other species adsorbed at the same type of active sites on the metal surface. The "saturation" effect will thus be explained by the increasing importance of the term $b_H P_H$ in the rate expression.

The influence of the reverse reaction $CH_4 \rightarrow C + H_2$ on the rates measured at high temperatures is thought to result from an approach to equilibrium conditions. The gasification does not stop completely because removal of methane from the reaction sites allows a certain distance to be kept from the equilibrium concentration. This distance will obviously depend on the contact times used. However, the amount of carbon gasified at a given temperature will depend on the value of the equilibrium methane concentration. As a result, the apparent activation energy would be expected to be close to the heat of reaction at high temperatures and is. Furthermore, at a given temperature, the equilibrium methane concentration will depend on the power 2 of the hydrogen pressure, and, if the partial pressure of hydrogen in the feed is decreased, the gasification rate will decrease proportionally. This too was observed (Table V).

C. Gasification by Steam

As in the case for carbon dioxide and hydrogen, steam gasification rates remained constant for long period of time, without deactivation being detected at intermediate values of carbon burnoff. The main products of the reaction were hydrogen and carbon dioxide. Carbon monoxide was detected only in small amounts, although its concentration was still in excess of that corresponding to the equilibrium of the water–gas shift reaction. Sufficient hydrogen was produced to ensure the metal was in the reduced state.

The kinetics of steam gasification are summarized in Table V. At temperatures higher than about 600°C the rates appeared to be affected by mass transfer.

In the lower-temperature zone, the value of E_a agreed quite well with a value reported for the steam gasification of nickel-doped graphite (Tamai *et al.*, 1977).

With this result in mind, the surface reaction was assumed to be the rate-controlling step at low temperatures. A mechanism can be envisaged where water dissociates at the metal surface, producing chemisorbed oxygen and hydrogen atoms, and that the oxygen subsequently reacts with carbon to produce carbon dioxide and hydrogen. The reaction rate should then depend on the partial pressure of steam and also on the relative coverages of the metal surface by oxygen and hydrogen. The fractional order observed agrees with this assumption.

There is considerable interest now developing in the gasification of carbon in the context of the methanation and steam re-forming reactions since McCarty *et al.* (1977; McCarty and Wise, 1979) identified carbonaceous intermediates that reacted readily with hydrogen. In the latter case, the overall reaction may be written as $RH + H_2O = CO + CO_2 + H_2 + CH_4 +$ etc., and attempts have been made to establish the importance of intermediate reactive carbonaceous species. The system is more complex than methanation, because gasification by hydrogen, carbon dioxide, and steam are possible. Careful studies by Gardner and Bartholomew (1981) and by Figueiredo (1981) have been reported, and more work can be expected.

IX. EXTENSION TO OTHER METALS

Some disagreement exists as to the relative efficiency of transition-metal catalysts for the gasification of coke. Walker *et al.* (1968) report the reactivity order

$$Fe > Co > Ni$$

whereas both Marsh and Adair (1975) and Tamai *et al.* (1977) report the order

$$Ni > Co > Fe$$

These differences extend to gasification by steam, hydrogen, and carbon dioxide and are thought to result from differences in the carbons and in the concentration and state of aggregation of the metal.

A recent article on studies of the gasification of carbon catalyzed by different metals indicates that the factors thought to be important for nickel are also important for other metals (Trimm, 1981). Thus, for example, Baker and Sherwood (1980) have shown that the catalytic effect of Ir or Rh depends on whether the metal or the metal oxide is present, which, in turn, depends on the reaction conditions. Again, Baker and Chludzinski (1981) have found that copper can catalyze the gasification of graphite above about 800°C, but that chromium, an important component of some alloys, is not a gasification catalyst.

It is regrettable that there have been few studies of the gasification of coke from alloys, because these results would be of great interest in connection with steam cracking. What is obvious is that the gasification of coke is determined by the catalytic action of the metal *on the surface*. The importance of this statement rests on the fact that the surface of an alloy need not necessarily be the same composition as the bulk. Thus, for example, the surface composition of Fe–Co, Fe–Ni, Ni–Cu, and Co–Ni alloys has been found to have a large effect on the gasification of graphite by steam in the presence of the alloys (Rao *et al.*, 1980). This surface composition, in turn, depends on the treatment to which the alloy has been subjected.

From a practical viewpoint, it seems better to prevent coke formation on reactor surfaces by coating with, for example, silica or alumina, and this was discussed earlier in this article. However, it would also be useful to know which materials favored gasification, because this could also reduce overall coke formation. Regrettably, the author suspects that those materials that favor gasification are the same as those that favor coke formation, and the net gain might be small. Nonetheless it would be valuable to carry out detailed studies of coke gasification on alloy surfaces.

Coke formation and gasification constitute a complex problem that is affected by many factors. In recent years it has been possible to begin to understand what happens in a few systems, but the complexity of the problem is such that the solution of one question may well lead to the posing of several more. As a result, this chapter gives an outline of some aspects more important in the context of applied pyrolysis, and it is hoped that the interested reader will be encouraged to explore individual difficulties in more detail where required.

REFERENCES

Badger, G. M. (1965). *Prog. Phys. Org. Chem.* **3**, 1.
Baird, T., Fryer, H. R., and Grant, B. (1974). *Carbon* **12**, 591.
Baird, R., Feates, F., and Harris, P. (1972). *Carbon* **10**, 93.
Baird, T., Fryer, J., and Grant, B. (1971). *Nature (London)* **233**, 329.
Baird, T., White, D., and Reid, M. (1980). *Electron Microsc., Proc. Eur. Congr., 7th*, London **1**, 300.
Baker, R. T. K., and Chludzinski, J. J. (1981). *Carbon* **19**, 75.
Baker, R. T. K., and Sherwood, R. D. (1980). *J. Catal.* **61**, 378.
Baker, R. T. K., and Waite, R. J. (1975). *J. Catal.* **37**, 101.
Baker, R. T. K., Barber, M., Harris, P., Feates, F., and Waite, R. (1972a). *J. Catal.* **26**, 51.
Baker, R. T. K., Feates, F. J., and Harris, P. (1972b). *Carbon* **10**, 93.
Baker, R. T. K., Harris, P. S., Thomas, R. B., and Waite, R. J. (1973). *J. Catal.* **30**, 86.
Baker, R. T. K., Gadsby, G. R., and Terry, S. (1975a). *Carbon* **13**, 245.
Baker, R. T. K., Gadsby, G. R., Thomas, R. B., and Waite, R. J. (1975b). *Carbon* **13**, 211.
Banarjee, B. C., Hirth, T. J., and Walker, P. L., Jr. (1961). *Nature (London)* **192**, 450.
Bennett, M., and Price, J. B. (1981). *J. Mater. Sci.* **16**, 170.
Bernardo, C. A. (1977). Ph.D. Thesis, Univ. of London, London.
Bernardo, C. A., and Trimm, D. L. (1976). *Carbon* **14**, 225.
Bertolini, J. C., Tardy, B., and Billy, J. (1981). *J. Chim. Phys. Phys.-Chim Biol.* **78**, 729.
Blair, G., and Presland, A. (1971). *Proc. Conf. Carbons Graphite, 3rd, London* S.C.1, p. 116.
Bokros, J. C. (1969). *Chem. Phys. Carbon* **5**, 1.
Bonzel, H. P., and Krebs, H. J. (1979). *Chem. Phys. Carbon* **5**, 19.
B.P. Trading Co. (1964). "Gas Making" London.
Brooks, J. D., and Taylor, G. H. (1968). *Chem. Phys. Carbon* **4**, 243.
Brown A. B., Emsley, A. M., and Hill, M. P. (1980). *Gas Chem. Nucl. React. Large Ind. Plant, Univ. Salford, England.*
Brown, D. E., Clark, J. T. K., McCarroll, J. J., and Suis, M. D. (1976). Ger. Patent 2,613,787
Buyanov, R. A., Chesnokov, V. V., Afanas'ev, A. D., and Babeuko, V. S. (1977). *Kinet. Catal. (USSR)* **18**, 839.
"Catalyst Handbook" (1970). Wolf Sci. Texts, London.
Chemical Society (1978). "Oxygen in the Metal and Gaseous Fuel Industries," Spec. Publ., No. 32. Chem. Soc. London.
Chesnokov, U. V., Buyanov, R. A., and Afanas'ev, A. D. (1979). *Kinet Catal. (USSR)* **20**, 390.
Conroy, J. S., Slysh, R. S., Murphy, D. B., and Kinney, C. R. (1957). *Proc. Carbon Conf., 3rd*, London p. 395.
Cooper, B. J., and Trimm, D. L. (1980). *J. Catal.* **62**, 35.
Cunningham, R., and Gwathmey, A. (1957). *Adv. Catal.* **9**, 25.

Dalmon, J. A., and Martin, G. A. (1978). Personal communication.
De Deken, J., Menon, P. G., Froment, G. F., and Haemers, G. (1981). *J. Catal.* **70**, 225.
Derbyshire, F. J., and Trimm, D. L. (1975). *Carbon* **7**, 1.
Dowden, D. A. (1978). *Spec. Period. Rep.*: "Catalysis." **2**, 1.
Ducros, R., Honsley, M., Aluct, M., and Cassuto, A. (1978). *Surf.* **71**, 433.
Escoubes, M., Quinson, J., and Eyraud, C. (1967). *Bull. Soc. Chim. Fr.* p. 2435.
Figueierdo, J. L. (1981). *Carbon* **19**, 148.
Figueiredo, J. L., and Trimm, D. L. (1975). *J. Catal.* **40**, 154.
Fitzer, E., Mueller, K., and Shaefer, W. (1971). *Chem. Phys. Carbon* **7**, 238.
Fryer, J., and Paal, Z. (1973). *Carbon* **11**, 665.
Gardner, D. C., and Bartholomew, C. H. (1981). *Ind. Eng. Chem., Fundam.* **20**, 229.
Graham, S. C., Homer, J. B., and Rosenfeld, J. L. J. (1975). *Proc. Eur. Symp. Combust., 2nd, Orleans, Fr.* p. 374.
Hamilton, J. C., and Blakeley, J. M. (1978). *J. Vac. Sci. Technol.* **15**, 559.
Holmen, A. (1981). Personal communication.
Karcher, W., and Glaude, P. (1971). *Carbon* **9**, 617.
Karu, A., and Beer, M. (1966). *J. Appl. Phys.* **37**, 2179.
Keep, C. W., Baker, R. T. K., and France, J. A. (1977). *J. Catal.* **41**, 232.
Kneale, B., and Ross, J. R. H. (1981). *Chem Commun.* p. 751.
La Cava, A. (1977). Ph.D. Thesis, Univ. of London, London.
La Cava, A., and Trimm, D. L. (1978). *Carbon* **16**, 505.
Lafitau, H., and Jacque, L. (1968). *Bull. Soc. Chim. Fr.* p. 4779.
Laurendeau, N. M. (1978). *Prog. Energy Combust. Sci.* **4**, 221.
Lobo, L. S., and Trimm, D. L. (1971). *Nature (London)* **234**, 15.
Lobo, L. S., and Trimm, D. L. (1973). *J. Catal.* **29**, 15.
Lobo, L. S., Trimm, D. L., and Figeriedo, J. L. (1973). *Proc. Int. Catal. Congr., 5th*, London p. 1125.
Long, R. H., Sze, M. C., and Under, H. (1975). *Met Prog.* p. 52.
Low, G. G., and Bell, A. T. (1979). *J. Catal.* **57**, 397.
McCarty, J. G., and Madix, R. J. (1977). *J. Catal.* **48**, 422.
McCarty, J. G., and Wise, H. (1979). *J. Catal.* **57**, 406.
McCarty, J. G., Wentrcek, P. R., and Wise, H. (1977). *Am. Chem. Soc., Div. Pet. Chem., Prepr. Chicago, Ill.* p. 1315.
McKee, D. W. (1974). *Carbon* **12**, 453.
Marsh, H., and Adair, R. R. (1975). *Carbon* **13**, 327.
Martin, M. T., and Hudson, J. B. (1978). *J. Vac. Sci. Technol.* **15**, 474.
Matsumoto, L. H., and Bennett, C. O. (1978). *J. Catal.* **53**, 331.
Moayeri, M. (1970). M.S. Thesis, Univ. of London, London.
Moeller, D. A., and Bartholomew, C. H. (1979). *Prepr. Pap.—Am. Chem. Soc., Div. Fuel Chem.* **25**, 54.
Murthy, P. S., and Edge, R. F. (1963). *Inst. Gas Eng., J.* **3**, 459.
Newman, B., Kroger, C., and Fingas, E. (1931). *Anorg. Chem.* **197**, 321.
Nishiyama, Y., and Tamai, Y. (1974). *J. Catal.* **33**, 98.
Nishiyama, Y., and Tamai, Y. (1976a). *Carbon* **14**, 13.
Nishiyama, Y., and Tamai, Y. (1976b). *J. Catal.* **45**, 1.
Nishiyama, Y., and Tamai, Y. (1980). *Chemtech* p. 680.
Oya, A., and Otani, S. (1979). *Carbon* **17**, 131.
Palmer, H. B., and Cullis, C. F. (1975). *Chem. Phys. Carbon* **1**, 266.
Pelissier, H. (1976). *Mater. Tech.* **64**, 195.
Presland, A. E. B., and Walker, P. L., Jr. (1969). *Carbon* **7**, 1.
Rao, V. U. S., Szirmae, A., and Fisher, R. M. (1980). *J. Catal.* **62**, **44**.
Rao, Y. K., and Taplan, B. P. (1972). *J. Met. Trans.* **3**, 2465.

Renaud, M., and Bonnetain, L. (1974). *J. Catal.* **33**, 160.
Robertson, S. D. (1969). *Nature (London)* **221**, 1044.
Robertson, S. D. (1970). *Carbon* **8**, 365.
Robertson, S. D. (1972). *Carbon* **10**, 221.
Rostrup-Nielsen, J. R. (1972). *J. Catal.* **27**, 343.
Rostrup-Nielsen, J. R. (1975). "Steam Reforming Catalysts." Teknisk Forlag A/S, Copenhagen.
Rostrup-Nielsen, J. R. (1979). *Symp. Sci. Catal., Sindri, India.*
Rostrup-Nielsen, J. R., and Tottrup, P. B. (1979). *Symp. Sci. Catal., Sindri, India.*
Rostrup-Nielsen, J. R., and Trimm, D. L. (1977). *J. Catal.* **48**, 155.
Sachtler, J. W. A., Kool, J. M., and Ponec, V. (1979). *J. Catal.* **56**, 284.
Shohachi, K., Egashira, M., and Katsuki, H. (1981). *J. Catal.* **68**, 237.
Slotboom, H. W., and Penninger, J. M. L. (1974a). *Erdoel Kohle, Erdgas, Petrochem.* **27**, 410.
Slotboom, H. W., and Penninger, J. M. L. (1974b). *Ind Eng. Chem., Process Des. Dev.* **13**, 296.
Tamai, Y., Nishiyama, Y., and Takahashi, M. (1968). *Carbon* **6**, 593.
Tamai, Y., Nishiyama, Y., and Takahashi, M. (1969). *Carbon* **7**, 209.
Tamai, Y., Watanabe, H., and Tomita, A. (1977). *Carbon* **15**, 103.
Tesner, I. A. (1973). *Symp. Faraday Soc.* No. 7, 104.
Tesner, P., Robinovich, E., Rafalkes, I., and Arefieva, E. (1970). *Carbon* **8**, 435.
Thompson, B. H., and Brooks, C. T. (1973). *Am. Chem. Soc. Natl. Meet., 165th, Dallas, Tex.*
Thuiller, M. A. (1976). *Mater Perform.* **15**, 9.
Toei, R., Kazuhiro, N., Yamada, K., and Okazaki, M. (1975). *J. Chem. Eng. Jpn.* **8**, 131.
Tomita, A., Yoshida, K., Nishiyama, Y., and Tamai, Y. (1972). *Carbon* **10**, 601.
Trimm, D. L. (1977). *Catal. Rev. Sci. Eng.* **16**, 155.
Trimm, D. L. (1981). *Catalysis* **4**, 210.
Trimm, D. L., and Turner, C. J. (1981). *J. Chem. Tech. Biotechnol.* **31**, 385.
Virk, P. S., Chambers, L. E., and Woebcke, H. N. (1974). In "Coal Gasification" L. G. Massey (ed.) Advances in Chemistry Series, No. 131, p. 237. Am. Chem. Soc., Washington, D. C.
Walker, P. L., Jr., Rusinko, F., and Austin, L. G. (1959). *Adv. Catal.* **11**, 133.
Walker, P. L., Jr., Austin, L. J., and Tietjen, J. J. (1965). *Chem. Phys. Carbon* **1**, 327.
Walker, P. L., Jr., Shelef, M., and Anderson, R. A. (1968). *Chem. Phys. Carbon* **4**, 287.
Walkov, W., Genkin, M. V., Sklyrov, A. V., and Voelter, J. (1977). *Z. Phys. Chem. (Liepzig)* **258**, 156.
Wang, N., Matula, R. A., and Farmer, R. C. (1981). *Symp. (Int.) Combust. [Proc.]* **18**, 1149.
Wentrcek, P. A., Wood, B. J., and Wise, H. (1976). *J. Catal.* **43**, 363.
Zdonik, S. B., Green, G. J., and Hallee, L. P. (1970). "Manufacturing Ethylene." Petroleum Publ. Co., Tulsa, Oklahoma.

10

Importance of Surface Reactions in Pyrolysis Units

LYLE F. ALBRIGHT and TOM CHUNG-HU TSAI

I. INTRODUCTION

The rapid escalation of petrochemical feedstock prices has caused ethylene plant designers to select, to an increased degree, highly selective cracking heaters. To achieve this objective, pyrolysis coils allowing operation at low hydrocarbon partial pressure—such as the tapered coils with increasing tube diameter and split coils with multiple tubes—have been found to be beneficial (de Blieck and Mol, 1975; Mathis, 1979). Pyrolysis coils using smaller diameter tubes have moreover resulted in higher selectivity toward production of olefins because of shorter residence time (Leftin *et al.*, 1976; Wallace, 1972). With smaller diameter tubes, surface reactions in pyrolysis units are relatively more important and have hence received greater attention from the designers. Recent evidence obtained

Pyrolysis: Theory and Industrial Practice

ISBN 0-12-048880-9

in both laboratory and industrial units indicates that these surface reactions are more important than it was thought. Surface reactions produce the following results.

1. Small amounts of ethylene, other olefins, and diolefins are destroyed (Dunkleman and Albright, 1976a,b).
2. At least some coke is produced (Baker, 1977; Baker and Harris, 1978; Cooper and Trimm, 1980; Dunkleman and Albright, 1976a,b; Tomita *et al.*, 1972).
3. Part of the coke is gasified. Apparently all CO and CO_2 produced occur as a result of surface reactions involving steam. For laboratory experiments in Vycor glass tubes, no CO or CO_2 was produced (Dunkleman and Albright, 1976a).
4. Carburization reactions weaken stainless steels and are a factor in eventual tube failure (Ibarra, 1980; Kane, 1980).
5. Surface roughening and corrosion occur as a result of repeated oxidizing–reducing and sulfiding–desulfiding steps (Tsai and Albright, 1976). Coke production at metal surfaces also destroys the integrity of the metal surface.

No definitive experiments have yet been reported as to the relative importance of the preceding results in industrial pyrolysis coils or in the transfer-line exchangers. Some quantitative information has been obtained in laboratory coils, however. In one such experiment (Dunkleman and Albright, 1976a), in which steam–ethane mixtures were pyrolyzed at 800°C and atmospheric pressure, ethylene yields were 80% and 90%, respectively, in Incoloy 800 and Vycor glass reactors. Surface reactions are much less important in Vycor glass reactors (Tsai and Albright, 1976).

Extrapolating these laboratory results to the industrial units involved the following problems, among others.

1. Surface-to-volume ratios of the laboratory coil were 4 to 20 times greater than those of industrial coils in the 1-in. to 5-in. range.
2. The ratio of surface to mass of reacting gases is inversely proportional to the total pressure of reactants. Because commercial units normally operate at 2 to 4 atm in the reaction zone (radiant section), the relative importance of surface reactions would be two to four times less because of the pressure effects as compared to the laboratory units operated at atmospheric pressure.

Another criticism of the laboratory investigation is that the coil was only operated for several hours. During the start-up period, the inner metal surface of the coil was apparently rather clean and thus fully exposed to the gaseous reaction mixture; but after 1 or 2 days of operation, the surface would be covered with coke. Because pure carbon is very inert, the importance of surface reactions is claimed to be small after the start-up period. Yet cokes in both the coils and the transfer-line exchangers (TLEs) contain significant amounts of iron, nickel, and chromium. These metals are known to promote coke formation and also decoking (or coke-gasification) reactions. Hence the coke on industrial furnaces is to some extent autocatalytic in relation to coke formation.

Surface reactions were assumed to be perhaps 80 times more important in the laboratory unit ($20 \times 4 = 80$ for correction of diameter and pressure effects, respectively). Then, instead of a 10% difference (90% − 80% = 10%) in ethylene yields, the decrease in ethylene yields because of surface reactions would be 0.125% (10%/80). For a unit whose nominal capacity is 1.0 billion pounds of ethylene per year, such a decrease is equivalent to an annual loss of about 1.25 million pounds of ethylene. Such a loss is a direct decrease in profits by at least $200,000, based on prices prevalent in the 1980s. Larger indirect losses also occur because the higher rates of coke formation result in the following: increased number of decokings (with subsequent loss in production and expensive decoking costs), lower energy efficiencies because of poor heat transfer, and more corrosion of coils. Surface reactions likely often account for 5 to 10% profit losses.

There is thus a real incentive to minimize surface reactions. Yet except for carburization, relatively little attention has been given until recently to developing approaches for minimizing or controlling these surface reactions.

II. MAJOR SURFACE REACTIONS DURING PYROLYSIS

At least seven surface reactions occur during most industrial pyrolyses. In a complex and as yet not completely understood manner, they all contribute directly or indirectly to coke formation.

A. Carburization of Surface

Carburization of the surface (i.e., formation of metal carbides) has two main effects. First, the carburization reactions lead to a chain of events that is instrumental in the eventual failure of furnace coils after about 3 to 6 years. The carburization seriously weakens the inner surfaces so that higher than normal stresses occur in the noncarburized portions of the metal. Creep often becomes critical because metal temperatures are frequently close to 1100°C (2012°F). Creep and weakening of the coil is accentuated by the cyclic operation of the furnace alternately in the pyrolysis and then decoking modes of operation (Ibarra, 1980; Kane, 1981).

The condition of the inside surface of the reactor tube has a marked effect on surface activity and therefore on the resistance to carburization (Clark, 1970). Clark (1972) reported that extruded surfaces of Incoloy 802 exposed to ethylene cracking service for over 12,000 picked up no identifiable amount of carbon. However, surfaces activated by overpickling to produce intergranular surface attack picked up carbon similar to that experienced by machined HK-40. The extruded surface is therefore apparently very beneficial for inhibiting carburization. There is thus a need to develop a surface-finishing technique that will cause low carbon-adsorption characteristics. The modified ASTM–HP40 alloy, Manaurite XA (25 wt% Cr, 35 wt% Ni, 5 wt% W, minor quantity of niobium, the rest Fe), developed and tested by Manior-Pompey in France, appears to be among the best alloys for pyrolysis coils currently available (*Hydrocarbon Processing*, 1980). This alloy is claimed to possess high carburization and oxidation resistance up to 1150°C. Supertherm, produced by the Abex Corporation, is basically an HP40 alloy but contains significant cobalt and titanium; it also seems highly promising.

The formation of metal carbides is also a precursor to the coke formed at or catalyzed by the metal well, as was discussed in detail in Chapter 9. The composition of the metal and more specifically the surface composition is a factor in the ability to carburize.

1. Up to about 2% silicon in Fe–Ni–Cr alloys minimizes carburization without serious loss in the strength of the metal (Kane, 1980). Presumably the silicon at or near the surface is oxidized, in part at least, to silica. Baker and Chludzinski (1980) and Brown *et al.* (1981) have recently reported that stainless-steel surfaces containing high concentrations of silica are subject to much less carburization and coking. The silica may be mixed or complexed in some manner with metal oxides. Another approach that may be related is to add a small amount of silanes in the feedstream (Clark and Foster, 1979; Goldman, 1966; Karcher and Glaude, 1971). These silanes obviously react with steam or oxygen to form silica, which probably collects on metal, forming a glasslike surface.

2. Chromium on the surface reacts with steam, forming a protective chromia coating that minimizes carburization (Kane, 1980). Interestingly, chromium diffuses much more rapidly than either nickel or iron. Diffusion of the chromium to the surface is rather rapid (Marek and Albright, 1981); because chromia does not diffuse rapidly, the surface becomes coated with it.

3. Aluminum on the surface reacts readily with steam or oxygen to form alumina that is also effective in minimizing carburization (Kane, 1980; Long *et al.*, 1975). Alumina can be obtained by employing alloys containing some aluminum, or by coating or impregnating the surface with aluminum as is done commercially by Alon Processing, Inc. of Tarentum, Pennsylvania.
4. Coating stainless steels with oxides of molybdenum, titanium, and tantulum also decreases the rate of coke deposition on the surfaces (Baker and Chludzinski, 1980).

Certain stainless steels are much better able to resist carburization, as explained in Chapter 16.

B. Coke Formation on Metal Surfaces

Most information reported in the literature relative to the role of surface reactions for the production of coke was obtained in the laboratory. Because coke deposits on the coils having skin temperatures up to 1100°C, and in the TLEs at temperatures as low as 400°C, a wide range of temperatures must be considered.

Direct measurement of the coking rate or the coke thickness deposited on the reactor surface in a commercial coil is a formidable task, if not an impossible one. Measuring the increase of the tube metal temperature as a function of a heater run length is often employed to derive coking-rate data in commercial coils. Via this route, however, reliable data are needed on the thermal conductivity and density of deposited coke. For various surfaces at 930 to 1300°C, Grisdale *et al.* (1951) reported the coke density of 129 lb/ft^3 (2065 kg/m^3) with thermal conductivity of 4.8 BTU/h-°F ft (8.3 W/m-°K). Coke thermal conductivity of 3.2 BTU/h-°F ft (5.5 W/m-°K) from ethane cracking and 3.7 BTU/h-°F ft (6.4 W/m-°K) from propane cracking have been reported by Schutt (1959) and Sundaram *et al.* (1979), respectively. Yet the thermal conductivity of the coke found in the TLEs is often considerably lower, often by a factor of several fold, than that of coke found in the pyrolysis coils.

Several hydrocarbons are known to be very effective as coke precursors; such hydrocarbons include acetylenic compounds, diolefins, aromatics, and olefins (Graff and Albright, 1982; Marek and Albright, 1981). Shah *et al.* (1976) reported that the "steady-state" coking rate for cracking *n*-octane within the temperature range of 750–800°C varies from about 0.5 to 3.0 mg/h/cm^2 of reactor surface depending on residence time. A study made on the coking rate in naphtha and gas oil heaters indicates that a typical steady-state coking rate in a commercial coil is within this range. The initial coking rate is higher than the steady-state coking rate, however.

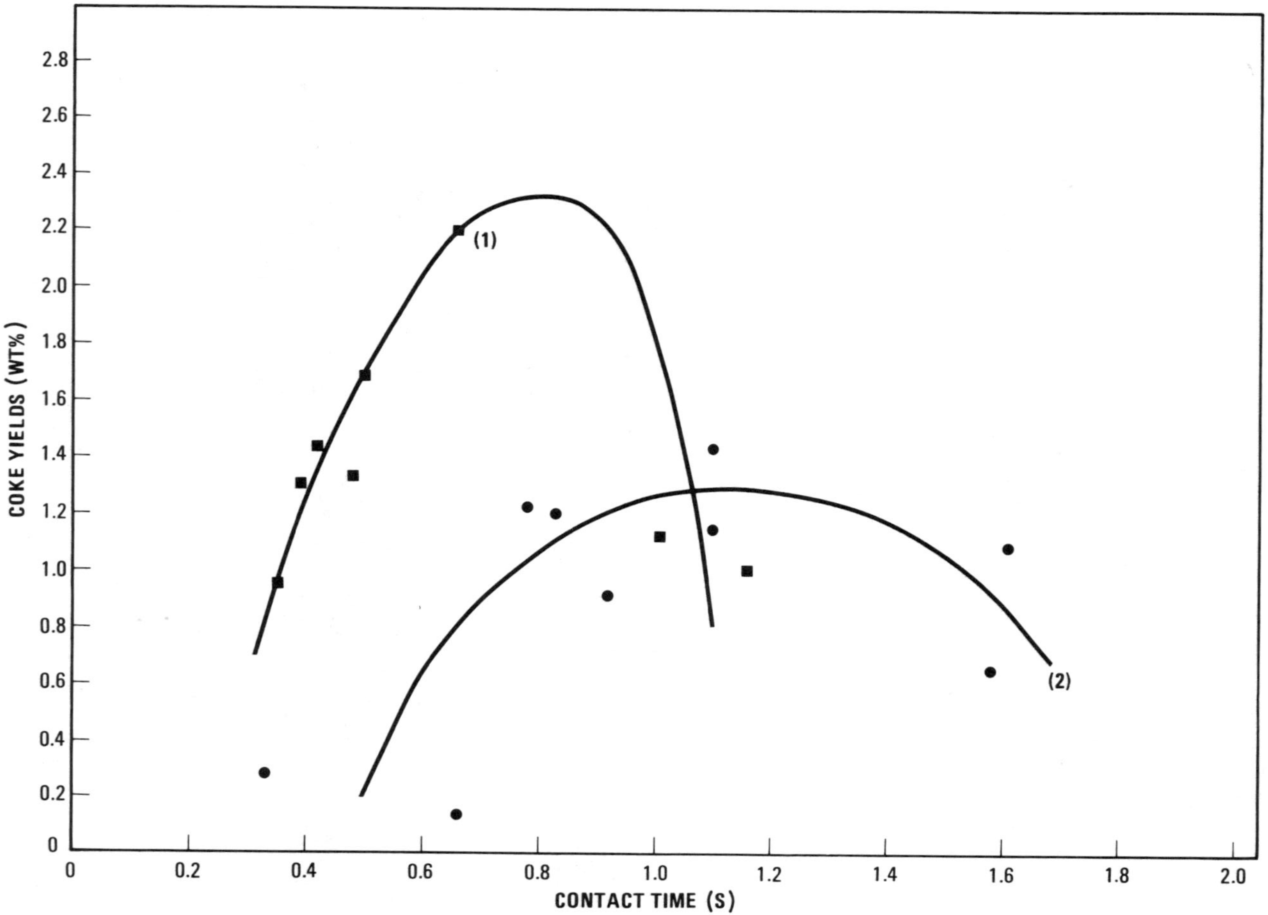

Fig. 1. Coke yields as a function of contact time in the pyrolysis of *n*-octane at 740°C in a quartz reactor. Curve (1), *n*-octane with 20 wt%

This is in agreement with the observation that the tube wall temperature of commercial units increases at a faster rate at the start of the run than at the end of the run.

As illustrated in Fig. 1, the coking rate versus contact time showed a maximum for both *n*-octane/steam and *n*-octane/ammonia cracking systems (Talisman and Shaburov, 1968). Ammonia, however, has a lower maximum. As the yield of lower olefins increases, the formation of coke and tar decreases. This phenomenon of maximum was also noted by Shah *et al.* (1976). The coking rates based on unit surface area in the reactor were considerably lower for coils with smaller S/V ratio (larger diameter), as compared to coils of greater S/V ratio. Coking is apparently mainly a result of surface effects, especially for coils with high S/V ratios.

Coking tendencies vary for different naphthas and gas oils (Newsome and Leftin, 1980; Nowak *et al.*, 1978). Quantitative correlation characterizing naphtha and gas oil coking tendency has been proposed by Leftin and Newsome (1979). The coking inhibition index (CII) is empirically predicted as follows.

$$\mathrm{CII} = 153.4 - [5.1\sqrt{\mathrm{BMCI}}(1 + \mathrm{S}/1.3) + 2\ \mathrm{API}]$$

where BMCI = (48640/VABP) + 473.7 (SG) − 456.8, VABP = volume average boiling point (°K), SG = specific gravity (15.6°C/15.6°C), S = weight percent sulfur, API = (141.5/SG) − 131.5.

In general, feedstocks with a CII of 10 or more have low coking tendencies.

Hydrotreating straight-run middle distillates often results in improved pyrolysis feedstocks that produce more olefins, more monocyclic aromatics, and less coke. During hydrotreating, polycyclic aromatics that are pronounced coke precursors are partially hydrogenated.

Types of Coke

At least three major families of cokes are formed during pyrolysis-type operations; these families are generally designated as filamentous, amorphous, and graphitic cokes (Baker and Harris, 1978), but various subtypes have also been proposed, as will be discussed here. Figure 2 shows pictures of several cokes as obtained when a scanning electron microscope was used to examine stainless-steel coupons contacted at 500 to 900°C with gaseous streams containing acetylene or ethylene, which are known to be coke precursors. These pictures emphasize the considerable differences in appearance of various cokes, as has been reported frequently in the past (Albright *et al.*, 1979; Graff and Albright, 1982; Grisdale *et al.*, 1951). Clearly the mechanisms of either coke formation or collection on the steel surfaces vary substantially.

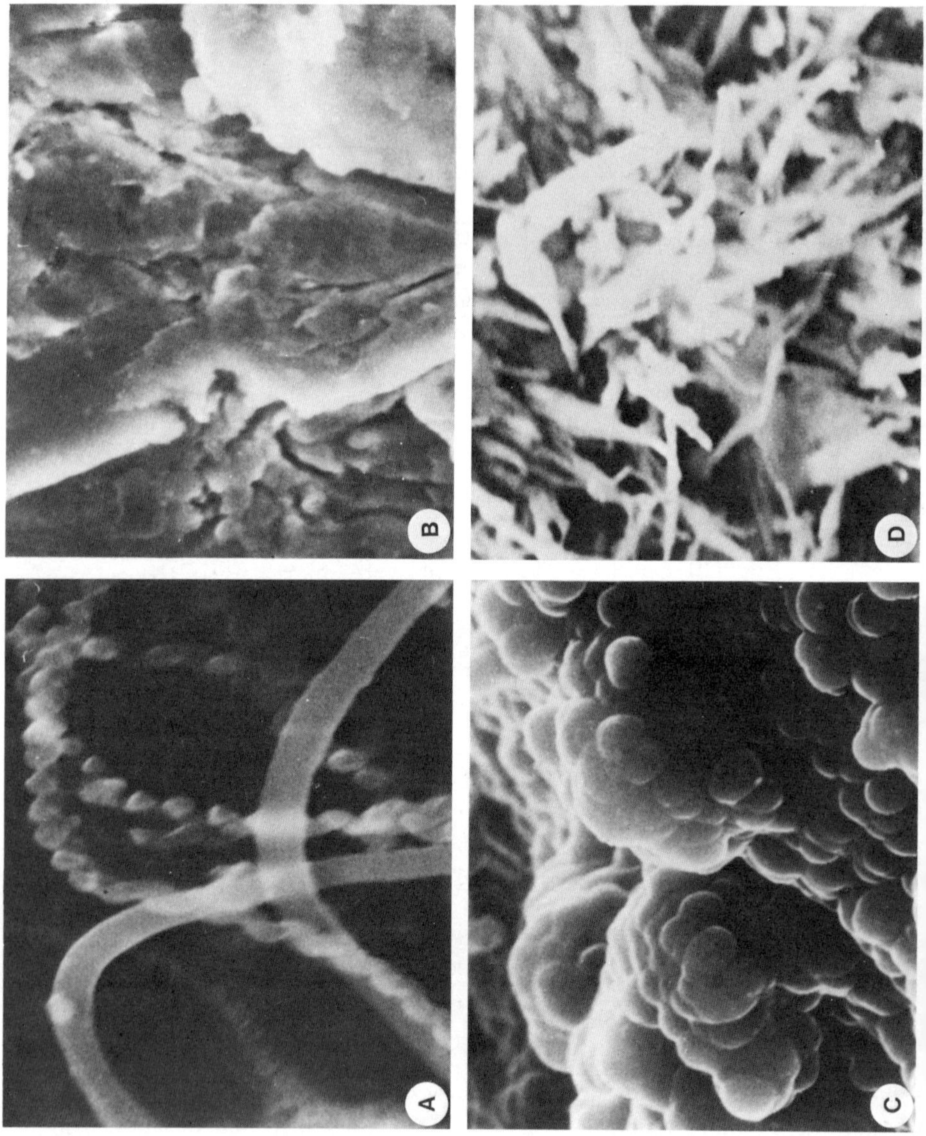
A
B
C
D

Fig. 2. Scanning electron micrographs showing wide variations in the appearance of several types of coke. (A) Filamentous coke; (B) amorphous coke; (C) globular (or spherical) coke; (D) arrowhead coke; (E) column and arrowhead coke; (F) tree-stump coke.

Amorphous coke (see Fig. 2B) has only a low or essentially zero metal content. Globular (or spherical) coke (Fig. 2C) is considered as a subtype of amorphous coke. Amorphous coke forms readily, especially at higher temperatures (e.g., 700 to 900°C) on all types of surfaces including stainless steels, aluminized (or alonized) stainless steels, and Vycor glass. On stainless steels, the amorphous coke is sometimes mixed with filamentous coke, with the result that some metal particles such as iron or nickel are detected in the coke.

Filamentous coke can be divided into at least several subtypes. Constant-diameter and braided (or ropelike) filaments are shown in Fig. 2A. Tree-stump coke (Fig. 2F) and a ribbon-like filament (Albright and McConnell, 1979; Marek and Albright, 1981) are also considered as varieties of filamentous coke. The column and arrowhead cokes shown in Figs. 2D and E contain metal particles, and presumably are formed by mechanisms related to those for filamentous coke production.

Graphitic-type coke with a high density and relatively high thermal conductivity is formed mainly at higher temperatures ($\geq$900°C).

Graff and Albright (1982) have postulated mass-transfer and overall chemical steps for the production of both filamentous and amorphous cokes, as summarized in Fig. 3. Amorphous coke is produced by a sequence that starts with gas-phase reactions (Albright and Yu, 1979; Brown and

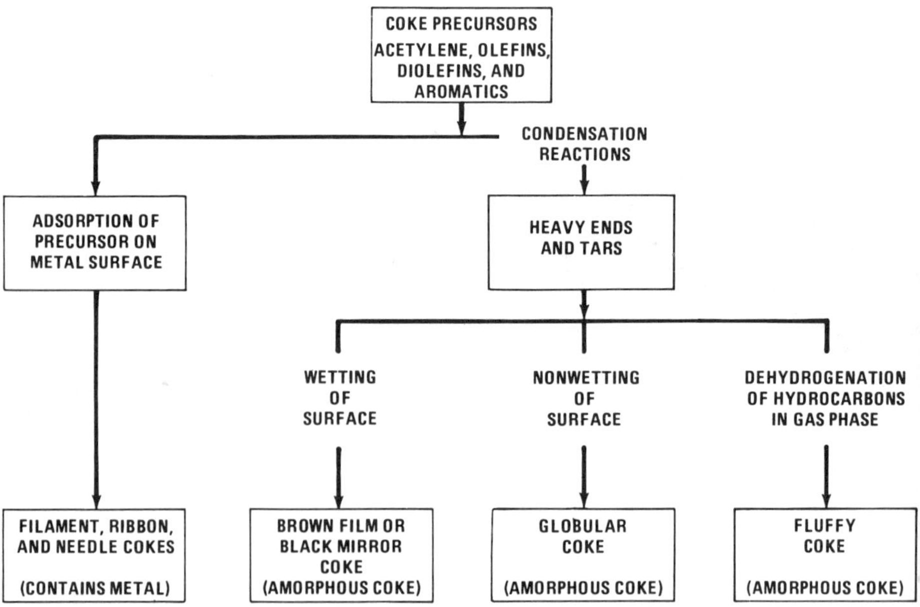

Fig. 3. Mechanism for production of filamentous and amorphous cokes.

Albright, 1976; Graff and Albright, 1982; Lahye *et al.*, 1977); acetylenic compounds and diolefins condense, forming aromatics; aromatics react, producing tar molecules; tar droplets impinge on the metal surfaces or are formed there; the droplets either wet or partially wet the solid surfaces; and finally the tar on the surface dehydrogenates, producing coke. In those cases in which the droplets only partially wet the surface, the coke formed has the appearance under the scanning electron microscope of a series of partially agglomerated spheres, designated here as globular coke (see Fig. 2C). This type of coke contains no metal particles unless it becomes mixed with surface-catalyzed coke, which will be discussed next. When the tar droplets wet the surface, they spread out to various degrees. In some cases, rather rough surfaces result, such as for the amorphous coke shown in Fig. 2B. In other cases, a smooth, almost mirror-like, surface is formed.

Filamentous coke is formed on stainless steel by surface-catalyzed reactions. Most if not all surface-catalyzed coke is formed from metal carbides such as nickel, iron, or cobalt carbides (Baker, 1977; Baker and Chludzinski, 1980; Baker and Harris, 1978; Bernardo and Lobo, 1975, 1976; Cooper and Trimm, 1980; Lobo and Trimm, 1973), as discussed in more detail in Chapter 9. Some general conclusions relative to filamentous coke have been reported (Graff and Albright, 1982; Marek and Albright, 1981).

1. This coke is formed on select surfaces including Fe–Ni–Cr stainless steels, in which case it contains nickel, chromium, and especially iron particles. It is seldom, if ever, formed on Vycor glass or aluminized surfaces such as aluminized (or alonized) Incoloy 800.
2. Acetylene, ethylene, and propylene readily form filamentous (or surface-catalyzed) coke.
3. At relatively low temperatures such as 500 to 600°C, most coke deposited on Incoloy 800 or other stainless steels is formed by surface-catalyzed reactions. At 800 and 900°C, the relative importance of surface-catalyzed coke is less, and that of coke formed by gas-phase reactions is greater. Iron, nickel, and chromium are, however, detected in the cokes found in commercial pyrolysis units, including both the pyrolysis coil and the transfer-line exchanger.
4. In all tests made to date, significantly more coke is deposited or formed on Incoloy 800 surfaces as compared to alonized (or aluminized) Incoloy 800 surfaces, and on decoked and oxidized Incoloy 800 surfaces as compared to new Incoloy 800 surfaces.

Filamentous coke is thought to be effective in filtering out or serving as collection sites for tar molecules or droplets formed in the gas phase.

In commercial pyrolysis furnaces, the coke appears to be primarily amorphous in character; most of it collects near the exit of the coil. In a naphtha furnace, the coke is often rather uniformly distributed and frequently has a laminated character when cross-sections are examined. The inner surface is generally rather shiny and contains metal particles, possibly indicating the presence of some filamentous coke.

The coke found in the TLE contains metal particles and is often magnetic, but the mechanism by which the metal enters this coke is not clear. Some of the metal could, of course, form as a result of surface-catalyzed reactions in the coil to the TLE because higher concentrations of metal are sometimes present on the gas-side portion of the coke. Three explanations have been considered.

1. Filamentous coke that contains metal particles is formed in the coil, then is fractured by the high-velocity gases, and is finally blown from the coil to the TLE, where it becomes trapped in tar droplets or deposits.
2. Pieces of metal are broken from the highly corroded surfaces in the coil; these pieces are also blown from the coil.
3. Metal sulfides are formed by reactions involving sulfur-containing compounds in the feedstream, as will be discussed in more detail later in this chapter. Some sulfides such as those of nickel are liquids at cracking temperatures and can be entrained and transported to the TLE by the gas stream.

Kinetics of Coke Formation

Significant coking-rate data have been reported. The coking data of Leftin and Newsome (1979; Newsome and Leftin, 1980) with naphthas indicate that coking in the bench-scale pyrolysis unit is catalytically controlled, with the hot tube walls furnishing the catalytic sites. This is particularly apparent in the initial stages of a pyrolysis run. Goossens *et al.* (1978a,b) also showed that coking is a kinetically controlled process. Examination of the deposited coke layer of a section of cracking tube clearly demonstrates the importance of the circumferential temperature distribution. Sundaram and Froment (1979) and Sundaram *et al.* (1979) have also reported significant kinetics information on coke formation in ethane and propane cracking heaters.

C. Oxidation and Reduction of Metal Surfaces

During pyrolysis, the inner surfaces of the stainless-steel coils in the furnace are oxidized in at least two ways, and then later reduced. The net result is that a repetitive oxidation–reduction sequence occurs.

Relative to oxidation of the inner surface, steam reacts with the stainless-steel surfaces at temperatures in the range of 800–1000°C; metal oxides and hydrogen are formed, resulting in considerable changes of the surface composition, as indicated, for example, by X-ray diffraction analyses. Preliminary information on such analyses is summarized next.

Steam treatment for 24 h at 800°C of relatively new surfaces of Incoloy 800 (which contains appreciable nickel and chromium and which is sometimes used in ethylene furnaces) resulted in a significant formation of NiO, Fe_3O_4, and hydrogen (Tsai and Albright, 1976). A major decrease occurred in the amount of the surface Fe–Ni–Cr alloy, as indicated by X-ray diffraction analysis. In addition, there was an increase in the Fe–Ni alloy and a decrease in Fe_2O_3. Somewhat comparable results occurred for stainless-steel 304. For nickel, NiO increased significantly.

Oxygen treatment of stainless steels at 700–900°C also resulted in a significant formation of metal oxides. For a new stainless steel, the kinetics of oxidation are high for several minutes but then decrease toward zero after perhaps 15–30 min. Such oxygen treatment is somewhat similar to that which occurs during decoking, as will be discussed later. Oxygen treatment of stainless-steel 304 for 12 h at 800°C resulted in a large increase of Fe_2O_3, Fe_3O_4, and Cr_2O_3 (Tsai and Albright, 1976). NiO increased only slightly, and the Fe–Ni–Cr alloy decreased by perhaps 50%. There was a large increase in the Fe concentration.

Metal oxides produced by steam or oxygen oxidation can be reduced with CO or hydrocarbons to levels essentially the same as before the oxidation (Tsai and Albright, 1976). On fairly new (and unroughened) surfaces, reductions are essentially complete at 700–900°C in about 30–60 min. Hydrogen also reduces some of the oxides but seems to be less effective than hydrocarbons or CO. Both oxidation and reduction reactions obviously occur concurrently on the inner surface of the coils while pyrolysis reactions are occuring.

Experimental data of Tsai and Albright (1976) indicate the approximate level of oxides on stainless-steel surfaces. For Incoloy 800 alloy, the levels obtained after rather complete oxidation were as follows: 33 mmol oxygen/m^2 after treatment of a new surface with oxygen; 17 mmol oxygen/m^2 after treatment of a fairly new surface with steam; 120–150 mmol oxygen/m^2 with oxygen treatment of a surface that had been oxidized and reduced several times. Oxidation–reduction sequence clearly roughened and changed the composition of the surface.

Considerable data indicate that partially oxidized stainless-steel surfaces promote the production of coke during pyrolysis. When a laboratory coil that has been decoked is returned to service for pyrolysis, coke formation is initially very fast until the surface is reduced to a considerable extent (Crynes and Albright, 1969; Dunkleman and Albright, 1976a; Emsley and Hill, 1977; Herriott *et al.*, 1972; Tomita *et al.*, 1972). Highly oxidized or somewhat reduced metal surfaces are relatively inert in relation to coke production, however, (Albright and Yu, 1979; Brown and Albright, 1976).

Almost no reactions occurred between highly oxidized surfaces and hydrocarbons. Once surface reactions started, however, and as the surface was partially reduced, the rates of surface oxidation first increased significantly and then passed through a maximum (Brown and Albright, 1976).

During pyrolysis, the level of metal oxides on the inner surface of a pyrolysis coil presumably approaches a more or less steady-state value after several hours of operation. This value depends on the operating conditions and on the feedstock used. Dunkleman and Albright (1976a) found that about 0.5 to 1.0 h was needed to obtain steady-state composition for the product stream whenever the feedstock to a laboratory reactor was switched between propane and ethane; inlet feeds and temperatures were stabilized within less than 5 min.

More experimental data are certainly needed with regard to the surface oxidations and subsequent reductions that occur in the coils of ethylene furnaces. Identification of the specific oxides formed, as well as the kinetics of producing and reducing each, would be most helpful in clarifying the surface chemistry.

D. Sulfiding and Desulfiding of Metal Surfaces

Hydrogen sulfide, light mercaptans such as methyl or ethyl mercaptan, thiophene, or dimethylsulfide are suitable sulfur-containing compounds that react readily with stainless-steel surfaces in ethylene furnaces to produce metal sulfides on the surface (Emsley and Hill, 1977; Tsai and Albright, 1976). Sulfur-containing gases are present in most if not all heavier feedstocks, including naphthas, gas oils, and crude oils. They are also sometimes added deliberately in minute quantities of about 35 to 200 wppm to ethane and propane feedstocks, which are essentially free of sulfur compounds. Antifoulants such as Nalco 262 and Nalco 5264 (Nalco Chemical Co., 1981) have been claimed to reduce furnace fouling by as much as 30%.

Reactions with these sulfur-containing gases are complicated and have not been completely clarified. Certainly the gases react with either oxidized or somewhat reduced metal surfaces to produce metal sulfides. Oxidized metal surfaces and H_2S react to produce metal sulfides on the surface, steam, sulfur dioxide, hydrogen, and sulfur; at least some H_2S decomposes to form sulfur and hydrogen. Sulfur also reacts with the metal surfaces to produce metal sulfides.

Treating a stainless-steel surface with H_2S, a mercaptan, or organic sulfide at 500–700°C produces in several minutes a fairly complete coating of metal sulfides. At 500°C and within several hours, a rather thick and effective coating is formed. At 850°C, H_2S attacks stainless steel rather rapidly; an Incoloy 800 coil became plugged with molten sulfides after several hours of treatment (Dunkleman and Albright, 1976a). No information was found in the literature as to the relative rates of formation of the several metal sulfides or as to the specific sulfides produced.

Sulfiding to a limited extent the inner surface of a coil that has just been decoked and is hence oxidized is often beneficial for subsequent pyrolyses of ethane and propane (Crynes and Albright, 1969; Dunkleman and Albright, 1976a). As a result, production of coke, carbon oxides, and hydrogen is significantly reduced while ethylene yields are improved.

When metal sulfides on the surface are treated with either steam or oxygen, they are slowly converted to metal oxides. When pure oxygen was contacted with a sulfide-coated surface in an Incoloy 800 tube (Tsai and Albright, 1976), most of the oxygen was initially absorbed; somewhat later, nearly pure sulfur dioxide was vented. During laboratory pyrolysis runs (Dunkleman and Albright, 1976a), the sulfide surfaces were slowly destroyed with propane–steam and ethane–steam feeds; during this time some metal oxides were formed on the surface.

E. Decoking Operations

During decoking of a pyrolysis coil, at least two oxidation reactions occur (Tsai and Albright, 1976). First, coke or metal carbides are oxidized to form carbon dioxide and some carbon monoxide. Second, the surface metal is oxidized, forming metal oxides. Steam or steam–air mixtures are used for oxidation. Steam–air decoking is normally recommended because it is somewhat faster and easier to control. Steam only is often used initially, and air is later added after decoking has started. In gas-cracking furnaces, the layer of coke can be very thick and hard, and may be stronger than the tube. When a coked tube is cooled before decoking, the metal—which contracts faster than coke—may become overstressed and crack. Partial decoking with steam generally weakens the coke sufficiently so that the tubes can be safely cooled.

To minimize damage to the metal, overheating with oxygen must be prevented because of the highly exothermic reactions. Temperatures for the steam–air decoking are often maintained in the 600 to 800°C range. However, the temperature for decoking with steam is generally close to 1000°C, in order to obtain a sufficiently fast rate of oxidation to decoke within a reasonable length of time. Besides temperature, variables of importance for decoking include the steam: air ratio and, to some extent, the total flow rates of gases. For a steam–air decoking operation, the steam-flow mass velocity per coil is generally in the range of 3.5 to 5.0 lb/s-ft^2 (17.1 to 24.4 kg/s-m^2) of cross-sectional area with an air: steam ratio in the range of 0.5–0.6. Higher steam- and air-flow mass velocities are used in smaller coils. Steam rates as high as 4.0 to 8.0 lb/s-ft^2 (19.5 to 39.1 kg/s-m^2) have, however, been reported for steam–air decoking (Taube, 1974). For steam decoking, the steam rate is usually about twice that for steam–air decoking. Steam velocities of 15 to 20 lb/s-ft^2 (73.2 to 97.6 kg/s-m^2) have been claimed to cause spalling of cokes from the reactor surfaces.

Relatively little quantitative data are available on the following factors.

1. Kinetics of decoking. Variables certainly include temperature; metal content of coke, which acts as a catalyst (Baker and Chludzinski, 1980; Turkdogan and Vinters, 1972); oxygen content of inlet gas stream; flow rate of gas; and type of coke. Graphitic-type coke is denser than amorphous coke and is more slowly oxidized. Filamentous coke is more closely packed on the shell of the filament than at the core; normally oxygen can penetrate and burn out the core before destroying the shell (Baker and Harris, 1978). Although coke formed on aluminized surfaces generally contains little iron or nickel (which catalyzes decoking operations), this coke seems to be oxidized fairly rapidly, possibly because it is highly amorphous.
2. When coke is partially oxidized, some of it probably breaks free of the metal surface. Such residual coke may be blown out of the coil by the fast-flowing gas stream. However, the importance of this phenomenon is as yet unknown.
3. Level of oxidation of the inner metallic surface at the end of decoking. Certain areas of the surface are certainly decoked before others. As a result, certain portions of the metal surfaces are probably oxidized to a greater extent than others.
4. Corrosive effects of repeated coking–decoking sequences. Figures 4A and B indicate that a single coking–decoking sequence employing only steam for decoking greatly roughens and corrodes the Incoloy 800 surface (Marek and Albright, 1981). Yet a similar sequence causes little roughening of an alonized Incoloy 800 surface. Decoking with air, however, resulted in a mottled appearance for the alonized Incoloy 800 surface but with no roughening, as shown in Figures 4C and D. EDAX analysis of the mottled surface did not indicate the reason for the discolorations. A coking–decoking sequence using steam and air on an Incoloy 800 surface also resulted in considerable roughening. Surface roughness of stainless steels likely increases to some extent, at least as the number of coking–decoking sequences increases.

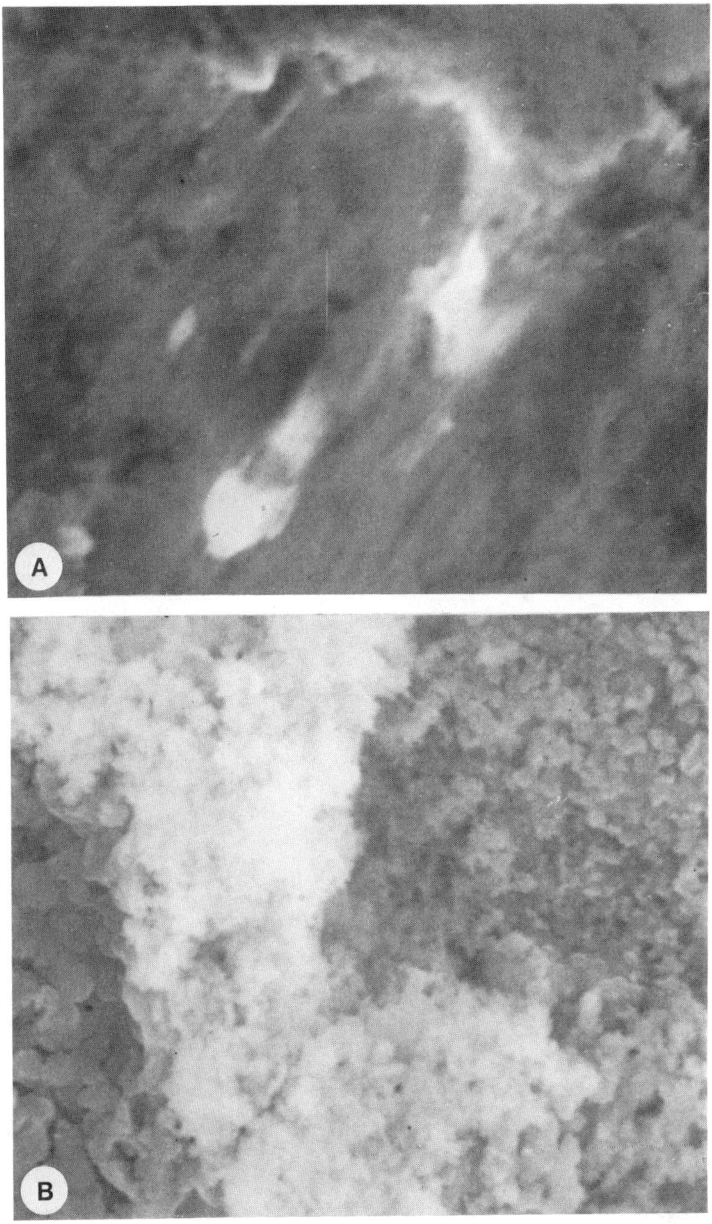

Fig. 4. Incoloy 800 and alonized Incoloy 800 surfaces: new and after decoking (or after burnoff). (A) New Incoloy 800; (B) Incoloy 800 after steam burnoff; (C) new alonized Incoloy 800; (D) alonized Incoloy 800 after O_2 burnoff.

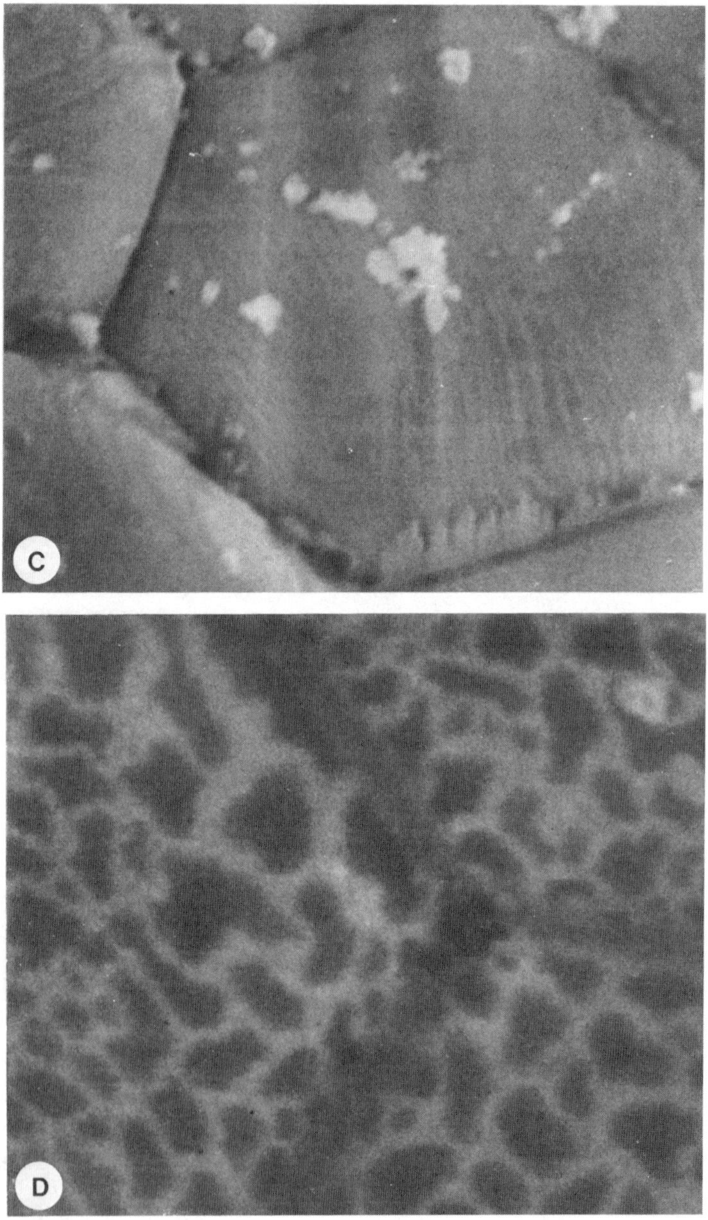

Fig. 4. (continued)

F. Effect of Surface on Kinetics of Pyrolysis

Crynes and Albright (1969) reported that propane conversions were slightly higher for pyrolysis runs in an oxidized coil as compared to runs in a sulfided coil. Dunkleman and Albright (1976a), however, found no significant conversion differences for ethane pyrolysis; they also investigated coils with different materials of construction. Presumably, any kinetic differences because of surface reactions are small.

One might speculate that during coke formation and as carbon–hydrogen bonds are broken, there would be an increase in hydrogen radicals (Albright, 1978). In addition, alkyl or hydrogen radicals may be destroyed in some manner at the surface. If such formation or destruction of free radicals occurs, apparently there is little net increase or decrease in free-radical concentrations, because otherwise the kinetics of pyrolysis would change significantly.

G. Corrosion in Coils of Pyrolysis Furnaces

The surface reactions already described are highly corrosive to the surface of stainless-steel coils. Coking often results in the removal of metal grains from the surface. The repeated sequence of oxidation–reduction, sulfiding–desulfiding, oxidation–sulfiding, and so on, also causes major changes in the integrity of the surface metal. Sulfiding at higher temperatures and high concentrations of H_2S is particularly corrosive. Tsai and Albright (1976) found, for example, that a stainless-steel 304 tube exposed to a sequence of several oxidations, reductions, and sulfidations was so roughened and corroded that it broke. There was no significant change of the outer diameter, but the inner diameter had decreased appreciably. The bottom wall was significantly thicker than the top one. The reason for this phenomenon is not known, but possibly molten sulfides from the top wall dropped to the bottom wall.

Some investigators have suggested that all or at least part of the steam to a pyrolysis unit be replaced with hydrogen (Chahvekilian, 1977), hydrogen sulfide (Tischer and Wing, 1973), or ammonia (Talisman and Shaburov, 1968). Such a replacement might not only significantly affect gas-phase reactions but would certainly affect surface reactions including corrosion; new materials of construction might be needed.

Because metal is incorporated in the coke and because the integrity of the metal surfaces changes as a result of oxidation, sulfiding, reduction, and so forth, it is not surprising that the composition of the inner surfaces of ethylene coils changes as the coil is used. Tsai and Albright (1976) found that the surface compositions of both Incoloy 800 and stainless-steel 304 changed greatly as a result of treatments that were comparable to some that occur in commercial units. Both stainless steel just mentioned were depleted almost entirely of nickel and chromium. An appreciable iron sulfide layer also was formed. Unfortunately, specific information has not yet become available on how surface compositions for industrial ethylene coils change with time, position in the coil, or operating conditions; such information would be helpful in developing improved design and operating conditions.

III. GENERAL CONCLUSIONS

As indicated, surface reactions during pyrolysis operations lead to several undesired effects, including destruction of ethylene product, production of some coke, carburization, and corrosion of the coils. Several promising approaches for reducing or controlling surface reactions have been reported.

1. Improved materials can be used for construction (Ibarra, 1980; *Hydrocarbon Processing*, 1980).
2. Specific coatings can be used on the surface of stainless steels. Aluminizing the surface and incorporating silica in the surface have given encouraging results, and additional applications of these techniques are likely in the near future. Addition of the oxides of molybdenum, titanium, or tantulum may also find applications (Baker and Chludzinski, 1980).
3. The feedstocks for pyrolysis units can be pretreated. For example, mild hydrogenation of naphthas and gas oils is beneficial in some cases (Nowak *et al.*, 1978); apparently such treatments destroy certain coke precursors.
4. Operating conditions can be improved and conditions in each coil of a given furnace can be better controlled during both pyrolysis and decoking.
5. Methods of sulfiding the coils of an ethylene furnace can be improved.
6. All or part of the steam normally used as diluent in pyrolysis reactions can be replaced with hydrogen (Chahvekilian, 1977), hydrogen sulfide (Tischer and Wing, 1973), or ammonia (Talisman and Shaburov, 1968).

Methods can probably be developed in the near future to reduce the level of undesired surface reactions; the time between decoking (or the heater run length) will be greatly lengthened. The need to shut down a unit for decoking may even be eventually eliminated.

Although the emphasis in this chapter has been on the surface reaction in ethylene furnaces, the information and general conclusions are also applicable to some extent for surface reactions in transfer-line exchangers of ethylene units, in other pyrolysis units, and in units for the production of vinyl chloride by dehydrochlorination of 1,2-dichloroethane. This conclusion for vinyl chloride units is based on the fact that the same coke precursors—namely acetylene, butadiene, and benzene—are present. The roles of HCl and elemental chlorine on the stainless steels and the cokes formed for the vinyl chloride units have not yet been reported; it is thought, however, that some key surface reactions occur because of metal chlorides that certainly form to some extent, at least on the metal surfaces.

REFERENCES

Albright, L. F. (1978). *Ind. Eng. Chem., Process Des. Dev.* **17**, 377.

Albright, L. F., and McConnell, C. F. (1979). *In* "Thermal Hydrocarbon Chemistry" (A. G. Oblad, H. G. Davis, and R. T. Eddinger, eds.), Advances in Chemistry Series, No. 183, pp. 205–224. Am. Chem. Soc., Washington, D.C.

Albright, L. F., and Yu, Y. H. C. (1979). *In* "Thermal Hydrocarbon Chemistry" (A. G. Oblad, H. G. Davis, and R. T. Eddinger, eds.), Advances in Chemistry Series, No. 183, pp. 193–203. Am. Chem. Soc., Washington, D.C.

Albright, L. F., McConnell, C. F., and Welther, K. (1979). *In* "Thermal Hydrocarbon Chemistry" (A. G. Oblad, H. G. Davis, and R. T. Eddinger, eds.), Advances Chemistry Series, No. 183, pp. 175–191. Am. Chem. Soc., Washington, D.C.

Baker, R. T. K. (1977). *Chem. Eng. Prog.* **73**(4), 97–99.

Baker, R. T. K., and Chludzinski, J. J. (1980). *J. Catal.* **64**, 464.

Baker, R. T. K., and Harris, P. S. (1978). *Chem. Phys. Carbon* **14**, 83–165.

Bernardo, C. A., and Lobo, L. S. (1975). *Catalysis* **37**, 267–278.

Bernardo, C. A., and Lobo, L. S. (1976), *Carbon* **14**, 287–288.

Brown, D. E., Clark, J. T. R., Foster, A. I., McCarroll, J. J., and Sims, J. L. (1981). Symposium of "Coke Formation on Catalysts or in Pyrolysis Furnaces," *Am. Chem. Soc. Meet., New York.*

Brown, S. M., and Albright, L. F. (1976). *In* "Industrial and Laboratory Pyrolyses" (L. F. Albright and B. L. Crynes, eds.), ACS Symposium Series, No. 32, pp. 296–310. Am. Chem. Soc., Washington, D.C.

Chahvekilian, E. (1977). *AIChE Nat. Meet., 83rd, Houston, Tex.*

Clark, J. T. K., and Foster, A. I. (1979). Br. Patent 1,552,284.

Clark, M. W. (1970). *NACE Natl. Conf.* Philadelphia, Pennsylvania.

Clark, M. W. (1972). *Pac. Chem. Eng. Congr. [Proc.]* **1**.

Cooper, G. J., and Trimm, D. L. (1980), *J. Catal.* **62**, 35–43.

Crynes, B. L., and Albright, L. F. (1969). *Ind. Eng. Chem., Process Des. Dev.* **8**, 25.

de Blieck, J., and Mol, A. (1975). *Pet. Int. (London)* **15**,(3), 50.

Dunkleman, J. J., and Albright, L. F. (1976a). *In* "Industrial and Laboratory Pyrolyses" (L. F. Albright and B. L. Crynes, eds.), ACS Symposium Series, No. 32, pp. 241–260. Am. Chem. Soc., Washington, D.C.

Dunkleman, J. J., and Albright, L. F. (1976b). *In* "Industrial and Laboratory Pyrolyses" (L. F. Albright and B. L. Crynes, eds.), ACS Symposium Series, No. 32, pp. 261–273. Am. Chem. Soc., Washington, D.C. Emsley, A. M., and Hill, M. P. (1977). *Carbon* **15**, 205–210.

Goldman, T. B. (1966). *World Pet. Nov.*, 51–52, 76.

Goossens, A. G., Dente, M., and Ranzi, E. (1978a). *Hydrocarbon Process.* **57**(9), 227.

Goossens, A. G., Dente, M., and Ranzi, E. (1978b). *Oil Gas J.* **76**(36), 89.
Graff, M. J., and Albright, L. F. (1982). *Carbon* **20**(4), 319.
Grisdale, R. D., Pfeister, A. C., and Von Roosbroeck, W. (1951). *Bell Syst. Tech. J.* **30**, 271.
Herriott, G. E., Eckert, R. E., and Albright, L. F. (1972). *AIChE J.* **18**, 84. *Hydrocarbon Processing* (1980). **59**(3), 136.
Ibarra, S. (1980). *Met. Prog.* Feb., 62–66.
Kane, R. H. (1981). *Corrosion* **37**(4), 188–199. *Int. Corros. Forum, Chicago, Ill.* Pap. No. 168.
Karcher, W., and Glaude, P. (1971). *Carbon* **9**, 617–625.
Lahye, J., Badie, P., and Ducret, J. (1977). *Carbon* **15**, 87.
Leftin, H. P., and Newsome, D. S. (1979). U.S. Patent 4,176,045 (to Pullman, Inc.).
Leftin, H. P., Newsome, D. S., and Wolff, T. J. (1976). *In* "Industrial and Laboratory Pyrolysis" (L. F. Albright and B. L. Crynes, eds.), ACS Symposium Series, No. 32, pp. 373–391. Am. Chem. Soc., Washington, D.C.
Lobo, L. S., and Trimm, D. L. (1973). *J. Catal.* **29**, 15–19.
Long, R. H., Sze, M. C., and Unger, H. (1975). *Met. Prog.* Feb., **107**(2), 52–53.
Marek, J. C., and Albright, L. F. (1981). Symposium of "Coke Formation on Catalysts or in Pyrolysis Furnaces," *Am. Chem. Soc. Meet. New York.*
Mathis, J. F. (1979). *Proc.—World Pet. Congr.* **10**.
Nalco Chemical Co. (1981). Product Bulletins G-262, G-5263, G5264, G5267, and G5268. Houston, Texas.
Newsome, D. S., and Leftin, H. P. (1980). *Chem. Eng. Prog.* **76**(5), 76–77.
Nowak, S., Keil, G., and Zimmerman, G. (1978). *Hung. J. Ind. Chem.* **6**, 105–114.
Schutt, H. C. (1959). *Chem. Eng. Prog.* **55**(1), 68.
Shah, Y. T., Stuart, E. B., and Sheth, K. D. (1976). *Ind. Eng. Chem., Process Des. Dev.* **15**, 518.
Sundaram, K. M., and Froment, G. F. (1979). *Chem. Eng. Sci.* **34**, 635–644.
Sundaram, K. M., Van Damme, P. S., and Froment, G. F. (1979). *AIChE Annu. Meet., 72nd, San Francisco, Calif.*
Talisman, L. V., and Shaburov, M. A. (1968). *Int. Chem. Eng.* **8**(1), 105.
Taube, A. T. (1974). *Hydrocarbon Process.* **53**(4), 151.
Tischer, L. G., and Wing, M. S. (1973). U.S. Patent 3,773,850.
Tomita, A., Yoshida, K., Nishiyama, Y., and Tamai, U. (1972). *Carbon* **10**, 601–611.
Tsai, C. H., and Albright, L. F. (1976). *In* "Industrial and Laboratory Pyrolyses" (L. F. Albright and B. L. Crynes, eds.), ACS Symposium Series, No. 32, pp. 274–295. Am. Chem. Soc., Washington, D.C.
Turkdogan, E. T., and Vinters, J. V. (1972). *Carbon* **10**, 97–111.
Wallace, B. A. (1972). U.S. Patent 3,671,198.

11

Laboratory Reactors for Pyrolysis Reactions

GUY-MARIE CÔME

Pyrolysis: Theory and Industrial Practice

ISBN 0-12-048880-9

I. INTRODUCTION

This chapter is devoted to laboratory reactors used for the kinetic study of thermal decomposition of hydrocarbons; it excludes photolytic, catalytic, and other nonthermal reactors. In Section II, some general problems connected with theory and use of reactors for kinetic studies are examined. Then, the six main types of pyrolysis reactors (batch, single-pulse shock tube, continuous-flow stirred tank, very low pressure pyrolysis, tubular, wall-less) are described along with their respective advantages and disadvantages (Sections III to VIII). In Section IX, neopentane pyrolysis has been selected as a convenient standard by which to compare the results produced by each reactor. A brief discussion about the errors arising in chemical kinetics and a survey dealing with other methods of kinetic interest are presented in Section X.

II. REACTORS: THEORY AND PRACTICE

Before the 1970s, most gas-phase pyrolyses of hydrocarbons in the laboratory had been carried out by the batch reactor, or static method. The use of open reactors for studying gas-phase pyrolyses in the laboratory is relatively recent, but their use is common in industrial plants.

Isothermal systems (in space and time) are advantageous for fundamental studies because temperature may be considered as an independent variable. If adequate heat transfer cannot be achieved, either because the evolution or absorption of heat due to reaction is too fast or because the reactor is too large, the reaction may be carried out adiabatically. Intermediate situations between the adiabatic and nonadiabatic cases may also be encountered.

In the case of nonisothermal reactors, in addition to the equations of mass conservation, an equation of energy conservation is to be written.

Furthermore, the equations of mass and energy conservation are closely coupled and thus the numerical calculation is more difficult. Globally, the treatment of experimental kinetic data, in order to get reaction rates, is subject to much more uncertainty than in the isothermal case. For these reasons, most laboratory reactors are designed to perform isothermally.

Gas flow through a tubular reactor implies a pressure drop along its length, and this effect becomes more significant if the flow rate is high. If there is a significant pressure drop, the equation of momentum conservation must be written beyond that of conservation of mass, and of energy, thus complicating the treatment of kinetic data. A commercial steam-cracking coil operates with a significant pressure drop, which is not the case for most laboratory reactors.

The degree of mixing of reactants and products in a vessel during a reaction is characterized by the residence time distribution (RTD) of molecules. The residence time of a given molecule in a reactor is the time spent by that molecule in the reactor. In a real reactor, molecules stay in the reactor for various times that are functions of the geometry and of the fluid dynamics of the reactor. There are two limiting cases of RTD.

In an ideal continuous-flow stirred tank reactor (CFSTR), mixing—achieved by diffusion, jets, or mechanical stirring—is sufficient to ensure uniform composition, pressure, and temperature in the whole reactor. The RTD of a CFSTR is a decreasing exponential, with a time constant equal to the reactor volume/volumetric flow rate.

In an ideal plug-flow reactor, there is negligible axial diffusion. Therefore, plug flow is characterized by an unique residence time for all the molecules. A plug flow is difficult to achieve on the laboratory scale because of the small Reynolds numbers involved. Therefore, for most laboratory tubular reactors, the gas flow is laminar. Furthermore, if the flow becomes turbulent, the assumption of plug flow becomes more nearly valid; but there may be a nonnegligible pressure drop along the reactor.

In an ideal stirred batch reactor, there is also only one residence time for all the molecules, which is simply the duration of the reaction itself.

The whole phenomenon leading to the existence of a RTD is usually called macromixing. If the reaction corresponds to first order, knowledge of the RTD is sufficient to calculate the reaction rate from experimental results. But if the rate law is not linear, it is necessary to know also the microscopic structure of the reaction mixture, characterized by the state of micromixing.

A reacting mixture, homogeneous from a macroscopic point of view, may exist in two limit states. In a microfluid, each molecule may enter into contact with any other molecule. In a macrofluid, on the contrary, there exist small segregation domains; the molecules of a given segregation domain stay together during their passage throughout the reactor and the macromixing concerns only the mixing of segregation domains themselves. Of course, a real fluid will commonly exhibit a partial segregation.

III. BATCH REACTOR

A. Characteristics

The main limitation of the batch reactor comes from the time required for turning valves and for the pressure and temperature to reach equilibrium values. Thus a batch reactor is suitable for slow reactions only, with reaction times greater than about 1 min. Borosilicate glass (Pyrex) and quartz are used for temperatures up to 500 and 1100°C, respectively. Other materials, such as stainless steel, nickel, and gold, have also been used. Temperature is kept constant by means of an electric furnace or a liquid thermostat. Usually, reactions in a batch reactor are conducted at subatmospheric pressures, but higher pressures may be used too. The volume of a batch reactor is limited by the dimensions of available thermostats, the time needed to admit the reactants, and the lack of temperature equilibration when the heat of reaction is important. Because of tubing necessary for gas introduction, pressure gauges, and sampling, batch reactors present dead volumes that, when they are greater than a few percent, introduce uncertainties in the kinetic measurements. Furthermore, a large vessel is desirable to minimize wall effects and to provide a large enough sample. Consequently, a few hundred cubic centimeters is a common size for batch reactors (Mulcahy, 1973).

B. Wall Effects

Uncertainties are introduced in the use of a batch reactor by the definition of zero time, by the existence of dead spaces, and by nonuniform temperature throughout the whole reactor; but the main hindrance concerns wall reactions. Wall effects on a homogeneous gas-phase reaction can be studied by two methods: changing the nature of the reactor walls, and changing the S/V ratio of the reactor.

The nature of a wall of course depends first on its material: Pyrex, quartz, gold, stainless steel, and so on. The surface of a given wall can be modified in several ways. A "clean" reactor is obtained by treating it sequentially with organic solvents to eliminate greases, and with hot oxygen to burn the carbonaceous materials deposited on the walls, and then simultaneous warming and pumping to desorb gases. A reactor is treated either with gases such as hydrogen, nitric oxide, and hydrogen sulfide, or with liquids such as hydrofluoric acid or phosphoric acid—always followed by pumping. Conditioning a reaction vessel is achieved by carrying out pyrolyses to large extents of reaction; it seems likely that conditioning involves deposition of a coating of carbonaceous material on the walls. Other coatings are obtained with lead oxide, potassium chloride, magnesium oxide, and magnesium perchlorate.

The relative extent of the walls of the reactor is increased by introducing loose packing of the same material into the reactor. Empty and packed reactors have typical S/V ratios ranging from about 1 cm^{-1} to 10 cm^{-1} or more, respectively.

Wall effects on the rate of hydrocarbon pyrolysis have been extensively studied by changing the nature and the relative extent of the reactor walls. Articles on these subjects are available by Purnell and Quinn (1967), Martin *et al.* (1976), and Volkan and April (1977). Further discussion of wall effects may be found in the next sections of this chapter and in Chapter 10 of this book, as well as in works mentioned in the reference list.

IV. SINGLE-PULSE SHOCK TUBE

The majority of kinetic studies of pyrolysis reactions by means of shock tubes have been carried out in fact by the single-pulse shock technique. From the point of view of reaction-rate measurements, a single-pulse shock tube may be considered as a batch reactor, but with reaction times of about 1 ms and temperatures in the range 1000–1800°K. Shock tubes have been operated from subatmospheric pressures up to a few atmospheres.

A. Conventional Shock Tube

A conventional shock tube is a straight tube, closed at both ends, containing gas at high pressure (driver) separated by a diaphragm from a gas at low pressure (reactant). When the diaphragm is ruptured, the driver gas expands into the low-pressure section, generating a shock wave that propagates into the low-pressure gas, heating and compressing it. The shock wave is then reflected from the closed end; this increases the temperature and pressure of the gas still further. Simultaneously, expansion waves, which lower temperature and pressure, travel into the high-pressure gas and are reflected from the end of the driver section. Repeated reflections of the pressure waves subject the reactant gas to a series of temperature pulses of slowly decreasing amplitude. This complex temperature history limits the value of the conventional shock tube for chemical kinetics investigations. Two modifications of the conventional shock tube to produce a single high-temperature pulse have been devised, as discussed next.

B. Conventional Single-Pulse Shock Tube

Glick *et al.* (1955) added to the shock tube, at the end of the high-pressure section, a large evacuated tank separated by a second diaphragm from the driver section. By breaking this diaphragm after the bursting of the first diaphragm, but before the arrival of any pressure waves, sonic flow into the evacuated tank is established and prevents any appreciable secondary temperature pulses. The bursting of the tank diaphragm also generates a strong expansion wave that cools the reacting gas sample. It is therefore possible to process a sample of gas with a single pulse of known temperature and duration. Because of the shortness of the temperature pulse, there is no appreciable heat transfer to the wall. The temperature behind the incident shock wave is substantially below the reflected-shock temperature, so the amount of reaction occurring before further heating by the reflected shock is very small and can generally be neglected. In the similar single-pulse shock tube of Bradley and Frend (1971), reactants are confined, at high dilution, between two valves, which are opened immediately before firing the shock. Thus the loss of time resolution due to different heating times is circumvented.

C. Unconventional Single-Pulse Shock Tube

In the design of Lifschitz *et al.* (1963) used by Tsang (1964) for pyrolysis studies, the dump tank was placed at the low-pressure side of the shock tube close to the diaphragm, instead of in its more conventional position at the end of the high-pressure section. The dump tank damps out the numerous reflected shock waves. The cooling rate in this design is higher than that in the preceding technique, because in the latter, only the reflected expansion wave from the driver section is utilized. The sampling port is placed near the end wall. Analyses are done by either gas chromatography or mass spectrometry.

D. Sources of Error

The shock temperature is computed from the incident or reflected shock velocities, measured by means of pressure transducers. The calculation is a straightforward application of the laws of conservation of mass, momentum, and energy; simplifying assumptions are made. The reaction time is deduced from the pressure–time history. As has been noticed, the reaction time varies along the test section and, therefore, the concentrations of the reaction products are not uniform. Thus a special sampling procedure is generally used.

The utility of the single-pulse shock tube as a tool for kinetics depends on the control of the reaction temperature (particularly the minimizing of secondary temperature pulses), the reaction time, the cooling rate to freeze the chemical reaction, minimizing secondary reactions to avoid complex reaction mechanisms, and the absence of surface effects. The errors arising from the finite cooling rate generally are negligible for reactions of sufficiently high activation energy. Bradley and Frend (1971) believe that reaction times of ~ 1 ms can be measured to $\pm 10\%$, the uncertainty being in determining the precise arrival time of the expansion wave; the uncertainty in the velocity measurements is much lower and corresponds to an error of $\pm 10°K$ in the temperature measurements. Precise thermodynamic data are necessary. Furthermore, special care has to be taken to maintain uniform temperature and pressure during the reaction period. Perturbations are introduced by the nonideal opening of the diaphragm and the interaction of the reflected shock wave with the boundary layer generated by the incident shock. Most authors claim that pyrolyses carried out in single-pulse shock tubes are free from surface effects, but recent results seem to indicate that this is not completely true (Lifschitz and Frenklach, 1977). In order to minimize secondary reactions and even to isolate the initiation process, Tsang (1964) used dilute reactants in an inert gas and a large excess of an inhibitor such as toluene.

In view of the difficulty of shock-tube temperature measurements, Tsang (1964) has developed and extensively used a comparative rate technique. It consists of decomposing two compounds with the same reflected shock wave and calculating the respective rate constants for unimolecular decomposition. If the rate parameters of one of these reactions—an internal standard—are well established, the corresponding values from the other process may be calculated.

V. CONTINUOUS-FLOW STIRRED TANK REACTOR (CFSTR)

A. Reactor Design

Mixing in the gaseous phase may be achieved in three ways: by mechanical stirring, by molecular or turbulent diffusion (the first process often being able to ensure the homogeneity of the mixture at low pressures), and finally by jets that create recycling streams.

The first CFSTR for studying gas-phase pyrolyses was designed by Mulcahy and Williams (1961). It is suitable for operation at pressures of about 1 to 50 torr and mean residence times of about 0.1 to 100 s. The reactants with or without carrier gas enter through a central spherical perforated "diffuser"; mixing occurs by a combination of convection and diffusion. Rondeau *et al.* (1976) have used spherical vessels with an internal baffle positioned in front of the capillary gas-flow inlet and for large space times, greater than 1 min. Matras and Villermaux (1973) and David and Matras (1975) have described a spherical jet-stirred reactor, for space times of around 1 s, in which stirring is achieved by means of four turbulent jets issuing from the nozzles of a cross-shaped injector. A similar reactor, but with a cylindrical geometry, has been designed by David *et al.* (1979).

B. Mixing

The degree of mixing in a CFSTR may be estimated by RTD experiments, temperature measurements, flow visualization, and, finally, studying the kinetics of a first-order reaction, or of any other reaction previously examined in other reactors, but especially in a batch reactor.

Mulcahy and Williams (1961) studied in their reactor the kinetics of pyrolysis of di-*t*-butyl peroxide at 430–550°K and pressures between 2 and 30 torr. Their results agree well with previous ones obtained in several studies by the static method. The authors concluded that the reactor functions as a perfectly stirred reactor. Two thermocouples located inside the reactor indicated in most cases temperature differences of less than 1.5°K but sometimes up to 4.5°K. Matras and Villermaux (1973) have established by RTD experiments that the reactor of Mulcahy and Williams (1961) contained a short-circuit flow path and thus was not always perfectly macromixed. Preferred flow paths have been made obvious by a visualization technique (Rondeau and André, 1975).

Residence time distribution and kinetic experiments (Rondeau *et al.*, 1976), as well as visualization experiments (Rondeau and André, 1975), have shown that the reactor of Rondeau *et al.* (1976) is well macromixed, and probably in a state not too far from micromixing.

Matras and Villermaux (1973) and David and Matras (1975) have defined three constraints of operation of their jet-stirred reactor: turbulence, sonic velocity, and internal recycling limitations. By RTD experiments and by studying the first-order decomposition of bicyclo[2.2.1]hept-2-ene these authors showed that good macromixing is obtained for reactor volumes ranging from a few to 50,000 cm^3, when the three constraints are met. However, Azay and Côme (1979a) showed that reactors built in agreement with Matras and Villermaux's criteria (1973) may exhibit temperature gradients as large as 80°K; the existence of strong temperature gradients was established, not only from temperature measurements *in situ*, but also from kinetic experiments. Azay and Côme (1979a) showed that the temperature gradient could be eliminated by a sufficient preheating of the reactants and/or of a carrier gas. David and Villermaux (1978) have attempted to determine the state of micromixing of the gas phase in their reactor. They studied the homogeneous catalytic dehydrogenation of isobutane by iodine in the presence of oxygen and concluded that the reactor works in a state of maximum mixedness.

The CFSTR designed by Matras and Villermaux (1973) and David and Matras (1975) appears to be very flexible. It can be operated from a few milliseconds to a few seconds, and built with volumes ranging from a few to 50,000 cm^3. When the reactor is constructed in agreement with the criteria, the RTD is very close to that of a perfect mixer. Furthermore, if an adequate preheating of input flow is achieved, temperature gradients in the reactor are negligible. Thus this reactor appears to be suitable for kinetic studies.

C. Non-Quasi-Stationary State Pyrolysis

It is possible to conduct experiments at steady-state flow, but during the induction period of a reaction—that is, during the establishment of quasi-stationary state for free-radical concentrations. By means of singular perturbation theory, a formula giving the length of the induction period associated with any reaction mechanism has been established (Côme, 1977). This allows choosing a priori experimental conditions, for a given reaction mechanism, for which a non-quasi-stationary state can be observed. Actually for reaction times of ~ 1 s, Marquaire and Côme (1977, 1978) observed the induction period of neopentane pyrolysis around 720°K, and Corbel *et al.* (1981) the one of ethane pyrolysis around 800°K. Such non-quasi-stationary state experiments give supplementary information about reaction mechanisms and free-radical reactivities. They allow determination of the initiation and the propagation rate constants of a chain reaction by measuring only the major reaction products.

D. Nonsteady-State Operations

A CFSTR may also be used at nonsteady state. A pulse of reactant(s) is injected into a carrier-gas stream and carried into the CFSTR, where it is partially converted to products. To follow the reaction, David and

Villermaux (1973) monitored the concentration signals by mass spectrometry at the inlet and outlet of the reactor. Baldwin *et al.* (1975) directed the reaction mixture into a gas chromatograph, where the amounts of reactants and products may be measured. Both techniques allow determination of reaction rates.

VI. VERY LOW PRESSURE PYROLYSIS (VLPP)

A. Characteristics

A VLPP reactor is in fact a CFSTR in which perfect mixing is achieved by molecular diffusion. However, the very special features of VLPP deserve attention. The first article dealing with VLPP was published by Benson and Spokes (1967). Later, Golden *et al.* (1973) gave a review of this technique, which will be largely the basis of discussion here. Since then, these researchers have continued to study many reactions by means of VLPP. Recently, another low-pressure pyrolysis technique has been developed (Davidson *et al.*, 1980).

In a VLPP experiment, a steady-state flow of reactant molecules passes into a thermostated Knudsen cell (i.e., at pressures up to 10^{-2} torr) through a short length of small-bore capillary tube, which prevents back diffusion from the reactor. Once in the cell, reactant molecules may decompose, or react with other products or with the walls, or finally escape from the cell. The mean free path of molecules within the cell exceeds the size of the exit aperture by about a factor of 10. Reactants and products effuse molecularly into the ionization chamber of a mass spectrometer where they are analyzed.

Under conditions of very low pressure, gas flow is chiefly free molecular flow such that energy transfer to reactant molecules is essentially by means of gas–wall interactions. Furthermore, the frequency of collisions in the gas phase between the products of the initial pyrolytic step and other species is very low, and thus secondary reactions are negligible in general.

VLPP has been operated from room temperature to 1400°K with a quartz reactor and to 1900°K with an alumina reactor, for pressures up to a few 10^{-2} torr. Typical flow rates range from 10^{14} to 10^{17} molecule s^{-1}. Reactor geometry and aperture size lead to mean residence times ranging from $\sim 10^{-3}$ to 1 s.

The calculation of a reaction rate from experimental results obtained in a VLPP experiment is a straightforward application of the law of mass conservation. However, it must be noticed that the rate of escape of a species from the reactor is given by the Knudsen equation—that is, it can be computed from easily known parameters of the experiment.

VLPP has been applied to the measurement of the rates of fast unimolecular reactions, very fast bimolecular and heterogeneous reactions, and energy transfer between excited molecules and surfaces or other molecules.

B. Unimolecular and Bimolecular Reactions

The measured rate constant k of a unimolecular reaction is a function of the collision frequency of the average molecule with the reactor walls. This happens because the frequency of wall collisions is not sufficient to maintain an equilibrium distribution of energy in the reacting molecules. Thus the measured rate constant k is lower than the constant k^{∞} obtained for an infinite wall-collision frequency. This is commonly referred to as the "falloff" of unimolecular rate constants. The relationship between k and k^{∞} is given by either Rice–Rampsperger–Kassel (RRK) or Rice–Rampsperger–Kassel–Marcus (RRKM) theories. By fitting the theoretical and experimental values of k, high-pressure activation energies can be deduced, if reasonable estimates of activation entropies can be made.

Primarily, VLPP had been designed to avoid bimolecular reactions, by working at low pressures and also short reaction times. Thus only "very fast" bimolecular events, such as radical–molecule or radical–radical processes, may be observed if "high" pressures and "large" mean residence times are used. Free radicals are produced by the thermal decomposition of a sensitizer, such as di-*t*-butyl peroxide.

Experiments carried out at pressures such that gas-phase collisions begin to compete with gas–wall collisions can be used to measure the relative effectiveness of walls and gases in the transfer of vibrational energy in the study of unimolecular reactions in the "falloff" region of pressures. Of course, the success of such experiments depends on the correctness of the assumptions used in RRKM theory.

C. Heterogeneous Reactions

Concerning heterogeneous reactions, the first point of question is as follows: in "normal" VLPP experiments, the pressure is so low that almost all collisions of reactants are with the walls, and thus it can be thought that observed reactions are heterogeneous. However, heterogeneous reactions are usually slow as compared with desorption processes, and therefore, only heterogeneous reactions with a collision efficiency in excess of about 10^{-4} (i.e., very fast heterogeneous processes) can be measured by VLPP.

VII. TUBULAR REACTOR

A. Characteristics

The main question connected with the use of the tubular reactor deals with the deviations from ideality. In the laboratory, it is very difficult to realize ideal plug flow, characterized by a Reynolds number greater than 10^4 and a length/diameter ratio greater than 100. Generally, the Reynolds number is less than 100 and the gas flow is laminar, with a parabolic profile. In the use of a laminar-flow tubular reactor, however, it is customary to assume that (*a*) the entering gas molecules warm up instantaneously to reaction temperature, (*b*) there is a single residence time, (*c*) the reactant mixture is quenched instantaneously at the outlet of the reactor, and (*d*) there is no pressure drop along the tube. Normally, the pressure drop is negligible in laminar-flow experiments, and measurements of pressures permit ready confirmation of this fact. But it is more difficult to correct the results of deviations from plug flow due to axial and radial diffusion, and of nonconstant temperature due to imperfect radial heat transfer.

B. Heat Transfer

In kinetic experiments, the temperature of the wall of the reactor is usually kept constant in time by means of thermal regulation devices, as well as in space by convenient adjustments of the furnace (see, e.g., Drillat *et al.*, 1962). Therefore, the temperature profile of the tube is flat with abrupt falls at the entrance and exit of the reactor. In general, however, the gas will pass some distance into the hot zone before reaching the wall temperature and again some distance out of it before sufficient cooling occurs for the reaction to stop. For a first-order reaction, with a rate constant following the Arrhenius law, Mulcahy and Pethard (1963) derived, by using simplifying assumptions, an expression giving the true value of the rate constant as a function of the "apparent" value, calculated by assuming thermal equilibration. Mulcahy (1973) points out the fact that thermal equilibration can be greatly accelerated by preheating a carrier gas. Furue and Pacey (1980) reexamined the question of heat transfer from the tube to the reacting gas. They derived a new expression, which was found to actually represent the first-order isomerization of cyclopropane in various cylindrical reactors. According to this expression, the fractional reduction in the measured rate constant is approximately $R^2/k\tau$, where R is the reactor radius, k the thermal diffusivity, and τ the average residence time.

C. Laminar Flow and Diffusion

Mulcahy and Pethard (1963) also examined the assumption that transport of the reagent molecules through the tube occurs only by simple piston flow. In fact, however, the molecules move not only by bulk flow, but also by diffusion down the concentration gradient produced by the reaction. Furthermore, the flow being laminar, there is no single residence time. Mulcahy and Pethard (1963) derived an expression giving the error committed by neglecting viscous flow and axial diffusion. Furue and Pacey (1980) found that this expression well represents their data on isomerization of cyclopropane. An upper limit to the fractional reduction of the measured rate constant would be 0.5 $k\tau$, k being the true rate constant and τ the mean residence time; this reduction is small. Villermaux (1971) examined the influence of diffusion occurring in an isothermal and laminar-flow tubular reactor either with a homogeneous or with a heterogeneous reaction. Lédé and Villermaux (1977) studied the same problem for simultaneous first-order homogeneous and heterogeneous reactions. They proposed methods for the determination of homogeneous reaction rate constants and found them suitable by studying the reaction of hydrogen atoms with ethane and propane (Lédé and Villermaux, 1978).

So far, discussion has been restricted to first-order or pseudo-first-order reactions. A more general approach is needed when the reaction is modeled by a mechanism comprising as many elementary processes as one deems necessary. The equations of mass, energy, and momentum are to be written and solved numerically.

D. Non-Quasi-Stationary State Pyrolysis

Tubular reactors have been used conventionally for studying pyrolysis reactions, at steady state and after the induction period of the reaction. However, as in the case of CFSTR, experiments can also be carried out during the induction period of the reaction. Actually, Pacey and Wimalasena (1978, 1980a,b) observed the induction period of neopentane pyrolysis at 822°K and of ethane pyrolysis at 900°K. In both cases, they were able to determine the initiation, propagation, and main termination rate constants. A priori calculation of the length of the induction period can be achieved by means of singular perturbation theory (Côme, 1979).

VIII. WALL-LESS REACTOR

Wall reactions in gas-phase kinetics always have been a source of controversy. That is the case not only for batch, stirred, and tubular reactors, but also even for shock-tube and VLPP experiments. Therefore, a wall-less reactor appears very attractive. Three wall-less reactors have been devised.

A. Reactor Design

The wall-less reactor, pioneered by Taylor *et al.* (1969), encompasses a flowing stream of hydrocarbon vapor inside a protective cylinder of inert flowing gas. The hydrocarbon enters the reactor at a low temperature and is brought to the reaction temperature by mixing with a zone of superheated gas. The flow is downward to avoid turbulence due to gravity effects. A cooled sampling tube is mounted on a three-dimensional microscope stage permitting movement to any part of the reaction zone. Products are analyzed by chromatography. By inserting a surface over which the gas stream may flow without significant resistance, it is possible to make a comparison of a reaction under both homogeneous and heterogeneous conditions (Hutchings *et al.*, 1969).

Because of the sampling probe, the reactor of Taylor *et al.* (1969) is open to the atmosphere at the downstream end. As paraffin pyrolyses are known to be very sensitive to trace amounts of oxygen, the rate measurements in this reactor could be in error. Thus Pratt and Rogers (1979a) designed another slightly modified reactor, to enable complete vacuum sealing against atmospheric pressure.

Taylor (1979) recognized that the primary disadvantage of the wall-less reactor, as he and his collaborators had used it, is the changing concentration due to diffusion as the injected reactant flows downstream. As a result of this effect, quantitative results are possible only with first-order reactions. To overcome this disadvantage, a new "homogeneous front" reactor has been developed by Taylor (1979). Gases I and II are brought to reaction temperature in the furnace before entering the reactor. Then they are distributed through two sets of alternative slits. The separate gases then pass into an open tube, where mixing does occur. When observations are made under wall-less (surface-free) conditions, the reaction time is limited to the time required for those reactants or products (which may be influenced by surface) to diffuse from the outer wall to center stream. The sampling procedure is the same as with the preceding reactor. Reaction times are around 1 s.

B. Treatment of Data

Beyond the experimental procedure, the main question regarding the wall-less reactor is about the reactor model for the treatment of data in order to get pertinent kinetic parameters.

The treatment of the experimental results by Taylor *et al.* (1969) does not take into account the laminar character of the flow, the radial and axial diffusion, and the temperature gradients. Moreover, Taylor *et al.* (1969) assume that all the studied pyrolyses are first-order reactions, which is contradictory with the results of Pratt and Rogers (1979a,b) in a wall-less reactor, with investigations performed by means of several other techniques, with the mechanisms deduced from our knowledge of free-radical chemistry and chemical kinetics, and with the fact that paraffin pyrolyses generally are strongly self-inhibited (the self-inhibition appears at small extents of reaction). Using a complete reaction mechanism and the corresponding rate laws could largely modify the interpretations and kinetic parameters of Taylor *et al.* (1969).

Pratt and Rogers (1979a) have used an entirely different procedure to treat their experimental data. First, they represent the reaction by a reaction mechanism comprising as many elementary processes as they consider necessary. Then they write the partial differential equations representing the conservation of material, energy, and momentum, with each equation including terms describing mass flow, diffusion, and chemical reaction. These equations are strongly coupled and nonlinear, and thus not easy to solve numerically. By means of convenient simplifying hypotheses (e.g., axial diffusion negligible compared to axial flow), these equations become of parabolic form, for which efficient numerical methods of solution have been described (e.g., in Patankar and Spalding, 1970). Pratt and Rogers (1979a) have modified the program of these authors in order to be able to use any reaction mechanism instead of global rate expressions.

C. Wall Effects

In their articles on the pyrolyses of ethane and propane, Pratt and Rogers (1979a,b) conclude that the observations made in the wall-less reactor (established and interpreted as already discussed) are in essential agreement with the results of most other studies. This proves that the body of knowledge accumulated on paraffin pyrolyses is not in general in serious error by reason of underrepresenting wall effects. If the conclusions of Taylor *et al.* (1969) were true, a very large part of the present data on hydrocarbon pyrolyses would be in error, despite the wide consistency of experimental and theoretical results (see also Section IX).

IX. NEOPENTANE PYROLYSIS

The pyrolysis of neopentane has been studied in most reactors previously mentioned: batch reactor (Engel *et al.*, 1957; Anderson and Benson, 1964; Baronnet and Niclause, 1969; Baronnet *et al.*, 1968, 1971, 1974; Halstead *et al.*, 1969; Franck and Sackett, 1969; Marshall *et al.*, 1976), single-pulse shock tube (Tsang, 1966; Bradley and West, 1976), continuous-flow stirred tank reactor (Rondeau *et al.*, 1976; Azay and Côme, 1979b), very low pressure pyrolysis (Baldwin *et al.*, 1979), tubular reactor (Pacey, 1973), non-quasi-stationary state pyrolysis (Marquaire and Côme, 1977, 1978; Pacey and Wimalasena, 1978, 1980a), wall-less reactor (Taylor *et al.*, 1969; Hutchings *et al.*, 1969; Pratt and Rogers, 1981). The reaction has been carried out in the temperature range 700–1300°K (i.e., over 600°K) and at global pressures from the Knudsen region to around 1 atm. Therefore, neopentane pyrolysis appears to be a good standard for comparing gas-phase pyrolysis reactors.

A. Stoichiometry

There is general agreement on the identity of the products of the reaction, on their primary or secondary nature, and on their major or minor importance (which of course depends on the temperature). Three articles at least, of a total of six (Halstead *et al.*, 1969; Baronnet *et al.*, 1971; Bradley and West, 1976; Azay and Côme, 1979b; Pacey, 1973; Taylor *et al.*, 1969), mention the following products: isobutylene, methane, hydrogen, ethane, 2-methyl-1-butene, ethylene, propene, 2-methyl-2-butene, isoprene, allene, propyne, isobutane; but most products have been found by nearly all the six authors.

B. Kinetics

The results of most studies of methane formation indicate a Rice–Herzfeld radical-chain mechanism of $\frac{3}{2}$ initial order, at any temperature and pressure sufficiently high for the rate constant of methyl radical recombination to be no longer in the "falloff" region.

Most authors also agree on the essentially homogeneous character of the reaction. Evidence for this is obtained by changing the nature and the relative extent of the reactor walls. Only Hutchings *et al.* (1969) found very different behaviors for the "homogeneous" and "heterogeneous" reactions.

C. Initiation Process

The Arrhenius parameters of the rate constant of the initiation process,

$$\text{neo-}C_5H_{12} \longrightarrow \cdot CH_3 + \cdot C(CH_3)_3$$

are reported in Table I, along with the numerical values of the rate constant at 821°K, which, of course, are achieved in most cases by a large extra-

TABLE I

Initiation Rate Constant of Neopentane Pyrolysis

log A (s^{-1})	E (kcal mol^{-1})	k (s^{-1}) $\times 10^5$ at 821°K	Temperature (°K)	References
17.6	84.0	1.7	703–743	Marquaire and Côme (1977, 1978)
18.2	85.8	2.3	713–823	Halstead *et al.* (1969)
16.8	82.0	0.9	723–803	Baronnet *et al.* (1968, 1971, 1974)
16.1	78.9	1.2	756–845	Marshall *et al.* (1976)
17.7	85.2	1.0	793–953	Pacey (1973)
—	—	1.7	821	Pacey and Wimalasena (1978)
17.3	82.0	2.9	908–1008	Azay and Côme (1979b)
16.9	80.5	2.9	923–1073	Taylor *et al.* (1969)
17.3	80.8	6.2	1000–1260	Baldwin *et al.* (1979)
16.5	80.4	1.3	1030–1300	Bradley and West (1976)
16.1	78.2	1.9	1070–1240	Tsang (1966)

polation (Table I). If log A and E are assumed to be constant throughout the temperature (T) interval, and if the same weights are given to all values of the table, the following mean values and 95% confidence intervals are obtained: log A (s^{-1}) $= 17.1 \pm 0.4$; E(kcal mol^{-1}) $= 81.8 \pm 1.5$; $k(s^{-1}, T = 821°K) = (2.2 \pm 0.9) \times 10^{-5}$.

The 95% confidence intervals correspond to variations of $\pm 2\%$ for both log A and E, variations clearly below the limits of precision of any kinetic method.

From this observation, two conclusions may be drawn. Because of the different experimental conditions used and the good consistency of the results, it would be difficult to claim superiority for one particular method. On the other hand, it is very unlikely that wall effects play an important role in neopentane pyrolysis.

Recently, using a wall-less reactor, Pratt and Rogers (1981) determined $k = 2.8 \times 10^{-3}$ at $T = 983°K$, a value much lower than $k = 8.2 \times 10^{-2}$ predicted on the basis of the mean values of A and E deduced from the table. This value thus appears surprisingly low.

X. SUMMARY COMPARISON OF REACTORS

As numerous reactors can be used for the kinetic study of pyrolysis reactions, a comparison is rather attractive. Three kinds of errors arise in the derivation of kinetic parameters from experimental results: errors in measurements, errors in reaction model, and errors in reactor model.

A. Experimental Errors

Stochastic errors of measurement affect the precision of parametric estimation, but not its validity. Furthermore, this type of error may be assessed from the scattering of results. Systematic errors of measurement are very difficult to detect a priori, but a comparison with other results makes them obvious. When consistent results are achieved by several different methods, it may be hoped—and supposed—that this is not due to artifacts.

B. Errors in Reaction Model

This type of error arises when reactions that do occur are neglected. For gas-phase reactions, either homogeneous elementary processes or wall reactions that really play a role in the reaction kinetics may be neglected. In some cases, secondary reactions are negligible and the reaction mechanism may be assumed to be simple. That is the case, for example, in VLPP, Tsang's experiments (1964, 1966), and non-quasi-stationary state pyrolysis; however, in VLPP the reaction model has to be complicated by an extensive recourse to RRKM theory. In most cases, indeed, complex modeling is necessary (Côme, 1982). Wall effects may occur in most reactors but are minimized in the wall-less reactor and the shock tube. In the other reactors, wall effects can be studied by changing the nature, the geometry, and the relative extent of the walls.

C. Errors in Reactor Model

Such errors stem from wrongly evaluated operating conditions (reaction and quenching time, reaction temperature, etc.) and from errors in the residence time distribution, in the micromixing state, and so on. From this point of view, the shock tube is a very complex reactor, whereas the batch reactor and also CFSTR are easily characterized and not too far from ideality.

D. Other Kinetic Methods

Besides the versatility of the reactor as a kinetic tool, the range of available experimental conditions must also be considered. In this respect, the discussion must be extended to other reactors or kinetic methods, which have not been included here for consideration. Let us enumerate some of these methods, which have also been proven valuable for the study of free-radical reactions (selected articles describe either the method or topics related to reactions of methyl radical, neopentane, ethane, etc.)

1. Toluene-carrier technique (Szwarc, 1950)
2. Aniline-carrier technique (Esteban *et al.*, 1963)
3. Rotating sector (Gomer and Kistiakowsky, 1951)
4. Very low pressure reactor (Baghal-Vayjooee *et al.*, 1978)
5. Molecular modulation spectrometry (Parkes *et al.*, 1976)
6. Free-radical buffer (Hiatt and Benson, 1972)
7. Isotope effect (Clark and Quinn, 1976)
8. Thermal sensitization (Szirovicza and Marta, 1975)
9. Photochemical sensitization (Furimsky and Laidler, 1972)

Many other methods are also of interest: flash photolysis, electrical discharge, plasma jet, pulsed lasers, molecular beams, and gas chromatographic reactor. See also the following: Albright and Crynes (1976), Ashmore *et al.* (1966), Banford and Tipper (1969), Benson (1960), Eyring *et al.* (1975), Friess *et al.* (1961), Laidler (1965), Melville and Gowenlock (1964), Nicholas (1976), Oblad *et al.* (1979), and Pratt (1969). Furthermore, "engineering" reactors are also to be considered because experimental results, whatever their origin, are to be integrated in our body of knowledge. It is epistemologically obvious that consistency is the master key in chemical kinetics, as in other sciences.

REFERENCES

Albright, L. F. and Crynes, B. L., eds. (1976). "Industrial and Laboratory Pyrolyses," ACS Symposium Series, No. 32. Am. Chem. Soc., Washington, D.C.

Anderson, K. H., and Benson, S. W. (1964). *J. Chem. Phys.* **40**, 3747.

Ashmore, P. G., Dainton, F. S., and Sugden, T. M., eds. (1966). "Photochemistry and Reaction Kinetics." Cambridge Univ. Press, London and New York.

Azay, P., and Côme, G. M. (1979a). *Ind. Eng. Chem., Process Des. Dev.* **18**, 754.

Azay, P., and Côme, G. M. (1979b). *Am. Chem. Soc., Natl. Meet., 177th; Chem. Soc. Jpn., Natl. Meet., 38th, Honolulu, Hawaii.*

Baghal-Vayjooee, M. H., Colussi, A. J., and Benson, S. W. (1978). *J. Am Chem. Soc.* **100**, 3214.

Baldwin, A. C., Davidson, I. M. T., and Howard, A. V. (1975). *J.C.S. Faraday Trans. I* **71**, 972.

Baldwin, A. C., Lewis, K. E., and Golden, D. M. (1979). *Int. J. Chem. Kinet.* **11**, 529.

Bamford, C. H., and Tipper, C. F. H., eds. (1969). "Comprehensive Chemical Kinetics. Vol. 1: The Practice of Kinetics." Elsevier, Amsterdam.

Baronnet, F., and Niclause, M. (1969). *C. R. Hebd. Séances Acad. Sci., Ser. C* **269**, 513.

Baronnet, F., Dzierzynski, M., Martin, R., and Niclause, M. (1968). *C. R. Hebd. Séances Acad. Sci., Ser. C* **267**, 937.

Baronnet, F., Dzierzynski, M., Côme, G. M., Martin, R., and Niclause, M. (1971). *Int. J. Chem. Kinet.* **3**, 197.

Baronnet, F., Côme, G. M., and Niclause, M. (1974). *J. Chim. Phys.* **71**, 1214.

Benson, S. W. (1960). "The Foundations of Chemical Kinetics." McGraw-Hill, New York.
Benson, S. W., and Spokes, G. N. (1967). *J. Am. Chem. Soc.* **89**, 2525.
Bradley, J. N., and Frend, M. A. (1971). *Trans. Faraday Soc.* **67**, 72.
Bradley, J. N., and West, K. O. (1976). *J.C.S. Faraday Trans. I* **72**, 8.
Clark, J. A., and Quinn, C. P. (1976). *J.C.S. Faraday Trans. I* **72**, 706.
Côme, G. M. (1977). *J. Phys. Chem.* **81**, 2560.
Côme, G. M. (1979). *Comput. Chem. Eng.* **3**, 603.
Côme, G. M. (1982). *In* "Comprehensive Chemical Kinetics" (C. H. Bamford and C. F. H. Tipper, eds.), Vol. 24, pp. 249–332. Elsevier, Amsterdam.
Corbel, S., Marquaire, P. M., and Côme, G. M. (1981). *Chem. Phys. Lett.* **80**, 34.
David, R., and Matras, D. (1975). *Can. J. Chem. Eng.* **53**, 297.
David, R., and Villermaux, J. (1973). *Can J. Chem. Eng.* **51**, 630.
David, R., and Villermaux, J. (1978). *J. Chim. Phys.* **75**, 656.
David, R., Houzelot, J. L., and Villermaux, J. (1979). *Chem. Eng. Sci.* **34**, 867.
Davidson, I. M. T., Dean, C. E., Lawrence, F. T., and Wood, I. T. (1980). *Int. Symp. Gas Kinet., 6th* p. 30.
Drillat, J., Tardieu De Maleyssie, J., and Foulatier, R. (1962). *Instrum. Lab.* **20**, 13.
Engel, J., Combe, A., Letort, M., and Niclause, M. (1957). *C. R. Hebd. Seances Acad. Sci.* **244**, 453.
Esteban, G. L., Kerr, J. A., and Trotman-Dickenson, A. F. (1963). *J. Chem. Soc.* pp. 3873, 3879.
Eyring, H., Henderson, D., and Jost, W., eds. (1975). "Physical Chemistry, an Advanced Treatise. Vols. VIA, VIB: Kinetics of Gas Reactions." Academic Press, New York.
Franck, D. J., and Sackett, W. M. (1969). *Geochim. Cosmochim. Acta* **33**, 811.
Friess, S. L., Lewis, E. S., and Weissberger, A., eds. (1961). "Technique of Organic Chemistry. Vol. 8, Parts I, II: Investigation of Rates and Mechanisms of Reactions." Wiley (Interscience), New York.
Furimsky, E., and Laidler, K. J. (1972). *Can. J. Chem.* **50**, 1115, 1123, 1129.
Furue, H., and Pacey, P. D. (1980). *J. Phys. Chem.* **84**, 3139.
Glick, H. S., Squire, W., and Hertzberg, A. (1955). *Symp. (Int.) Combust. [Proc.]* **5**, 393.
Golden, D. M., Spokes, G. N., and Benson, S. W. (1973). *Angew. Chem., Int. Ed. Engl.* **12**, 534.
Gomer, R., and Kistiakowsky, G. B. (1951). *J. Chem. Phys.* **19**, 85.
Halstead, M. P., Konar, R. S., Leathard, D. A., Marshall, R. M., and Purnell, J. H. (1969). *Proc. R. Soc. London, Ser. A* **310**, 525.
Hiatt, R., and Benson, S. W. (1972). *J. Am. Chem. Soc.* **94**, 25.
Hutchings, D. A., Taylor, J. E., and Frech, K. J. (1969). *J. Phys. Chem.* **73**, 3167.
Laidler, K. J. (1965). "Chemical Kinetics." McGraw-Hill, New York.
Lédé, J., and Villermaux, J. (1977). *J. Chim. Phys.* **74**, 459.
Lédé, J., and Villermaux, J. (1978). *Can J. Chem.* **56**, 392.
Lifschitz, A., and Frenklach, M. (1977). *J. Chem. Phys.* **67**, 2803.
Lifschitz, A., Bauer, S. H., and Resler, E. L. (1963). *J. Chem. Phys.* **38**, 2056.
Marquaire, P. M., and Côme, G. M. (1977). *Int. Symp. Gas Kinet., 5th*, p. 2.
Marquaire, P. M., and Côme, G. M. (1978). *React. Kinet. Catal. Lett.* **9**, 165, 171.
Marshall, R. M., Purnell, J. H., and Storey, P. D. (1976). *J.C.S. Faraday Trans. I* **72**, 85.
Martin, R., Niclause, M., and Scacchi, G. (1976). *In* "Industrial and Laboratory Pyrolyses" (L. F. Albright and B. L. Crynes, eds.), ACS Symposium Series, No. 32, p. 37. Am. Chem. Soc., Washington, D.C.
Matras, D., and Villermaux, J. (1973). *Chem. Eng. Sci.* **28**, 129.
Melville, H. W., and Gowenlock, B. G. (1964). "Experimental Methods in Gas Reactions." Macmillan, New York.
Mulcahy, M. F. R. (1973). "Gas Kinetics." Nelson, London.
Mulcahy, M. F. R., and Pethard, M. R. (1963). *Aust. J. Chem.* **16**, 527.

Mulcahy, M. F. R., and Williams, D. J. (1961). *Aust. J. Chem.* **14**, 534.
Nicholas, J. (1976). "Chemical Kinetics: A Modern Survey of Gas Reactions." Harper, New York.
Oblad, A. G., Davis, H. G., and Eddinger, R. T., eds. (1979). "Thermal Hydrocarbon Chemistry," Advances in Chemistry. Series, No. 183. Am. Chem. Soc., Washington, D.C.
Pacey, P. D. (1973). *Can. J. Chem.* **51**, 2415.
Pacey, P. D., and Wimalasena, J. H. (1978). *Chem. Phys. Lett.* **53**, **593**.
Pacey, P. D., and Wimalasena, J. H. (1980a). *J. Phys. Chem.* **84**, 2221.
Pacey, P. D., and Wimalasena, J. H. (1980b). *Chem. Phys. Lett.* **76**, 433.
Parkes, D. A., Paul, D. M., and Quinn C. P. (1976). *J.C.S. Faraday Trans. I* **72**, 1935.
Patankar, S. V., and Spalding, D. B. (1970). "Heat and Mass Transfer in Boundary Layers. A General Calculation Procedure." Intertext, New York.
Pratt, G. L. (1969). "Gas Kinetics." Wiley, New York.
Pratt, G. L., and Rogers, D. (1979a). *J.C.S. Faraday Trans. I* **75**, 1089.
Pratt, G. L., and Rogers, D. (1979b). *J.C.S. Faraday Trans. I* **75**, 1101.
Pratt, G. L., and Rogers, D. (1981). *J.C.S. Faraday Trans. I* **77**, 2751.
Purnell, J. H., and Quinn, C. P. (1967). *In* "Photochemistry and Reaction Kinetics" (P. G. Ashmore, F. S. Dainton, and T. M. Sugden, eds.), p. 330 Cambridge Univ. Press, London and New York.
Rondeau, J. A., and André, J. C. (1975). *Rev. Sci. Instrum.* **46**, 692.
Rondeau, J. A., Large, J. F., and Côme, G. M. (1976). *In* "Industrial and Laboratory Pyrolyses" (L. F. Albright and B. L. Crynes, eds.), ACS Symposium Series, No. 32, p. 131. Am. Chem. Soc., Washington, D.C.
Szirovicza, L., and Marta, F. (1975). *React. Kinet. Catal. Lett.* **3**, 9.
Szwarc, M. (1950). *Chem. Rev.* **47**, 75.
Taylor, J. E. (1979). *NBS Spec. Publ. (U.S.)* No. 561, 1353.
Taylor, J. E., Hutchings, D. A., and Frech, K. J. (1969). *J. Am. Chem. Soc.* **91**, 2215.
Tsang, W. (1964). *J. Chem. Phys.* **40**, 1498.
Tsang, W. (1966). *J. Chem. Phys.* **44**, 4283.
Villermaux, J. (1971). *Int. J. Heat Mass Transfer* **14**, 1963.
Volkan, A. G., and April, G. C. (1977). *Ind. Eng. Chem., Process Des. Dev.* 16, 429.

12

Pyrolysis of Petroleum Liquids: Naphthas to Crudes

SIEGFRIED NOWAK and HUBERT GÜNSCHEL

I. INTRODUCTION

Hydrocarbon pyrolysis is economically one of the most important reactions that has been increasingly used industrially since the mid 1950s. It is the key process for modern petrochemistry and will remain so in the

Pyrolysis: Theory and Industrial Practice

ISBN 0-12-048880-9

future. In a single stage of processing, important basic chemicals are formed such as ethylene, propylene, butadiene, and C_6–C_8 aromatics. Presently the main product, ethylene, represents the quantitatively most important chemical substance based on crude oil, and it is the base stock for 30% of all petrochemicals produced. The historical development of hydrocarbon pyrolysis is characterized by (*a*) continuously growing capacities of pyrolysis plants; (*b*) increase of space–time yields in the overall process; (*c*) an increase of the ethylene selectivity; (*d*) an improvement in energetic efficiency, especially by a better utilization of the waste heat; (*e*) the extension of running times; (*f*) a wider range of feedstock flexibility; and (*g*) the expansion of the feedstock basis toward high-boiling fractions. These advances and increases in performance were first of all due to constructional and technological improvements of the cracking furnace and its equipment. In the years to come the development will continue in the directions indicated. Production capacities, however, will be raised not so much by enlarging the plants as by increasing the number of plants.

Unfortunately the high level of technological advancement is not matched by an equally high level of knowledge of the scientific fundamentals and correlations of the reactions taking place in the overall process. Without underestimating the pioneering work performed by Rice and Herzfeld (1934) and Kossiakoff and Rice (1934), today the cracking process is far from being scientifically described. Therefore, the theoretical aspect of pyrolysis must be given further consideration.

In comparison to the technical advances, the cause for the slow theoretical increase of knowledge lies in the nature of the reactions. The pyrolytic decomposition of hydrocarbons is one of the most complicated reactions of hydrocarbon chemistry. This can be seen clearly in the following characteristics.

1. Feedstocks and final products are multicomponent mixtures.
2. The overall reaction takes place according to a mechanism with consecutive and parallel partial reactions.
3. Arising intermediate compounds such as radicals are ephemeral and experimentally difficult to detect.
4. The radicals initiate reaction chains.
5. Light olefinic reaction products inhibit further decomposition; impurities in the feed and wall effects can affect the rate and direction of the decomposition.
6. The reaction is a homogeneous endothermic gas-phase reaction.
7. The process is not isothermal, isobaric, or isochoric.
8. The conversions occur at high temperatures ($>600°C$).
9. The reaction is kinetically controlled and after reaction times of less than 0.5 s, it must be terminated by quenching.

II. PYROLYSIS OF NAPHTHA FRACTIONS

Straight-run naphtha is currently the most significant feedstock for commercial pyrolysis. Its share in the available overall base feedstocks rose continuously until the mid 1970s, while in the same period that of liquefied petroleum gases, C_2 up to C_4 (LPG), gradually diminished. The cracking of liquid hydrocarbon fractions became superior to LPG cracking to such an extent that (*a*) higher yields of olefins were achieved, (*b*) a considerable amount of associated products could be provided for appropriate chemical utilization, and (*c*) LPG were not available at each location and in the amounts required.

In 1975 about 90% of the ethylene produced was obtained from naphtha in Europe. Recently, because of the worldwide shortage of crude oil, the share of heavy crude oil fractions (kerosene, atmospheric gas oil, vacuum gas oil) has risen continuously, but without endangering the dominant role of straight-run naphtha.

Naphtha is a complex mixture of hydrocarbons with different structures and molecular configurations. Because of its specific data (distillation characteristics, distribution of molecular weight, evaporation enthalpy, etc.), the tubes for the pyrolysis furnaces display characteristic design features.

A. Industrial Naphtha Cracking (Steam Cracking)

From among all the techniques suggested for the thermal cracking, the tube furnace—also called the steam cracker—has proved best. The oil cracking is accomplished in a dynamic system under nonisothermal and nonisobaric conditions within a flame-heated tube in the presence of steam. First, in that section of the tube called the convection zone the steam–hydrocarbon mixture is heated to temperatures of about 600°C, to be subsequently cracked within the tubular coils of the radiant zone at temperatures up to 900°C. The coil lengths in the radiant zone amount to about 90–120 m, the diameter of the coils varies from 8 to 10 cm, and the residence times of the reaction mixture in the radiant zone vary from 0.2 to 0.5 s. By developing ever more heat-resistant nickel–chromium steel tubes, the heat-flux density of the tube material could be increased to 350,000 kJ/m^2 h. The specific heat requirement lies at about 5000 kJ/kg naphtha and steam. The cracked-gas mixture, emerging with a velocity of 200 m/s and having a temperature of up to 900°C, must be cooled immediately to 350 to 400°C directly at the furnace outlet. This is indirectly affected in quench boilers, the sensible heat being gained as high-pressure steam. The cooling velocities must be very high—they lie in the order of

10^4 °C/s. In the quench boiler, the temperature must be higher than the hydrocarbon dew point in order to prevent fouling by deposits. A further cooling of the cracked gas to about 280°C is effected by injecting quenching oil. The subsequent oil-scrubbing column (topping fractionation) brings about another cooling to about 100°C. After that the mixture of about 30°C, which is only partially gaseous, goes through air and water coolers and into precipitator containers, where the steam condensate and liquid hydrocarbons are separated from the light cracked gas. After stripping, the process water is recycled for steam generation. The hydrocarbon condensates provide for the required reflux for the head of the topping column. The remaining cracked gas is separated into its components according to the principle of low-temperature, medium-pressure distillation in a joined gas-separation plant.

B. Process Conditions and Numerical Values of Cracking Severity

The yields of cracked products are a function of the process parameters. Precise knowledge of process parameters is useful for both the design and the economical operation of a plant. Consequently, fixed and variable parameters are distinguished. The former are given by the designer, whereas the latter, serving to control the plant, can be freely set, within limits.

The most important design parameters are (*a*) the overall tube length of the furnace, (*b*) the tube diameter, (*c*) the maximum heat flux, (*d*) the maximum tube wall temperature, (*e*) the maximum hydrocarbon and process steam throughput, and (*f*) the boiling range of the base stock.

The most important control parameters are (*a*) temperatures alongside the radiant zone (T), (*b*) the residence times at cracking temperatures (τ), (*c*) the partial pressures of hydrocarbons (P), and (*d*) the nature of the components of the feedstock (N).

A description of the nature of the feedstock by a numerical value is most difficult. The introduction of the UOP-K factor, as well as the BMCI factor (U.S. Bureau of Mines correlation index) have proved useful. Also, certain measuring values from the PONA analysis (i.e., the distribution of *n*-paraffins, isoparaffins, olefins, naphthenes, and aromatics), as well as the branching degree of dearomatized fractions, defined as the quotient from C numbers,

$$x = \frac{C_{\text{methyl}}}{C_{\text{total}}} \tag{1}$$

are well suited for arriving at a rough characterization of a complex fraction used industrially.

The UOP-K factor, a universally accepted index for the chemical nature of hydrocarbon fractions, is defined as follows (Watson and Nelson, 1935):

$$\text{UOP-K} = 1.216 \frac{\sqrt[3]{T_s}}{d} \tag{2}$$

where T_s = median molar boiling point (°K), and d = specific gravity at 15.6°C.

Other relations for estimating UOP-K factors of heavy hydrocarbon distillates were published by Woodle (1980).

The BMCI value can be calculated by the equation (Green *et al.*, 1975):

$$\text{BMCI} = \frac{87552}{\text{VABP}} + 473.7(d) - 456.8 \tag{3}$$

where VABP = volumetric average boiling point (R) and d = specific gravity at 15.6°C.

The use of these factors is discussed in subsequent sections. For a given naphtha the cracking severity, and in connection with it the yield (y) of a final product, is a function of the time–temperature–pressure response alongside the reaction coil.

$$y = \mathrm{f}(\tau, T, P) \tag{4}$$

If the cracking-coil geometry is known, instead of τ also the throughput of naphtha and diluent steam can be given. A first useful empirical value W for the time–temperature relation was given by Linden and Reid (1959). They found out experimentally that certain value pairs of furnace-outlet temperature and mean residence time lead to nearly the same cracking severity (conversion level), if the relation

$$W = T\,\tau^{0.06} \tag{5}$$

is fulfilled, thus the same feedstock and the same hydrogen partial pressure must be given. For high ethylene yields, however, shorter residence times and higher temperatures are more advantageous (Balbig and Lange, 1975) (Fig. 1).

According to the proposal from Zdonik *et al.* (1967a,b), the integral effect of the time–temperature response alongside the reaction coil can also be well expressed by the numerical value KIF (kinetic cracking-severity function), and it can be used as a measurement for the cracking severity.

$$\text{KIF} = \int_0^{\tau} k_5\, d\tau = \ln \frac{C_0}{C} = 2.3 \log \frac{1}{1 - x} \tag{6}$$

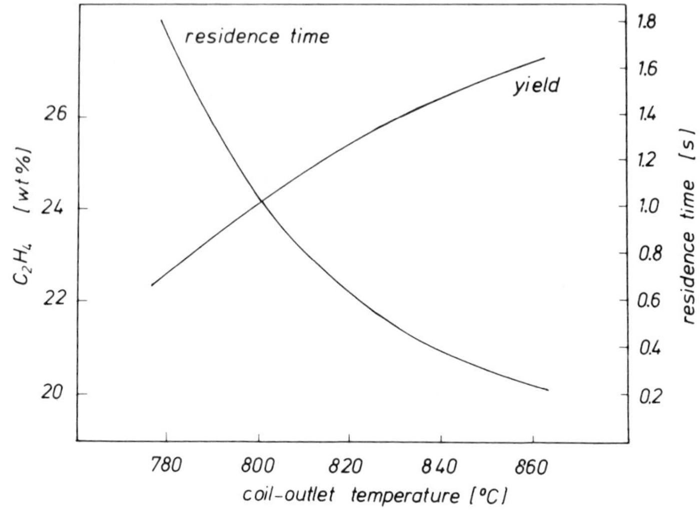

Fig. 1. Yield of ethylene in relation to coil-outlet temperature and residence time (Balbig and Lange, 1975). Fixed parameters: KIF, coil-outlet pressure, dilution ratio, feed, and feed rate.

where k_5 = the decomposition constant of the guiding component *n*-pentane, C_0 resp. c = concentrations of the guiding component in the feedstock resp. in the effluent (mol/mol), and x = conversion level of the guiding component. For example, in a conversion of *n*-pentane up to 90%, a KIF value of 2.3 resulted from a calculation according to the formula in Eq. (6). The value of k_5 is temperature dependent and it can be calculated according to the Arrhenius equation, if both the activation energy and the frequency factor are known. A repeatedly branched isoparaffin with a molecular weight of about 100 should have a cracking response corresponding to that of an average molecule of "naphtha." Although such a compound is better suited as a guiding component for determining the KIF value, it is more difficult to analyze it in the reaction mixture. Typical dependencies of the yields on the KIF value will be shown later on.

An overall cracking-severity function (OKSF), based likewise on the overall decomposition kinetics, was derived by Illés *et al.* (1978; Illés and Horvath, 1975). It can be calculated by means of the following relation:

$$\mathrm{OKSF} = \int_0^\tau \left(\frac{C_0}{E_e}\right)^{n-1} k\, d\tau \tag{7}$$

where C_0 = concentration of the feedstock at the tube inlet (mol/mol),

n = overall kinetic reaction order at a given pyrolysis temperature, E_e = effective expansion of the feed hydrocarbon mixture at the conversion level x, and k = decomposition constant of the overall reaction $((\mathrm{mol/mol})^{1-n}\,\mathrm{s}^{-1})$.

The function is formally as well as physically analogous to the KIF value, yet it considers the reaction orders deviating from value $n = 1$ in case of high conversions. The yields of the reaction products depend on the OKSF value, as shown in Section II,D,1.

An integral expression considering not only the residence time and temperature, but also the hydrocarbon partial pressure, was developed by Witt and Wall (1977). They combine these three parameters into a simple value called "molecular collision parameter" (MCP).

$$\mathrm{MCP} = \int_0^{\tau} \frac{(HCPP)^2}{T^{1.5}}\, d\tau \tag{8}$$

where $HCPP$ = hydrocarbon partial pressure (psi), τ = residence time (s), and T = absolute temperature (°K).

A low hydrocarbon partial pressure can be reached by means of a high steam dilution or a low overall pressure in the cracking coil.

The MCP value is a measure for the collisions of the hydrocarbon molecules. It is independent of the cracking-coil geometry. The rates of bimolecular reactions rise with the MCP value. The smaller the MCP value is, the larger the ethylene and butadiene yields at a constant propylene/ethylene ratio and the smaller the methane and pyrolysis fuel yields are. By means of the MCP value, yields from a single feedstock but different reactors can be compared with one another.

C. Influence and Significance of Cracking Parameters

In order to ascertain significant process parameters from applicable process parameters including local measuring points alongside the pyrolysis coil, tests according to an elaborated plan must be run and the data evaluated. Respective investigations on a bench-scale unit were carried out by Günschel and Nowak (1973) using a naphtha-cracking furnace. In this instance a test plan created by Stove and Meyer (1966) was used. The furnace permitted settings over a wide throughput, pressure, and temperature range. By a sectionwise heating of the cracking coil, temperature profiles from 600 to 900°C could be set.

Table I gives yield alterations for selected compounds, with parameter settings varying in the given limits. From Table I one can see that increased

TABLE I

Shifts in Yield versus Changing of Parameter Settings[a]

	Change of parameter settings: range						
Products	ΔT_E 400–450°C	ΔT_1 640–700°C	ΔT_2 750–780°C	ΔT_A 800–840°C	ΔP_A 2–3 atm	Δm_N 4–6 kg/h	ΔV(st/N) 0.50–0.75
Ethylene	−0.02	+0.25	+1.05	+2.38	−1.15	−0.68	−0.02
Propylene	0	+1.41	−3.90	+0.27	−0.43	+2.67	+1.47
Butenes	+0.11	−0.48	−0.65	−1.83	−0.69	+1.57	+0.31
Butadiene	−0.02	+0.15	−0.85	+0.15	−0.38	+0.75	+0.23
Benzene	−0.24	+0.21	−1.40	+3.55	+1.70	−1.13	+0.11

[a] Günschel and Nowak (1973).

[b] Abbreviations: T_E, temperature at coil inlet; T_1, temperature at 25% coil length; T_2, temperature at 50% coil length; T_A, temperature at coil outlet; P_A, pressure at coil outlet; m_N, naphtha throughput; V(st/N), ratio of steam to naphtha (g/g); +, increase of yield (wt%); −, decrease of yield (wt%).

yields of the following products can be obtained: ethylene, if the temperature is raised at the coil outlet; propylene, if the throughput of naphtha is increased; and benzene, if the pressure and temperature settings at the coil outlet are raised. Even if the results cannot be conveyed to other pyrolysis plants, they nevertheless provide qualitative information on how the parameter settings influence the yields. In the same way, other dependent target quantities—such as operating expenses, energy consumptions, running times, and profits—can be brought into correlation with independent variable process parameters. In this connection the effects resulting from resetting the measured values on the target quantities of interest can be estimated. Finding those process parameters that are really independent of one another, however, has turned out to be a problem.

D. Characteristic Product Yields of Naphtha Pyrolysis

1. *Yield Dependence on the Cracking Severity*

If a naphtha–steam mixture is fed to a pyrolysis tube at constant temperature and pressure, the naphtha conversion increases alongside the tube.

Figure 2 shows the product yields for a light naphtha plotted against the cracking severity (KIF) (Balbig and Lange, 1975). These types of curves

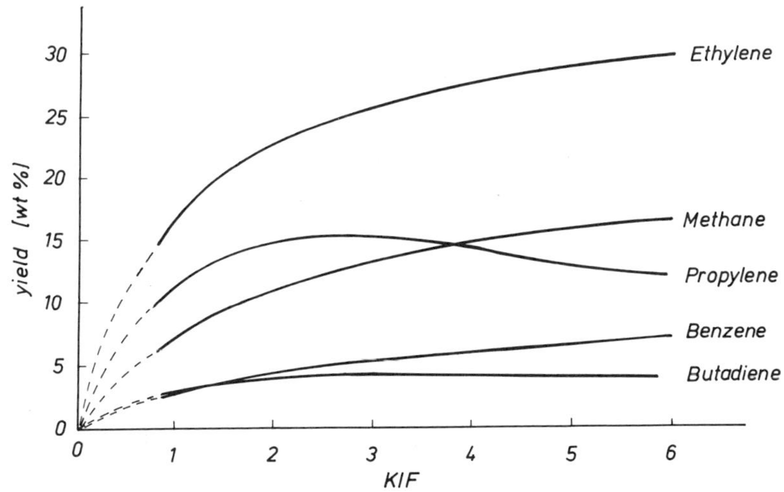

Fig. 2. Product curves as a function of KIF at fixed temperature and pressure, naphtha. (From Balbig and Lange, 1975.)

depend on the nature of the feedstock, on the pressure, and particularly on the cracking temperature. Analogous dependencies exist between the yields and the OKSF value (Illés and Horvath, 1975).

A correlation of the hydrogen and methane yield curves form convex curves as achieved by means of the function

$$y = a(1 - e^{b\ \mathrm{OKSF}}) \tag{9}$$

whereas a correlation of olefin and diolefin yield curves is obtained by using the function

$$y = a(e^{b\ \mathrm{OKSF}} - e^{c\ \mathrm{OSKF}}) \tag{10}$$

The decrease of hydrocarbon components i of the feed naphtha, too, is affected exponentially according to the approximation

$$y_i = a\ e^{b\ \mathrm{OSKF}} \tag{11}$$

In these functions the symbols a, b, and c denote empirical constants and

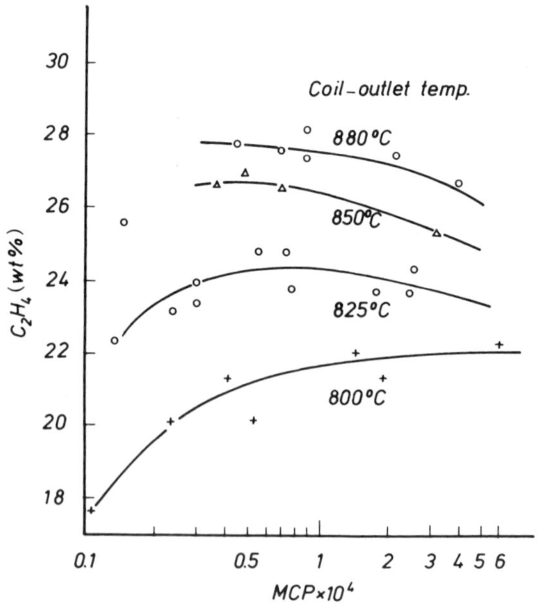

Fig. 3. Ethylene yield plotted against molecular collision parameter (MCP). (Anonymous, 1978*b*. Reproduced with permission from *Oil Gas J.*)

TABLE II

Influence on Yields by Decreasing MCP[a]

Product	Influence on yields at falling MCP value	
	High cracking severity	Low cracking severity
Methane	−	−
Ethane	−	−
Ethylene	+	−
Methane/ethylene	−	−
Propylene	+	−
1,3-Butadiene	+	−
BTX[b]	−	−

[a] Lohr and Dittmann (1977a, 1978). Reproduced with permission from *Oil Gas J.*

[b] Benzene, toluene, and xylene.

the symbol y_i, the yields in wt%. In Fig. 3 the ethylene yield is plotted against the MCP value (Anonymous, 1978b).

With respect to a favorable yield pattern, low MCP values at high cracking severities or high MCP values at low cracking severities are advantageous, which is illustrated by the survey in Table II (Lohr and Dittmann, 1977a).

2. *Yield Dependencies on the Nature of Feedstock*

The olefin potential of a hydrocarbon fraction correlates with the paraffin content. Whereas from normal paraffins ethylene is preferentially formed, from repeatedly branched isoparaffins predominantly propylene is obtained. Base stocks with a relatively high content of naphthene and a small content of paraffin are interesting for the production of diolefins. Butadiene and ethylene in equimolecular amounts are formed in a primary reaction during the decomposition of cyclohexane, and butadiene in smaller amounts in a secondary reaction is formed by the dimerization of ethylene. These dependencies are illustrated in Table III showing yields from different naphthas (de Blieck *et al.*, 1971).

The following naphthas are listed: column 1, primary reference fuel; column 2, a highly naphthenic naphtha; column 3, an isoparaffinic naphtha; column 4, a high naphthenic naphtha rich in aromatics. The overall value of the products obtained falls from naphtha 1 to naphtha 4. Eisenlohr *et al.* (1967) tested different naphthas in an industrial pyrolysis plant, as well.

TABLE III

Influence of PONA Analysis on Product Distribution[a]

	Naphtha type[b]			
	1	2	3	4
Density	0.720	0.710	0.712	0.727
PONA				
nP	41	30	29	20
iP	34	31	40.5	31
N	20	36	23.5	33
A	5	3	6.5	16
Plant yields (incl. C_2H_4 cracking)				
CH_4 + others	14.2	13.7	13.9	13.4
C_2H_4	27.6	26.6	26.8	25.4
C_3 fraction	17.6	17.0	17.1	16.1
other C_4	6.9	7.5	7.1	7.2
C_4H_6	4.2	4.6	4.3	4.4
Gasoline	25.5	28.1	26.0	27.5
Fuel oil	4.0	2.5	4.8	6.0
Relative product value	104.0	103.0	102.5	100.0

[a] de Blieck *et al.* (1971). Copyright by *Hydrocarbon Processing*, reprinted with permission.

[b] Full-range naphtha: 40–180°C. Severity: Propylene/ethylene ratio, 0.64. 1, standard PONA analysis; 2, high in naphthenic components; 3, high in isoparaffinic components; 4, high in naphthenic and aromatic components.

Table IV shows the internal substances and most important characteristics of the different naphthas. The light naphthas 1–3 are distillates of different boiling ranges but contain different paraffins and naphthenes.

Figure 4 shows the result of thermal cracking under comparable conditions (i.e., constant reactor outlet temperature, constant residence time, and constant naphtha/steam ratio). The outlet temperature was varied, with adherence to the characteristics of the temperature pattern. The figure shows the dependence of the ethylene yield plotted against the cracking severity, defined as the sum of the cracking gas $< C_3$. Naphtha 1, with the highest content of *n*-paraffins and the lowest content of aromatics, provides the highest yields of ethylene.

If by means of the given formulas (see Section II,B) from both the mean boiling points and the density, the UOP-K factor and the BMCI value are calculated for different base stock fractions, proportional and inversely proportional dependencies on the olefin yield will result.

TABLE IV

Nature and Composition of Different Naphtha Types for Short-Residence-Time Cracking[a]

<table>
<tr><th rowspan="2">Characteristics</th><th colspan="10">Naphtha type</th></tr>
<tr><th>1</th><th>2</th><th>3</th><th>4</th><th>5</th><th>6</th><th>7</th><th>8[b]</th><th>9[b]</th><th>10[b]</th></tr>
<tr><td>Density</td><td>0.675</td><td>0.685</td><td>0.695</td><td>0.761</td><td>0.723</td><td>0.681</td><td>0.793</td><td>0.700</td><td>0.679</td><td>0.710</td></tr>
<tr><td>Boiling range (°C)</td><td>35–145</td><td>35–155</td><td>35–165</td><td>142–191</td><td>38–193</td><td>52–162</td><td>152–220</td><td>35–140</td><td>40–163</td><td>51–176</td></tr>
<tr><td>H Content (wt%)</td><td>15.9</td><td>15.7</td><td>15.4</td><td>14.6</td><td>15.0</td><td>15.6</td><td>13.9</td><td>15.5</td><td>15.6</td><td>—</td></tr>
<tr><td>PONA (wt%)</td><td></td><td></td><td></td><td></td><td></td><td></td><td></td><td></td><td></td><td></td></tr>
<tr><td>nP</td><td>42</td><td>39</td><td>37</td><td rowspan="2">60</td><td>34</td><td>26</td><td rowspan="2">56</td><td>38</td><td>39.5</td><td>29</td></tr>
<tr><td>iP</td><td>41</td><td>40</td><td>38</td><td>35</td><td>67</td><td>34</td><td>42</td><td>40.5</td></tr>
<tr><td>N</td><td>13</td><td>15</td><td>17</td><td>23</td><td>20</td><td>5</td><td>37</td><td>20</td><td>16</td><td>23.5</td></tr>
<tr><td>A</td><td>4</td><td>6</td><td>8</td><td>17</td><td>11</td><td>2</td><td>7</td><td>8</td><td>2.5</td><td>6.5</td></tr>
</table>

[a] Eisenlohr *et al.* (1967). Reproduced with permission from *Erdoel Kohle*.
[b] Added by the authors.

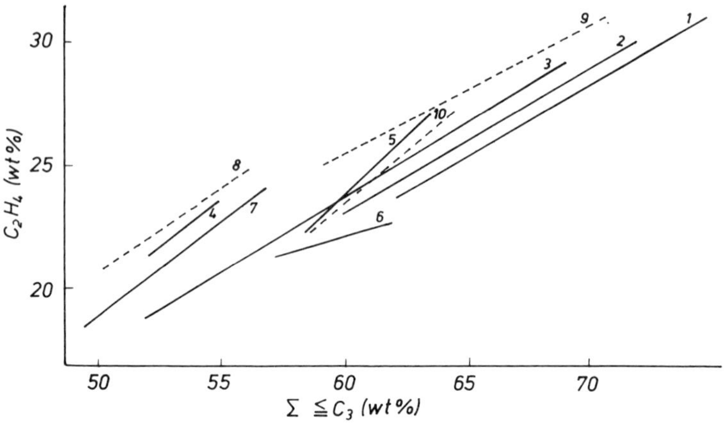

Fig. 4. Yield of ethylene for different naphtha types (designated by numbers; see Table IV). (From Eisenlohr *et al.*, 1967. Reproduced with permission from *Erdoel Kohle*.)

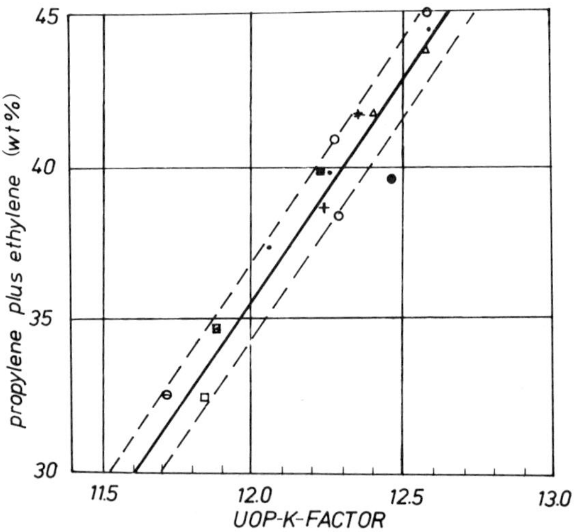

Fig. 5. Yields of ethylene and propylene in relation to UOP-K factor of naphtha at a conversion from 95 to 99%. Note: $\int k\,dt \sim 3-5$.

In Fig. 5 the correlation between the sum of ethylene plus propylene, and the UOP factor ascertained for a limited range of cracking severities is shown. The linearity and the small scattering of points from the commercial pyrolysis of different naphthas indicate the usefulness of the UOP factor for predicting the yields of olefins for different feedstocks.

A plot of ethylene–propylene yields against the BMCI value of different feedstocks is shown in Fig. 6. The smaller the BMCI value is, the better olefin yields are to be expected.

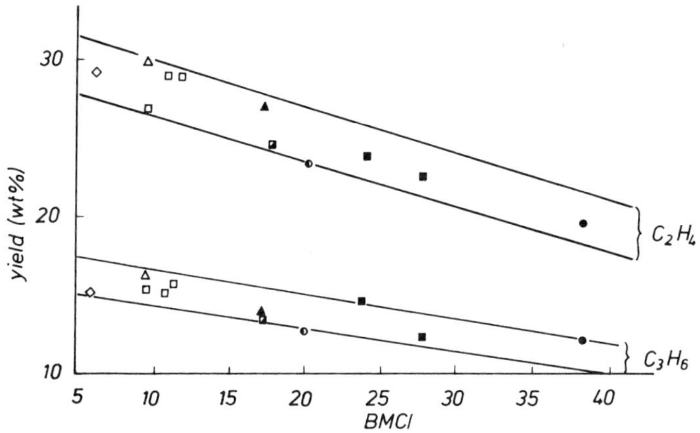

Fig. 6. Yield of ethylene and propylene in relation to BMCI of different virgin and hydrogenated feeds. ●, VGO; ◐, partially hydrogenated VGO; ■, AGO; ◪, partially hydrogenated AGO; □, hydrogenated AGO; ▲, kerosene; △, hydrogenated kerosene; ◇, light naphtha. (From Günschel and Nowak, 1980.)

E. New Developments in Steam-Cracking Technology

For the production of small olefins by means of pyrolysis, the steam-cracking tube has maintained its dominating role mainly because of simple design, easy maintenance, and simple control. Other well-known furnace designs that have resulted from further developments are the gradation pyrolysis furnace of Selas Corp., short-residence-time (SRT) heater of Lummus Co., small-diameter (SD) heater of KTI Co., ultraselective conversion (USC) heater of Stone and Webster Engineering, and the short-time cracking unit of Lurgi G.m.b.H.

In the new designs, the findings show that a rapid rise in temperature, low hydrocarbon partial pressure, and short residence times lead to higher yields and longer running times.

The Linde Selas A.G. developed the LSCC cracker (Linde Selas combined coil) (Lohr and Schwab, 1979). Its peculiarity is the splitting of the cracking coils in the first section of the radiant zone into two parallel tubular coils with smaller diameters. By this means a larger S/V ratio was provided. In the second section the two tubular coils are united to provide one coil with a larger diameter, thereby reducing flow resistance in order to bring about a smaller pressure drop. By splitting into the smaller diameter coils, an improved heat transfer—and consequently, a quicker heat-up of the tubular coil of LSCC design in a furnace of conventional construction, that is, with coils of constant diameter—resulted in improvements of the ethylene yield of 0.9 wt% as well as a reduction of the pyrolysis oil of 1.2 wt% without any capacity loss.

Likewise, high olefin yields at the expense of undesired by-products are obtained by the pyrolysis in the millisecond furnace designed by the firms Kellogg (United States) and Idemitsu Petrochemical Co. (Japan) (Prescott, 1975). The tube furnace operates at residence times of only 0.03 to 0.1 s and temperatures from 900 to 930°C. The ethylene-yield improvements are claimed to be 2.8 to 5.8% absolute. The furnace has no coils but several parallel and vertically arranged tubes of small diameter. Heating is achieved by burners at both sides along the single tube lines. The tubes are fixed at the lower end of the radiating section. Their thermal expansion is possible only upward.

In a pilot plant of 25 kt/a ethylene, besides straight-run naphtha, kerosene and light gas oil were also converted similarly with good success. Typical yields for a full-range naphtha and a light gas oil are shown in Table V (Sajus, 1978; Kunugi *et al.*, 1978).

Further advancements of steam cracking led to transfer-line quench cracking (TLC) (Goossens *et al.*, 1976). Here, by adding crackable hydrocarbons to the hot reactor-outlet mixture, a portion of the heat to be dissipated is utilized as cracking heat. Besides improved heat economy, higher quantities of desired products are obtained.

The demands for a further increase of heat transfer in the cracking zone led to the considerations of installing ceramic tubes or utilizing the heat

TABLE V

Typical Yields of Full-Range Naphtha and Light Gas Oil Obtained by Millisecond Pyrolysis[a]

Conditions and products	Feed	
	Full-range naphtha	Light gas oil
Cracking condition		
Outlet temperature (°C)	927	899
Steam/oil (wt. ratio)	0.5	0.6
Contact time (s)	0.05	0.05
Product yield (wt%)		
Hydrogen	1.2	0.8
Methane	15.2	10.6
Ethylene	33.0	28.0
Propylene	11.6	12.8
Butadiene	4.7	5.3
$HC_5 \geqq C_5$	26.5	36.2

[a] Sajus (1978).

of partial combustion by adding small amounts of oxygen to the feedstock fraction. Also, preheating the combustion air and reducing the excessive air in furnace burners led to better energy efficiency. The tense situation on the feedstock market and a return to worldwide shortages of energy will initiate new incentives for further improvements of pyrolysis technology.

III. PYROLYSIS OF HEAVY HYDROCARBON FRACTIONS

A. General Feedstock Trends

In all parts of the world the development of the production of olefins is determined to a large extent by the feedstocks available. Although at present the use of LPG—especially in the United States and the Soviet Union—as well as light naphthas in most European countries is dominating, the worldwide trend toward the processing of high-boiling hydrocarbon fractions is nevertheless already evident.

The rising demand for light olefins (i.e., C_2–C_4 olefins) and the growing shortage of conventional pyrolysis feeds require an increased use of high-boiling crude oil fractions for the production of olefins (Sajus, 1978; Kunugi *et al.*, 1978; Walley, 1977). The experience today in processing straight-run gas oils suggests that the problem of converting gas oils to light olefins can be considered as being resolved technically, but not economically or scientifically.

Analogously to naphtha hydrocarbons, the ethylene potential of hydrocarbons boiling above 180°C depends on the C/H ratio as well as the amounts of *n*-paraffins, isoparaffins, naphthenes, and aromatics.

B. Characterization of Heavy Crude Oil Fractions

In contrast to naphtha fractions, heavy hydrocarbon fractions have another structure. Characteristic differences exist in the content of cyclic compounds and in the increased occurrence of bicyclic and polycyclic naphthenes and aromatics, which have a significant effect on the yield pattern. Whereas in naphtha fractions the cyclic hydrocarbon/paraffin ratio amounts to 30:70, in gas oils it increases to ratios of more than 50:50, and in vacuum gas oils it reaches ratios of 70:30.

In Fig. 7 the increase of the aromatics content with rising boiling range in Romashkino crude oil fractions (Soviet Union) is illustrated. Whereas the content of total aromatics rises with increasing molecular weight, the alkylbenzenes rise to a maximum. A peculiarity of gas-oil fractions, besides

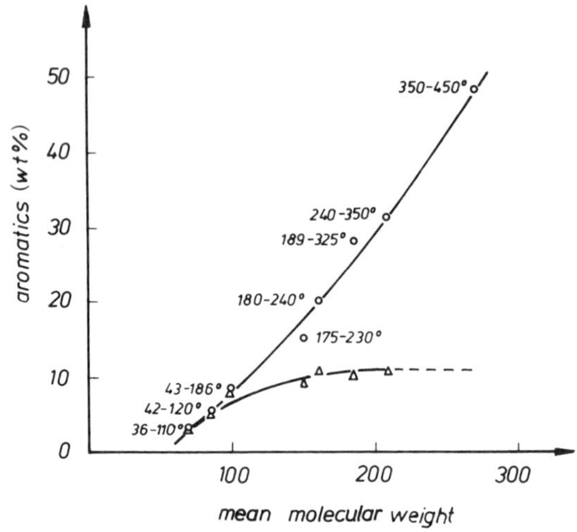

Fig. 7. Overall aromatics (○) and alkylbenzenes (△) in petroleum fractions in relation to the mean molecular weight. [From Nowak *et al.*, 1979. Reproduced with permission of *Chem. Tech. Leipzig*).]

the rising content of aromatics, is the high content of sulfur compounds. In comparison with naphtha fractions, the sulfur content is higher by a factor of 10. The sulfur is almost completely bonded in aromatic structures (Zdonik *et al.*, 1976). A comprehensive study on quantity and nature of sulfur compounds in Soviet crude oils and hydrocarbon fractions is published in Obolentzev (1967).

A general relation between the density (ρ) and the sulfur content (wt%) is given by the equation:

$$\rho = 0.0087\ (\%S)^2 + 0.0607\ (\%S) + 0.7857 \tag{12}$$

The sulfide sulfur (Y), depending on the overall sulfur (x), follows the linear equation

$$Y = 0.1821x - 0.0211 \tag{13}$$

C. Peculiarities of the Pyrolysis of Gas-Oil Fractions

Because of their structural composition, the cracking behavior of high-boiling hydrocarbon fractions differs essentially from that of light hydrocarbons. Differences exist not only in the pattern, but also in the process conditions and the design of the pyrolysis plant itself.

Unlike the cracking of naphtha, the pyrolysis of gas oils yields smaller yields of light olefins and a greater amount of by-products. In particular, a high output of pyrolysis oil, a fraction boiling above 200°C, must be taken into account. If, for instance, about 3 to 3.3 tons of naphtha are necessary for producing 1 ton of ethylene, the production of the source quantity of ethylene from gas-oil fractions requires about 4.5 tons of feedstock.

Table VI shows typical yields for different feeds obtained by the same cracking severity (Lohr and Dittmann, 1977b). With rising boiling range of feeds, the yields of olefins decrease while the C_5 fraction increases. Both a high production of fuel oil and a high coke formation are characteristic of pyrolysis of gas oil.

An excellent survey about the product distribution in the pyrolysis of atmospheric gas oil (AGO) and vacuum gas oil (VGO) for different cracking severities is provided in the literature (Lohr and Dittmann, 1977b; Hatch and Matar, 1978; Zdonik *et al.*, 1967, 1975; Zdonik and Hayward, 1975; Virk *et al.*, 1974). The product distribution is dependent on the composition of feedstocks and their specific properties.

The deteriorating cracking behavior of gas oil expressed by small yields of olefins and high formation of gum and coke is due to the high aromatics content of gas oils. Especially the formation of gum and coke, which is higher than that which occurs in naphtha pyrolysis, renders the process more complicated and leads to shorter running times of the pyrolysis plant. Coke deposits in both the coil and the quench cooler not only greatly impede heat transfer but also bring about superheating and consequently shorter lives of coils.

TABLE VI

Typical Yields Obtained by the Same Cracking Severity for Different Feeds[a]

	Light naphtha	Heavy naphtha	Light gas oil	Heavy gas oil
Boiling range, °C	36–110	40–164	176–343	335–515
Product yields (wt%)				
Aromatics content	2.6	12.1	24.1	48.0
Methane	17.4	13.8	11.6	8.9
Ethylene	31.0	25.5	24.1	18.9
C_3 Fractions	18.8	15.3	14.3	13.9
C_4 Fractions	10.0	8.3	8.4	9.7
Pyrolysis gasoline	14.4	26.9	18.1	19.0
Pyrolysis fuel oil	2.0	5.1	18.9	24.4

[a] Lohr and Dittmann (1977b). Reproduced with permission from *Oil Gas J.*

1. *Coke Formation of Heavy Hydrocarbon Fractions*

Although a large number of publications are devoted to coke formation in the pyrolysis of hydrocarbon fractions (Hatch and Matar, 1978; Albright, 1978; Albright and Yu, 1979; Dunkleman, 1976; Lehaye and Prado, 1975; Badie, 1975; Lehaye *et al.*, 1977; Magaril, 1973, 1976; Magaril and Beresina, 1975, 1976; Zychlinski, 1972; Spengler *et al.*, 1973; Ghaly and Crynes, 1976; Aarna, *et al.*, 1978; Goossens *et al.*, 1978; Kopinke *et al.*, 1981), only a few studies have systematically investigated the dependence of coke formation on the boiling range of pyrolysis feeds (Albright, 1978; Albright and Yu, 1979; Dunkleman, 1976; Badie, 1975; Magaril and Beresina, 1976; Zychlinski, 1972; Magaril, 1973; Kopinke *et al.*, 1981).

Interesting investigations on this problem have been performed by Kopinke *et al.* (1981). Unlike the other publications, theirs concentrates on coke formation for various hydrocarbon fractions in the reactor as well as in the quench cooler. The separate recording of primary coke

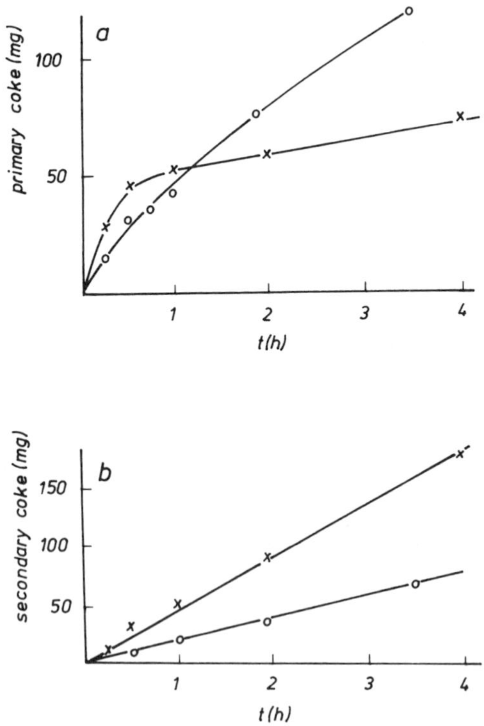

Fig. 8. Formation of coke in relation to running time. ○, Naphtha (0.01 wt% S) at 810°C; ×, gas oil (0.43 wt% S) at 780°C. (From Kopinke *et al.*, 1981.)

(coke in reactor) and secondary coke (coke in the quench cooler) provide interesting correlations between the process parameters and the coke-formation behavior of naphtha and gas oil. The pyrolysis of naphtha at $T = 810°C$, and of gas oil at 780°C, with a residence time of 0.35 s and at a steam/hydrocarbon ratio of 0.7 led to surprising results concerning the formation of primary and secondary coke (Figs. 8 and 9). The results show that both primary and secondary coke formation depend on process conditions and moreover to a large extent on boiling range as well as composition of the feedstocks. The change in primary and secondary coke formation for hydrocarbon fractions of different boiling points is illustrated in Fig. 10.

While the coke deposits in the pyrolysis reactor decline with rising boiling range (Fig. 10a), the coke formation in the quench cooler increases by a factor of three or four for gas-oil pyrolysis compared with naphtha (Fig. 10b).

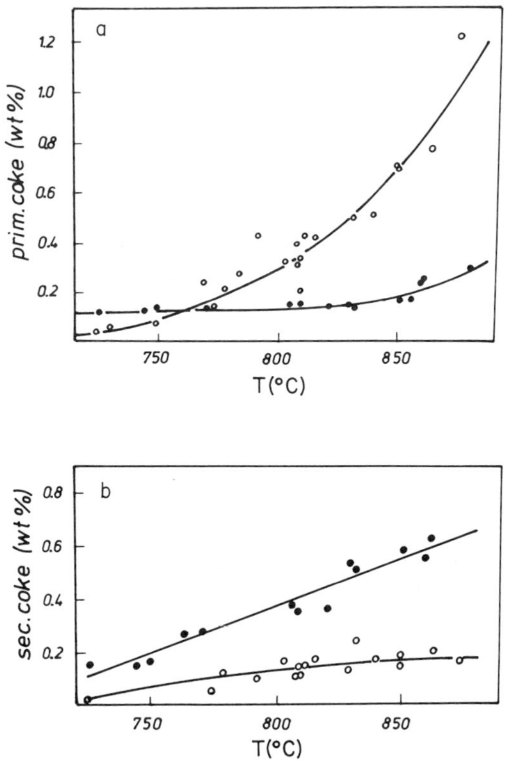

Fig. 9. Formation of coke in relation to coil-outlet temperature. ○, Naphtha; ●, gas oil. (From Kopinke *et al.*, 1981.)

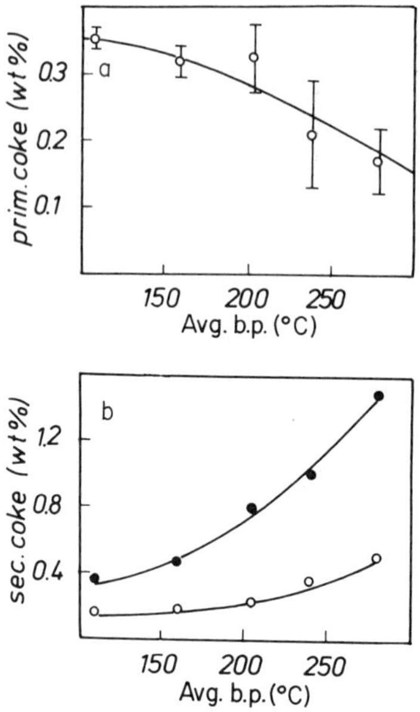

Fig. 10. Formation of coke in relation to average boiling point of feed. ○, Naphtha; ●, gas oil. (From Kopinke *et al.*, 1981.)

Results of coke formation for different hydrocarbon fraction show in Table VII that the aromatics content in the pyrolysis feeds exerts the greatest influence on secondary coke formation.

The present results on coke formation for different feeds suggest that with a rising boiling range of the feed, the coke formation increasingly shifts from pyrolysis reactor toward the quench unit. The considerable increase of secondary coke in gas-oil pyrolysis is due to the increasing content of polycyclic aromatics.

In order to produce a high yield of ethylene and to counteract the increased formation of gum and coke in pyrolysis of gas oil, the pyrolysis is performed with a relatively high steam/hydrocarbon ratio. This not only

TABLE VII

Coke Formation from Different Feedstocks[a,b]

Feedstock	Primary coke (wt%)	Secondary coke (wt%)
Straight-run naphtha	0.35	0.17
Full-range gas oil	0.23	0.34
Hydrocatalytically dearomatized gas oil	0.15	0.13

[a] Kopinke *et al.* (1981).

[b] Conditions: temperature, 810°C; residence time, 0.35 s; H_2O/HC ratio, 0.7.

requires greater energy but also leads to a reduction in reactor capacity. Because heat recovery is only partly achieved, an increase of the steam/hydrocarbon ratio is economically limited. In order to obtain shorter residence times and to keep secondary reactions within limits, reaction coils used in the pyrolysis of gas oil have a larger diameter but a smaller length than those used in naphtha-cracking furnaces. Because of the high throughput, the heating surfaces of the reaction coils must withstand extraordinarily high heat-flow densities, which implies a qualitatively high requirement of reactor material. These are the major differences between the naphtha and gas oil technical plants. The yield of by-products also increases considerably with rising boiling range of feedstocks. This requires more extensive chemical use of products, such as allene, C_4 olefins, isoprene, aromatic hydrocarbons, (e.g., benzene, toluene, styrene, indene, naphthalene, acenaphthenes, phenanthrene, etc.). The pyrolysis of gas oil, therefore, requires higher investment and processing costs. The investment costs for the thermal cracking of AGO and VGO are about 8–15% higher than those for commercial cracking of naphtha (Mikulla *et al.*, 1978). The technical difficulties and the higher costs of the pyrolysis of heavy hydrocarbon fractions are the most important reasons preventing the producers from using high-boiling hydrocarbon fractions on a larger scale. Unlike the situation in naphtha cracking, few data are available from commercial gas-oil cracking plants.

2. *Cracking Behavior of Gas-Oil Fractions*

Many investigations on the cracking behavior of gas oil and fractions with different structural groups in the same boiling range have shown that with rising boiling range the deteriorating cracking behavior cannot be attributed so much to the increasing mean molecular weight as to the increasing content of aromatic hydrocarbons. This is especially so for bicyclic and polycyclic feedstocks, which are dangerous in that they produce solid deposits in the coil as well as in the quench cooler of the cracking plant.

In order to test the influence of the content of aromatics in feedstocks with the same boiling range on the results of thermal cracking, five gas-oil fractions (240° to 360°C, aromatics content 28%, 22%, 14%, 8%, and 2%) were produced by Nowak *et al.* (1975a, 1979). For this purpose the aromatic hydrocarbons were separated from straight-run gas oil by solvent extraction using dimethylformamide (DMF), and both raffinate and extract were remixed to obtain the defined fractions for pyrolysis. Figure 11 contains typical cracking results. From these it was possible to calculate the yields of ethylene, pyrolysis fuel oil (PFO), and coke from both the saturated and unsaturated portions of gas oil (Table VIII).

Pyrolysis fuel oil (9 wt%) is formed from saturated HCs by condensation or polymerization of lighter intermediates. The aromatics in gas oil

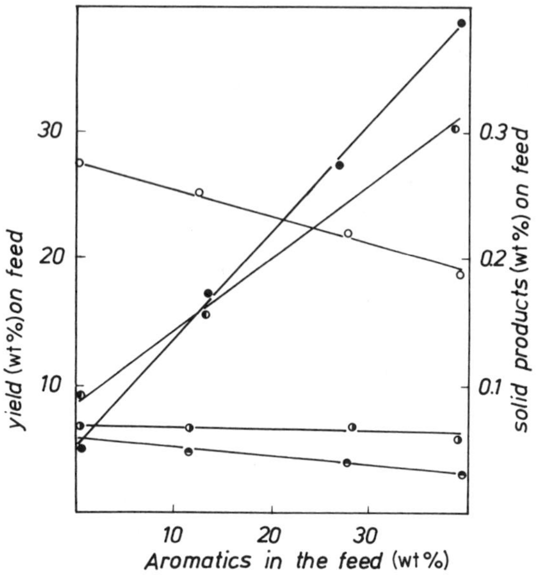

Fig. 11. Yields in relation to the content of aromatics in gas-oil fractions. ○, Ethylene; ●, coke; ◐, fuel; ◑, BTX-aromatics; ◓, butadiene. Conditions: temperature, 770°C; residence time, 0.25 s; H_2O/HC ratio, 0.8. (From Nowak *et al.*, 1975a.)

TABLE VIII

Yields from Saturated and Unsaturated Gas-Oil Hydrocarbons[a]

Feed	Yield (wt%)		
	C_2H_4	PFO	Coke
Saturated HCs	27	9	0.05
Unsaturated HCs	7	31	0.40

[a] Günschel and Nowak (1980).

provide only 7 wt% ethylene. The formation of coke from aromatics was higher by the factor eight than from paraffins and naphthenes.

The results show that the cracked-gas formation is essentially determined by the content of aromatics. The decline of the aromatics from 28 to 2 wt% leads to a higher yield of cracked gas, especially ethylene, and a smaller amount of fuel oil. The formation of coke is significantly reduced with declining content of aromatics.

Studies by other authors led also to the result that aromatic hydrocarbons are less suitable feeds for ethylene production.

In a very interesting article by Green *et al.* (1975), the cracking behavior of differently structured C_{16} hydrocarbons—which represent compounds with a mean molecular weight of a gas-oil fraction—was calculated by means of a semiempirical method.

The C_2H_4 potential of C_{16} hydrocarbon types contained in gas-oil fractions is highly variable. Thus from hexadecane an ethylene yield of about 33 wt% has been calculated, whereas for tricyclic aromatic hydrocarbons, the results revealed from 12 to only 4.3 wt% ethylene. Correspondingly, the yields of liquid hydrocarbons in the effluent change from about 17.2 to 90.3 wt%. Moreover, together with the increase in carbon content and in the number of rings in the molecule, the coke formation increases. These results give unambiguous evidence of how heavily the cracking results depend on the hydrogen content and the chemical composition of the gas-oil fractions.

These results on the pyrolysis of high-boiling hydrocarbon fractions indicate that the cause for the increasing deterioration of the cracking behavior of heavy crude oil fractions is primarily affected by the changed structural groups and particularly by the rising content of aromatics, but not by the boiling range.

From all these facts the conclusion was drawn either to remove the disturbing aromatics from the medium distillates or to convert them into products having a more favorable cracking behavior.

D. Trends in the Pyrolysis of Heavy Crude Oil Fractions

During the last 5 to 10 years many publications and patents have come out describing and claiming AGO as well as VGO, including crude oil, as feeds for the production of olefins. The most essential trends in the field of the pyrolysis of heavy hydrocarbon fractions can be summarized as follows.

1. Technological improvement of available plants and development of new cracking technologies for processing heavy crude oil fractions
2. The changing of the composition of heavy crude oil fractions and their adaptation to conventional pyrolysis technologies
3. Higher feedstock flexibility of cracking plants
4. New cracking technologies

1. *Pretreatment of Heavy Hydrocarbon Fractions*

The modification of heavy crude oil fractions as feeds for pyrolysis is essentially directed toward removing the heteroelements, sulfur and nitrogen, as well as separating or converting the disturbing polycyclic aromatics (Zdonik *et al.*, 1976; Green and Zdonik, 1968; Davis *et al.*, 1972; Ger. Offen., 1972; Soonawala and Den Broeder, 1976; Wernicke and Kreuter,

1978; Becker *et al.*, 1978; GDR Patent Application, 1980; Rhoe *et al.*, 1979; Eigenson *et al.*, 1970; Stork, 1977; Wernicke and Kreuter, 1979; Goetzmann *et al.*, 1979; Korosi *et al.*, 1979) by using proper pretreating processes, including the following.

1. Physical separation of aromatic hydrocarbons such as by adsorption, solvent extraction, or extractive distillation
2. Hydrocatalytic pretreatment methods such as hydrorefining, partial hydrogenation, or total hydrogenation to convert aromatic hydrocarbons into naphthenes, and bonded sulfur and nitrogen into H_2S and NH_3

a. Dearomatization by Means of Selective Solvents. Among the adsorptive, distilling, and extractive methods for separating aromatics, the latter have received most attention, although no technique operating according to this principle has yet been developed for obtaining aromatics-poor pyrolysis feedstocks. However, it was proved by means of bench-scale and pilot plants that a raffinate poor in aromatics provides higher feedstock-specific yields of final products in pyrolysis. As selective solvents, DMF, N-methylpyrrolidone, phenol, furfural, aniline, ammonia, and methylamine were applied separately as mixtures, and partly mixed with water (Magaril, 1976; Magaril and Beresina, 1975; Nowak *et al.*, 1975a; Muchina *et al.*, 1974; Spirkin *et al.*, 1972; Kubicka, 1974; Barton *et al.*, 1968; Riad *et al.*, 1969; Soonawala and Den Broeder, 1976; Eigenson *et al.*, 1968, 1970). Eigenson *et al.* (1968, 1970) give information about the yields as well as the data of raffinates and extracts obtained during treatment of a straight-run crude oil fraction of the boiling range 180 to 350°C with DMF and a specially treated pyridine solution. The raffinates can be processed without any difficulties and thereby olefin yields analogous to those in naphtha pyrolysis are achieved.

Muchina *et al.* (1974), who likewise carried out investigations on the dearomatization of crude oil fractions by means of DMF, obtained in the case of a DMF/feedstock ratio of 3:1, for instance, a reduction of the aromatics content from 28 to 6%; in the case of sulfur they obtained a decrease to 0.19 wt%. In the case of smaller DMF/feedstock ratios, higher concentrations of aromatics and sulfur remain in the raffinate. Although a selective removal of bicyclic and polycyclic aromatics should be desirable, their separation is brought about to the same extent as for monocyclic compounds. Thus the distribution coefficients are independent of the number of aromatic cycles in the molecules.

The pyrolysis of such gas-oil fractions poor in aromatics, in comparison with straight-run gas oil, gave an increase in the yield of ethylene from 22.5 to 28 wt%, a decrease in pyrolysis oil, and a coke formation lower by a factor of four (Zdonik *et al.*, 1976).

Analogous results have also been obtained by Eigenson *et al.* (1968) and Kubicka (1974), who also performed aromatics extraction with the purpose of using the extract rich in aromatics for producing high-octane gasolines and C_6 to C_8 aromatics.

These investigations have shown that the extraction of aromatics by means of solvents represents a suitable method for pretreating pyrolysis feeds from the gas-oil boiling range. The restrictions occurring during the industrial use of this method are essentially economic in nature. The costs of dearomatization in using this method are obviously so high that the industrial realization cannot be expected soon.

b. Hydrocatalytic Pretreatment Methods. The significant role that high-boiling hydrocarbon fractions already play as feedstocks for pyrolysis in many countries is also evidenced by the noteworthy increase in publications and patent registrations with regard to hydrocatalytic pretreatment during the last 10 years. For the desulfurization and removal or depletion of the disturbing aromatic hydrocarbons from gas-oil fractions, the well-known and already commercially introduced hydrocatalytic techniques such as hydrofining, hydrogenation, and hydrocracking are well suited.

First of all hydrocatalytic refining should be discussed, because in regard to the economics involved it can be considered the simplest catalytic hydrogen treatment of crude oil distillates. In connection with pyrolysis, this pretreatment method is of interest for several reasons.

1. Organically bonded sulfur, nitrogen, and oxygen are converted in the process to H_2S, NH_3, and H_2O.
2. In the hydrogen recycle gas, H_2S and NH_3 can be more effectively and more cheaply scrubbed than in the cracked gas of pyrolysis (elimination of regenerative scrubbing susceptible to fouling).
3. The low sulfur and nitrogen content in the liquid pyrolysis products increases the value in use of the latter and brings about an improved gasoline and fuel oil.
4. The reaction products of the pyrolysis are altogether richer in hydrogen, because no hydrogen from the feedstock is required for removing the heteroelements.
5. The lives of the furnace tubes are longer, because the defects introduced by sulfurizing the cracking-coil material are reduced and the coking of the tubes is delayed.
6. Fewer corrosion problems initiated by acid gases in the gas-separation plant occur.

By means of catalytic hydrofining alone, the difficulties that arise during the pyrolysis of high-boiling crude oil fractions—above all, those due to the formation and deposit of cokelike products—cannot be overcome. The expectations linked with the pretreatment of pyrolysis feed products can only be partially fulfilled using the method of catalytic hydrofining. Consequently, the technological significance of this method for pretreating pyrolysis feedstock products is considered only in connection with other pretreatment methods.

The essential aim of hydrocatalytic pretreatment of high-boiling hydrocarbon fractions as feedstocks for pyrolysis is in reducing the high aromatics content.

In catalytic hydrogenation the aromatics are converted into more easily crackable naphthenes having the same cycle structure. This means that in contrast to the dearomatization by means of the solvent extraction, the valuable hydrocarbons are almost wholly preserved as pyrolysis feedstock. The method aiming at prehydrogenation is characterized by the following conditions: pressure, 10 to 25 MPa; temperature, 280 to 450°C; gas/feed ratio, 1000 to 1500 m^3/m^3; space velocity, 1.5 to 5.0 v/vh. Combinations of nickel–tungsten oxides and nickel–molybdenum oxides on Al_2O_3 carriers are used as catalysts. By such means the content of aromatics of a straight-run gas oil was decreased from 20–25% to only 2–5% in a single-step operation (Becker *et al.*, 1978)

A shift of the working pressure into the medium range is possible only by employing a two-step operation (with preceding hydrofining), with the use of noble metal-containing catalysts. The development of sulfur-resistant catalysts permitting single-step hydrogenation under medium pressure is being intensively studied. Processes for the selective hydrogenation of bi- and polycyclic aromatics in a mixture with monocyclic aromatics are of current interest and are already being developed (Ger. (East) Pat. Appl., 1980; Wernicke and Kreuter, 1979; Korosi *et al.*, 1979).

Systematic studies on the hydrogenation of individual, differently structured aromatic hydrocarbons led to the following results.

1. In accordance with investigations of Virk *et al.* (1974), the delocalization energy of electrons is determined for the velocity of the conversion of aromatics into naphthenes and partially hydrogenated aromatics, respectively.
2. Alkyl-substituted aromatics of the same aromatic structure are more easily hydrogenated with lower degrees of alkylation.
3. Indans and tetralins are preferentially hydrogenated in comparison to alkylated benzenes.

This state of affairs is represented in Fig. 12 and Table IX, in which the relative hydrogenation rates for individual aromatic hydrocarbons are given (Nowak *et al.*, 1979).

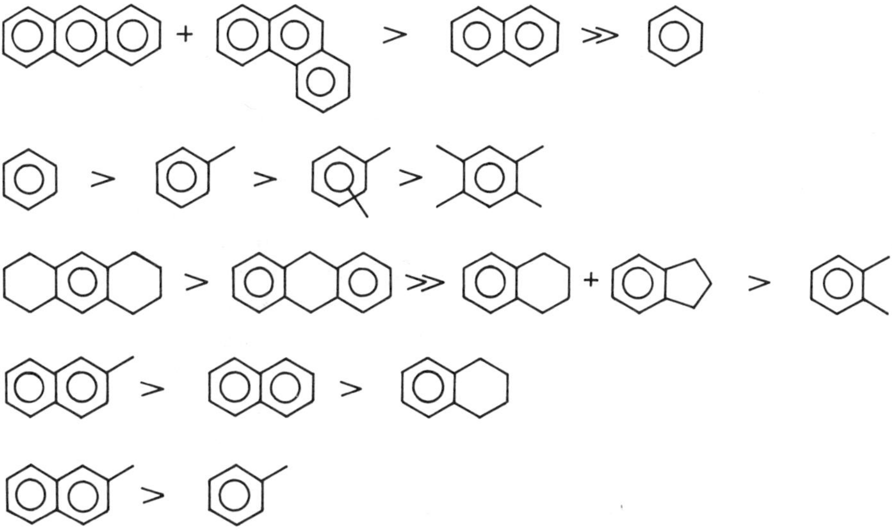

Fig. 12. Relative rates of hydrogenation of different hydrocarbon mixtures on Ni catalyst. [From Nowak *et al.*, 1979. Reproduced with permission from *Chem. Tech.* (*Leipzig*).]

TABLE IX

Relative Reaction-Rate Constants ($k_{rel.}$) of the Catalytic Hydrogenation of Aromatic Hydrocarbons in the Presence of a Ni–Al_2O_3 Catalyst under Competitive Conditions[a]

Aromatics	$k_{rel.}$
Durene	0.22
o-Xylene	0.48
Toluene	0.78
Benzene	1.00
Indan	1.05
Tetralin	1.05
2-Methylnaphthalene	1.08
9,10-Dihydroanthracene	1.10
Naphthalene	1.13
Anthracene	1.15
Phenanthrene	1.15

[a] From Nowak *et al.* (1979). Reproduced with permission from *Chem. Tech.* (*Leipzig*).

Although the hydrogenation rates of aromatic hydrocarbons are dependent on the carrier catalyst used, as well as on the reaction conditions, the given sequence of the relative reaction rates remains unaffected.

In conformity with this sequence, the aromatics content in partially hydrogenated fractions decreases with rising boiling range, as shown in Fig. 13, but increases in contrast to the content of aromatics in straight-run fractions.

From these figures as well as from the relative reaction rates, one can deduce that in the case of partial hydrogenation, the residual aromatics in the gas oil consist mainly of monocyclic compounds (alkylbenzenes). Starting from these findings, the attempt was made to saturate or crack the polycyclic aromatics selectively into alkylbenzenes that remain within the fractions (Ger. (East) Pat. Appl., 1980).

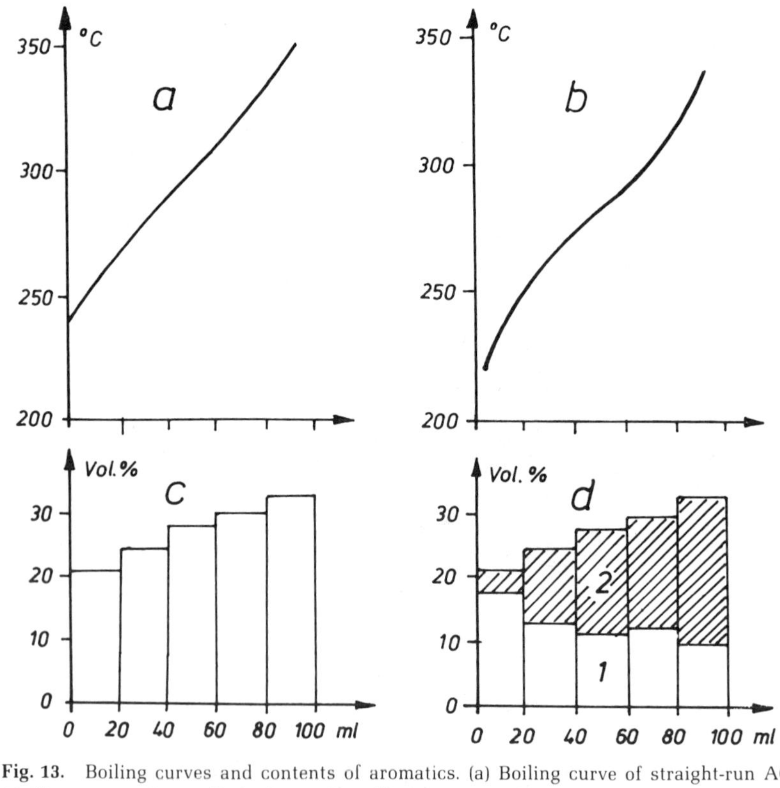

Fig. 13. Boiling curves and contents of aromatics. (a) Boiling curve of straight-run AGO. (b) Boiling curve of partially hydrocatalytically dearomatized AGO. (c) Aromatic contents in distillates of straight-run AGO. (d) Aromatic contents in distillates of hydrocatalytically pretreated AGO. 1, Aromatics remaining; 2, converted aromatics. [From Nowak *et al.*, 1979. Reproduced with permission from *Chem. Tech. (Leipzig)*.]

1. *Catalytic Hydrocracking*

An alternative conversion process to catalytic hydrogenation is catalytic hydrocracking. Whereas in hydrogenation the conversion of aromatics into naphthenes is the main reaction, in hydrocracking the diminution of molecular weight is the aim, and dearomatization is of less importance. Primarily the polycyclic structures among the aromatics are degraded, and the arising intermediary cycloalkenoaromatics are partially cracked into alkyl-substituted monocyclic aromatics. These reactions and the conversion of naphthenes into paraffins are the objectives sought from the viewpoint of subsequent pyrolysis. The cracking and isomerizations of *n*-paraffins proceeding simultaneously in the process are undesirable, not only because the cracking consumes hydrogen but also because in the subsequent pyrolysis the isoparaffins provide lower ethylene yields than *n*-paraffins.

Recent studies have favored hydrocracking over hydrogenation, obviously because by hydrocracking both engine fuels (gasoline, diesel fuel) and pyrolysis feedstocks can be produced. According to investigations performed by Stork (1977), hydrocracked residues from cracking of VGOs are particularly well suited as pyrolysis feedstocks. Stork found that such residues have a high content of *n*-paraffins and few aromatics.

2. *Pyrolysis of Hydrocatalytically Pretreated Gas-Oil Fractions*

The hydrogenation of gas-oil fractions and their subsequent pyrolysis have been investigated and described in detail by Korosi *et al.* (1979) and Nowak *et al.* (1975a, 1979).

Table X contains typical process conditions and test results of the pyrolysis of straight-run and hydrogenated hydrocarbon fractions. The results give rise to the following observations.

1. Hydrocatalytically dearomatized gas-oil fractions provide a yield pattern analogous to that of naphtha.
2. In comparison with straight-run naphthas, hydrogenated gas-oil fractions provide higher yields of monocyclic aromatic hydrocarbons, particularly benzene and toluene.
3. Under similar conditions hydrocatalytically dearomatized gas oils yield a little more PFO than naphtha, but much less than straight-run gas oils.
4. In pyrolysis of largely dearomatized feedstocks, the yield of fuel oil, gum, and coke is lower than in the case of partially hydrogenated and straight-run gas oils.

These investigations reveal that the well-known difficulties in the pyrolysis of high-boiling hydrocarbon fractions can be avoided through an appropriate hydrocatalytic pretreatment. This provides real possibilities

TABLE X

Results of the Pyrolysis of Different Feeds[a,b]

Conditions and results	Feed: Straight-run naphtha	Straight-run gas oil	Partially hydrogenated gas oil	Totally hydrogenated gas oil
Mean molecular weight	100	190	187	181
Boiling range (°C)	40 . . . 180	190 . . . 330	172 . . . 320	165 . . . 320
Aromatics	8	27	15	1
Naphthenes	20	26	30	45
Paraffins	72	47	54	54
Reaction products (wt% related to feed)				
Cracked gas $\leq C_4$	73.6	58.8	62	70
Ethylene	26	21	25	25
Propylene	16	14	14	16
Butadiene	5	2	4	6
Pyrolysis gasoline (C_5 to 200°C)	21.7	22.9	27.7	23.8
Benzene Toluene	11	6	8	16
Pyrolysis fuel oil ($\geq$200°C)	4	17	9	5.9
Coke and tar				
in the reactor coil	0.4	0.2	0.2	0.1
in the quench cooler	0.3	1.1	0.7	0.2

[a] Nowak *et al.* (1979). Reproduced with permission from *Chem. Tech.* (*Leipzig*).

[b] Conditions: temperature profile, 700–830°C; residence time, 0.2–1.0 s; H_2O/HC, 0.5–3.0 g/g.

of replacing naphtha fractions by pretreated high-boiling hydrocarbon fractions for olefin production. Economic advantages are particularly apparent if such crude oil fractions (e.g., VGO) are pretreated and used as pyrolysis feeds, for which the price difference is considerable compared to naphtha fractions.

Investigations on the reaction behavior of alkylbenzenes under pyrolysis conditions did not reveal any negative effects on the pyrolysis process (Ger. (East) Pat. Appl., 1980). Alkylbenzenes were preferentially dehydrogenated and dealkylated respectively. Table XI demonstrates these facts.

The pyrolysis of partially hydrogenated gas-oil fractions with a residual aromatics content of 6 to 10 wt% provided high yields of light olefins and BTX; undesirable second reactions such as coke deposits and the formation of pyrolysis oil were mostly avoided.

The advantage of partial hydrogenation consists especially in reducing the amount of hydrogen needed for pretreatment, compared with total hydrogenation or hydrocracking.

3. *Hydrocracker Products as Feeds for Olefin Production*

At the end of the 1970s, some publications and patents came out claiming hydrocracker products as feeds for the production of olefins (Zdonik *et al.*, 1976; Offen, 1976; Wernicke and Kreuter, 1978; Rhoe *et al.*, 1979; Stork, 1977; Wernicke and Kreuter, 1979; Goetzmann *et al.*, 1979).

TABLE XI

Reaction Behavior of Individual Aromatics in Gas-Oil Cracking[a]

Aromatics	Hydrogenated AGO with 15% monocyclic aromatics (yield from 100%)	Hydrogenated AGO without added aromatics (yield from 85%)	ΔYield
Benzene	5.6	4.6	1.0
Toluene[b]	6.1	2.6	3.5
Ethylbenzene[b]	1.4	0.4	1.0
m/p-Xylene[b]	3.1	0.5	2.6
o-Xylene[b]	2.2	0.2	2.0
Styrene	2.9	0.9	2.0
Mesitylene[b]	1.5	0	1.5
Durene[b]	1.1	0	1.1
Polycyclic aromatics	5.9	6.1	0
Total			14.7

[a] Ger. (East) Pat. Appl. (1980).
[b] 2.5% of each labeled aromatic was added.

An interesting conception for using VGOs for the production of olefins was proposed by Lummus at the 10th World Petroleum Congress in Bucharest (Rhoe *et al.*, 1979). Through the LC-fining of an atmospheric residue, products appropriate as feeds for the production of olefins resulted.

The hydrocracking cuts boiling above 350°C that are normally recycled into the hydrocracking process represent, because of their high paraffin content, an especially suitable feedstock for producing olefins. Rhoe *et al.* (1979) also present a survey of the reaction products obtained from different hydrocracking fractions. The yields of ethylene clearly correspond to the aromatics content of the pyrolyzed hydrocracking fractions. Thus both the light naphtha fraction C_5 (85°C) and the hydrocracking residue produce analogous yields of ethylene (27 wt%). For the medium hydrocracking fraction, the yields of ethylene are lower (17–20 wt%). Analogous results were also obtained according to the hydroconverter process of the Linde A. G. (Wernicke and Kreuter, 1979; Kreuter, 1978), which had been particularly designed for gaining heavy feeds for the production of olefins. The process mentioned is single staged, operating at 380°C and 7 MPa.

The yields of ethylene from hydrocracking residues increase in direct proportion to the extent of hydrocracking. The yields of ethylene, including ethane recycling, as well as the respective yield of PFO, are represented in Fig. 14 (Kreuter, 1978).

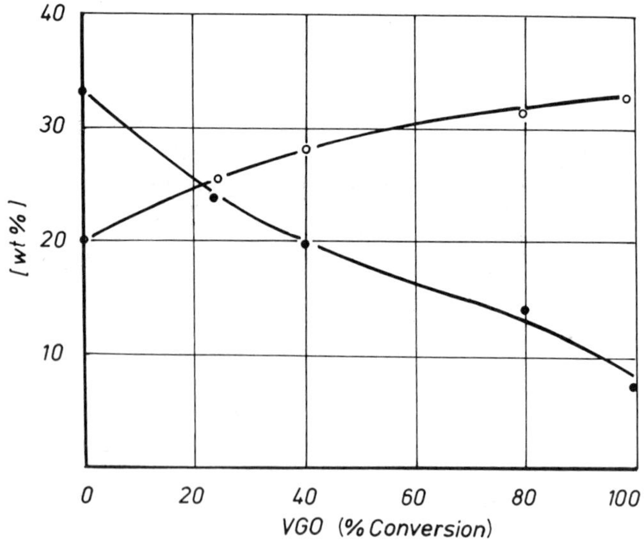

Fig. 14. Steam cracker yields from VGO hydroconverter residue. ○, Ethylene; ●, pyrolysis fuel oil. (From Kreuter, 1978. Reproduced with permission of Linde Aktiengesellschaft.)

For a cracking degree of 50%, about 2×10^6 tons of VGO are required in order to produce 300,000 tons of ethylene. Moreover, the pyrolysis of hydrocracking residues is expected to be possible in conventional naphtha-cracking plants.

Although these results can be judged positively with respect to the yield pattern, so far there have been only very general data on the cracking behavior of hydrocracking fractions and practically none on carbon formation. Despite this situation, the method of hydrocracking offers several attractive possibilities of gaining usable feeds from heavy crude oil fractions and from residues for the production of olefins.

For the production of olefins the use of light hydrocracking products obtained in conventional hydrocracking plants is possible only within limits—or with difficulties. The main cause for the problems is the high content of branched compounds leading to low yields of ethylene. With regard to coke formation in the pyrolysis of hydrocracking products, either no statements or only contradictory ones exist. Coke deposits corresponding to those occurring in the pyrolysis of straight-run gas oil are to be expected.

In order to obtain appropriate feedstocks for the production of olefins through hydrocracking, the process preferably directed toward producing high-octane gasoline and diesel fuel must be necessarily modified in regard to both the catalyst used and the operational procedure. Changes are found with regard to hydrocracking severity, which should lie between hydrocracking and hydrogenation, and which can be achieved by means of such catalysts having a high hydrogenating activity but a low cracking and isomerization activity. Besides removing the undesirable heteroelements, the catalyst should selectively crack the polycyclic aromatics and hydrogenate the cracking products.

From the available results on the pyrolysis of hydrocracking products obtained under medium- and high-pressure conditions, one can conclude that such hydrocarbon fractions are particularly suited for pyrolysis, especially those obtained under high-pressure conditions ($>$15 MPa). The survey given in Table XII presents typical yields of olefins obtained by feeding hydrocracking effluents and hydrocracking residues from high-pressure plants (20 MPa) (Günschel and Nowak, 1980).

Obviously the technical and economic characteristics for the production of light olefins depend on the hydrocracking conditions as well as on the nature of gas oils fed—that is, the boiling range and composition—the conditions and results in pyrolysis itself, and also the utilization of by-products arising during the pyrolysis.

Among the numerous possibilities of processing heavy hydrocarbon fractions, the development is evidently determined by the hydrocatalytic

TABLE XII

Structural Analysis of Technical Hydrocracking Fractions in the Gas-Oil Boiling Range and Typical Yields of Pyrolysis[a]

	Fraction			
	Hydrogenated gas oil	Hydrocracking effluent (untreated)	Hydrocracking effluent (hydrogenated)	Atmospheric gas oil
Characteristics				
Boiling range (°C)	142–310	108–311	100–329	189–318
Molecular weight	180	150	169.9	219
Composition (wt%)				
Paraffins	53.6	57.7	~60	37.4
Naphthenes	41.3	30.3	~39	35.3
Aromatics (total)	5.1	12	1	27.3
Alkylbenzenes	0.2	9	1	4.5
Yields				
Cracked gases (wt%)	71.3	67.6	69.0	69.1
Methane	14.7	13.8	13.0	12.8
Ethylene	25.4	23.3	26.8	21.4
Propylene	14.9	14.9	16.0	13.6
Butadiene	5.3	4.9	5.0	4.4
Pyrolysis gasoline (wt%)	26.4	24.3	24.1	23.6
Benzene	10.2	8.2	9.0	2.5
Toluene	6.6	4.9	4.7	—
Xylene	1.2	1.2	0.8	—
Pyrolysis fuel oil (>200°C)	2.3	8.0	6.8	17.2

[a] Günschel and Nowak (1980).

pretreatment. Among the hydrocatalytic pretreatment methods discussed, partial and total hydrogenation must be given precedence because these feeds not only result in high yields of light olefins and a low coke formation, but they require the least consumption of hydrogen.

E. Pyrolysis of Individual Components from Gas-Oil Fractions

The composition of structural groups in gas-oil fractions is substantially changed by pretreatment methods. Contrasting straight-run fractions with hydrogenated fractions, the content of naphthenic hydrocarbon increases, which fundamentally influences the products distribution during pyrolysis. Besides summaries and qualitative descriptions of the pyrolysis of pre-

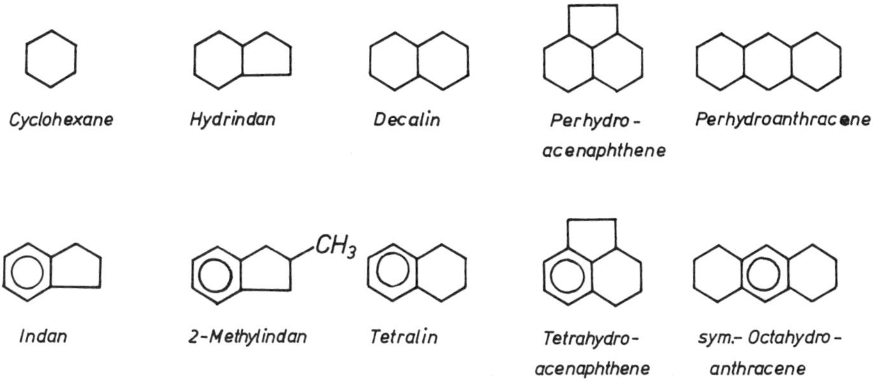

Fig. 15. Cyclic hydrocarbons for pyrolysis.

treated crude oil fractions, in the literature there has been virtually no reliable information on the cracking behavior of polycyclic naphthenes and partially hydrogenated aromatics. Correlations between the structure of alicyclic hydrocarbons in the gas-oil boiling range, the process conditions, and the reaction products arising during the pyrolysis have been mainly investigated (Nowak *et al.*, 1975a,b; Korosi *et al.*, 1979; Bajus and Vesely, 1976; Penninger and Slootboom, 1973; Bredael and Rietvelde, 1979).

In Fig. 15 some totally and partially hydrogenated polycyclic hydrocarbons are shown, which in a temperature ranging from 700 to 830°C, a residence time ranging from 0.3 to 0.5 s, and mixed with *n*-heptane, were submitted to pyrolysis (Nowak *et al.*, 1975a, 1979). The results can be summarized as follows.

1. The decomposition rates of polycyclic naphthenes are altogether higher, whereas those of the cycloalkenoaromatics are lower than the decomposition rates of the *n*-heptane (see Fig. 16).
2. Paraffins and monocyclic and polycyclic naphthenes basically lead to other product distributions. The survey in Table XIII shows how different hydrocarbon types affect the qualitative yield pattern.
3. Cycloalkenoaromatics and polycyclic naphthenes differ not only with respect to their decomposition rates but also with respect to their products. Thus, in the pyrolysis of bicyclic naphthenes, surprisingly high yields of monocyclic aromatics were found. It is interesting to note that the structure of polycyclic naphthenes exerts only a small influence on the quantity and nature of monocyclic aromatics. In

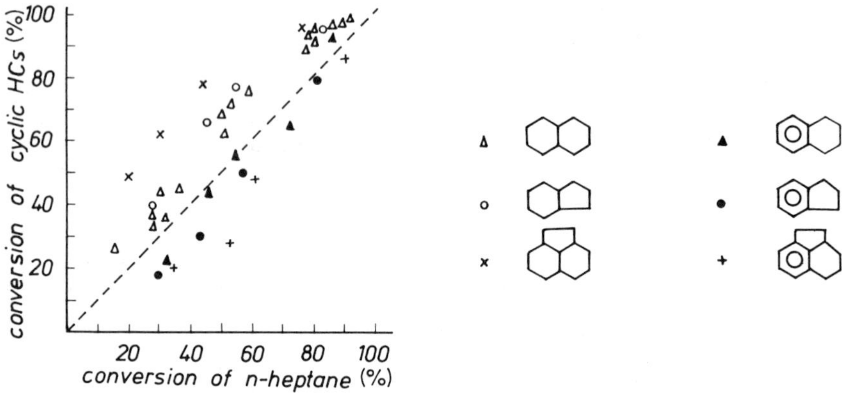

Fig. 16. Percentage conversion of cyclic hydrocarbons (HCs) in presence of *n*-heptane. (From Nowak *et al.*, 1975b.)

TABLE XIII

Comparison of Yields Obtained in Using Paraffins, Monocyclic and Polycyclic Naphthenes as Feeds[a,b]

Yield	mN	pN
C_2–C_4 Olefins	~P	<P
Butadiene	>P	~P
Liquid product $\geq C_5$	~P	>P
Benzene	~P	≫P

[a] Nowak *et al.* (1975b).
[b] mN, Monocyclic naphthenes; pN, polycyclic naphthenes; P, paraffins.

contrast to the cycloalkenoaromatics, the polycyclic naphthenes deliver a considerably higher contribution toward the production of olefins (Fig. 17).

The small yields of C_6–C_9 aromatics from cycloalkenoaromatics are due to their preferred dehydrogenation by pyrolysis, and consequently their conversion into the respective condensed aromatic hydrocarbon. This result has been confirmed by systematic experiments with Tetralin (see Fig. 18).

Cyclic HC in n-Heptane		Crack-gas $\leq C_4$	Olefins $C_2 + C_3$	Liq. prod. $\geq C_5$	Aromatics $C_6 - C_9$
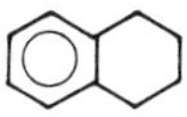	30%	45.3	24.6	54.7	2.9
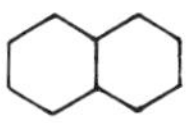	30%	52.3	27.7	47.7	14.7
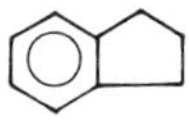	30%	40.6	23.1	59.4	2.2
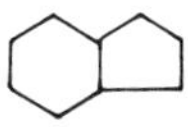	30%	60.7	33.2	39.3	12.6

Fig. 17. Yield distribution in relation to different cyclic hydrocarbons (HC). [From Nowak *et al.*, 1979. Reproduced with permission from *Chem. Tech.* (Leipzig).]

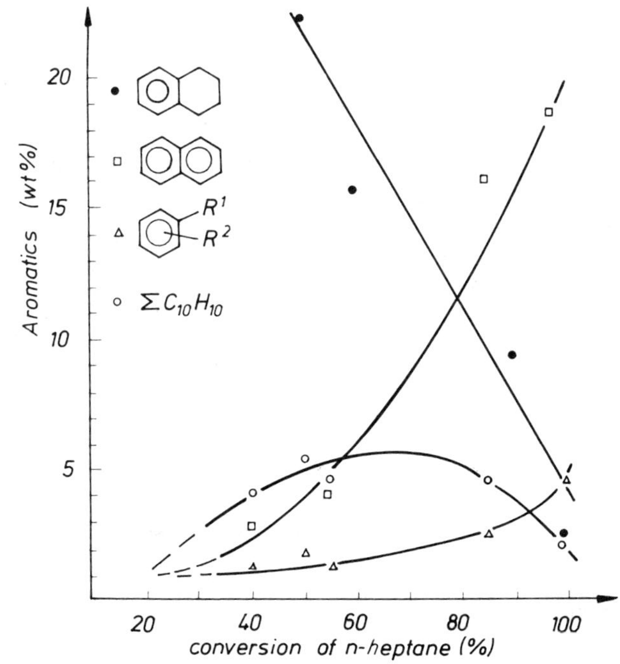

Fig. 18. Products of pyrolysis of Tetralin in presence of *n*-heptane. [From Nowak *et al.*, 1979. Reproduced with permission from *Chem. Tech.* (Leipzig).]

These findings on the pyrolysis of polycyclic hydrocarbons confirm the results achieved with pretreated heavy hydrocarbon fractions.

The results as a whole permit the conclusion that the pyrolysis of hydrocatalytically pretreated gas-oil fractions can be regarded as an alternative method for producing light olefins and aromatic hydrocarbons from naphtha fractions. This method offers the possibility of petrochemically using high-boiling hydrocarbon fractions poor in hydrogen, in addition to replacing naphtha, which is not sufficiently available.

IV. NEW ROUTES IN OLEFINS PRODUCTION

The development of new pyrolysis technologies deviating from the sound tubular steam-cracking technology is determined to a large extent by the availability of conventional pyrolysis feedstocks and the development of prices. The numerous investigations pursue the aim of converting high-boiling hydrocarbon fractions including crude oil and residue oils without any pretreating into light olefins and BTX aromatics, with high yields being provided and no difficulties met. The advances and principles in the development of new techniques available thus far give rise to the following classification.

1. Pyrolysis in the presence of gaseous heat carriers (combustion gases, superheated steam, etc.)
2. Pyrolysis by means of solid recycling heat carriers (coke or inorganic substances)
3. Pyrolysis in the presence of hydrogen under pressure
4. Production of olefins by means of heterogeneous catalysts

The majority of the results quoted in the literature stem from bench-scale and pilot plants. A number of technological and economic problems have not been solved satisfactorily to permit the realization of large-scale production.

A. Pyrolysis in the Presence of Gaseous Heat Carriers (Homogeneous Pyrolysis)

The pyrolysis of hydrocarbons in the presence of highly heated gaseous carriers is an interesting and widely known method for obtaining light olefins and acetylene from different feedstocks. Gases or "steam" heated to high temperatures, or combustion products of a portion of a feedstock or of an external hydrocarbon source, are used as heat carriers. The use of very high reaction temperatures (between 1200 and 2000°C) and short residence times (between 0.005 and 0.01 s) are characteristic for this technique. Process conditions thus positively influence the production of final products and retard the progress of secondary reactions.

For obvious reasons this technique was also used for obtaining light olefins from high-boiling hydrocarbon fractions and crude oil, respectively.

Among the newer developments of flame cracking and partial oxidation for the production of olefins from crude oil, based on the developments of the firms Texas Eastman, SBA-Kellogg, Montecatini, and Hoechst (Bredael and Rietvelde, 1979; Acim *et al.*, 1958; Barry, 1960; Krekeler *et al.*, 1959; Kamptner and Krause, 1966), the autothermal pyrolysis of Dow Chemical Co. (Anonymous, 1977) is given particular attention. The heat required for the reaction is obtained by burning a portion of the feedstock with oxygen and with added hydrogen. The process operates at reaction temperatures between 1200 and 2000°C; the residence time is about 0.1 s. When using crude oil, yields of ethylene between 22 and 27 wt% are obtained. The C_2H_4/C_3H_6 ratio is 3 or 4:1, and the C_2H_4/C_2H_2 ratio is 10 or 15:1.

The commercial realization of the production of light olefins according to this technique meets with a number of difficulties, among them the following.

1. The design of the pyrolysis reactor and its equipment (burner, combustion chamber) for producing the gaseous heat carriers
2. Problems of mixing the gaseous heat carriers with the pyrolysis feedstock, because studies have shown that the mixing time must be considerably smaller than the reaction time of the hydrocarbons
3. The need for extensive downstream operations to separate the reaction products from the combustion gases
4. The yields of tar and fuel oil, especially in the pyrolysis of high-boiling hydrocarbon fractions and crude oil

An attractive method of obtaining olefins is crude oil pyrolysis in the presence of superheated steam as a heat carrier, developed by the firms Kureha and Union Carbide. This development has also become known under the name ACR (advanced cracking reactor) (Hosoi and Keisler, 1975).

The first results on a pyrolysis process operating with steam heated to 2000°C were reported by Gomi and Araki (1971) and Wilkinson and Gomi (1974). The steam is produced in a regenerating furnace filled with ceramic pebbles (ZrO_2, Al_2O_3). The steam/feedstock ratio is about 2:3. In the pyrolysis of different crude oils, ethylene yields from 20 to 30 wt% and acetylene yields between 2 and 17 wt% are obtained at temperatures in the reaction zone from about 950 to 1150°C with residence times from 0.005 to 0.01 s. The process developed further is expected to be used industrially in the 1980s (Hosoi and Keisler, 1975; Wett, 1972; Ishikawa and Keister, 1978; Anonymous, 1978). The relatively high yields of both pitch and fuel oil are negative attendant phenomena of this technique. Also, the economy of the process is decisively determined by the regional possibilities of utilizing the acetylene produced.

B. Pyrolysis in the Presence of Solid Heat Carriers

Another interesting method of converting heavy feedstocks or crude oil into light olefins is the processes that recycle solid heat carriers such as coke, sand, and ceramic materials. The thermal contact processes, such as the fluidized-bed reactor presented by Lurgi (Schmalfeld, 1963a,b and the thermal contact processes devised in the Soviet Union (Tmenov, 1972; Tmenov *et al.*, 1968; Gorislavets *et al.*, 1977; Lavrovski and Brodski, 1957)—all of which operate according to the principle just indicated—have not been developed further for various reasons. An exception is the development in Japan. As a result of extensive research and developmental works, Kunugi *et al.* (1978) and Kunii and Kunugi (1975) presented a fluidized-bed reactor operating with coke particles as heat carriers. The process is run at temperatures between 700 and 850°C and at residence times between 1 and 3 s. The yields of ethylene obtained under these conditions for the pyrolysis of crude oil amount to about 18–20 wt%. Problems result from a high content of sulfur, which is concentrated in the cracked gas up to 2.2%, in the naphtha fraction to 0.5%, and in the other reaction products, such as coke, to about 6%. The relatively high yields of by-products such as fuel oil, pitch, and other high-boiling reaction products whose economic and effective utilization are problematic, as well as the reliability of control of the complex fluidized-bed reactor technology, influence the further development and industrial use of this process.

C. Pyrolysis in the Presence of Hydrogen

The technique of hydropyrolysis, developed by the French firms Pierrefitte-Auby in cooperation with S. Heurtey and Naphtha-Chimie S. A. (Barre *et al.*, 1976; Anonymous, 1975, 1976), operates at temperatures between 800 and 900°C, at pressures of 10 to 30 atm, and at a residence time of about 0.1 s. This process uses hydrogen as a makeup gas.

By combining a high reaction temperature and very short residence times, with hydrogen, high-boiling crude oil fractions can be converted into light olefins, bringing about high yields of the latter without any noticeable problems. Hydrogen plays a particular role in this process. According to fundamental studies performed by Kunugi *et al.* (1967) on the acceleration of the cracking reaction in the presence of hydrogen, this element not only favors the cracking of hydrocarbons, it also inhibits coking. Although it is widely known that applying pressure negatively affects the production of olefins, the use of very short residence times minimizes the disadvantages brought about by the pressure. On the other hand, the application of pressure brings about advantages, because some energy costs for gas compression can be saved. Furthermore, hydropyrolysis allows different feedstocks (naphtha or gas oils) with high yields of ethylene to convert equally without any technological alterations of this technique. Thus in recycling ethane and propylene into the pyrolysis process, about 45 wt% ethylene is obtained during the cracking of naphtha. In gas-oil pyrolysis, the ethylene yields are considerably less on a once-through basis, and amount in the case of naphtha to only about 11%.

Despite the marked advantages of this technique, it has not yet been possible to develop this process for industrial use. The very complicated high pressure–hydrogen technology of this process, as well as the lack of assurance of stable operation, give rise to considerable technological problems. The provision and consumption of expensive hydrogen (about 50 to 100 m^3/ton of feedstock), as well as the extensive separation of hydrogen from the reaction products, are to be regarded as factors whose solution exerts a strong influence on the further development and introduction of this technique into practice.

D. Pyrolysis in the Presence of Catalysts

A direction of research in the field of olefin production that has aroused increasing interest during recent years, from both the theoretical and the practical points of view, is the conversion of hydrocarbons in the presence of catalysts. The publications and patents dealing with this problem pursue the following goals.

1. An increase of selectivity in the production of olefins
2. A reduction in the specific energy requirement of the pyrolysis by lowering the reaction temperature and saving energy
3. An increase in the flexibility of the pyrolysis process in regard to different reaction products
4. The provision of processing possibilities for heavy hydrocarbon fractions

The catalysts claimed in patents for catalytic pyrolysis comprise a large range of elements in the periodic table. Basically, the catalysts can be arranged in three large groups.

1. Amorphous and crystalline aluminosilicates in connection with conventional cracking catalysts
2. Alkaline or alkaline earth–aluminate compounds
3. Different metal oxides on different catalyst carriers such as alumina, pumice, and amorphous and crystalline (zeolite) aluminosilicates.

By patent law especially the metal oxides of iron, chromium, cobalt, molybdenum, vanadium, and nickel were protected. Moreover, catalysts containing noble metals have been patented. Research teams that have become known through studies in this field are predominantly in the Soviet Union (Adel'son *et al.*, 1965, 1972, 1975a,b,c; Pauskin *et al.*, 1968, 1970; Lysych *et al.*, 1969, 1970; Talisman *et al.*, 1961a,b), Japan (Tomita *et al.*, 1971; Tomita and Yamaguchi, 1974; Tomita and Kitagawa, 1977), France (Senes *et al.*, 1969, 1972; Senes and Pittier, 1972; Lhonore *et al.*, 1969), the United States, and Denmark (Bogart *et al.*, 1972; Rostrup-Nielsen, 1973; Fischer *et al.*, 1973; Wrisberg *et al.*, 1969). These studies so far are almost exclusively in the stage of basic research. Central to these investigations are the development of suitable catalysts, the study of the mechanism of heterogeneous decomposition, and the interaction between substrate and catalyst. The high yields of ethylene (between 30 and 40 wt%) brought about by different catalysts at temperatures between 700 and 750°C appear to justify the intensive research work in this field. Unresolved questions are the insufficient time stability of the catalysts and the required regeneration due to coke deposits. In contrast to homogeneous pyrolysis—in which, as a rule, an ethylene/propylene ratio of 2:1 is obtained—in catalytic pyrolysis this ratio changes toward the value of 1:1.

Mention must be made of the introduction of a technique (THR process) by Tomita and Kitagawa (1977) of Japan, using a BeO, CaO, or SrO—Al_2O_3 catalyst for the production of ethylene from high-boiling hydrocarbon fractions. With a catalyst of the composition $Ca_{12}Al_{14}O_{33}$ and $Ca_3Al_2O_3$, an ethylene yield of 41 wt% is obtained from a hydrocarbon fraction of the boiling range of 200–400°C at temperatures of about 720°C and a H_2O/HC ratio of 3:1. The advantage of this technique, which is already at a pilot scale, is the insensitivity of the catalyst to sulfur and heavy metals of the feedstock. Furthermore, no carbon formation has been observed on this catalyst, obviously because it is well known that alkaline earth compounds show gasification activity to carbon.

Although this process by changing the catalyst composition can also be used for obtaining hydrogen, syngas, and fuel gas, information is as yet available on its large-scale realization.

V. CONCLUSIONS

Developments in the sphere of olefin production are characterized by (*a*) increased use of heavy hydrocarbon fractions and the development of new techniques and cracking technologies, (*b*) an increase in the flexibility of pyrolysis plants with respect to processing different feedstocks, (*c*) an increase in selectivity with respect to the yield of ethylene, (*d*) an energetically more effective accomplishment of the pyrolysis process, (*e*) longer running times of pyrolysis plants, and (*f*) lower investment and processing costs.

A simultaneous fulfillment of all these requirements and the realization of a cracking technology meeting all these needs appear to be impossible from the viewpoint of science and technology. Depending on the regional and specific conditions a producer is faced with, the objective can only be to reach an acceptable compromise.

The advances achieved in the field of olefin production are a result not so much of new scientific results as they are of constructional and technological improvements of the tube reactor and its equipment (quench cooler, etc.). As a result of this development, the well-known furnace designs for high-temperature, short-time pyrolysis mentioned in an earlier section, as well as new techniques devised to operate in the presence of gaseous and solid heat carriers respectively or in the presence of hydrogen, came into being. Although an increase in the efficiency of the cracking-coil reactor with regard to the rise of selectivity and throughput meets with technological limitations (heat-flux density, rapid heat transfer, heat-resisting materials of coils, etc.), the constructional possibilities of those technologies deviating from the tube reactor are not yet exhausted. All the studies directed toward the direct pyrolysis of heavy hydrocarbon fractions reflect a more or less advanced stage of development, but thus far these projects have not come into commercial use, with the exception of autothermal pyrolysis. In spite of different techniques, all these developments have in common (*a*) high yields of fuel oil, gum, and coke, respectively, (*b*) problems resulting from the presence of heteroelements and organic metal compounds in the feedstock during the pyrolysis, and (*c*) the production of a large yield pattern and for that reason high expenses for separating and commercially utilizing the by-products obtained. However, the production of a large assortment of organic basic chemicals must by no means be judged a completely negative development. For instance, the pyrolysis of crude oil in areas far from conventional refineries can be of great significance for developing a chemical industry in such isolated areas.

There are still quite a number of unresolved technological and economic problems with the direct pyrolysis of heavy crude oil fractions, so the pretreatment methods for producing suitable feedstock for olefins from crude oil fractions have a good chance to be realized in practice. On the one hand, the well-known difficulties in the pyrolysis of high-boiling hydrocarbon fractions poor in hydrogen are removed by the pretreatment; but on the other hand, the pyrolysis can be performed without any expensive changes by means of the proven tubular cracking technology. Nevertheless, the fact remains that pretreating represents an additional step within the process, affecting the economy of olefin production.

Direct pyrolysis in improved tubular furnace (SRT, USC, LSCC, Millisecond P.)	Pretreatment of feedstock (extraction of aromatics, hydrogenation, hydrocracking)	Pyrolysis in presence of gaseous carriers (combustion gases, superheated steam)	Direct pyrolysis in presence of hydrogen, at normal and elevated pressures	Pyrolysis in presence of solid recycling heat carriers (coke, ceramic pebbles)	Pyrolysis in presence of heterogeneous catalysts
	↓				
	Pyrolysis in tubular furnace				

Fig. 19. Developments in olefin production from heavy crude oil fractions.

At present it appears that the techniques shown in Fig. 19 are well suited for the production of olefins on the basis of heavy hydrocarbon fractions. Which of these techniques will succeed in the long run to be commercially used on a large scale is dependent on many factors, such as technological progress, the feedstock situation, local conditions, the utilization of final and by-products, and requirements for control of environmental pollution.

REFERENCES

Aarna, A. J., Zirgakov, J. H., and Soone, J. C. (1978). *Zh. Prikl. Khim. (Leningrad)* **51**, 1417–1418.

Acim, C. A., Reid, T. F., and Schrader, R. J. (1958). *Chem. Eng. Prog.* **54**(1), 41–48.

Adel'son, S. V., Bulavina, G. M., Visuyakova, T. M., Rutman, A. M., Somov, A. K., Liakumovic, A. G., Lupin, A. R., Micurov, A. M., Rutman, G. I., and Tarov, U. T. (1965). USSR Patent 219,593.

Adel'son, S. V., Smirnov, A. P., Gurevic, V. I., Rudyk, E. M., Nikonov, V. I., Pauskin, J. M., and Ivankovaskii, B. L. (1972). USSR Patent 410,073.

Adel'son, S. V., Lysych, O. V., and Pauskin, J. M. (1975a). *Neftekhimiya* **15**, 836–841.

Adel'son, S. V., Rudyk, E. M., Smirnov, A. P., Gerevic, V. J., and Pavlyuchin, V. N. (1975b). *Khim. Tekhnol. Topl. Masel* 8, 2.

Adel'son, S. V., Nikonov, V. I., Ivankovskii, B. L., Smirnov, A. P., Rudyk, E. M., and Gurevic, V. I. (1975c). USSR Patent 523,133.

Albright, L. F. (1978). *Ind. Eng. Chem., Process Des. Dev.* **17**, 377.

Albright, L. F., and Yu, Y. H. C. (1978). *Am. Chem. Soc.*, (Washington, DC) Div. Petrol. Chem. Preprints, **23**, 218.

Anonymous (1975). *Chem. Eng.* (N.Y.) **82**(16), 50.

Anonymous (1976). *Chem. Eng. News* **54**(44), 19.

Anonymous (1977). *Chem. Eng. News* **55**(48), 6.

Anonymous (1978a). *Chem. Eng. News* **56**(13), 29–30.

Anonymous (1978b). *Oil Gas J.* **76**(18), 75–76.

Badie, P. (1975). Ph.D. Thesis, Univ. Strasbourg.

Bajus, M., and Vesely, V. (1976). *Ropa Uhlie* **18**, 356.

Balbig, D., and Lange, H. (1975). Internal Report, VEB Chemieanlagenbau, Leipzig.

Barre, C., Chahvekilian, E., and Dumon, R. (1976). *Hydrocarbon Process.* **55**(11), 176–178.

Barry, M. I. (1960). *Chem. Eng. Prog.* **56**(1), 39.

Barton, R., McCormick, R. H., and Fenske, M. R. (1968). *Ind. Eng. Chem., Process Des. Dev.* **7**, 366.

Becker, K., Bindernagel, K., Grossmann, M., Guenschel, H., Nowak, S., Nette, W., Pechstein, G., Pfeiffer, F., Prag, M. *et al.* (1978). Ger. (East) 131,474 and 131,475.

Bogart, M. J. P., Nielsen, A., Wrisberg, J., and Rostrup-Nielsen, J. (1972). Ger. Offen. 2,227,769.

Bredael, P., and Rietvelde, D. *Fuel* **58**, 215.

Davis, P. T., Glover, T. G., and Jones, J. R. (1972a). Ger. Offen. 2,164,951.

Davis, P. T., Glover, T. G., and Jones, J. R. (1972b). Ger. Offen. 2,254,131.

de Blieck, J. L., Cijfer, H. J., and Jungerhans, R. R. J. (1971). *Hydrocarbon Process.* **50**(9), 177–182.

Dunkleman, J. J. (1976). Ph.D. Thesis, Purdue Univ., Lafayette, Indiana.

Eigenson, A. S., Kritschko, A. A., Kozig, B. L., Pestrikov, S. W., Sodikova, S. R., Ivanova, L. S., Skudina, J. J., and Ajasjan, G. N. (1968). *Khim. Tekhnol. Topl. Masel* **9**, 7.

Eigenson, A. S., Sokov, I. F., Putilova, S. D., Pau, G. M., and Kusnetzov, A. P. (1970). *Khim. Tekhnol. Topl. Masel* **7**, 1.

Eisenlohr, K. H., Naumburg, K., and Zengel, H. G. (1967). *Erdoel Kohle, Erdgas, Petrochem.* **20**, 82–89.

Fischer, F. H., Rostrup-Nielsen, J., and Wrisberg, J. H. (1973). Ger. Offen. 2,340,904.

Ger. Offen. (1967). 1,668,774 (to Universal Oil Products Co.).

Ger. Offen. (1972). 2,254,131 (to Lummus Co.).

Ger. (East) Pat. Appl. (1980). Regist. No. 223,007. Acad. Sci. GDR, Berlin.

Ghaly, M. A., and Crynes, B. L. (1976). *In* "Industrial and Laboratory Pyrolyses" (L. F. Albright and B. L. Crynes, eds.), ACS Symposium Series, No. 32, pp. 218–241. Am. Chem. Soc., Washington, D.C.

Goetzmann, S., Kreuter, W., and Wernicke, H. J. (1979). *Hydrocarbon Process.* **58**(6), 109.

Gomi, S., and Araki, S. (1971). *Proc.—World Pet. Congr., 8th, Moscow* **4**, 371–378.

Goosens, A. G., Westerduin, R. F., and Mol. A. (1975). *Erdoel Kohle, Erdgas, Petrochem.* **28**, 471–478.

Gossens, A. G., Dente, M., and Ranzi, E. (1978). *Hydrocarbon Process,* **58**(10), 227–237.

Gorislavets, S. P., Tmenov, D. N., and Maiorov, W. J. (1977). "Pirolis Uglewodorodnovo sirja," p. 179. Nauk. Dumka, Kiev.

Green, E. J., and Zdonik, S. B. (1968), Fr. 1,600,622; (1969), Ger, Offen. 1,922,665.

Green, E. J., Zdonik, S. B., and Hallee, L. P. (1975). *Hydrocarbon Process.* **54**, 164.

Günschel, H., and Nowak, S. (1973). *Alkene '73, Bilateral Colloq. GDR/CSSR, Marianske Lazne, Czech.*

Günschel, H., and Nowak, S. (1980). Unpublished data, 210C. Acad. Sci. GDR, Berlin.

Hatch, L. F., and Matar, S. (1978). *Hydrocarbon Process.* **57**, 129–139.

Hosoi, T., and Keisler, H. G. (1975). *Chem. Eng. Prog.* **71**(11), 63–67.

Illés, V., and Horvath, A. (1975). *Hung. J. Ind. Chem.* **3**, 369–390.

Illés, V., and Horvath, A. (1978). *Erdoel Kohle, Erdgas, Petrochem.* **31**, 42.

Illés, V., Szalai, O., Csermely, A., and Otto, A. (1978). *Chem. Tech. (Leipzig)* **30**, 122–126.

Ishikawa, T., and Keister, R. G. (1978). *Hydrocarbon Process.* **58**(12), 108.

Kamptner, H. K., and Krause, W. R. (1966). *Chem. Eng. (N.Y.)* **73**(4), 80–98.

Kopinke, F., Poizel, E., Bach, G., Zimmerman, G., and Nowak, S. (1981). *Erdoel Kohle, Erdgas. Petrochem.* **34**, 204–208.

Korosi, A., Virk, P. S., and Woebcke, H. N. (1979). *Erdoel Kohle, Erdgas, Petrochem.* **32**, 473.

Kossiakoff, A., and Rice, F. O. (1934). *J. Am. Chem. Soc.* **56**, 590.

Krekeler, H., Wertz, R., and Pechthold, H. (1959). *Erdoel Kohle, Erdgas, Petrochem.* **12**, 358.

Kreuter, W. (1978). *Linde Olefines Symp., Munich* Sect. A, pp. 1–5.

Kubicka, R. (1974). *Ropa Uhlie* **16**, 204–215.

Kunii, D., Kunugi, T., *et al.*, (1975). *Proc.—World Pet. Congr., 9th, Tokyo* **5**, 137–143.

Kunugi, T., Tominaga, H., and Abiko, S. (1967). *Proc.—World Pet. Congr., 7th, Mexico City* **5**, 239–245.

Kunugi, T., Kunii, D., and Kobayashi, A. (1978). *Trends Avail, Feedst. Compos., World Conf. Future Sources Org. Raw Mater., "Chemrawn 1," Toronto*, pp. 449–456.

Lavrovski, K. P., and Brodski, A. M. (1957). *Khim. Nauka Promst.* **2**, 189–196.

Lehaye, J., and Prado, G. (1975). *J. Chim. Phys.* **72**, 483–489.

Lehaye, J., Badie, P., and Duret, J. (1977). *Carbon* **15**, 87–93.

Lhonore, P., Quibel, J., and Senes, M. (1969). U.S. Patent 3,624,176.

Linden, H. R., and Reid, J. M. (1959). *Chem. Eng. Prog.* **55**(3), 71.

Lohr, B., and Dittmann, H. (1977a). *Oil Gas J.* **75**(29), 78–85.

Lohr, B., and Dittmann, H. (1977b). *Oil Gas J.* **75**(27), 53–58.

Lohr, B., and Dittmann, H. (1978). *Linde Ber. Tech. Wiss.* **43**, 9–22.

Lohr, B., and Schwab, W. (1979). *Chem.-Anlagen Verfahren* **3**, 27–36.

Lysych, O. V., Pauskin, J. M., Adel'son, S. V., and Arbitman, S. V. (1969). *Gazov. Promst.* **14**(11), 46–49.

Lysych, O. V., Pauskin, J. M., Adel'son, S. V., and Panova, I. J. (1970). *Gazov. Promst.* **15**(3), 44–47.

Magaril, R. Z. (1973). "Formation of Carbon during Thermal Transformation of Individual Hydrocarbons and Petroleum Products" Izd. Khimiya, Moscow.

Magaril, R. Z. (1976). "Theoretical Principles of Chemical Petroleum Refining Processes, p. 87. Izd. Khimiya, Moscow.

Magaril, R. S., and Beresina, S. N. (1975). *Khim. Tekhnol. Topl. Masel* 6, 5–8.

Magaril, R. S., and Beresina, S. N. (1976). *Zh. Fiz. Khim.* **9**, 2282–2285.

Mikulla, K. D., Bölt, H., and Richter, H. (1978). *Linde Olefines Symp., Munich* Sect. U, p. 5.

Muchina, T. N., Furer, C. M., and Leschina, G. F. (1974). "Production of Unsaturated olefins," pp. 27–33. NIISS, Moscow.

Nowak, S., Keil, G., Günschel, H., and Pechstein, G. (1975a). *Proc.—World Pet. Congr. Tokyo*, **5**, 145–153.

Nowak, S., Keil, G., and Zimmermann, G. (1975b). *Hung. J. Ind. Chem.* **6**, 105.

Nowak, S., Günschel, H., and Zimmermann, G. (1979). *Chem. Tech.* (*Leipzig*) **31**, 496–500.

Obolentzev, R. D. (1967). *Proc.—World Pet. Congr., 7th, Mexico City* **9**, 109–116.

Offen, M. J. (1976). *Hydrocarbon Process.* **55**(10), 123.

Pauskin, J. M., Liakumovic, A. G., Adel'son, S. V., Nikiforov, P. A., and Lysych, O. V. (1968). *Khim. Promst.* (*Moscow*) No. 11, 811.

Pauskin, J. M., Adel'son, S. V., Smirnov, A. P., and Rudyk, E. M. (1970). *Neftepererab. Neftekhim.* (*Moscow*) No. 7, 24–25.

Penninger, M. L., and Slootboom, H. W. (1973). *Rec. Trav. Chim. Pays-Bas* **92**, 513–530.

Prescott, J. H. (1975). *Chem. Eng.* **82**(14), 52–53.

Rhoe, A., Fernandes-Baujin, I. M., Neuwirth, O., and Dorn, R. K. (1979). *Proc.—World Pet. Congr., 10th, Bucarest*, **4**, 185–193.

Riad, M. A., Mahmoud, B. A., Hafiz, M. Y., and Abdou, I. K. (1969). *J. Inst. Pet.* **55**, 568.

Rice, F. O., and Herzfeld, K. F. (1934). *J. Am. Chem. Soc.* **56**, 284–289.

Rostrup-Nielsen, J. (1973). Ger. Offen. Patent 2,310,464.

Sajus, L. (1978). *Trends Avail. Feedst. Compos., World Conf. Future Sources Org. Raw Mater.*, "*Chemrawn 1*," *Toronto*, p. 421.

Schmalfeld, P. (1963a). *Proc.—World Pet. Congr., 6th*, Frankfurt-am Main, **4**, 303–309.

Schmalfeld, P. (1969b). *Erdoel Kohle Erdgas, Petrochem.*, **16**, 534–537.

Senes, M., and Pittier, M. (1974). Fr. Demaude **2**, 187, 408.

Senes, M., Lhonore, P., and Quibel, J. (1969). Ger. Offen. Patent 1,927,061.

Senes, M., Lhonore, P., and Quibel, J. (1972). Fr. Demaude **2**, 186, 452.

Soonawala, H. D., and Den Broeder, S. E. (1976). Ger. Offen. 2,601,875 and 2,603,142.

Spengler, G., Wallenhorst, F., and Kern, J. (1973). *Erdoel Kohle, Erdgas, Petrochem.* **26**, 82.

Spirkin, V. G., Bataev, P. S., Chernova, K. S., and Kozlov, A. I. (1972). USSR Patent 349,674.

Stork, K. (1977). *Proc. Lummus-Symp. Konvers.-tech. Raffinerien, Wiesbaden.*

Stove, E. A., and Meyer, R. P. (1966). *Ind. Eng. Chem.* **58**, 36–40.

Talisman, L. V., Fomina, V. I., and Astrina, A. D. (1961a). USSR Patent 141,481.

Talisman, L. V., Fomina, V. I., Astrina, A. D., Stepanov, A. G., and Cernuchina, V. G. (1961b). USSR Patent 154,248.

Tmenov, D. N. (1972). Ph.D. Thesis, Inst. Neftekhim. Sint., Akad. Nauk, Moscow.

Tmenov, D. N., Aliev, W. S., and Mamedov, C. T. (1968). *Khim. Tekhn. Topl. Masel 7*, 7–10.

Tomita, T., and Kitagawa, M. (1977). *Chem. Eng. Tech.* **49**, 469.

Tomita, T., and Yamaguchi, Y., (1974). Ger Offen. Patent 2,431,983.

Tomita, T., Kikuchi, K., and Skamoto, T. (1971). Ger. Offen. Patent 2,158,073.

Virk, P. S., Chambers, L. E., and Woebcke, H. N. (1974). *Adv. Chem. Ser.* No. 131, 237.

Watson, K. M., and Nelson, E. F. (1935). *Ind. Eng. Chem.* **27**, 1460.

Wernicke, H. J., and Kreuter, W. (1978). Ger. Offen. 2,721,504.

Wernicke, H. J., and Kreuter, W. (1979). *Hydrocarbon Process.* **58**(10), 137–142.

Wett, T. (1972). *Oil Gas J.* **70**(6), 76.

Wilkinson, L. A., and Gomi, S. (1974). *Hydrocarbon Process.* **53**(5), 109–111.

Witt. R., and Wall, F. (1977). *AIChE Meet.* (Nov.), *New York.* 13–17.

Wolley, K. H. (1977). *Chem. Ind. (Düsseldorf)* **24**, 17.

Woodle, R. A. (1980). *Hydrocarbon Process.* **59**, 171.

Wrisberg, J., Andersen, K. J., and Mogensen, K. (1969). Br. Patents 1,305,568 and 1,306,087.

Zdonik, S. B., and Hayward, G. L. (1975). *Hydrocarbon Process.* **54**(8), 95–98.

Zdonik, S. B., Green, E. J., and Hallee, L. P. (1967a). *Oil Gas* J. **65**(26), 96–101.

Zdonik, S. B., Green, E. J., and Hallee, L. P. (1967b). *Oil Gas* J. **65**(34), 86–88.

Zdonik, S. B., Hayward, G. L., Fishtine, C. H., and Feduske, F. C. (1975). *Hydrocarbon Process.* **54**(12), 111.

Zdonik, S. B., Hayward, G. L., Fishtine, C. H., and Feduske, F. C. (1976). *Hydrocarbon Process.* **55**(1), 149–154.

Zychlinski, W. (1972). Ph.D. Thesis, Univ. Leipzig.

13

Pyrolysis Furnace Design: Conventional and Novel

LAWRENCE L. ROSS

I. INTRODUCTION

A. Feedstock Flexibility

Developers of pyrolysis reactor technology are today faced with the formidable task of operating in an environment of uncertainty with respect to feedstock availability. This uncertainty is aggravated by potential

Pyrolysis: Theory and Industrial Practice

ISBN 0-12-048880-9

instability in world politics. Although petrochemical producers have been able to bid successfully for feedstocks in the fuels market, the ability to use a variety of different materials to produce ethylene has become increasingly important (Wett, 1980). This situation is evidenced by the new American facilities that were designed for heavier feedstocks but are now confronted by economics favoring natural gas liquids available from the North Sea and the Middle East. It is obvious that olefins manufacturers will have to be flexible regarding feedstock selection in order to optimize profitability, particularly in the present over-capacity market.

B. Reactor Design

Constraints on downstream facilities, rather than the reactor, tend to limit the flexibility of operation. A flexible plant will cost more than a plant designed to handle the poorest feed, because not all equipment is limited by the poorest feed, nonetheless all equipment has to be sized for the limiting case. The flexibility required of a reactor design is therefore dictated by the extent to which additional capital charges for a plant can be justified.

The most widely employed system for producing olefins is by thermal cracking of hydrocarbons in fired tubular furnaces. The large modern plants under design and construction use this method. In general, the optimal residence times in tubular reactors decrease with increasing heaviness of feedstock. Shorter residence times are achieved using smaller diameter tubes and/or higher tube-metal temperatures. Both the number and cost of such furnaces are higher than the furnaces with larger tube diameters, which are adequate for cracking of lighter feedstock. As a result, if excessive flexibility is initially installed to accommodate heavy feedstocks but is not fully utilized, considerable nonproductive investment will have been made.

To the extent that economic factors dictate that progressively heavier feedstocks be used, two shortcomings of conventional tubular technology are becoming evident. First, coking rates in both the furnace tubes and the effluent-quench zones are greatly increased, ultimately to the point of precluding the use of tubular technology. Second, increased tail gas and heavy liquid by-products add to the capital cost of downstream-handling facilities. Designers of tubular reactors have attempted to correct these shortcomings by developing significant innovations in configuration and processing to yield shorter residence times and lower partial pressures of hydrocarbons, conditions that allow increased selectivity of olefins. The designers have been aided by improved processes for the upgrading of heavy feedstocks to be used in production of olefins (Nahas, 1980; Offen,

1976). In parallel with these developments, numerous novel concepts have been proposed to exploit more fully the operating regimes that have been precluded by conventional tubular technology, and to deal with coking problems.

The prime objectives of this chapter are to review the state of the art with respect to technology for conventional tubular reactors and to provide a general overview of novel concepts of reactor design now being explored.

II. CONVENTIONAL REACTOR TECHNOLOGY

A. Introduction

Indirectly fired tubular reactors hold a dominant position in olefins production. This type of reactor is expected to be developed continually and to be used for pyrolysis for many years to come. A schematic diagram of a typical reactor–furnace combination is shown in Fig. 1. The reactants enter the furnace countercurrent to the flue-gas flow and pass through a

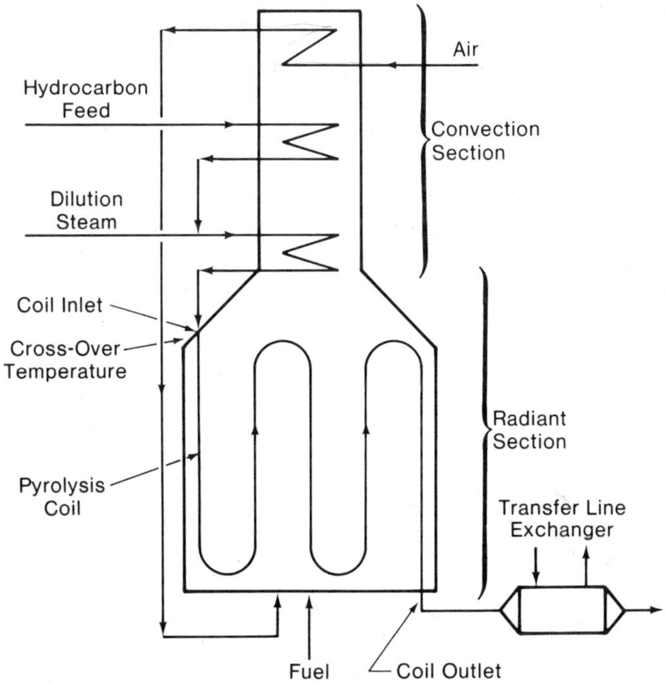

Fig. 1. Diagram of a typical pyrolysis furnace.

convection-heating section and a radiant coil. The convection section is shielded from direct radiation from burners by several rows of high-alloy bare shielding tubes. Extended-surface tubes are generally used in the convection section in order to increase the heat-transfer rate on the flue-gas side. Most fireboxes are designed so that radiation is received on both sides of the tubes, exposing them uniformly to radiant energy. Because of the high operating temperatures in the firebox and reactor, both the effluent and the flue gases are at high enthalpy levels. A significant fraction of the usable heat from the flue gas can be recovered in the convection section by preheating the feed and by generating steam in economizer coils. Heat recovery from the reactor effluent is normally accomplished by exchange with boiling water followed by direct quench with water or oil.

B. Design Criteria

The trend in pyrolysis is toward higher temperatures and shorter residence times, particularly for distillate feedstocks. This trend has been accelerated by the increased use of gas-oil feedstocks. For these hydrogen-deficient feeds, short residence time is the prime means of maximizing olefins yields and run lengths. Added to the demands for high temperature and low coil volume are the needs to minimize buildup of coke in the coil and to quench rapidly. Quenching must avoid fouling and recover as much high-level heat as economically justified.

As discussed earlier, a key consideration in design is flexibility in use of feedstock. Although special circumstances may warrant a wide range of flexibility, a more common practice is to design for dual-feedstock flexibility (e.g., ethane–LPG, LPG–naphtha, and naphtha–atmospheric gas oil (AGO). Table I compares approximate maximum throughputs of hydrocarbon for the same coil (Lohr and Schwab, 1979). Relative throughputs are sensitive to coil geometry and to peak tube-skin temperature. For example, for an

TABLE I

Relative Hydrocarbon Throughputs of Cracking Coils

Ethane	60
Propane	85
Butane	90
Naphtha	100
Kerosene	110
AGO	115

TABLE II

Typical Once-Through Conversions for Hydrocarbon Pyrolysis

Feed	Conversion (%)	Typical effluent yields[b] (wt%)	
		C_2H_4	C_3H_6
Ethane	60.6	49.1	1.1
Propane	91	35.5	14.4
Naphtha (81° API)	92[a]	31.3	17.1
AGO[c] (27 BMCI)	90[a]	24.1	12.9
VGO[c] (42 BMCI)	94[a]	22.7	10.9

[a] Conversion based on cracking-severity index (Ross and Shu, 1979).
[b] Yields from generalized pyrolysis models: ethane, propane (Ross and Shu, 1977); naphtha, gas oils (Shu and Ross, 1979).
[c] AGO, Atmospheric gas oil; VGO, vacuum gas oil.

equivalent ethane conversion of 60%, Ross and Shu (1977) show a relative throughput of about 0.8 as compared with the value of 0.7 computed from Table I.

For a given feedstock, the main variables affecting yields are feedstock conversion, residence time, and partial pressure of hydrocarbon. A maximum yield of total olefins is achieved at moderate conversion, whereas maximum yield of ethylene requires the highest practical conversion limited by coking in the reactor or by the transfer-line exchanger (TLE). By-product values usually set the optimum conversion level. Short residence time and low partial pressure of hydrocarbon enhance yields of olefin at all levels of severity. Typical once-through conversions as a function of type of feedstock are summarized in Table II.

After selection of conversion level, the designer has to devise a furnace coil and establish operating conditions that optimize yields and costs of capital and energy. Constraints include minimum steam ratio from a fouling standpoint, a maximum allowable tube-metal temperature, and a coil-outlet pressure above atmospheric.

C. Operating Conditions

The feedstock, level of conversion, and economic criteria allow determination of an optimum value of the partial pressure of hydrocarbon in the coil that corresponds to a specific combination of coil-outlet pressure and

TABLE III

Nominal Reactor Steam-Dilution Ratios

Feed	Nominal hydrocarbon partial pressure at outlet (psia)	Nominal hydrocarbon partial pressure at outlet (10^6 Pa)	Corresponding steam ratio (weight basis) at 25 psia (0.17×10^6 Pa)
Ethane	20–18	0.14–0.12	0.25–0.40
Propane	18–15	0.12–0.10	0.30–0.50
Naphtha	14–11	0.097–0.076	0.50–0.80
Gas oil	9–8	0.062–0.055	0.80–1.00

dilution-steam ratio. A minimum pressure for the coil is generally indicated, with pressures above atmospheric being required to prevent leakage of air into the suction of the cracked-gas compressor. Typical minimum coil-outlet pressures, which depend on the equipment between the furnace and the compressors, are 20–30 psia ($0.14–0.21 \times 10^6$ Pa).

Operational and capital costs for utilities associated with dilution steam, high-pressure steam, cooling water, and fired duty decrease with decreasing dilution by steam. For a fixed production of ethylene, a decrease in dilution steam requires additional feedstock to compensate for reduced yields of olefins. The degree to which this incremental cost in feedstock offsets the savings in utility costs depends on the economic criteria employed—for example, the extent of credit for by-products. The increased costs associated with the shorter furnace and run lengths of TLEs at reduced steam dilutions must also be established and included in the analysis. General guidelines for steam dilution for different feeds are shown in Table III.

D. Furnace Coil Design

1. *Objectives*

Variables that come into play when comparing designs of tubular reactors include mass velocity, heat flux, and configuration. The effects of changes in these parameters on product distribution, tube-wall temperature, and cost can be determined by means of rigorous reactor simulation (Pang *et al.*, 1979; Rase, 1977; de Blieck and Goossens, 1971). Because a minimal pressure drop is desired, the mass velocity can usually be varied only within narrow limits. In addition, it is logical to exploit tube metallurgy to its fullest extent. These factors are discussed next with respect to current design practice.

2. Coil Geometry

The designer of a radiant coil must consider tube orientation (vertical versus horizontal), staggered versus single-row arrangement, tube spacing, and top versus bottom outlet for the furnace. In addition, the coil may have a constant or expanding diameter (sometimes called a swaged coil), or it can be split (sometimes called a branched or combined coil).

With a single exception, tube cross-sections are circular. The exception is Mitsubishi (1972b), which has developed and is commercially exploiting an oval cross-section. Because the radiant intensity on a tube is a maximum on the sides facing the burners, an oval tube allows a higher permissible heat-transfer rate than a circular tube. Although the pressure drop is increased, the net result is a higher processing capacity per coil. High tube costs make this concept unattractive when applied to the entire coil, so that Mitsubishi currently employs oval tubes only in the front—that is, on the hot end of its designs.

The constant-diameter coil represents the historical configuration. The trend toward short-residence-time cracking of naphtha and heavier feedstocks has resulted in a dramatic shift to short, small-diameter coils. The effect of coil diameter on residence time and yield for given constraints on tube-skin temperatures is illustrated in Figs. 2 and 3 (Chambers and Potter, 1974). Figure 3 shows an increase of about 2 wt% in the yield of ethylene from naphtha cracking when the tube diameter is decreased from 4 to 2 in.

Kellogg has taken the concept of a constant-diameter coil to its limiting configuration—namely, a single straight-tube pass of 1 to 2 in. (2.54–5.08 cm) in diameter. Although other companies have expressed interest in this approach, Kellogg is the only one offering it commercially under the name of "millisecond furnace" (Ennis *et al.*, 1975). At residence times below 0.1 s this furnace provides olefins selectivity for liquid feedstocks that are unattainable by designs giving longer residence times. Its economic edge, however, is reduced if there is a favorable market for by-products such as aromatics. Note should be made that the low processing capacity of small-diameter coils results in a significant increase in the number of passes per furnace. This increase requires technological innovations for both the feed and the quench sections of the reactor. Uniform distribution of flow through parallel passes can be achieved by use of restriction orifices. To take full advantage of the short residence time offered by a small-diameter

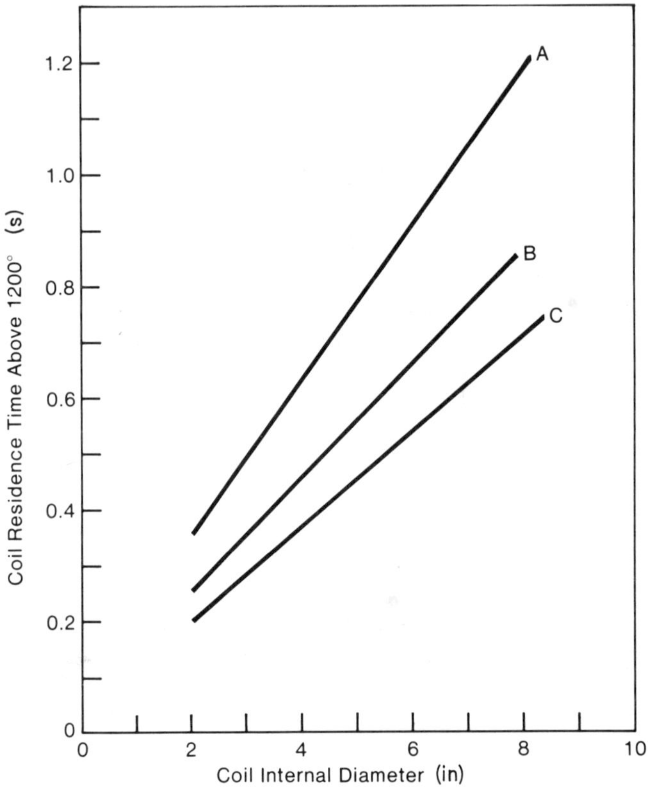

Fig. 2. Residence time plotted against coil internal diameter, based on constant mass velocity and severity. Results given for three different constraints on tube-skin temperature: A, 1700°F (926.7°C); B, 1800°F (982.2°C); C, 1900°F (1037.8°C). (1 in. = 2.54 cm; 1200°F = 648.9°C.) (From Chambers and Potter, 1974; reproduced with permission of *Hydrocarbon Process.*)

coil, rapid quenching is mandatory. Kellogg has considered double-pipe heat exchangers on each pass, as well as a lead-bath exchanger. Although the lead bath would appear to offer much promise, problems of lead contamination and mechanical vibration have to be considered.

3. *Expanding and Split Coils*

By increasing the diameter along the coil, coil capacity is increased while maintaining the same yield pattern and furnace run length. This increase is also achieved by use of a split coil wherein smaller diameter coils are joined in groups (normally pairs) to larger diameter coils.

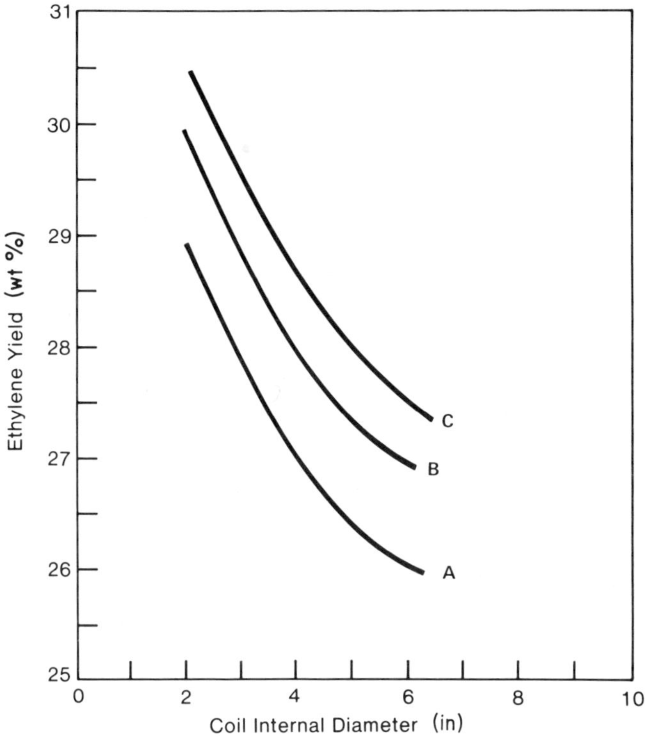

Fig. 3. Coil ethylene yield plotted against tube internal diameter, based on constant mass velocity, cracking severity, hydrocarbon partial pressure, and feed composition. Results given for three different constraints on tube-skin temperature: A, 1700°F; B, 1800°F; C, 1900°F. (From Chambers and Potter, 1974; reproduced with permission of *Hydrocarbon Process.*)

The effects are illustrated in Fig. 4, which qualitatively compares temperature and pressure profiles of split, expanding, and constant-diameter coil at the same throughputs. The split coil exhibits a steeper temperature profile at the coil inlet and a flatter profile toward the coil outlet. This pattern results in lower outlet and tube-skin temperatures. The expanding coil has a relatively high mass velocity at the inlet, which increases the heat-transfer coefficient on the inner wall of the coil and thus lowers the tube-skin temperature. The smaller diameter tubing at the inlet of a split coil provides an increased S/V ratio, which increases the heat absorption at the inlet relative to that for a constant-diameter coil. Early versions of the split-coil design embodied a constant mass velocity throughout the coil. Later designs, however, have decreased mass velocity along the coil similar to an expanding coil. This factor further enhances

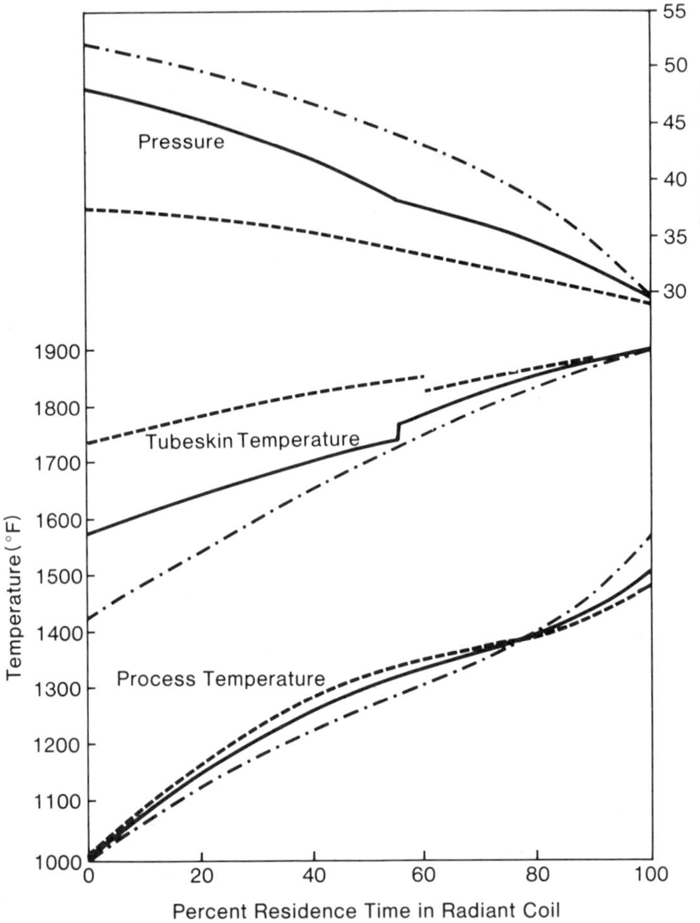

Fig. 4. Temperature and pressure profiles for three different types of radiant coils at the same throughputs. -----, Split coil; ——, expanding coil; -·-·-·-, straight coil. (1000°F = 537.8°C; 30 psia = 0.21 × 10^6 Pa; 55 psia = 0.38 × 10^6 Pa.)

the split-coil performance with regard to increased heat transfer at the coil-inlet region.

Both expanding and split coils take more advantage of limitations on maximum tube-skin temperature than do single-diameter coils. More of the coil operates at close to its upper limit, thereby decreasing the coil size for a given throughput and duty. In addition, by decreasing the mass velocity along the coil, the pressure drop near the coil outlet is reduced. This reduction allows a lower overall partial pressure of the hydrocarbon and improves selectivity of the olefins.

Although split-coil designs are gaining favor over the expanding-coil design, the results of a recent comparison of optimal design of coils for a specific gas-oil feed and specific economic parameters tended to be inconclusive (Ross, 1977). Very flat optima were found. Both designs produced virtually identical optima based on the same ethylene production. When comparisons were based on the same throughput, the split coil exhibited a small but significant advantage. The advantage, however, of utilizing a split coil with its low pressure drop must be balanced against the fact that it requires at least twice the number of inlet-tube passes per furnace than does the expanding coil.

4. *Orientation and Tube Spacing*

Vertical coils evolved in the early 1960s by overcoming the high-temperature support problems inherent in horizontal coils. The vertical configuration also permits higher temperatures and severities, and is now an industry standard. There is controversy as to the number of parallel tube rows. Should there be a single or a double-staggered row? Advantages of the double row are that manifolding is simpler, and the design permits use of a specially designed four-hole fitting to connect four coil passes to a single TLE. These advantages must be balanced against the lower processing capacity per coil in the double-staggered-row arrangement. This lower capacity stems from a maldistribution of the heat flux (American Petroleum Institute, 1978), which gives a lower processing capacity for a given limitation on the peak tube-skin temperature. Alternatively, for the same throughput, a single-row arrangement could accommodate a higher cracking severity.

Mitsui (Yamamota and Nakase, 1977) offers an interesting variation of the two-pass, double-staggered row. As seen in Fig. 5, each of three passes crisscrosses from one row to the next on long-radius return bends. Another set of three passes in mirror image allows six passes to be collected in one TLE. This concept lends itself to a large number of parallel passes.

A split-coil design combines the advantages of both the single and double-row arrangements (Lohr and Schwab, 1979). In the first part of the coil, where the tube-skin temperatures are low, the double-row arrangement is applied to make the coil more compact. In the second part of the coil, where the tube-skin temperatures are high, the single-row arrangement gives a more uniform temperature distribution at the tube circumference.

5. *Top/Bottom Outlet*

The radiant-coil exit can be at the top or bottom of the firebox. Most designs have a top radiant inlet that for an even number of tubes results in a top outlet and for an odd number of tubes results in a bottom outlet.

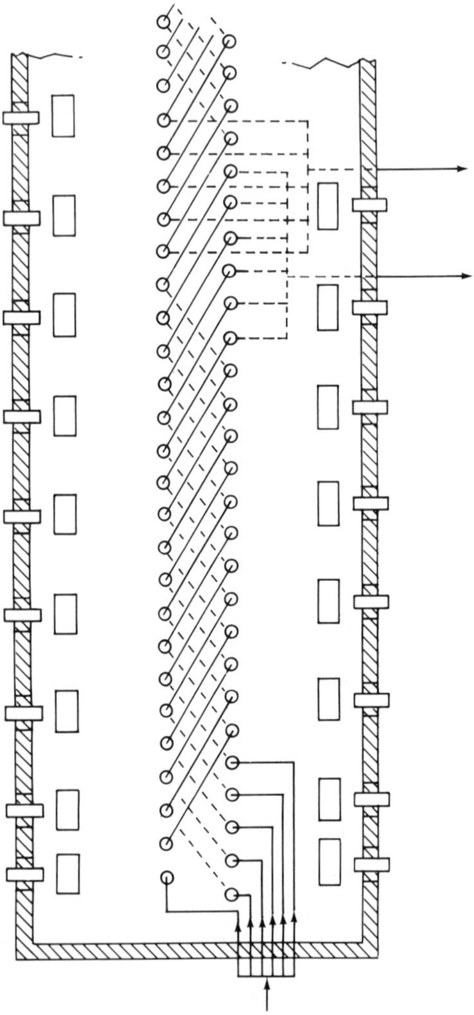

Fig. 5. Plan view of Mitsui six-pass coil. (Adapted from U.S. Patent 4,002,149—1977.)

From the standpoint of mechanical design a top outlet is easier to support, although it is not an overriding factor in furnace design. A top outlet permits a close-coupled TLE with a very short transfer line. Although it minimizes plot space, it requires more structure for supporting the TLE and steam-drum combination at a high elevation.

The widely used bottom outlet is easier to maintain, especially for a horizontal TLE. On the other hand, a bottom outlet with a vertical-upflow TLE elevates the cracked gas to a point best suited for an oil-quench fitting.

The residence time in a conventional bottom-outlet transfer line is of the order of 0.03 to 0.04 s. This time would be excessive for a millisecond radiant-coil design, which has a total residence time of less than 0.1 s.

6. *Firebox Arrangement*

The heat-flux and tube-skin temperature profile along the radiant coil may be varied in a given firebox by adjusting burner size and location, and furnace setting. For a given cracking duty, harder firing at the coil inlet reduces the tube-skin temperature at the coil outlet. Alternatively, this mode of firing permits the use of a shorter radiant coil for a given constraint on the peak tube-skin temperature.

The degree to which the firing pattern can be altered in a conventional firebox is small. Figure 6, from a recent patent by Linde (Dorner, 1977), represents an attempt to optimize the firing profile by dramatically altering both the configuration of the coil layout and the firebox. Although the design is unorthodox when compared to a conventional furnace, it has many attractive features. Inlet tubes are situated around the inside walls of the firebox where they are subject to intense firing from side-wall burners. The outlet tubes are located in the central portion of the box where

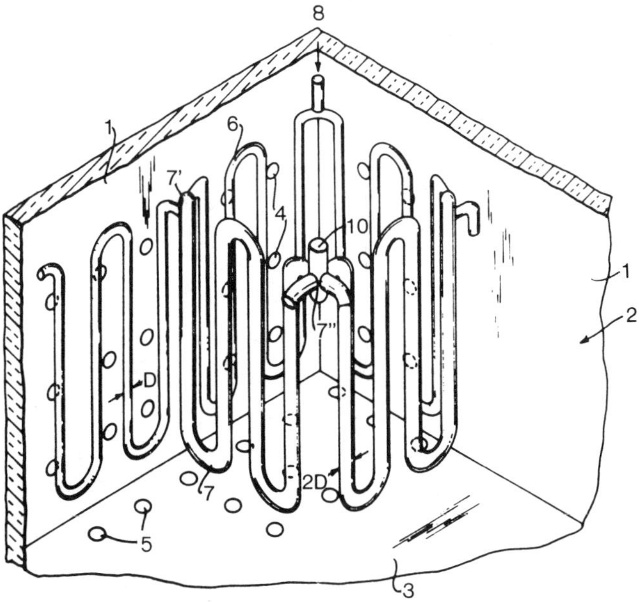

Fig. 6. Linde firebox configuration. 1, Tube furnace; 2,3, combustion chamber; 4,5, burners; 6,7, tubes; 8, manifold feed system; 10, coil exit. (For a description of those numbered items that do not appear here, see U.S. Patent 4,008,128—1977; Dorner, 1977.)

heat generation is from floor burners. As can be seen, the layout is a perfect fit for a split-coil design. In addition it lends itself to a four-coil TLE—a rather nice bonus.

Multizone furnaces can provide improvements in the firing pattern. Foster Wheeler has attempted this by using vertical refractory walls to create two or three fired zones within a single firebox (Demarest and Kraus, 1970). Kellogg (Wallace and Frankhanel, 1968) has also proposed a three-zone furnace.

Complete flexibility in profile firing is achieved by constructing completely independent radiant zones. Foster Wheeler applied this concept in the 1960s. More recently, Stone and Webster (Woebcke, 1974) and Lummus (Cross, 1972) have also proposed independent radiant-zone configurations. Although both groups use two zones, Stone and Webster shows a horizontal-coil arrangement whereas Lummus shows a vertical coil. The high cost of independent zones probably outweighs the potential improvements in coil capacity and product selectivity.

E. Tube Materials

Temperature limitations of commonly used tubing materials for pyrolysis furnaces are summarized in Table IV. Kobe Steel in Japan produces a modified HK-40 that is supposed to have superior resistance to carburization. Other alloys similar to HP-45 are Manaurite 36XS, containing tungsten and chromium, and Thermax 63, containing tungsten and titanium. These materials have somewhat greater strength than HP, but superior resistance to carburization has not yet been demonstrated. This lack of improved resistance, plus higher cost, have prevented wider use of these alloys in cracking furnaces.

Another tube material being tried in cracking furnaces is a Cr–Ni–Al steel alloy produced by Alonizing (Albright *et al.*, 1979). In this process, the base material is exposed to vaporized aluminum at 930°C to form an

TABLE IV

Temperature Limitations of Tubing Materials

	Weight percent		Maximum temperature (°C) at end of run
	Cr	Ni	
Incoloy 800	21	32	990–1010
HK-40	25	20	1050–1065
Incoloy 802	21	32	1050–1065
HP-45	25	35	1100–1120

alloy of 50% aluminum at the surface. The advantages for pyrolysis tubing are claimed to be protection against carburization and reduced rate of coking. Although several furnaces have Alonized tubes in service, it is too early to assess their performance, and further testing will be needed to demonstrate their effectiveness at the higher temperatures used with HP material.

Several years ago, International Materials Company tried to initiate a program to test silicon carbide tubing for pyrolysis (Glover, 1974). Although this material, which has a melting point above 2650°C, has been used for many years in autothermal flame reactors, its major disadvantage for tubular reactors is its brittleness. Although the application of fiber reinforcing may overcome this problem, this material is presently very costly and has to be considered as in the early stage of development.

F. Processing Innovations

1. *Hydropyrolysis*

Hydropyrolysis represents one of the major processing innovations in the last 20 years and was pioneered in bench-scale equipment in the early 1970s by Pierrefitte Auby. It was tested in prototype in a unit installed in 1976. It combines features of conventional steam pyrolysis and thermal hydrocracking. Hydrogen is used as a diluent and as a reactant with levels of about 40 mol % in the effluent. Energy required for recycling of hydrogen is minimized by operation at a reactor pressure of 140–650 psig, preferably about 305 psig. Although it was well known that conventional hydrocracking processes produced mainly paraffins, Pierrefitte Auby found that by operating at high temperature (700 to 900°C) and short residence time (0.01 to 0.1 s), it obtained relatively high yields of olefins. The short residence times also drastically reduced formation of tar and coke.

Pierrefitte Auby's 1974 patent (Chahvekillian and Plichon, 1974) shows operating conditions and yields for several feeds. Compared to conventional pyrolysis, the yields of methane, ethane, and propane are higher; on the other hand, ethylene, and C_4 and heavier yields are lower. If recycled ethane and propane are steam-cracked in conventional furnaces, however, ultimate yields of ethylene are higher than in conventional reactors. Hydropyrolysis also has the capability of cracking undesired by-products, such as propylene, C_4s, and gasoline fractions. Theoretically, ethylene yields of over 50% from liquid feeds can be attained by maximizing recycle cracking of by-products.

Several advantages for hydropyrolysis are that it (*a*) upgrades feed potential for olefins by making up for hydrogen deficit, which upgrading is particularly advantageous for low-hydrogen feeds such as gas-oils; (*b*) reduces coking by hydrogenating precursors of the coke; (*c*) recycles olefinic by-products or can even use olefinic raw materials; and (*d*) reduces cracked-gas compression horsepower by operating under pressure.

Disadvantages include low yields of butadiene and the need for makeup hydrogen [1000 to 2000 SCF (standard cubic feet)/bbl, or 28.3 to 56.6 m^3/bbl] at high severity. Overall it appears that hydropyrolysis of naphthas may not be competitive with steam cracking as long as the by-products from steam cracking can be sold at prevailing market prices. There is, however, a commercial chance for the use of feeds of vacuum gas oil (VGO).

2. *Other Concepts*

Tube-metal temperature is the limiting factor in pushing a pyrolysis coil to its maximum capability. This constraint is especially true for short-residence-time cracking, where reduced coil volume results in increased heat flux. One way to reduce heat flux is to inject hot steam into the coil. This procedure is referred to as hot-diluent cracking. As an example, assume a gas-oil cracker with a dilution-steam weight ratio of 0.8. If one half of the total dilution steam is superheated to 1200°C and mixed with the furnace-inlet gas at 600°C, the resulting mix temperature will be 700°C, which corresponds to a reduction of about 16% in radiant-coil duty. The upper limit in the percentage of diluent steam to be used as hot diluent depends on the minimum steam requirement for vaporizing the hydrocarbon feed before the final hydrocarbon–steam superheat coil. In addition to producing a more optimal radiant-coil temperature profile, hot-diluent injection results in a shorter (smaller surface area) coil, which produces a more favorable partial pressure profile for the hydrocarbon.

Various proposals have been made to inject feedstock or a product stream at some intermediate point in the furnace/transfer-line system. Injected hydrocarbon has varied from highly aromatic recycled products to portions of the cracked feedstock. It appears that in the case of injection of unsaturates, rapid coking will result unless the injection point is well toward the coil outlet.

An Esso patent carries the concept of intermediate-feed injection to its ultimate (Kivlen, 1971). Here preheated fresh feed, and/or steam if desired, is injected at spaced locations through the reactor. This pattern results in fresh feed being contacted with at least partially converted feed and products from a preceding zone. The reasoning behind this processing scheme is the claim that faster reaction rates and better yield selectivities are achieved by cracking fresh feed with an admixture of hot products. Although data cited for gas-oil cracking show a 17% increase in ethylene yield, a good proportion at the expense of propylene, this alteration in product selectivity would appear to be primarily due to increased feed conversion rather than to the distributed feed injection.

The extent to which the transfer line connecting the reactor outlet and quench exchanger can accommodate cracking is discussed in an ICI patent (Dyer and Alexander, 1977). Although the patent claims an enhancement of yield from deferred quenching, the enhancement appears to be primarily the result of increased conversion of feed. KTI (Goossens and Westerduin, 1975) claims a deleterious effect of transfer-line cracking and recommends a direct prequench of 50 to 100°C with a colder gas as the most convenient way to solve the problem. The stated degradation of effluent yields appears to be unrealistically high (5 and 21% for ethylene and propylene, respectively). KTI states that once the effluent has been prequenched it becomes advantageous to allow "reaction soaking" in the transfer line.

Another innovation proposed in a Monsanto patent (Cruse, 1971) is to inject additional hydrocarbon into the transfer line for processing and/or quenching. It cites an example of cracking a wide-boiling condensate (40–400°C) with varying amounts of condensate being injected into the transfer line. The results show a substantial increase in yield of aromatics at the expense of more valuable olefins. In spite of the degradation of product value, the patent claims a net advantage because of the increase in reactor capacity for a given fired duty. Even this advantage is open to question as the incremental transfer-line processing lowers the temperature into the TLE, thereby reducing make of high-pressure steam.

G. Furnace Coking

1. *Mechanisms*

As yet there is no theoretical foundation for prediction of coking in furnace tubes, in the transfer line, or in the TLE tubes themselves. Coke buildup on furnace-tube walls is believed to stem from unsaturated species that diffuse through the gas-film boundary layer to the high-temperature tube wall. Here they undergo dehydrogenation reactions leading ultimately to coke formation. Tube materials, particularly iron and nickel, are believed to catalyze coking reactions. This observation is supported by data showing that sulfur-free feedstocks coke at a faster rate than those containing sulfur. It is suggested that the role played by sulfur is to poison the active catalytic sites. The rate of coking decreases with run length because of coverage and deactivation of the surface by the coke layer (Newsome and Leftin, 1979). Metals might still diffuse through the coke layer to provide active sites for continued coking (Kivlen and Koszman, 1971).

There is a controversy about whether the rate of coking is limited by mass transfer through the gas film or by reaction rate at the surface. Some investigators (Fernandez-Baujin and Solomon, 1976; Chambers and Potter,

1974) argue that the reaction-velocity constant at the high temperature of the surface is much greater than the mass-transfer coefficient for the coke precursor. As a result, coking rate would be mass-transfer controlled. Experimental evidence shows, however, that coking diminishes after an initial high rate and that rates are virtually proportional to reactor S/V ratio. Those observations lead to a reaction-controlled mechanism.

Extensive research on surface reactions involved in light-hydrocarbon pyrolysis has been carried out at Purdue University (Tsai and Albright, 1976). Metal oxides produced during steam–air decoking were found to promote coke formation. During pyrolysis, hydrogen reduces these surface oxides, desulfurizes coke, and reacts with coke to produce methane. In addition to coke being formed from precursors such as olefins, it also could be produced by disproportionation of carbon monoxide to carbon dioxide and carbon.

In lieu of a priori prediction methods for coil-coking rates (or lengths of furnace runs), most furnace designers have developed proprietary empirical correlations weighted heavily with respect to commercial operating experience (Chen and Maddock, 1973; Mol, 1974). Another approach that is being pursued by several investigators (Leftin and Newsome, 1979; Marteno *et al.*, 1979) involves the translation of bench-scale coking data to commercial-scale conditions. The translation factor is typically obtained by comparing coking rates for several feedstocks under both bench and commercial operations.

The specific area of the furnace–TLE combination that limits length of run depends on the nature of the feedstock. For gas cracking, there is general agreement that run length is limited by coke buildup on the inlet tubesheet of the TLE, which causes tube blockage. The complex interaction of heat, mass, and momentum transfer in this region precludes a simple theoretical understanding of the mechanism. For naphtha feeds, runs are generally terminated by high temperature of the radiant-tube wall or fouling at the outlet end of the TLE. The latter effect becomes more important with increasing heaviness of the feedstock. Fouling at the TLE outlet is due to condensation of heavy ends when the cracked gas is cooled below its dew point. This consideration is embodied in commercial practice in that gas-cracker TLE effluents can be cooled well below 200°C whereas outlet temperatures for naphthas are maintained above 320°C. TLE-outlet temperatures for AGOs must be maintained above 540°C, whereas TLEs are not practical with heavy VGOs.

2. Fouling Inhibition

Many ways of reducing rates of coking have been proposed and/or patented. Surface coatings containing aluminum are claimed to reduce

TABLE V

Proposed Fouling Inhibitors

Inhibitor	Where added	Assignee	Reference
Nitrogen oxides	Feed	Dow	Schultz (1967)
Phosphorus and/or bismuth	Feed	Exxon	Koszman (1970b)
TBC, Phenylene diamine	Effluent	Marathon	Marhartz (1968)
Alkali metal salts	Effluent	Jefferson/Texaco	King and Jones (1971)

coking and carburization. A Shell patent (Nap *et al.*, 1974) for an aluminum coating claims that the coating is suitable for process temperatures up to 1100°C. Laboratory experiments are cited in which run lengths were quadrupled by use of the coating of aluminum.

Various inhibitors are claimed to reduce coking. Some are added to control tube fouling, whereas others are added to the effluent to reduce fouling in the TLE or quench zones. A sampling of the patent literature is listed in Table V.

Exxon (Koszman, 1970a) has found that arsenic greatly accelerates coking and so advocates its removal from pyrolysis feedstocks. Exxon also claims that sulfur contents in feed in excess of 400 ppm tends to increase coking rates in furnace tubes.

Mechanical devices to improve length of run include a basket installed in the inlet cone of a TLE (Schneider *et al.*, 1975) and an effluent-manifold baffle claimed to eliminate spiral flow (Pamphilis and Kivlen, 1970).

3. *Decoking Methods*

Several methods are used for removing coke. The most common are off-line steam–air decoking of furnace tubes, and mechanical or hydrojet cleaning of the TLEs. This procedure normally requires shutting down the furnace with attendant thermal stresses and a downtime of 8 h or more. An improvement is to steam–air decoke through the TLE with hydrojet cleaning after several cycles—say three. In this procedure the TLE is only partially decoked because the temperature of the boiling water on the shell side is below the coke-ignition temperature. It should be noted that Mitsubishi (1972a) has a TLE with spiral tubes that can be heated to high temperatures without water in the shell. That allows more complete decoking with steam and air.

A further improvement is on-line decoking with steam only (Wallace, 1972). In this operation the furnace–TLE combination stays on line, and the decoking effluent enters the recovery system along with the cracked gas. With steam only, decoking is by the water–gas reaction, $C + H_2O = CO + H_2$. This reaction is slower than burning with air and requires high temperature to achieve a reasonable rate of coke removal. The reaction is said to be catalyzed by iron and nickel, which diffuse into the coke layer from the tube surface. Sulfur tends to poison these catalysts and slows down the reaction. Another reaction mechanism for coke removal, much slower than the water–gas reaction, is hydrogenation, $C + 2\,H_2 = CH_4$. A temperature in excess of 1040°C is needed to achieve effective coke removal by this reaction.

Exxon has several patents for decoking of furnace tubes. In one of them (Cahn *et al.*, 1968), hydrocarbon feed is replaced by a water–steam mixture. One or more tubes are steam-decoked on line while the remaining tubes are maintained in normal cracking service. In an improvement (Kivlen and Koszman, 1971), hydrogen is added to the decoking steam. This method is claimed to remove sulfur from the coke and allow the water–gas reaction to proceed more rapidly. Decoking rates are presented for coil-outlet temperatures up to 1040°C. In a related approach, Exxon (Kivlen and Struth, 1972) and Monsanto (Oliver, 1969) have patents for decoking liquid furnaces by intermittent switching to ethane feed. The hydrogen produced in ethane pyrolysis presumably desulfurizes the coke. In Monsanto's concept, rather frequent switching of feeds is proposed, about 4 h in every 100.

Mitsui has a patent (Ozawa *et al.*, 1973a) for on-line decoking that involves adding sand to the feedstock. This method is supposed to remove coke by erosion, not only from the furnace tubes but the TLE as well. No examples are given, and the procedure appears to be more of an interesting idea rather than an experimentally proven concept.

III. NONCONVENTIONAL REACTOR TECHNOLOGY

A. Introduction

Over the past 20 years many reactor concepts have been proposed to overcome limitations inherent in fired tubular reactors. They include autothermal (partial oxidation, hot diluent), cyclic, molten bath, and heterogeneous—which encompasses fixed bed, moving bed, fluidized bed, molten spray, and catalytic. Commercial success of these concepts has mainly been decided by related technological breakthroughs as well as by changing economics and feedstock availability.

B. Autothermal Reactor

1. *Description*

The autothermal type of reactor operates in the short-residence-time regime and may be classified as either partial oxidation (PO) or hot diluent (HD) with respect to mode of operation. In the PO mode, the hydrocarbon feed and diluent are mixed with oxygen or other oxidizing gas in a proportion such that in burning the hydrocarbon is partially oxidized. The heat thus generated supplies the sensible and heat-of-reaction requirements for the unoxidized portion of the feed. The HD mode is basically a two-chamber refinement of the PO operation. Hot diluent is first generated by oxidation of a fuel, which is then mixed with the hydrocarbon to be cracked. Because of the high reaction rate, the admixing of these streams must be carried out with high intensity and at high velocities, usually equal to or greater than sonic.

2. *Historical Perspective*

At present, the most fully developed and commercially viable reactors are those that have been developed by the Kureha/Chiyoda/Union Carbide (K/C/U) joint venture and by Dow. Their unchallenged positions stem not so much from patents on the basic reactor concept, which has been around for many years, but rather from innovations relating to heavy-feed processing with special emphasis on crude-oil cracking. Early contenders were primarily interested in evolving a process that would supersede the classic partial-oxidation processes for the formation from light-hydrocarbon feedstocks (Schrader, 1957; Kamptner *et al.*, 1966) of products with high acetylene/ethylene ratios. This goal left the door open for K/C/U and Dow to develop technology on feed atomization, high-intensity reactant/hot-diluent mixing, and product quenching—patentable concepts that are critical in the processing of crude oil. The attraction of crude-oil cracking to a chemical company is that it provides an opportunity to integrate backward to a basic raw material and eliminate price uncertainties of buying naphtha and gas-oil cuts from refiners (Wing, 1980).

3. *Kureha/Chiyoda/Carbide Technology*

In 1973 Kureha, Chiyoda, and Union Carbide completed an agreement for the joint development of technology on an advanced cracking reactor (ACR). This joint venture has led to an extensive investigation of ACR technology through a series of research-and-development programs (Davis and Keister, 1976).

Initial data were obtained in Kureha's 2–5 ton/day test facility. In the Kureha/Chiyoda reactor (Gomi and Araki, 1971), superheated steam (up to 2000°C) was generated by means of a pair of regenerative-type furnaces and was continuously introduced into the top part of the pyrolysis reactors. A pilot facility at Nishiki, Japan, was used by the K/C/U joint venture to test improvements in ACR technology (Hosoi and Keister, 1975). Here the high-temperature steam regenerators were replaced by generation via burning of oxygen and fuel to produce combustion gases of about 2000°C. This energy source was supplemented by superheated steam generated externally from the reaction zone. Details on these developments are found in several patents issued between 1968 and 1976 (Sogawa, 1968a,b; Hosoi *et al.*, 1974; Albright and Keller, 1976; Washimi *et al.*, 1972). To shift the ethylene/acetylene yield ratios of the original program into the regime needed for olefins production, studies of optimal operating conditions were carried out (Hosoi and Keister, 1975). The reactor residence time was increased to the range of 20 ms and steam/feed ratios decreased to levels that balanced the increase in steam temperature. Some typical yields are shown in Table VI. Although most of the data are at an ethylene/acetylene

TABLE VI

Yield Patterns in ACR Reactor[a,b]

	Arabian light	Light naphtha	Light gas oil	Vacuum gas oil
C_2H_4/C_2H_2 wt. ratio	7.54	9.99	7.45	6.48
Residence time (s)	15	16	15	16
Products				
C_2H_2	4.21	4.34	4.49	4.34
C_2H_4	31.78	43.62	33.55	28.14
C_2H_6	1.68	3.14	1.83	1.57
C_3H_4	2.36	1.79	1.83	2.10
C_3H_6	6.09	11.47	7.04	5.62
C_3H_8	—	0.37	0.16	0.07
C_4H_4	0.68	0.19	0.32	0.72
C_4H_6	2.80	3.43	3.29	3.08
C_4H_8	0.06	1.65	0.64	0.36
C as CO, CO_2, H_2S	4.14	1.97	1.58	3.14
C_5 s—160°F (71.1°C)	1.26	1.52	0.94	1.14
Pyrogasoline	4.51	9.81	32.68	39.27
C_0—430°F (221.1°C)	5.10			
430–650°F (221.1–343.3°C)	16.39			
≥650°F (343.3°C)	7.71			

[a] Hosoi and Keister (1975).
[b] Yields given in weight percent.

yield ratio equal to 8, some economic trends toward lower severity are indicated.

As part of the process development, heat recovery became a significant part of the economics. With conventional cracking, high-level heat recovery in a TLE is accepted practice. Early in the studies, the conventional TLE was abandoned because of the potential fouling problem at the high-temperature inlet. The other option of removing the sensible heat with stabilized quench-oil liquid, as performed in Kureha's semicommerical plant, required extremely large equipment and achieved extremely poor heat-transfer efficiency. To compensate for this inequity, the Ozaki quench cooler (OQC) heat-recovery system was developed (Ozawa *et al.*, 1973b; Anonymous, 1977a). Tests on the cooler, which is a combination of direct and indirect quench cooling, indicate that it will remain in a nonfouled condition at the operating temperatures required to generate high-pressure steam.

Recently Union Carbide reported results of a detailed experimental program on the interaction of fluid dynamics, vaporization, and kinetics as they influence scaling of the ACR pilot data to a commercial size (Kearns's *et al.*, 1979). Specific details on the reactor itself, including a description of hydrodynamic considerations used to optimize its configuration, are documented in a recent patent (Kamm *et al.*, 1979), from which Fig. 7 was taken.

According to Union Carbide, the results of these test programs, supplemented by data from Kureha's commercial reactor, have allowed scaling up to a 100 million pounds/year ethylene reactor (45.5×10^6 kg/year). In order to minimize the technical risk and complete development work, however, they will carry out tests on a $15 million ACR prototype unit presently being completed at Seadrift, Texas. This unit will also be used to evaluate long-term operability and to provide the basis for a full commercial-scale unit by the mid 1980s.

4. *Dow Technology*

In late 1977, Dow disclosed that it had developed a process to produce olefins directly from crude oil (Anonymous, 1977b). Development work was started in 1973 that led to start-up of a pilot unit in 1977. Dow has stated that it plans to use this technology and not conventional naphtha cracking for its second American cracker planned for the Gulf Coast in the mid 1980s. The process is described as partial-combustion cracking. Details are proprietary, and apparently no patent has been issued. However, based, on limited information released by Dow, the process appears to be a two-chamber hot-diluent type similar to that of Kureha/Chiyoda/Union Carbide. One source (Anonymous, 1978a) states that Dow's process departs

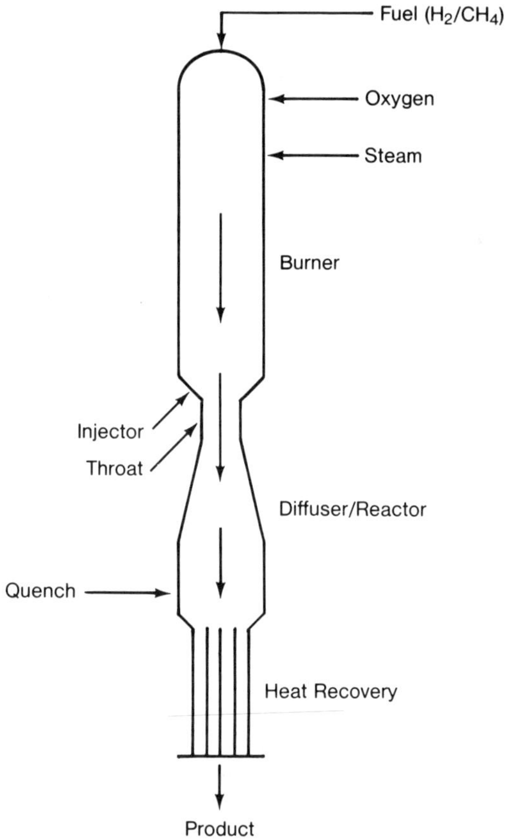

Fig. 7. Diagram of ACR autothermal reaction. (Adapted from Kamm *et al.*, 1979.)

from that of K/C/U in that cracking occurs in a hydrogen atmosphere, perhaps pointing to a reactor combining the hot-diluent cracking and hydropyrolysis technologies. A more recent publication (Wing, 1980) provides detailed yield structures for a whole crude and two heavy gas-oil cuts.

5. *Crude-Cracking Consensus*

There is no question that Union Carbides and Dow's technology represents a significant shift in temperature level and residence-time operation. By precisely controlling the variables, extremely high selectivity of the most valuable products can be achieved. Although the prime incentive for the development of this technology is crude-oil cracking, Union Carbide claims an economic advantage over conventional tubular technology even for distillate cracking (Anonymous, 1978b). The major concern in crude-oil cracking is its high yield of tar and pitch. Carbide recently reported the start of construction of what it called the first commercial-scale plant for the manufacture of carbon fibers produced from petroleum pitch (Anonymous, 1980). According to Carbide, developing applications for this product cover a "multitude of high-volume industries." There may be a limit, however, to the amount of this product the market could absorb.

The consensus of many commercial suppliers of cracking technology, primarily of tubular configuration, is that, even if proven on a commercial scale, a pronounced shift to crude cracking is unlikely. Dow and Carbide are investing millions, however, in the hope of bringing the concept to commercial fruition. If successful, it could claim a significant share of the olefins market. More significantly, by allowing petrochemical complexes to be built in areas where oil production occurs but refineries are not needed, it could be a decisive factor in changing the way basic chemicals are supplied worldwide (Wing, 1980).

C. Cyclic Reactor

1. *Description*

A cyclic-reactor system consists of a pair or pairs of furnaces lined with insulation. The inside of each furnace, except for a small combustion zone, is filled with a refractory mass of specialized configuration. The furnaces are operated in pairs on a regenerative cycle. Heat is removed from the refractory mass by pyrolysis of preheated feed in the presence of a suitable diluent, usually steam. Heat is restored to the refractory mass by combustion of fuel with oxidizing agent, normally air. Although the refractory inadvertently provides a high-coking surface, the resulting coke is burned off during the thermal-regeneration portion of the cycle. A complete cycle consists of a cracking step and a combustion step in the forward direction, followed by a cracking step and combustion step in the reverse direction. The cycle time may be of the order of several minutes. Reactor residence times are nominally about 0.5 s.

2. *Status*

These reactors have been used commercially in the production of acetylene—primarily from light hydrocarbons. Their potential for processing heavier feedstocks and providing high ethylene/acetylene ratios remains doubtful, however, in the light of successful technological breakthroughs leading to the present generation of autothermal reactors. It is ironic that the cyclic nature of the operation, which was the key to the reactor's early triumphs, now represents its chronic weakness. Cyclic operation requires complex equipment and results in high operating costs.

In addition, the transient nature of the cracking part of the cycle (i.e., refractory-mass temperature decreasing with time) makes cracking conditions difficult to optimize. Finally, heavy feedstocks introduce the problem of rapid fouling of the refractory mass.

The present literature survey did not reveal significant breakthroughs that could signal a revival of interest. Design innovations have centered mainly on the refractory-mass configuration and feed distribution. With respect to crude-cracking applications, three companies—Krupp-Koppers, Paccal (Wett, 1973), and Azote et Products Chimiques SA (1974)—have some experience but no proven process ready to license.

D. Molten-Bed Reactor

1. *Description*

Through its high S/V ratio, a molten-bed reactor is able to access the desirable high-temperature, short-residence-time operating regime. The process employs a molten metal or salt as heat carrier. The reactants and diluents (dispersed phase) pass through the molten medium (continuous phase) where cracking takes place. Regeneration of the thermally exhausted and fouled melt can be accomplished by burning with fuel and suitable oxidizing agent, normally air.

The concept has many attractive features. The process is continuous, and temperature control to maintain the thermal regime is quite simple. A wide range of feedstocks can be processed even in the absence of a diluent. The carbon formed during pyrolysis is removed continuously from the reaction zone. Finally, the reactor lends itself naturally to a combination reactor–quench configuration capable of generating high-pressure steam.

Extensive prototype testing is needed to resolve basic problems before the molten-bed reactor can be fully exploited. In cases where the molten bath is relatively static (the normal configuration) and hydrocarbon feed is bubbled through it, the high superficial velocities required can lead to severe vibration, slugging, and backmixing. Potential entrainment problems can result in carryover of minute concentrations of metals that will adversely affect the operation of downstream equipment. In spite of these problems, several companies show continuing interest in this concept. Emphasis is on commercializing the process for the "bottom of the barrel" feedstocks (heavy resids, VGOs).

2. *Mitsui*

Mitsui has proposed several designs in a series of patents. The earliest design (Yoshida *et al.*, 1973) is based on a simple two-chamber system wherein gas lifts, provided by the process and combustion streams, move the melt between the combustion and reaction chambers. Possible problems are gas leakage between the two chambers and nonuniform dispersion of carbonaceous matter in the melt. The latter problem can lead to excessive buildup of coke in the reaction chamber.

The features of an improved reactor by Mitsui (Ozawa *et al.*, 1973b, 1974) include indirect heating of the melt, bath circulation by diluent gas, and a fluidized-bed quench system capable of generating high-pressure steam. Although no operating data are cited, the reactor is designed for less than 0.5 s residence time and cracking temperatures between 700 and 1100°C. The coke-removal feature is quite novel. Coke, generated in the reactor melt, is carried into the quench zone by virtue of its lower density. The fluidizing action pulverizes the coke, which leaves with the cracked gas and is recovered outside of the reactor.

3. *Molten Metal/Tubular Hybrid*

A 1972 Japanese patent (Ozawa *et al.*, 1972) illustrates an interesting combination of tubular and molten-bath reactor concepts. Here molten metal, heated from 700 to 1000°C in a fired furnace, is combined with preheated feed. The feed, in intimate contact with the molten metal, is cracked and products are then quenched and separated from the melt. Owing to density differences, coke produced in the reaction and quenching zones floats to the surface of the molten metal in the tower where it is separated and removed. High-pressure steam is generated in the quench tower. The reactor appears to offer promise for high Conradson-carbon feeds, in that the cleansing action of the molten metal (moving in annular flow) prevents reactor fouling. Molten-metal injection also tends to optimize the cracking-temperature profile in a manner analogous to hot-diluent injection discussed previously.

Yamaguchi *et al.* (1979) present typical product patterns for various feedstocks at atmospheric pressure. The melt is not believed to provide catalytic action for the cracking reactions. Rather, the difference in product patterns found between the presence and absence of melt is inferred to result primarily from changes in the temperature distribution caused by the presence of the melt. The increased yields of hydrogen and carbon oxides are believed to be due to the catalytic effect that the melt imparts in the water–gas reaction.

4. *Exxon*

Patents in 1974 and 1975 (Dugan and Higgins, 1974; Dugan *et al.*, 1975) illustrate a concept investigated by Exxon. Cracking is carried out in a regenerable alkali-metal-carbonate molten medium containing a glass-forming oxide, such as an oxide of boron. The residence time can be as low as 0.01 s, and with the aid of a diluent, hydrocarbon partial pressures can range from 0.1 to 50 atm. With hydrogen as the diluent, the process could prove compatible with hydropyrolysis. The carbonaceous materials, which are formed and dispersed in the molten medium, are gasified by contacting the melt with a gaseous stream containing oxygen, steam, or carbon dioxide at temperatures up to 1650°C. The composition of this

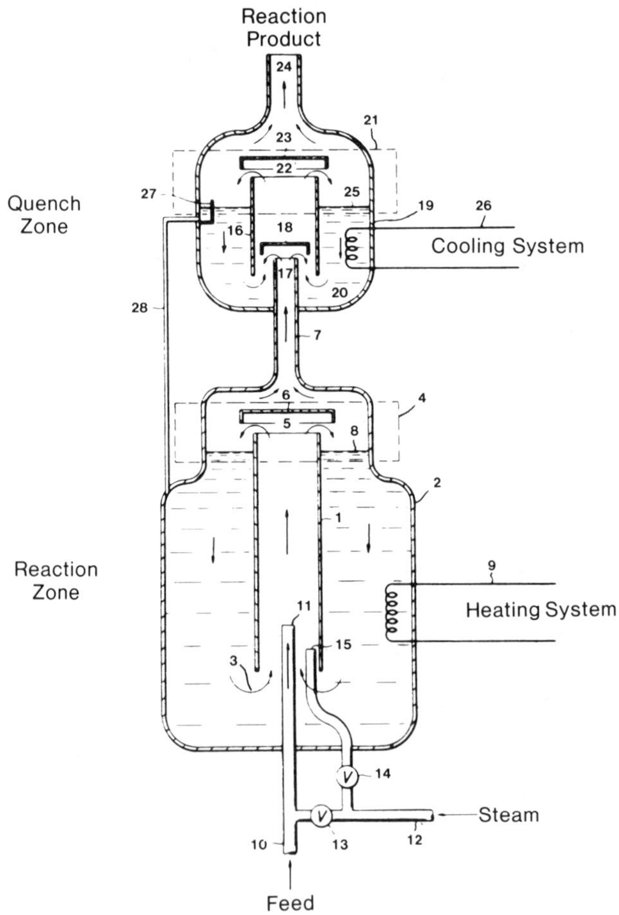

Fig. 8. Diagram of Shell molten-bed reactor and quench. 1, Forced-circulation riser; 2, reactor vessel; 4, hot-product separation; 6, gas–liquid separator; 7, transition zone; 16, cooling chamber; 19, cooling vessel; 21, cooled-product separation; 23, gas–liquid separator. (For a description of those numbered items that do not appear here, see U.S. Patent 4,012,457; from Bredeson *et al.*, 1977.)

stream is dictated by the environmental requirements on the exiting flue gas.

This process represents an attempt to tie together novel molten-bed concepts with proved petroleum catalytic-cracking technology. It is especially well suited for low-value, high-sulfur feeds such as crude bottoms, which are not readily upgraded to attractive petroleum products. The use of glass-forming oxides in the melt (up to about 20%) is the key to uniform dispersion of the carbonaceous material, as well as to acceleration of the desulfurization process. Sulfur, rather than being emitted as hydrogen sulfide, is retained in the melt and by suitable melt regeneration is converted to sulfur dioxide, which can be recovered by additional processing. The melt also leaches out sulfur from the carbonaceous material, further decreasing the hydrogen sulfide emissions.

Typical yield data presented, although confirming the olefins potential of the process, show high carbon oxide production, which might impose an excessive processing burden downstream of the reactor. A high coke-make is also evident. Whether all the coke is consumed during gasification depends on the energy requirements needed to heat-balance the system.

5. *Shell*

A combination molten-bed reactor and quench concept patented recently by Shell (Bredeson *et al.*, 1977) is shown in Fig. 8. It is similar to Mitsui's earlier concept, except that both hydrocarbon feed and diluent provide the gas lift needed for bed recirculation. In addition, the use of a molten-metal quench system enables the return of entrained metal to the reaction chamber. The patent claims that circulation of the melt overcomes the severe vibration and slugging characteristics inherent in a relatively static pool.

E. Heterogeneous Reactor

1. *Overview*

This technology has long been a mainstay of the petroleum industry. Much of it can be readily adapted to olefins production. As a result, the time between conceptual design and commercial exploitation may be much shorter than that required for the more novel concepts described previously. With respect to production of olefins, the key to the success of a heterogeneous reactor is not necessarily better selectivity of the product. Rather, through its reactor–regenerator sequence, it should be able to accommodate heavy coke-depositing feedstocks at high cracking severities—a capability denied conventional tubular reactors.

2. *Description*

In a heterogeneous reactor, gaseous feed (continuous phase) is cracked in the presence of a solid or liquid (dispersed phase). With few exceptions, the endothermic heat requirements are supplied either by the dispersed phase (heat carrier) or generated by the process stream (autothermal heat supply). Reactors may for convenience be divided into four broad categories depending on the state of the dispersed phase—fixed bed, moving bed, fluidized bed, and molten spray.

3. *Fixed Bed*

Fixed-bed reactors are the heterogeneous analog of the homogeneous cyclic reactors described previously. They are not suitable for high-fouling service, which would be the case with heavy feedstocks. Their chronic weakness lies in the complex equipment and high operating cost associated with cyclic operation.

4. *Moving Bed*

The moving-bed reactor is a natural adaptation of the catalytic-riser process to thermal cracking for olefins. The capability for continuous removal of coke permits operation at severities above those possible in conventional tubular crackers. Early work by BASF (Krekeler *et al.*, 1959) applicable to heavy-feed cracking led to a demonstration unit that was started up in 1970 and abandoned after only a few months of trial operation.

More recently, Gulf has expressed interest in this moving-bed concept (McKinney *et al.*, 1977). Its process, which is illustrated in Fig. 9, was tested in bench equipment to collect and establish yields and product qualities. The process is currently being codeveloped with Stone and Webster, as the TRC (Thermal Regenerative Cracking) process, with funds committed for a $15 million program to demonstrate the process with commercial scale components (Ellis *et al.*, 1982).

In the process, a stream of catalytically inert, hot solids is mixed with preheated feed and a diluent, normally steam. A steam/hydrocarbon ratio of 0.2 is cited as satisfactory. Weight ratios of solids to feed are between 10:1 and 30:1. The resulting mixture flows up through a riser at an average temperature between 700 and 1400°C and an average pressure between 5 and 50 psig ($0.034–0.34 \times 10^6$ Pa). Riser residence times are between 0.05 and 0.5 s. Gaseous products are quenched and separated from the solids.

The hot solids provide a surface for deposition of coke and a vehicle for its removal. The entrained coke-coated solids leaving the riser are passed to a burner to burn off the coke and to supply heat requirements for the next pass. The solids do not normally contain sufficient coke for heating requirements, and so supplementary fuel must be supplied.

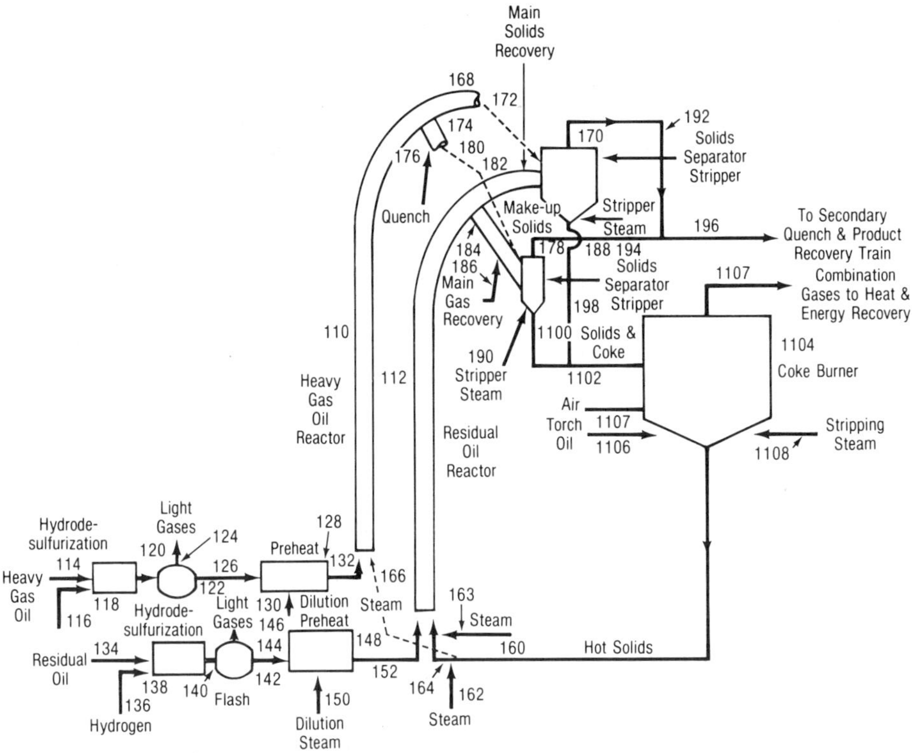

Fig. 9. Diagram of Gulf moving-bed reactor. (For a description of the numbers in the figure, see U.S. Patent 4,061,562; from McKinney *et al.*, 1977.)

One of the major attractions of the process is its compatibility with hydrodesulfurization processes. Sulfur in the feed tends to concentrate in the coke, and it leaves in the regenerator flue gas as sulfur oxide emissions. As a result, the feed must be desulfurized to the extent needed to meet environmental standards. Hydrotreating for sulfur removal is not necessarily a debit in that it also upgrades the cracking performance of the feedstock. As illustrated in Gulf's patent, much of the hydrogen consumption can be recovered via increased hydrogen yields in the riser cracker. The patent claims that the system can recover more than half of the hydrogen consumed by the feed during hydrodesulfurization. It should also be noted that the unrecovered hydrogen is not lost, but rather has helped to upgrade the slate of cracking product.

As seen in Fig. 9, the reactor system can process distillate-heavy gas oils and hydrodesulfurized residual oils in parallel. This procedure can be accomplished by either blending or independent processing. The latter approach is preferred as it enables each feedstock to be cracked at its optimum conditions. It also provides flexibility in desulfurizing to meet total sulfur oxide emission standards allowed in the composite flue gas.

Potential advantages of the TRC process over conventional tubular technology have been summarized by Ellis *et al.*, 1982. They include:

1. No practical limit to severity.
2. Feedstocks can be extended to include vacuum gas oils residuums and crudes as well as heavy synthetic feedstocks.
3. Reduced fresh feed consumption by recycle cracking of undesired coproducts.
4. Lower dilution steam ratio.
5. Widest range of process fuels.
6. Feed flexibility.

5. *Fluidized Bed*

Over the past 20 years several attempts have been made to commercially exploit fluidized-bed reactors for olefins production using heavy feedstocks. Early contenders—such as BASF (Steinhofer *et al.*, 1963; Nonnenmacher *et al.*, 1966), Ube Industries (Matsunami *et al.*, 1970; Suzukawa *et al.*, 1970), and Lurgi (Schmolfeld, 1963)—are currently not active. Although the patent literature reveals new contenders, only one group, Mitsubishi/Chiyoda, appears to be advanced in its development program.

The process is illustrated in Fig. 10 (Ueda *et al.*, 1977). Basically, preheated oil feed and diluent are fed to a fluidized bed consisting of inert refractory and coke particles. Cracking takes place at a temperature of about 780°C and a residence time of about 0.4 to 0.8 s. The overhead is quenched and products separated and recovered. The coke goes to the regenerator bed for partial oxidation.

Cyclone and gravitational separators are used to recover bed particulate matter and recycle it back to the reactor system. Another patent (Nakagawa *et al.*, 1977) deals with the problem of controlling the size distribution and amount of coke in the reaction system. Here the entire bed consists of coke particles.

6. *Molten Spray*

A 1965 Japanese patent (Minami, 1965) describes a novel regenerable molten-spray reactor. The processing scheme is similar to that of a moving-particulate-bed reactor and regenerator system except that molten spray and process gas are contacted countercurrently in the reactor. The patent claims an improvement over the use of a moving-solid particulate bed in

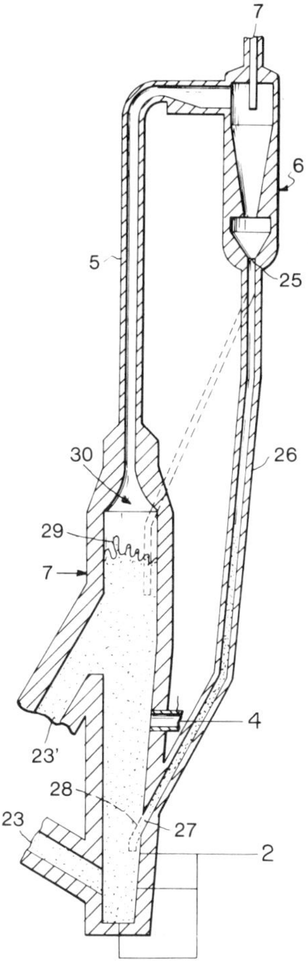

Fig. 10. Diagram of fluidized-bed reactor. 2, Steam nozzle; 4, feed nozzle; 6, cyclone dust collector; 23, duct from regenerator; 23′, duct to regenerator; 26, particle-return duct. (For a description of those numbered items that do not appear here, see U.S. Patent 4,049,540; from Ueda *et al.*, 1977.)

that problems related to clogging, particle attrition, and uniformity of particle size are eliminated.

Basically, pressurized molten metal is atomized and falls through the reactor, contacting upward-flowing reactants. The molten metal is regenerated by hot combustion gases in a gas-lift regenerator. The hot flue gas is used to generate high-temperature steam. Means are provided to recycle a portion of the exhausted flue gas to the regenerator. The steam is used as a hot diluent to raise the temperature of the hydrocarbon entering the reactor, while the reactor-effluent heat is recovered by exchange against incoming feed and a conventional quench cooler.

F. Catalytic Reactor

1. *Status*

Although thermal cracking is a highly successful source of olefins, much effort has been directed at developing a competitive catalytic route.

2. *Problem Areas*

The role of a cracking catalyst is threefold. It (*a*) lowers the reaction temperature, (*b*) accelerates the reaction rate at a given temperature, and (*c*) alters product selectivity.

Considerations of thermodynamic equilibrium dictate more favorable olefins production at higher rather than lower temperatures. Faster reaction rates in catalytic reactors require shorter residence times and higher heat-transfer rates—conditions not easily achieved in fixed or moving beds. Under these conditions fluidized beds suffer from catalyst attrition and backmixing problems. Gross backmixing is detrimental to olefins selectivity. It is apparent that, in order to overcome these disadvantages, catalytic reactors can only compete by offering intrinsically higher product selectivities—by accelerating the olefin-producing reactions relative to those leading to undesirable products and coke formation. Finally, coking is a much more serious problem in catalytic reactors in that it dramatically reduces catalytic activity. In order to maintain high catalyst activity, a much shorter reaction–regeneration cycle or a much larger reactor is required.

3. *Fluor, Haldor/Topsoe*

The state of the art can be illustrated by the efforts of Fluor and Haldor/Topsoe (Andersen *et al.*, 1975; Wrisberg *et al.*, 1973, 1974) to develop a cracking catalyst that is resistant to coke laydown. The amorphous zirconia catalyst containing from 0.1 to 10% alkali metal oxide has been used to crack naphtha and heavier feedstocks at temperatures between 700 and 900°C and at pressures between 1 and 5 atm ($0.1–0.5 \times 10^6$ Pa). Steam ratios between 0.1 and 10 are cited. Typical yield data are illustrated in Table VIII.

Although the naphtha yields appear to be attractive, the high olefins yield from the gas oil appears to be due to the high steam dilution rather than from catalytic enhancement. Note that the 4.6 steam ratio has produced a rather unacceptable 14.5% yield of carbon oxides. The success of this catalyst and others like it rests on its ability to inhibit fouling. Runs of 290 h on gas-oil feed have shown very little fouling. If this fouling inhibition can be demonstrated on a pilot or larger scale, the catalyst could be a viable commercial contender.

TABLE VII

Catalytic Cracking Yields[a]

	Feed		
	Naphtha	Gas oil	
Conditions			
Specific gravity	0.71	0.86	0.86
Steam ratio (wt.)	0.6	1.4	4.6
Temperature out (°C)	850	790	790
Pressure out (atm)	1.5	3.5	4.0
Ultimate yields (wt%)			
Hydrogen	0.88	1.2	2.5
Carbon oxides	0.99	4.5	14.5
Methane	13.0	9.9	18.7
Ethylene	33.2	26.8	34.0
Propylene	18.7	17.2	16.6
Butadiene	4.9	5.8	5.5
Butenes	5.7	7.1	3.5

[a] Andersen *et al.* (1975).

4. *Mobil*

A promising route to making ethylene from coal is found in Mobil's methanol-to-gasoline process, which uses a ZSM-5 class zeolite catalyst (Wett, 1980; Berry, 1980). By stopping the reaction at the proper time, the zeolite can also be used to produce light olefins and aromatics.

REFERENCES

Albright, C. W., and Keller, C. G. (1976). U.S. Patent 3,959,401 (to Union Carbide).

Albright, L. F., Mc Connell, C. F., and Welther, K. (1979). In "Thermal Hydrocarbon Chemistry" (A. G. Oblad, H. G. Davis, and R. T. Eddinger, eds.), Advances in Chemistry Series, No. 183, pp. 175–191. Am. Chem. Soc., Washington, D.C.

American Petroleum Institute (1978). "Recommended Practice for Calculation of Heat Tube Thickness in Petroleum Refineries," RP-530. Am. Pet. Inst., Washington, D.C.

Andersen, K. J., Rostrup-Nielsen, J., and Wrisberg, J. (1975). U.S. Patent 3,872,179 (to Haldor Topsoe A/S).

Anonymous (1977a). *Chem. Week* **121**(13), 39.

Anonymous (1977b). *Chem. Eng.* **84**(26), 65.

Anonymous (1978a). *Chem. Week* **122**(5), 39.

Anonymous (1978b). *Chem. Age* **116**(3061), 14.

Anonymous (1980). *Chem. Mark. Rep.* **217**(9), 3, 18.

Azote et Products Chimiques SA (1974). "Onia–GEGI Process, Production of Ethylene and Propylene by Crude Oil Steam Cracking." Toulouse, France.

Berry, R. I. (1980). *Chem. Eng.* **87**(8), 86–88.
Bredeson, L. D., McCullough, G. R., and Wylie, O. L. (1977). U.S. Patent 4,012,457 (to Shell Dev. Corp.).
Cahn, R. P., Kivlen, J. A., and Thurlow, D. K. (1968). U.S. Patent 3,365,387 (to Esso Res. & Eng. Co.).
Chahvekillian, E., and Plichon, J. M. (1974). U.S. Patent 3,842,138 (to Pierrefitte-Auby, Paris).
Chambers, L. E., and Potter, W. S. (1974). *Hydrocarbon Process.* **53**(1), 121–126.
Chen, J., and Maddock, M. J. (1973). *Hydrocarbon Process.* **52**(5), 147–150.
Cross, A. (1972). U.S. Patent 3,667,429 (to Lummus Co.).
Cruse, C. M. (1971). U.S. Patent 3,579,438 (to Monsanto Co.).
Davis, H. G., and Keister, R. G. (1976). *In* "Industrial and Laboratory Pyrolyses" (L. F. Albright and B. L. Crynes, eds.), ACS Symposium Series, No. 32, pp. 392–411. Am. Chem. Soc., Washington, D.C.
de Blieck, J. L., and Goossens, A. G. (1971). *Hydrocarbon Process.* **50**(3), 76.
Demarest, K. D., and Kraus, M. (1970). *Chem. Eng. Prog., Symp. Ser.* **66**, 103.
Dorner, A. (1977). U.S. Patent 4,008,128 (to Linde AG).
Dugan, J. J., and Higgins, J. P. (1974). U.S. Patent 3,852,188 (to Exxon).
Dugan, J. J., Pasternak, I. S., and Yao, K. C. (1975). U.S. Patent 3,876,527 (to Exxon).
Dyer, P., and Alexander, D. C. (1977). U.S. Patent 4,021,501 (to I.C.I., Ltd.).
Ellis, A. F., Gartside, R. J., and Bowen, P. B. (1982). *Chem. Eng. Progress* **78**(3), 66–69.
Ennis, B. P., Boyd, H. B., and Orriss, R. (1975). *CHEMTECH* pp. 693–699.
Fernandez-Baujin, J. M., and Solomon, S. M. (1976). *In* "Industrial and Laboratory Pyrolyses" (L. F. Albright and B. L. Crynes, eds.), ACS Symposium Series, No. 32, pp. 345–372. Am. Chem. Soc., Washington, D.C.
Glover, W. A. (1974). Personal correspondence with T. F. Newkirk, International Materials Co., Darlington, Massachusetts.
Gomi, S., and Araki, S. (1971). *World Pet. Congr., 8th, Moscow* **4**, pp. 371–378.
Goossens, A. G., Westerduin, R. F., and Mol, A. (1975). *Erdoel Kohle, Erdgas, Petrochem.* **28**, 471–476.
Hosoi, T., and Keister, H. G. (1975). *Chem. Eng. Prog.* **71**(11), 63–67.
Hosoi, T., Kanbayashi, M., Washimi, K., and Gomi, S. (1974). U.S. Patent 3,855,339.
Kamm, G. R., Milks, D., Kearns, J. D., Britt, H. I., and Khavarian, C. P. (1979). U.S. Patent 4,136,015 (to Union Carbide).
Kamptner, H. K., Krause, W. R., and Schilken, H. P. (1966). *Hydrocarbon Process.* **45**(4), 187–193.
Kearns, J. D., Milks, D., and Kamm, G. R. (1979). *In* "Thermal Hydrocarbon Chemistry" (A. G. Oblad, H. G. Davis, and R. T. Eddinger, eds.), Advances in Chemistry Series, No. 183, pp. 107–128. Am. Chem. Soc., Washington, D.C.
King, J. E., Jr., and Jones, S. T. (1971). U.S. Patent 3,617,478 (to Jefferson Chemical Co., Inc.).
Kivlen, J. A. (1971). U.S. Patent 3,579,601 (to Esso Res. & Eng. Co.).
Kivlen, J. A., and Koszman, I. (1971). U.S. Patent 3,557,241 (to Esso Res. & Eng. Co.).
Kivlen, J. A., Struth, B. W., and Weiss, C. P. (1972). U.S. Patent 3,641,190 (to Esso Res. & Eng. Co.).
Koszman, I. (1970a). U.S. Patent 3,546,316 (to Esso Res. & Eng. Co.).
Koszman, I. (1970b). U.S. Patent 3,531,394 (to Esso Res. & Eng. Co.).
Krekeler, H., Kirschner, T., Kirschner, W., Relken, G., and Waellnitz, O. (1959). U.S. Patent 2,911,454 (to Hoechst).
Leftin, H. P., and Newsome, D. S. (1979). U.S. Patent 4,176,045.
Lohr, B., and Schwab, W. (1979). *In* "Thermal Hydrocarbon Chemistry" (A. G. Oblad, H. G. Davis, and R. T. Eddinger, eds.), Advances in Chemistry Series, No. 183, pp. 153–174. Am. Chem. Soc., Washington, D.C.
McKinney, J. D., Sebulsky, R. T., and Wynne, F. E., Jr. *et al.* (1977). U.S. Patent 4,061,562 (to Gulf).
Marhartz, F. C. (1968). U.S. Patent 3,366,702 (to Marathon Oil).
Marteno, A. J., Glass, J., and Gaultier, J. L. (1979). *Hydrocarbon Process.* **58**(4), 199–202.

Matsunami, I., Suxukawa, Y., Nawata, G., and Kono, H. (1970). *Hydrocarbon Process.* **49**(11), 121–126.
Minami, S. (1965). U.S. Patent 3,192,018 (to Hitachi).
Mitsubishi Petrochemical Co., Ltd. (1972a). "Mitsubishi Quench System for Ethylene Plants." Tokyo.
Mitsubishi Petrochemical Co., Ltd. (1972b). "Mitsubishi Thermal Cracking Process." Tokyo.
Mol, A. (1974). *Hydrocarbon Process.* **53**(7), 115–118.
Nahas, R. S. (1980). *Oil Gas J.* **78**(35), 73–81.
Nakagawa, T., Matsuda, N., Mitani, K., and Tanemoto, T. (1977). U.S. Patent 4,049,541 (to Chiyoda).
Nap, C., Van Haarlem, A., and Emile, O. H. M. (1974). U.S. Patent 3,827,967 (to Shell Oil Corp.).
Newsome, D. S., and Leftin, H. P. (1979). *AIChE Annu. Meet., 72nd, San Francisco, Calif.* Pap. 21d.
Nonnenmacher, H., Engelbach, H., and Sinn, R. (1966). U.S. Patent 3,238,271 (to BASF).
Offen, M. J. (1976). *Hydrocarbon Process.* **55**(10), 123–127.
Oliver, G. D. (1969). U.S. Patent 3,433,731 (to Monsanto Co.).
Ozawa, T., Uyeda, M., Hashimoto, R., and Suzuki, S. (1972). U.S. Patent 3,696,166 (to Mitsui).
Ozawa, T., Uyeda, M., Hashimoto, R., and Suzuki, S. (1973a). U.S. Patent 3,764,634 (to Mitsui).
Ozawa, T., Uyeda, M., and Suzuki, S. (1973b). U.S. Patent 3,718,708 (to Mitsui).
Ozawa, T., Uyeda, M., and Suzuki, S. (1974). U.S. Patent 3,795,490 (to Mitsui).
Pamphilis, N. C., and Kivlen, J. A. (1970). U.S. Patent 3,527,832 (to Esso Res. & Eng. Co.).
Pang, K. H., Shu, W. R., and Koenig, M. I. (1979). *In* "Computer Applications to Chemical Engineering" (R. G. Squires and G. V. Reklaitis, eds.), Advances in Chemistry 124, pp. 333–402. Am. Chem. Soc., Washington, D.C.
Rase, H. F. (1977). "Chemical Reactor Design for Process Plants," Vol. 1, pp. 420–489; Vol. 2, pp. 13–31. Wiley, New York.
Ross, L. L. (1977). C. F. Braun Project Report, Feb. 16. Alhambra, California.
Ross, L. L., and Shu, W. R. (1977). *Oil Gas J.* **75**(43), 58.
Ross, L. L., and Shu, W. R. (1979). *In* "Thermal Hydrocarbon Chemistry" (A. G. Oblad, H. G. Davis, and R. T. Eddinger, eds.), Advances in Chemistry Series, No. 183, pp. 129–152. Am. Chem. Soc., Washington, D.C.
Schmolfeld, P. (1963). *Hydrocarbon Process. Pet. Refiner* **42**(7), 145–148.
Schneider, H., Hussner, R. G., and Fischer, R. E. (1975). U.S. Patent 3,880,621 (to Deutsche Texaco AG).
Schrader, R. J. (1957). U.S. Patent 2,790,838 (to Eastman Kodak Co.).
Schultz, J. A. (1967). U.S. Patent 3,308,056 (to Dow Chem. Co.).
Shu, W. R., and Ross, L. L. (1979). *Oil Gas J.* **77**(36), 72.
Sogawa, T. (1968a). U.S. Patent 3,408,417 (to Kureha).
Sogawa, T. (1968b). U.S. Patent 3,419,632 (to Kureha).
Steinhofer, A., Frey, O., and Nonnenmacher, H. (1963). *Hydrocarbon Process. Pet. Refiner* **42**(7), 119–124.
Suzukawa, Y., Kono, H., Terai, K., and Niwa, T. (1970). U.S. Patent 3,551,513 (to Ube).
Tsai, C. H., and Albright, L. F. (1976). *In* "Industrial and Laboratory Pyrolyses" (L. F. Albright and B. L. Crynes, eds.), ACS Symposium Series, No. 32, pp. 274–295. Am. Chem. Soc., Washington, D.C.
Ueda, T., Yoshioka, S., Shohji, Y., Tanaka, T., Kikuchi, Y., Fujita, N., Kaneko, N., and Tanemota, T. (1977). U.S. Patent 4,049,540 (to Chiyoda Chem. Co. and Mitsubishi Petrochem. Co.).
Wallace, B. A. (1972). U.S. Patent 3,671,198 (to Pullman, Inc.).
Wallace, B. A., and Frankhanel, M. O. (1968). U.S. Patent 3,405,691 (to Pullman, Inc.).
Washimi, K., Kanbayashi, M., Kanai, T., and Fujimori, A. (1972). U.S. Patent 3,674,679 (to Kurehaphgoa).

Wett, T. (1973). *Oil Gas J.* **71**(48), 73–75.

Wett, T. (1980). *Oil Gas J.* **78**(35), 65–71.

Wing, M. (1980). *CHEMTECH* pp. 20–22.

Woebcke, H. N. (1974). U.S. Patent 3,820,955 (to Stone and Webster Eng. Corp.).

Wrisberg, J., Andersen, K. J., and Morgensen, E. (1973). U.S. Patent 3,725,495 (to Haldor Topsoe A/S).

Wrisberg, J., Holte, G., Andersen, K. J., and Mogensen, E. (1974). U.S. Patent 3,839,485 (to Haldor Topsoe A/S and Fluor Corp.).

Yamaguchi, F., Sakai, A., Yoshitake, M., and Saegusa, H. (1979). *Hydrocarbon Process.* **58**(9), 167–172.

Yamamoto, E., and Nakase, N. (1977). U.S. Patent 4,002,149 (to Mitsui).

Yoshida, F., Akamatsu, K., and Hara, N. (1973). U.S. Patent 3,729,297 (to Mitsui).

14

Economic Considerations of the Design and Operation of Conventional Pyrolysis Furnaces

J. C. HEIMAN

I. INTRODUCTION

High-temperature pyrolysis of hydrocarbons has been practiced for over 40 years. The tubular furnace is the single most important piece of equipment in the traditional ethylene plant, because the product-yield distribution from it determines the downstream-plant design and operation.

Since the 1950s there has been a trend toward increasing the pyrolysis temperature and reducing residence time in the cracking coil. A better conceptual understanding of the mechanism has permitted improvements to be made in the design of the radiant heat-transfer section of furnaces, resulting in better control of reaction-zone severity. With this reduction of residence time and the higher temperatures required for yield improvements, the heat flux necessary to provide the heat of reaction has risen significantly.

Pyrolysis: Theory and Industrial Practice

ISBN 0-12-048880-9

Control of the pyrolysis reaction has become much more important because of these changes, but this control is very difficult and complex. Once the yield pattern, severity level, steam/hydrocarbon ratio, and outlet pressure have been established, the unfixed variables can be said to be tube dimensions, flow rate, inlet temperature, residence time, and tube-metal temperature. The dependent variables become outlet temperature, pressure drop, furnace box temperature, and heat flux. Limiting factors that must be considered include the maximum coil-metal temperature, maximum allowable furnace box temperature, and the maximum coil-pressure drop.

By looking at all of these design variables and the effect they have both on each other and on the operation of the process, the economic impact can be derived and placed in perspective with the rest of the plant.

II. FURNACE-DESIGN CONSIDERATIONS

The current economic environment has intensified efforts to maximize total olefin yields. The modern designer must often incorporate extreme flexibility into furnace design to allow the thermal cracking of feedstocks that may vary from light petroleum gases to heavier gas oils. The burden of economic optimization falls first on the designer to make this flexibility available, and then on the operations personnel to realize the required operating conditions.

To appreciate the complexity of furnace-design criteria, one must first understand the multiple relationships that exist among the various design variables. An in-depth discussion of these variables is presented by Schutt (1947). When designing an ethylene plant, the yield pattern must be determined for a particular feedstock. The products encompass a range of compounds from methane to high-molecular-weight tars. Their relative proportions will vary significantly depending on the feedstock composition and the pyrolysis conditions. Zdonik *et al.* (1974) have given an analysis of feedstocks and their typical cracking characteristics. With *n*-butane as the feedstock, the effluents at three conversion levels are shown in Table I. These reproducible data were obtained in small experimental miniplants. The "yield patterns," thus determined, are the basis for subsequent plant design. One way to measure pyrolysis for different naphtha or gas-oil feedstocks is to utilize a parameter called the kinetic severity function (KSF; Chambers and Potter, 1974). This function is the equivalent *n*-pentane conversion, defined as the disappearance of *n*-pentane via a first-order reaction. For example, if a given feedstock is cracked at a kinetic severity of 2.3, the time–temperature conditions in the reactor are

Table I

Pyrolysis of *n*-Butane

	Conversion of feed (%)		
	75	85	95
Conditions			
Coil-outlet temperature (°F)	1405	1440	1510
Coil-outlet pressure (psig)	15	15	15
Steam dilution (mol/mol)	0.5	0.5	0.5
Yield (wt%)			
H_2	0.4	0.5	0.6
CH_4	16.1	19.5	24.1
C_2H_2	0.1	0.2	0.4
C_2H_4	20.0	24.4	29.7
C_2H_6	6.3	7.2	7.4
C_3H_6	21.8	21.8	17.4
C_3H_8	0.4	0.4	0.5
C_4+	34.9	26.0	19.9

such that *n*-pentane would be 90% converted. The KSF correlates the degree of reaction of the feedstock and is quite useful for comparing yield data and for designing and evaluating cracking-coil performance. A detailed explanation of KSF and its relationship to the various pyrolysis variables is given by Zdonik *et al.* (1970). Figure 1 shows the effect of KSF on the pyrolysis of a typical feedstock.

The severity function can be related to the time and temperature profiles of the furnace coil, and a specific KSF can be obtained for various combinations of temperatures and residence times (see Fig. 2). The KSF incorporates both time and temperature and, more importantly, is consistent with the kinetics. When one includes the economic considerations of both ethylene and by-product yields, then consideration of product values must be used in choosing the design KSF. Once a furnace has been designed and built, however, the yield patterns can be varied quite significantly by changing variables to alter the specific KSF. The plant operator often changes KSF by adjusting the coil-outlet temperature. For a given equivalent time value, the logarithm of KSF is proportional to the reciprocal of the absolute temperature at the coil outlet; and for a constant coil-outlet temperature, KSF is directly proportional to the equivalent time. The KSF is thus virtually a unique function of the coil-outlet temperature, once the furnace is designed and built. At fixed outlet temperatures and at normal

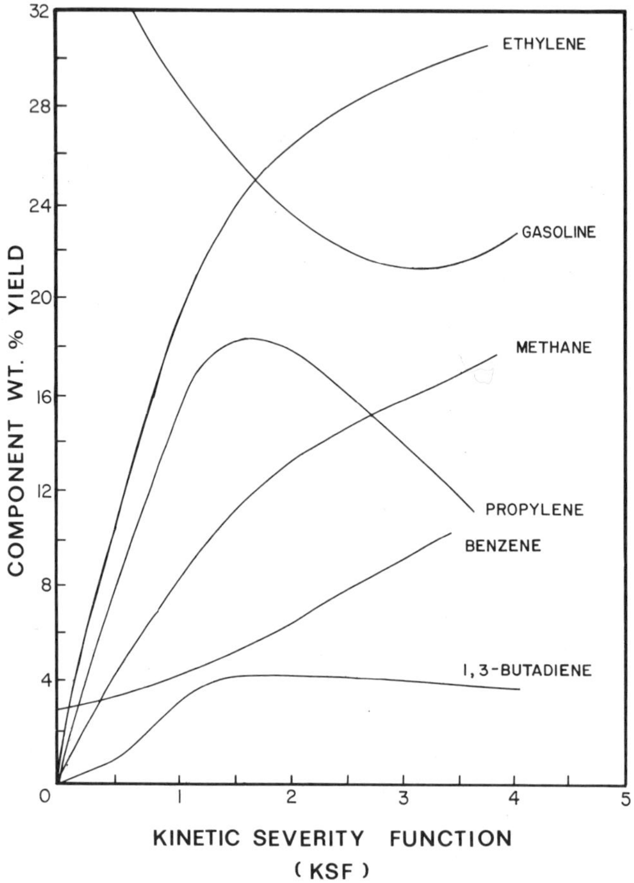

Fig. 1. Component yield as a function of KSF. (Reproduced with permission from *Hydrocarbon Processing*).

changes in feed rate and steam/hydrocarbon ratios, there are only small changes of severity. Thus, for a specific feedstock being cracked at a specific pressure level, the severity levels at other outlet temperatures can be estimated as a function of equivalent time, as shown in Fig. 3.

A definite relationship exists among the residence time, hydrocarbon partial pressure, and conversion of the feedstock—which determines product yield. A study (*Oil and Gas Journal*, 1967) of these effects on naphtha pyrolysis showed the advantages of using shorter residence times. At a specific hydrocarbon partial pressure and conversion level, lower residence times result in higher ethylene yields. Valuable by-product yields such as butadiene and acetylene are also enhanced. At a specific

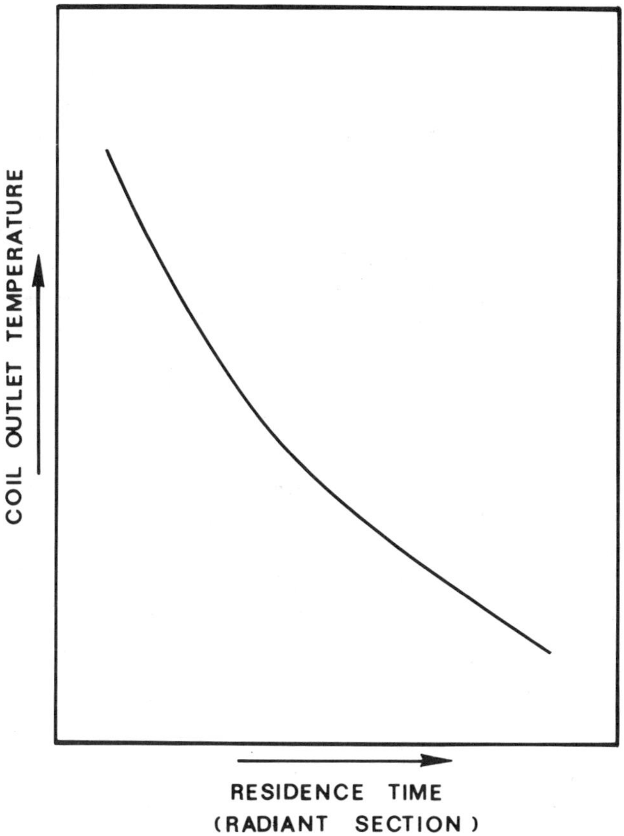

Fig. 2. Coil-outlet temperature as a function of residence time (constant KSF).

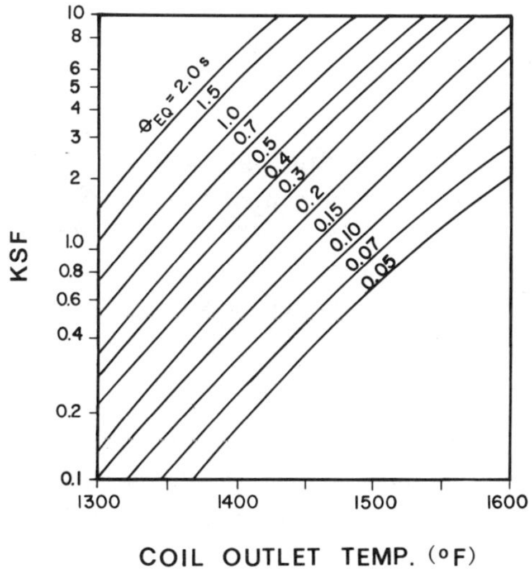

Fig. 3. Severity as a function of coil-outlet temperature at constant equivalent time (Zdonik *et al.*, 1970).

residence time and conversion level, a lower hydrocarbon partial pressure improves ethylene yield. The ability to raise ethylene yield by lowering the coil residence time and the hydrocarbon partial pressure is, however, quite limited.

The average hydrocarbon partial pressure is a function of the total pressure at the coil outlet, pressure drop in the coil, and the molar ratio of steam to hydrocarbon. The residence time is inversely proportional to the mass velocity, and any increase in mass velocity is accompanied by an increase in the coil-pressure drop. Higher hydrocarbon partial pressures in the coil occur with higher pressure drops (and higher inlet pressures). Any gain in the olefin yield caused by the resulting small reduction in residence time (brought about by increasing the mass velocity) may be more than offset by the simultaneous rise in the hydrocarbon partial pressure. The hydrocarbon partial pressure can also be lowered by raising the steam/hydrocarbon ratio, but this is limited by steam availability. At the low coil-outlet pressures normally run, a decrease of the hydrocarbon partial pressure of about 8% would require an increase in steam flow of some 20%. Thus the residence time and hydrocarbon partial pressure, though they significantly affect the ethylene yield, cannot be varied to any noticeable degree.

As will be discussed in the control section of this chapter, the design of the furnace must be made flexible enough to handle variation in feedstock and desired yield patterns, but once the different design variables are set, the effect of the basic operating parameters is minimized.

III. COKE FORMATION AND DECOKING

Pyrolysis always involves the complete dehydrogenation of some compounds and the slow deposition of coke or carbon on the internal surface of the coil. This coke inhibits heat transfer from the tube wall to the process gas and increases the tube-wall temperatures as the coke layer thickens. This coke buildup also increases the coil-pressure drop and thus has a negative influence on the yield pattern by increasing the hydrocarbon partial pressure.

The various mechanisms of coking, as well as the effect and interaction with tube metal, and subsequent tube degeneration are discussed in other chapters, but coke formation has a major effect on overall furnace economics and is discussed here.

The accumulation of coke deposits ultimately necessitates removal by decoking. When the tube walls reach their maximum allowable temperature, shutdown of the furnace is necessary. According to Goossens *et al.*

(1978), during a typical onstream cycle the tube-skin temperature rises 100°C, resulting (for the same throughput) in a 1–2% decrease in firebox efficiency with an increase in fuel consumption of about 5%. For a 1 billion pounds/year ethylene plant operating on naphtha, this would correspond to an increase of almost 9 million BTU/h/furnace, which is normally provided by extra fuel gas.

Decoking is accomplished by removing the coil from service and using a steam–air mixture to burn the carbon out of the coil. The steam–air mixture is heated to 900–1000°C and is slowly reacted with the carbon to produce hydrogen and carbon monoxide. The amount of air added is gradually increased as the amount of carbon decreases, until the decoking is complete. If too much air is added, causing too rapid a rate of decoking, fires can result in "hot spots," with holes subsequently burned in the tube. On severely coked coils, it is not uncommon for decoking to take 72 h. This process is costly because both steam and fuel are consumed. Furthermore, both coking and decoking slowly remove metal from the surfaces of the coils, leading to pitting and erosion. In some modern ethylene plants, approximately 10% spare furnace capacity is used to eliminate onstream downtime and resulting production losses; the extra furnace capacity, however, increases capital costs.

The advantages of reducing coke formation in pyrolysis coils are obvious, and coking can be minimized by careful control of the coil-metal temperature, steam/hydrocarbon ratio, and proper coil-outlet conditions. In data presented by Mol (1974), it can be seen that if the coil-outlet temperature is kept below 970°C, run lengths of over 100 days can be achieved before decoking is necessary.

Waste-heat recovery in the transfer-line exchanger (TLE) is also affected by coking. Coke laid down in the TLE reduces the heat transfer to steam generation; efficiency losses that occur over a month or two can add up to large energy expenditures. The TLE also eventually has to be shut down and mechanically cleaned.

In designing a furnace, there is a need to consider carefully the type and layout of the furnace burners and to optimize the diameter and length of the cracking coil for a given feedstock and a given severity level. Heavier feedstocks, in general, result in higher coking rates.

IV. WASTE-HEAT RECOVERY

In earlier plants, energy conservation was not nearly as important a factor as it is today, and the design and operation of these earlier furnaces reflect this fact. With the current increasing fuel costs, the success or failure of an ethylene plant requires minimal energy consumption. Because the energy required to crack the feedstock to lighter products is roughly fixed, the ability to recover expended energy becomes critical. Energy recoveries of 91–93% are typical in the modern-day ethylene plant, and major efforts are continually being made to improve these recoveries. Bailey and Wall (1978) have reported on some of the measures taken to recover energy expended in the cracking furnace.

Pyrolysis heat is recovered mainly in the convection section of the furnace and in the TLE. The off-gases to the furnace stack are at very high temperatures, and heat-transfer coils in the stack and upper chambers of the furnace are used to preheat the feedstock by convection heating to the "crossover temperature." The objective is to heat the feed to a maximum temperature but below the point at which cracking is initiated. This temperature ranges from 550 to 700°C, depending on the feedstock. To reduce the stack-gas heat content even more, some plants preheat boiler feed water in additional coils located in the stack, or may even generate low-pressure steam used in other areas of the plant.

The TLE is used to stop the pyrolysis reaction by a sudden and dramatic drop of several hundred degrees in temperature as the process materials leave the cracking coil. At the same time, high-pressure steam is generated (475–600 psig) as an internal energy source or exported from the plant as a by-product credit.

Typically, the pyrolysis gas leaving the TLEs is in the range of 375–500°C for a naphtha feed plant, but as high as 500–600°C for a gas-oil feed plant. The allowable outlet temperature depends on the amount of coke deposited in the TLEs. For processes necessitating a high TLE-outlet temperature, additional recovery of heat can be accomplished by direct injection of quench oil into the cracked gas directly downstream of the TLEs. The temperature of the cracked gas is reduced to 200–220°C, hence stopping all degradation reactions.

After the pyrolysis gas and quench oil are separated in a fractionation column, the gas at 100–110°C can be further cooled in air- or water-cooled exchangers, or by direct water quenching. The analysis and design procedure for quench towers is reported by Picciotti (1977). Hot quench water after separation from the heavy pyrolysis gasoline is sent to various heat exchangers throughout the plant. These efforts result in improved heat recovery, removal of the heavier tars and residuals, and preparation of the product gas for compression. Cooling procedures employed by different producers vary widely; apparently increased engineering effort and additional capital equipment investment will still be justified in the future for even better fuel efficiency.

V. CONTROL OF FURNACE OPERATION

Besides the design of the pyrolysis furnace, correct operation is the most important way of ensuring proper yields and good economical production. Without question the most significant variable in pyrolysis is the temperature. The coil-outlet, tube-wall, firebox, crossover, and transfer-line temperatures all have a noticeable effect on the overall economics of furnace operation. Increased severity causes increased coking in the coil, which increases coil-pressure drop and reduces heat transfer to the process. This increase in coking of the coil will eventually result in increased coking of the TLE because some coke or tar particles formed in the coil are caught by condensation in the TLE. Subsequent onstream runtime of the furnace will be lessened. Overheating and more frequent decoking also reduce coil life, necessitating large capital outlays for retubing a furnace. Overheating—that is, the operation of the coil at skin temperatures in excess of the desired value—leads to increased coking rate on the coil walls. This rate determines the run length, tube-service life, and also selectivity because of the higher pressure drop in the coil.

The accuracy and number of thermocouples used in a furnace are often not adequate. Many furnaces employ thermocouples to measure temperatures only at the inlet and outlet of the coil, at one or two positions in the firebox, and maybe at most one or two positions on the coil. Older plants rarely utilize spare thermocouples to make measurements along the coil. The ability to determine the location of the crossover temperature and coil-skin temperatures are very useful in maximizing coil-cracking performance. Without this information, furnaces are often operated either too hot, resulting in overcracking; too cold, resulting in undercracking; or with extreme heat fluxes, resulting in excessive coke buildups.

A typical thermocouple at the coil outlet will operate at $\pm 2°C$ accuracy when new. With use at these high temperatures, oxides form in the junction of the wires, which often produce errors as high as $\pm 10°C$. Also, as coke forms on the inside of the thermowell, an insulating layer causes a reduction of the read temperature. This error, coupled with a thermocouple error, can cause total errors of as much as 20°C. Such an error can lead to severe overheating, faster coking rates, overcracking of the feedstock, and significantly poorer yield patterns. Therefore, use of spare thermocouples and replacement of all high-temperature thermocouples every 6 months are almost always justified and tend to result in much improved operation.

For a 1 billion pounds/year ethylene plant running on a full-range naphtha, a deviation of 10°C from the desired coil-outlet temperature would typically result in a loss of $1.5 million/year. Such a loss is based on a combination of ethylene loss or increased coking, and subsequent shorter onstream time, depending on whether the temperature deviation is low or high.

The steam-flow rate affects to some extent both residence time and hydrocarbon partial pressure; the actual flow value is much less important as compared to the coil-outlet temperature. If the hydrocarbon flow rate is in error by 5%, however, the resulting change in partial pressure and resulting yield pattern will be quite significant in affecting the product composition. By using onstream analyzers and a good maintenance program of instrument recalibration, the operator can minimize flow errors.

VI. FURNACE ECONOMIC CONSIDERATIONS

The pyrolysis section is considered the heart of an ethylene plant. All the products of the plant are formed here, whereas the remainder of the plant serves to separate and purify these products. Because of the efficiency of fuel utilization, pyrolysis yield patterns, and feedstock consumption, the pyrolysis furnace has the greatest influence on the economics of the plant. Freiling and Simone (1973) have enumerated several examples of how these variables can be manipulated to gain an economic advantage depending on the market for the various products. Goossens *et al.* (1978) reported that in a typical 1 billion pounds/year ethylene plant, a 0.8 wt% increase in ethylene selectivity could correspond to a net profit gain of $1 million.

Fuel utilization is paramount in an ethylene plant. A typical furnace fuel consumption for the production of ethylene from ethane in a modern 1 billion pounds/year plant is over 7000 BTU/pound of ethylene. In a gas-oil plant, this number approaches 14,000. The value of fuel gas, although it varies from place to place, is often over $3.75 per million BTUs. With increasing fuel costs and a process that consumes billions of BTUs every year, there is still the need for even better heat recovery.

A particular study by Huang and Hwang (1980), conducted for several olefin plants, indicated a poor understanding of the furnace compressor-chilling train coupling relationship. The results showed excessive losses of ethylene to the tail gas and excessive recycle gas to the compressor, causing a 5% waste of compressor horsepower. With higher methane content in the furnace effluent, the compressor horsepower load is increased. The methane content is a function of feedstock composition, cracking severity, and coking condition. The excessive loss in the tail gas caused by poor ethylene refrigeration was traced to the operation of the demethanizer. The compressor suction pressure influences furnace yields, and discharge pressure influences the partial condensation of the separation train. Obviously, minimal deviation from plant-design conditions is of key importance. Increased coking also compounds the problems.

VII. SUMMARY AND RECOMMENDATIONS

The economics of a pyrolysis plant are by no means simple, and ultimately the optimal operating conditions depend on all factors previously mentioned. However, one can definitely conclude that because of the current energy situation and because an ethylene plant is severely energy intensive, the pyrolysis section of the plant has the greatest impact on economics relative to both energy expended and potential waste-heat recovery. More engineering effort and capital expenditure are recommended to decrease energy demands still further. Waste-heat recoveries of 95–96% are believed possible, but the degree of necessity will determine what level of effort and capital costs are deemed justified.

Additional research is definitely needed in the areas of coking and coking prevention in order to determine the real mechanisms of coking and the interrelationship of metal loss and coke inhibitors. Areas currently being investigated include alumina inpregnation and silica treatment of coils. It is predicted that major breakthroughs will occur in this general area of research.

REFERENCES

Bailey, T., and Wall, F. M. (1978). *Chem. Eng. Prog.* **74**(7), 45–50.

Chambers, L. E., and Potter, W. S. (1974). *Hydrocarbon Process.* **53**(1), 121–126.

Freiling, J. F., and Simone, A. A. (1973). *Pet. Petrochem. Int.* **13**(3), 70–76.

Goossens, A. G., Dente, M., and Ranzi, E. (1978). *Hydrocarbon Process.* **57**(9), 227–236.

Huang, W., and Hwang, J. J. (1980). *Hydrocarbon Process.* **59**(12), 105–109.

Mol, A. (1974). *Hydrocarbon Process.* **53**(7), 115–118.

Oil and Gas Journal (1967). **65**(27), 76–78.

Picciotti, M. (1977). *Hydrocarbon Process.* **56**(6), 163; **56**(9), 179.

Schutt, H. C. (1947). *Chem. Eng. Prog.* **43**(3), 103–116.

Zdonik, S. B., Green, E. J., and Hallee, L. P. (1970). "Manufacturing Ethylene." Petroleum Publ. Co., Tulsa, Oklahoma.

Zdonik, S. B., Bassler, E. J., and Hallee, L. P. (1974). *Hydrocarbon Process.* **53**(2), 73–81.

15

Production and Economics in the Recovery of Ethylene, Other Olefins, and Aromatics from Pyrolysis Systems

STANLEY B. ZDONIK

Pyrolysis: Theory and Industrial Practice

ISBN 0-12-048880-9

I. HISTORY OF ETHYLENE AND THE PETROCHEMICAL INDUSTRY

In the last 20 years, the petrochemicals industry has grown very rapidly, and until the last decade the greatest activity for this industry was centered in the United States. Today, petrochemicals have attracted the interest of many nations, and at the present growth rate on a worldwide basis, the center of petrochemical activity is moving from North America to Europe. A parallel can be drawn between a large portion of the petrochemical industry and the ethylene industry. Ethylene, which is used as a building block for many chemical derivatives and intermediates, has created product lines that have rivaled the volume and value of many major chemicals derived from nonpetroleum sources.

The ethylene-producing industry, like many other industries, has gone through a number of cycles. In the mid-1950s, some ethylene plants that were cracking naphtha feedstocks in Europe were primarily interested in converting a relatively inexpensive naphtha into ethylene. The propylene and C_4's products, in many cases, were burned with the residue gas as plant fuel, and the pyrolysis gasoline was blended into a large gasoline pool without hydrotreating.

In the early 1960s, as ethylene plants became larger in capacity, the disposition of the by-products became an important part of plant economics. The propylene produced was directed toward the production of derivatives such as polypropylene, isopropanol, acrylonitrile, and cumene. Butadiene was separated from the butenes. The pyrolysis gasoline, because of its large quantity, now required hydrotreating before blending into a gasoline pool. Hydrotreating of the C_6–C_8 fraction of the pyrolysis gasoline was also required if a BTX fraction was to be separated into benzene, toluene, and mixed xylenes. Up to the mid-1960s, many ethylene plants that cracked naphtha feedstocks operated at propylene/ethylene weight ratios of 0.6 to 0.7 in order to satisfy the demand for propylene.

In the late 1960s, the supply of propylene exceeded demand, and operators of ethylene plants began to look for methods of reducing the production of propylene. Supply of naphtha feedstocks also began to become scarce in certain locations, and the price of naphtha increased faster than the supply. Some ethylene-plant designers were developing cracking furnaces that were capable of higher once-through ethylene yields than previously possible with a given feedstock. They achieved propylene/ethylene weight ratios in the range of 0.4 to 0.5 and consumed less feedstock for a given production of ethylene. To achieve such furnace-yield patterns, the feedstock was cracked with a relatively short residence time and a high coil-outlet temperature. Such furnace designs currently have wide acceptance.

In recent years, the petrochemical industry has had to face many problems. Many operators have begun consideration of alternative feedstocks and methods to improve thermal efficiencies of plants. The days of cheap, abundant feedstocks and fuels are gone. Moreover, because of uncertainty in supply of feedstocks and because of cost, owners of some recently completed plants for naphtha and/or gas-oil feedstocks are considering substitution of portions of the design feedstocks with ethane, propane, or butanes. By removing certain bottlenecks in the plant, additional capacity is achievable. The incentives in adding this flexibility in feedstocks are their availability and cost.

Another problem that has not yet been felt or fully evaluated by producers of petrochemicals in the United States is the competition from new projects being constructed or planned in the Middle East where feedstocks are available and, currently, at a relatively low price.

In spite of economic problems facing the petrochemical industry, new projects are being planned on a more deliberate basis. The petrochemical industry, now valued in the multibillion-dollar range, rivals the refining industry in capital expenditures and will probably continue to expand in the future.

II. PETROCHEMICAL CENTERS

No petrochemical plant can survive without an assured supply of fairly uniform feedstock. The worldwide feedstock and energy crisis has created severe problems for some, and new opportunities and markets for those nations with available feedstocks. It has been customary in the past to ship feedstocks from the point of production to other locations where they are converted into petrochemical intermediates and finished products. The finished products are then sent throughout the world. The flow direction may change in the future. Hydrocarbon feedstocks may be converted into petrochemical intermediates or finished products at the location of origin of the feedstocks and the products transported to petrochemical centers such as the United States, Japan, and some European countries.

III. THE PETROCHEMICAL PLANT

A. Simple Version

A petrochemical plant that produces basic raw materials may take various forms. Some are relatively simple, such as shown in Fig. 1, and may consist only of an ethylene unit with supporting off-site facilities.

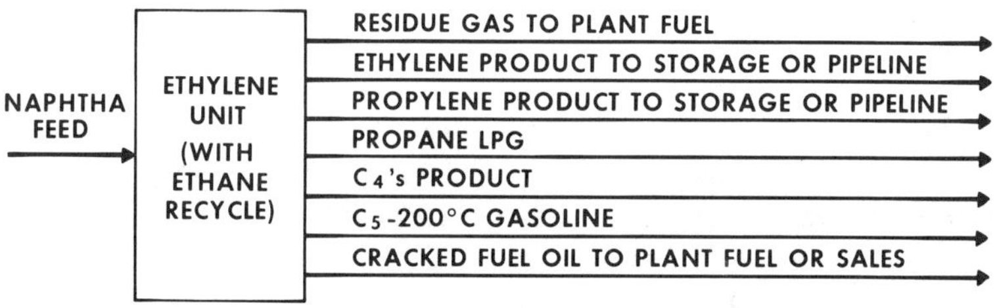

SUPPORTING OFFSITE FACILITIES

1. FEED AND PRODUCT TANKAGE
2. BOILER FEED WATER TREATMENT SYSTEM
3. STEAM GENERATION SYSTEM
4. COOLING WATER SYSTEM
5. POWER DISTRIBUTION SYSTEM
6. FUEL GAS AND FUEL OIL SYSTEM
7. PLANT AIR SYSTEM
8. INERT GAS SYSTEM
9. WASTE LIQUIDS TREATMENT SYSTEM
10. VAPOR RELIEF SYSTEM AND FLARE
11. FEED AND PRODUCT PIPELINES
12. DOCKS FOR TANKERS (OPTIONAL)
13. SPURS FOR RAILROAD SHIPMENT

Fig. 1. Flow diagram of a simple petrochemical complex.

If the plant uses naphtha as a feedstock, it will produce a residue gas that is consumed within the plant and an ethylene product that is sent directly either to a consuming plant across battery limits or to a product pipeline that may be several hundred miles in length. The ethane stream produced may be cracked in recycle or sent to fuel. The propylene product may be produced as chemical grade or polymer grade. If polymer-grade propylene is produced, the propane stream may be cracked in recycle or sold as liquefied petroleum gas (LPG). The liquid C_4's product can be fractionated to separate the butadiene from the butenes. The pyrolysis gasoline is sold for the production of motor gasoline to refiners or chemical plants with hydrogenation and extraction facilities for the recovery of benzene, toluene, and xylenes. The cracked fuel oil may be consumed as plant fuel or sold as product.

The naphtha feedstock consumed by a petrochemical complex is produced in a topping unit in a refinery. It is then delivered to the petrochemical plant by pipeline or tanker depending on the location of the refinery. In the latter case, considerable feedstock tankage is required to assure a steady supply of naphtha for the cracking furnaces. Product tankage is also an important part of such an installation if contract commitments are to be met. Utilities systems such as steam, cooling water, fuel, and power are also important and must be capable of meeting instantaneous demands imposed by the process plant.

B. Petrochemical Complex

A petrochemical plant may be quite complex and consist of an ethylene unit and numerous satellite units, which will add flexibility and profits to the operation as they produce a wider variety of products. Figure 2 illustrates the numerous combinations of processing units that may be added initially or at some time in the future. This plant can vary the ratio of olefins to aromatics over a wide range.

Relative to capital cost, the ethylene unit represents a major portion of the total capital cost for a petrochemical plant. Table I is a typical representation of capital cost for a petrochemical plant containing a 450,000 MTA ethylene unit cracking naphtha feedstock. As indicated by Example 1, the ethylene unit accounts for approximately 77% of the total plant cost; and in Example 2, the ethylene unit with satellite units accounts for approximately 42%. The ethylene unit obviously represents a large proportion of the total capital cost of a petrochemical plant, is responsible for most of the products, and is hence the most important unit in any petrochemical complex.

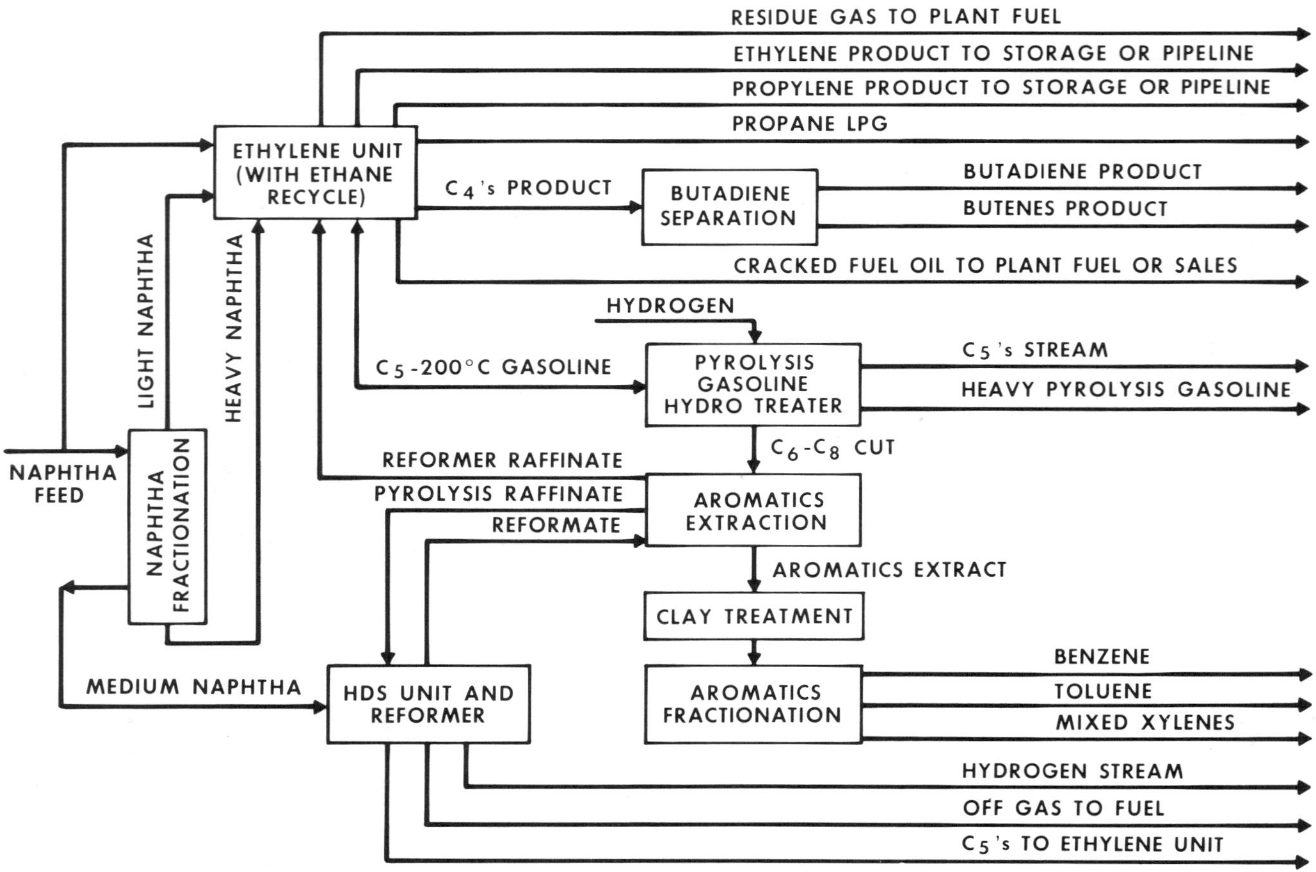

Fig. 2. Flow diagram of a petrochemical complex for olefins and aromatics production.

TABLE I

Typical Capital Cost for a Petrochemical Plant

	Capital cost (millions of dollars)	% Total capital cost
Example 1		
450,000 MTA Ethylene unit[a]	180.0	77.0
Off-site facilities	54.0	23.0
Total plant cost	234.0	100.0
Example 2		
450,000 MTA Ethylene unit[a]	180.0	42.0
Satellite units	105.0	25.0
Off-site facilities	143.0	33.0
Total plant cost	428.0	100.0

[a] Naphtha feedstock with ethane recycle.

IV. ETHYLENE FUTURE GROWTH

The future of the market for ethylene and by-product olefins will depend on a number of related factors. By-products include propylene, butadiene, butylenes, aromatics and/or pyrolysis gasoline, and cracked fuel oil. The relative quantities of these by-products will vary with the feedstocks used and the cracking severity employed. The plant economics are greatly influenced by whether chemical values or fuel values are or can be assigned to the by-products. Other factors are escalation as it affects equipment cost, costs of construction materials and labor, availability of feedstocks, environmental problems, and the availability of capital to invest in new facilities.

Ethylene-production capacity in the United States has generally been in step with demand, and producers are currently planning their expansions carefully so as not to exceed demand by a wide margin. In the mid-1960s and early 1970s, expansion programs were geared to making available to the market approximately 6 to 7% additional capacity each year. The future growth rate for ethylene in the next 10 years will likely expand by only 3 to 4% per year. Rising prices will be partially responsible for the expected slowdown.

Ethylene prices in the United States in the first quarter of 1980 averaged 24.5¢/pound (54¢/kg), and in Europe approximately 36¢/pound (79.4¢/kg) (*Chemical Week*, 1980). Indications are that the price of ethylene could

TABLE II

Free-World Ethylene Supply[a]

Location	Year				
	1976	1978	1980	1982	1984
North America	12.5	14.5	17.5	19.0	20.0
Western Europe	11.5	13.0	15.5	16.0	16.5
Asia and Pacific	5.5	6.0	6.5	7.5	8.0
All others	1.0	1.5	2.0	2.5	3.5

[a] Given in millions of tons.

increase at an annual rate of 10–15% per year over the next 5 years. Ethylene prices, by 1985, could average 40–50¢/pound (88–110¢/kg).

Table II presents the free-world ethylene supply or capacity in millions of tons from 1976 to estimates through 1984 (Grimes, 1980); Table III presents similar information for ethylene demand for the same time period (Grimes, 1980). A word of caution is in order because many ethylene plants are designed with some overcapacity and operate at partial capacity for several years before full plant output is realized. Compilers of production and consumption statistics, unless they are aware of this fact, are at a disadvantage in accurately presenting figures on usage and production of ethylene.

Some capacity that has been added recently will, in many cases, replace existing capacity that is now obsolete or energy inefficient. Some ethylene producers are building new large units to replace as many as two or three small obsolete units. In some cases, when the new and more efficient capacity is placed in operation, the existing capacity will be used for standby capacity.

TABLE III

Free-World Ethylene Demand[a]

Location	Year				
	1976	1978	1980	1982	1984
North America	11.0	13.0	15.0	17.0	19.0
Western Europe	10.0	11.0	12.5	13.0	15.0
Asia and Pacific	4.0	5.0	6.0	7.0	7.5
All others	1.0	1.0	1.0	1.5	2.5

[a] Given in millions of tons.

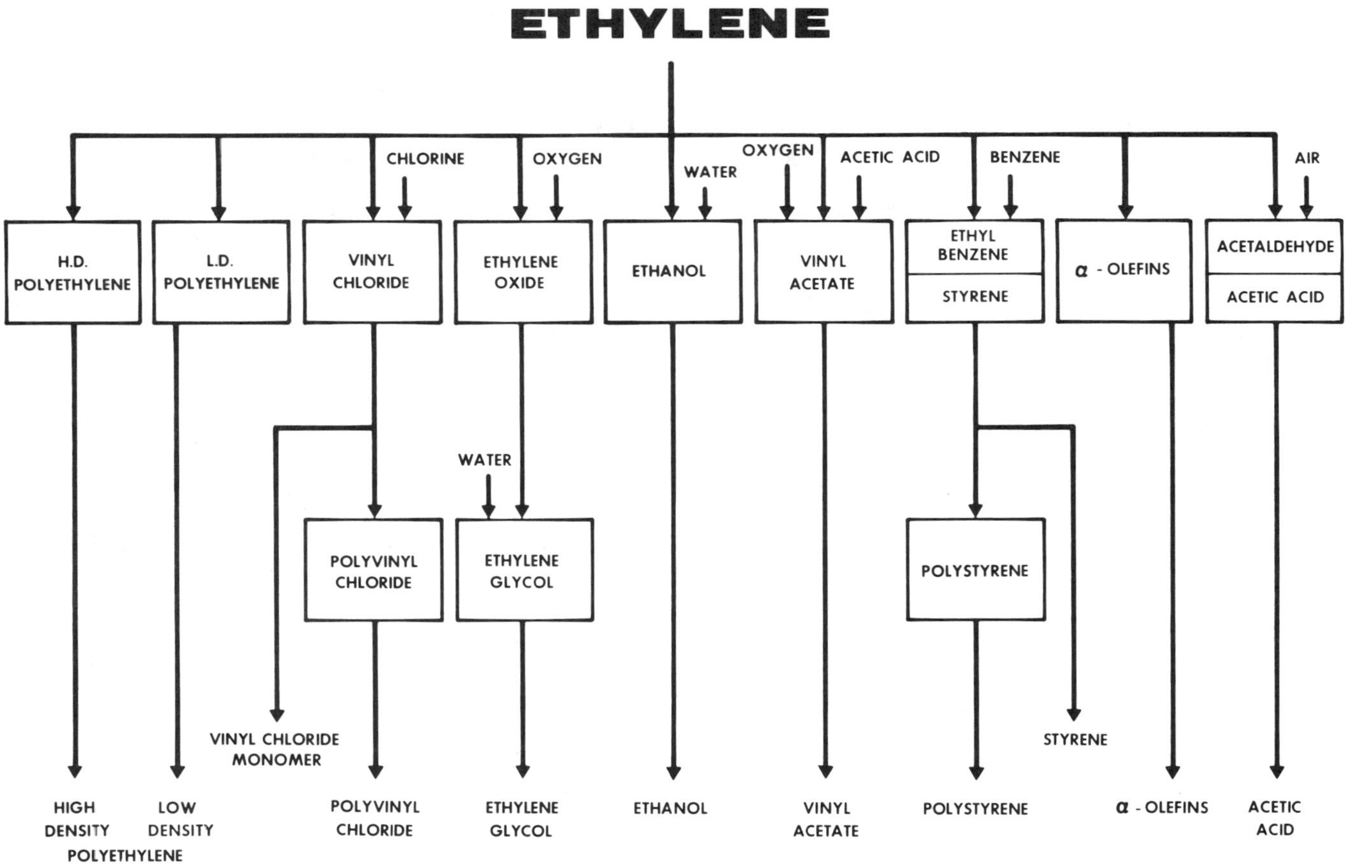

Fig. 3. Major derivatives of ethylene.

V. ETHYLENE DERIVATIVES

Because ethylene producers plan new plants based to a considerable extent on expected future demand, comments on factors that create this demand are in order. Figure 3 shows the major derivatives of ethylene, of which polyethylene, vinyl chloride, ethylene oxide, styrene, and ethyl alcohol are just a few. Those expected to show moderate increases in growth in the next few years are polyethylene, vinyl chloride, vinyl acetate, and ethylene oxide.

Although ethylene may be considered to be the principal product from an olefins plant, the by-products are also of great importance relative to the overall economics. Propylene uses have not developed as rapidly as those of ethylene. Figure 4 shows the most important derivatives of propylene.

Figure 5 illustrates some uses of butadiene, an important by-product, and Fig. 6 presents a few uses of the butylenes. If a refinery is close by, the butylenes may be consumed in alkylation or catalytic-polymerization units. Figures 7, 8, and 9 depict many of the uses of benzene, toluene, and xylenes.

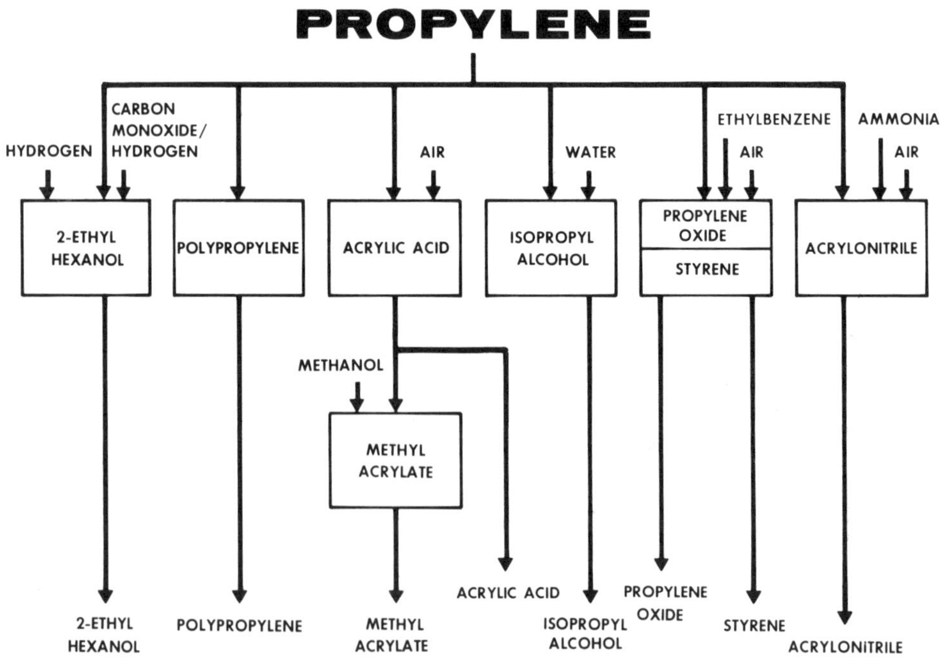

Fig. 4. Major derivatives of propylene.

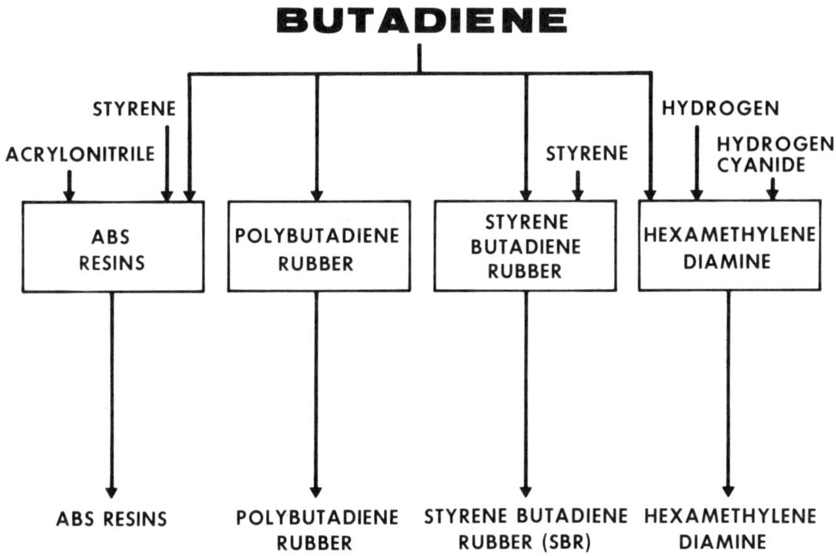

Fig. 5. Major derivatives of butadiene.

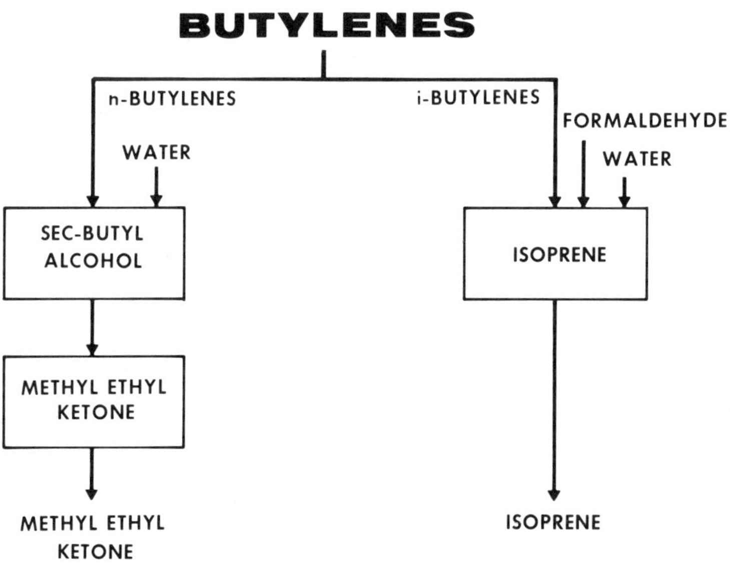

Fig. 6. Major derivatives of butylenes.

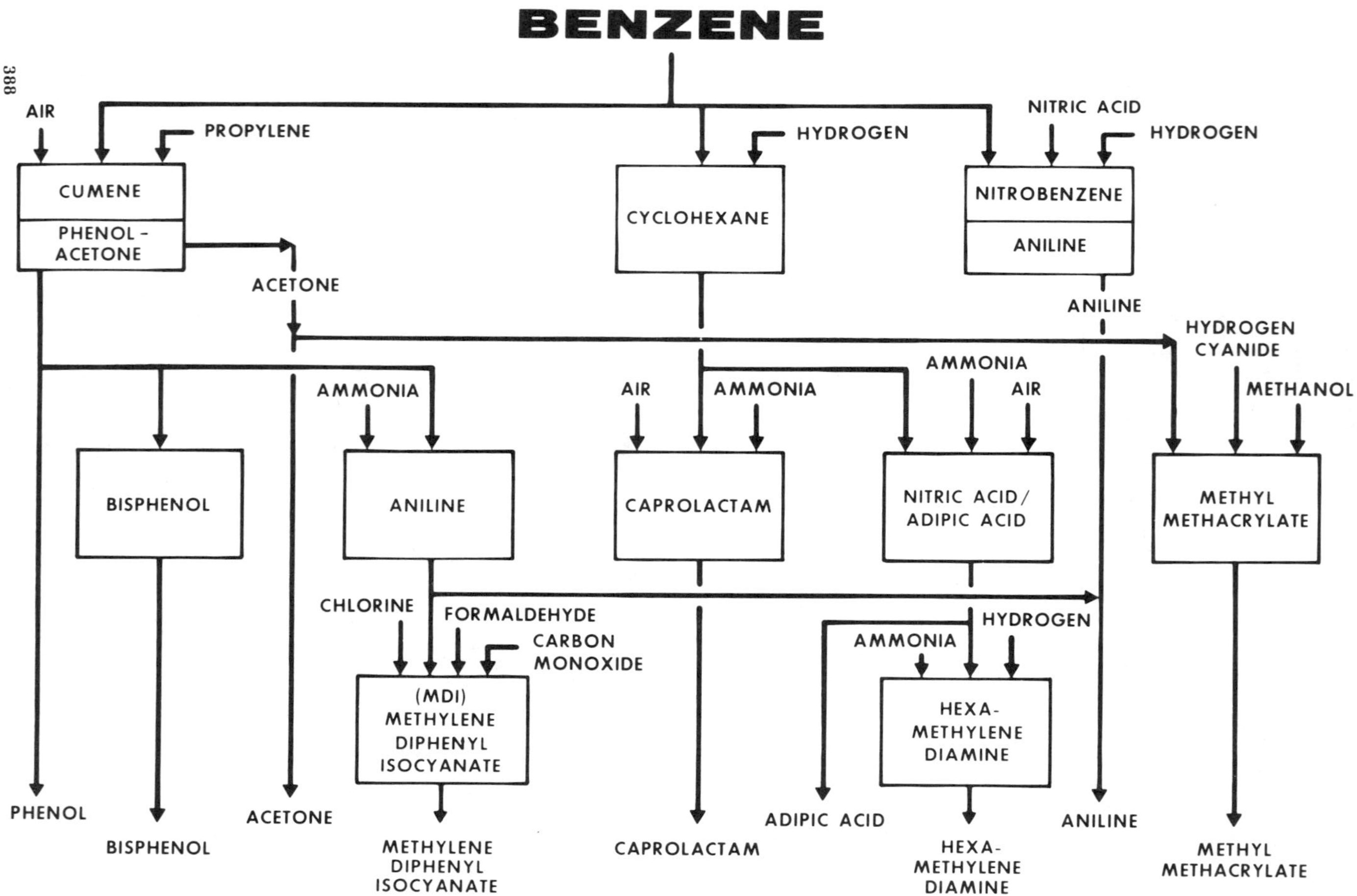

Fig. 7. Major derivatives of benzene.

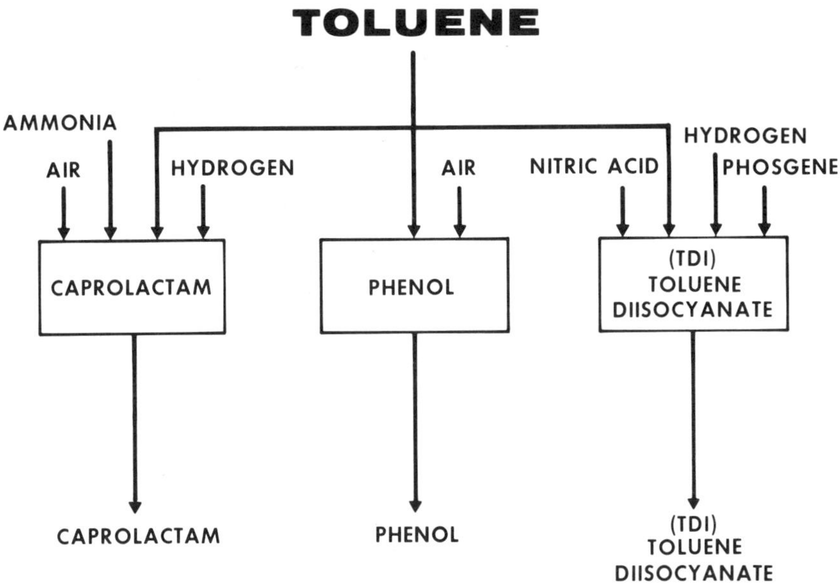

Fig. 8. Major derivatives of toluene.

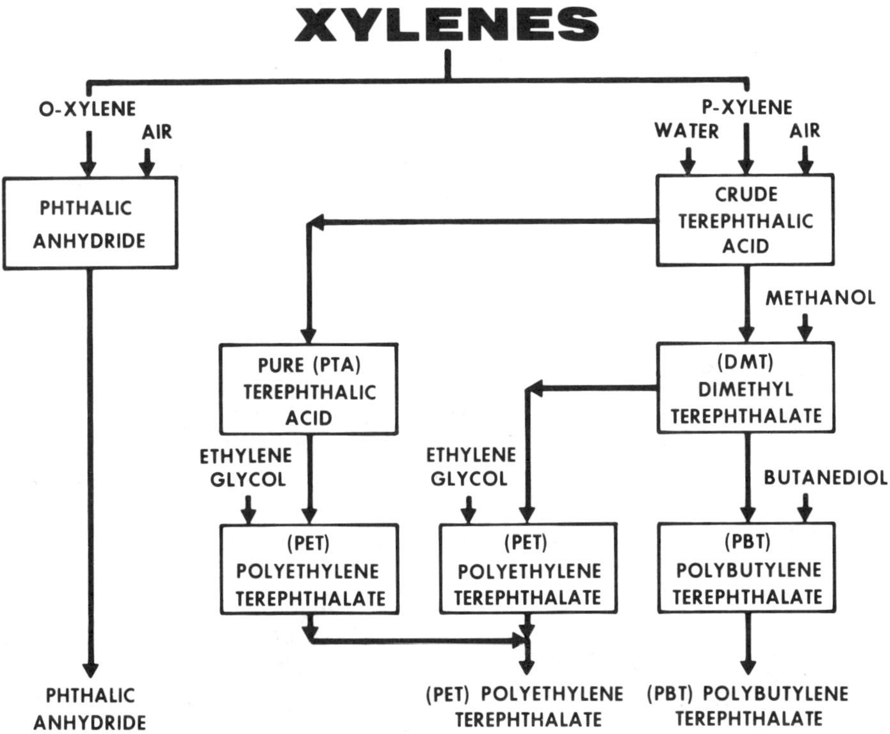

Fig. 9. Major derivatives of xylenes.

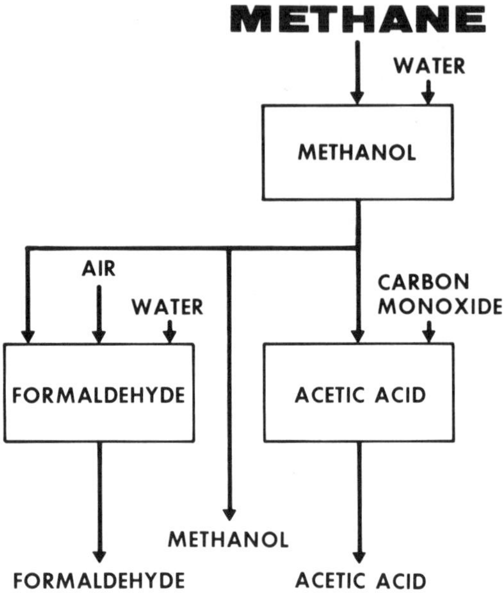

Fig. 10. Major derivatives of methane.

Methane mixed with hydrogen is contained in the residue gas. The residue gas is normally used as fuel gas for the cracking furnaces. When hydrogen is recovered, however, it is also feasible to produce methane. The methane can be used as a feedstock to produce hydrogen by re-forming, or can be used to produce petrochemicals as shown in Fig. 10.

VI. ETHYLENE-PLANT FEEDSTOCKS

A. Feedstock Selection

The project planner devotes considerable time and money during the early phases of a project to selecting the best feedstock or feedstocks to satisfy the basis of design for the mix of products. For light-hydrocarbon feedstocks, price and availability are of major importance. For naphthas and gas oils, although price and availability are of importance, feedstock composition is of equal importance; all naphthas are not alike in their capacity to produce ethylene at a given cracking severity. Sometimes it is economical to pay a few cents a gallon more for a better feedstock.

Ethylene, used in the first commercial production of polyethylene in 1939, was obtained by the dehydration of ethyl alcohol. In the early 1940s, much ethylene was recovered from refinery gases in the United States and some from coke-oven gases in Europe. Pyrolytic cracking was still in its infancy, and some ethane, propane, naphtha, and gas oil were cracked either at relatively low conversion or at low severity in the United States. Very little cracking was practiced in Europe. By the mid-1950s, pyrolytic cracking in the United States and Europe began to attract attention. The Japanese olefins industry was just starting to develop with naphtha as the basic feedstock.

If ethylene capacity is to be increased in the next decade, new and reliable sources of feedstocks must be established. The point cannot be emphasized too strongly that the selection and evaluation of the feedstock or feedstocks to be used in the production of ethylene is one of the most important considerations in beginning a new ethylene-plant project. The yield of ethylene, the quantity and composition of the by-products, and the plant design vary with the properties of the hydrocarbon material selected as a feedstock. Each feedstock must be evaluated on the basis of long-term availability, cost, and by-product market demand. If ethylene is the only product desired, then ethane would be the preferred feedstock because by-product production is minimal. If a complete slate of products is desired, then a liquid feedstock, such as naphtha or gas oil, would be preferable.

B. Refinery Gas

When refinery gas is available, it can be an attractive feedstock for ethylene production because appreciable quantities of ethylene can be recovered directly. Ethane and propane can also be recovered and used as cracking stock for more ethylene production. If a refiner will sell his refinery gas on a long-term basis at a fuel-gas value, ethylene can be competitively produced from this source relative to ethane from natural-gas sources. With refinery gas as a feedstock, the ethylene plant must be located in or near the refinery.

C. Ethane

Ethane is one of the more desirable feedstocks for the production of ethylene and is usually separated from natural gas in low-temperature-extraction plants. It is not universally available as a feedstock for all ethylene producers. In spite of the current energy crisis in the United States, ethane continues to be an important feedstock. Presently, the gas liquids have a cost advantage, because prices for crude oil have increased more rapidly than those for gas liquids. Ethane is usually cracked at a conversion level of between 60 and 65% per pass, with an ultimate ethylene yield of approximately 81 wt% at 60% conversion, and with a minimum production of by-products. When by-products have no value other than as fuel, ethane is an ideal feedstock. Table IV gives yield data for typical once-through cracking for several feedstocks and shows a conversion of 60% per pass for ethane.

D. Propane

In the United States propane has been more widely available geographically than ethane, but its use as a feedstock has logically been considered only in cases of attractive prices. It is normally cracked at a conversion level of 65–93% per pass, depending on the desired ratio of propylene to ethylene. At a conversion level of 93% per pass, the ultimate yield of

TABLE IV

Typical Yield Data for Light Hydrocarbons[a,b]

	Feedstock			
	Ethane (60%)[c]	Propane (93%)	n-Butane (96%)	Isobutane (80%)
H_2	3.72	1.56	1.49	1.08
CH_4	3.47	23.65	19.30	16.56
C_2H_2	0.42	0.77	1.07	0.72
C_2H_4	48.82	41.42	40.59	5.65
C_2H_6	40.00	3.48	3.82	0.88
C_3H_4	0.20	1.09	1.07	2.34
C_3H_6	0.99	12.88	13.64	26.35
C_3H_8	0.03	7.00	0.48	0.38
C_4H_6	1.33	2.82	4.13	1.49
i-C_4H_8				19.60
C_4C_8	0.25	0.89	1.92	—
i-C_4H_{10}				20.00
n-C_4H_{10}			4.00	0.20
C_5's	0.46	1.37	2.12	2.35
BTX	0.31	3.07	5.25	2.40
C_6–C_8(Nonaromatics)	—	—	—	—
C_9–400°F	—	—	0.10	—
Fuel oil	—	—	1.02	—
Total	100.00	100.00	100.00	100.00

[a] Pure hydrocarbons.
[b] Product yields given as weight percent.
[c] Percentage conversion per pass.

ethylene is approximately 47 wt%. This yield includes the ethylene produced by recycling the unconverted propane and the ethane formed, but not the propylene. By-products are produced to a greater extent with propane than with ethane. Depending on conversion level, the ultimate yield of propylene can be varied from about 27 wt% at 75% conversion to 16 wt% at 90% conversion. Table IV presents typical once-through yield data for propane at 93% conversion per pass.

E. Normal Butane

Normal butane can be cracked economically for the production of ethylene and propylene when it is available at a competitive price. It is used in relatively few commercial installations in the United States and western Europe, although more recently it has been gaining favor. From the standpoint of maximum utilization of feedstock, normal butane should be cracked at the highest practical conversion per pass because unconverted normal butane dilutes the C_4's product, making the recovery of butadiene more difficult. Conversion of about 90–96% in a single pass can be realized in a well-designed pyrolysis coil without encountering excessive coke formation. When cracking normal butane at a conversion of 96% per pass, the once-through furnace yields are approximately 40 wt% ethylene and 14 wt% propylene. Table IV presents typical once-through yield data for normal butane at a conversion of 96% per pass.

F. Isobutane

Isobutane is not commonly used as feedstock in the production of ethylene because ethylene yields are comparatively low. At an isobutane conversion of 90% per pass, the ethylene ultimate yield is about 11 wt% when the ethane produced is cracked on recycle. Isobutane is attractive if propylene and isobutylene are desired products. At elevated pressure and low conversion, good yields of isobutylene are attained; Table IV presents typical once-through yield data for 80% conversion per pass.

G. Naphtha

For many years naphtha feedstocks have been widely used for the production of olefins and aromatics in western Europe and Japan. In these locations, ethane and propane were not available in sufficient quantities to satisfy the needs of olefin producers. Naphtha was available, either locally or by importation, at reasonable prices, and has become the base feedstock for most western European and Japanese ethylene plants.

In the United States, until recently, ethane and propane feedstocks satisfied the major needs of olefin manufacturers. More recently, some naphthas and refinery raffinates have been selected as feedstocks for olefins manufacture in the United States. Olefin manufacturers in the United States must compete with refiners and gas-transmission companies who use naphtha fractions for the production of premium gasolines and substitute natural gas. The price established for a naphtha fraction is based on its competitive value as a fuel or as a feedstock for the production of olefins.

One of the advantages naphtha feedstocks have over gaseous feedstocks is the wide spectrum of possible co-products. By varying the cracking severity, the ratio of propylene to ethylene on a once-through basis may be changed from about 0.40 at high severity to about 0.75 at low severity. Additionally, butadiene and BTX (benzene, toluene, and xylene) production may also be varied within limits.

TABLE V

Typical Once-through Yields from Liquid Feedstocks at Low and High Severity[a]

Feedstock characteristics	Kuwait naphtha	Kuwait AGO	Es Sider vacuum distillate
Specific gravity (°API)	67.0	38.6	30.0
Specific gravity (15.5/15.5°C)	0.713	0.832	0.876
Boiling range (°C)	32–171	232–327	299–538
Watson characterization factor	12.30	11.98	12.21
Hydrogen (wt%)	15.2	13.7	13.0
Aromatics (wt%)	7	24	28
Yields (wt%)			
CH_4	10.3 (15.0)[b]	8.0 (13.7)[b]	6.6 (9.4)[b]
C_2H_4	25.8 (31.3)	19.5 (26.0)	19.4 (23.0)
C_2H_6	3.3 (3.4)	3.3 (3.0)	2.8 (3.0)
C_3H_6	16.0 (12.1)	14.0 (9.0)	13.9 (13.7)
C_4H_6	4.5 (4.2)	4.5 (4.2)	5.0 (6.3)
C_4H_8	7.9 (2.8)	6.4 (2.0)	7.0 (4.9)
BTX	10.0 (13.0)	10.7 (12.6)	18.9 (16.9) (BTX and C_5–200°C combined)
C_5–200°C (other than BTX)	17.0 (9.0)	10.0 (8.0)	
Fuel oil	3.0 (6.0)	21.8 (19.0)	25.0 (21.0)
H_2, C_2H_2, C_3H_4, C_3H_8	2.2 (3.2)	1.8 (2.5)	1.4 (1.8)
Total	100.0 (100.0)	100.0 (100.0)	100.0 (100.0)

[a] Residence time approximately 0.3 s.

[b] Two figures are given for yields weight percent from each feedstock. The first is for a low cracking severity; the second, in parentheses, is for a high cracking severity.

Naphtha feedstocks are mixtures of many components and are normally characterized by their specific gravity, ASTM boiling range, C/H ratio, and PONA analysis (which indicates the distribution of normal paraffins, isoparaffins, olefins, naphthenes, and aromatics). Normal paraffins are especially suited for the production of ethylene, whereas branched-chain paraffins (isoparaffins) yield more propylene. A naphtha with a high content of naphthenes and a low content of paraffins produces ethylene but favors the production of C_4 hydrocarbons and high yields of butadiene. Naphthas normally contain very few olefins but generally from 3 to 20 vol% aromatics. Aromatics are very refractory and remain essentially unconverted in the pyrolysis furnaces. Table V presents typical once-through yield data for a Kuwait full-range naphtha at low and high severity.

H. Atmospheric Gas Oil

Atmospheric gas oil (AGO) is now being cracked in a number of ethylene plants as an alternative liquid feedstock to naphtha for the production of ethylene. While not generally recognized, gas oil cracking is an old art. Gas oil was cracked for the production of olefins and aromatics in the 1930s and early 1940s in the United States. However, it lost favor as a feedstock and was replaced subsequently by readily available and inexpensive ethane and propane. Economic considerations were primarily responsible for this trend. In recent years, however, changes have occurred in feedstock availability for petrochemical plants. Depending on the price at which naphtha becomes available in a competitive market for either gasoline or synthetic natural gas production, it may very well develop that olefins producers will again find it economical to use gas oil as a feedstock. The price of AGO as an olefins feedstock will be determined by its value as a heating oil for domestic and industrial uses, as catalytic cracking feedstock, or possibly as a feedstock for manufactured gas. Table V presents typical once-through yield data for Kuwait AGO at low and high severity.

The cracked fuel oil can be used internally to some extent in the ethylene plant for some of the energy for the cracking furnaces and most of the energy for the steam boilers. Some of the cracked fuel oil may be blended with No. 6 fuel oil when the pool of No. 6 fuel oil is large. The two types of oils may not always be compatible, however. The total >400°F fraction may be split into a 400–650°F distillate fuel oil and a >650°F heavy-residual fuel oil. This separation may be advantageous to some marketers of fuel-oil products. Cracked fuel oil can also sometimes be used as a feedstock for carbon black production, but the quality of the cracked fuel oil varies from plant to plant relative to carbon black production. Recent investigations have indicated that when hydrogenated, the light fraction of the cracked fuel oil produces a superior jet fuel-oil product.

I. Vacuum Gas Oil

When light-hydrocarbon liquids, naphthas, and AGOs become less available, heavier liquids must be considered as alternative feedstocks for the production of olefins and aromatics. Vacuum gas oil (VGO) is currently attracting interest and a few ethylene plants are presently cracking VGO on a regular basis (Green, 1969). The technology is available to pyrolyze VGO, recover heat from the furnace effluent, and fractionate prior to compression. (All VGOs boiling in the range of approximately 600 to 1000°F are not alike, however, and some are more preferred as olefin feedstock

TABLE VI

Pyrolysis Yields from a Typical Hydrogenated Coal Liquid

	Hydrogenated coal liquid		Light Kuwait gas oil	
Feedstock characteristics				
Specific gravity (60/60°F; 15.6/15.6°C)	0.895		0.828	
Distillation (°F)				
Initial b.p.	363	(183.9°C)	379	(192.8°C)
50% Evaporation	545	(285°C)	559	(292.8°C)
End point	658	(347.8°C)	732	(388.9°C)
H_2 (wt%)	12.76		13.70	
Furnace effluent (wt%)				
Hydrogen	1.03		0.60	
Methane	11.66		12.00	
Acetylene	0.49		0.39	
Ethylene	18.83		22.66	
Ethane	2.40		3.96	
Propadiene, propyne	0.58		0.41	
Propylene	7.83		13.02	
Propane	0.21		0.34	
1,3-Butadiene	4.68		4.82	
Other C_4's	1.63		3.12	
C_5's	2.56		2.72	
C_6–C_8 Aromatics	22.34		12.33	
Other C_6–400°F (204.4°C)	3.30		3.11	
Fuel oil	22.46		20.52	
Total	100.00		100.00	

than others.) Some VGOs cause excessive fouling in the radiant-coil and effluent exchangers, whereas others are much preferred as olefin feedstocks.

Most VGOs must be hydrodesulfurized to reduce corrosion and acid gas-removal requirements in the ethylene plant. More severe hydrotreating increases the hydrogen content of the feedstock and provides a better olefin potential while significantly reducing cracked-fuel-oil production. Table V presents typical once-through yield data for Es Sider vacuum distillate at low and high severity of cracking.

J. Feedstocks Derived from Coal and Oil Shale

Future feedstocks may be produced from the liquefaction and/or gasification of coal or shale oil. The liquids produced likely will be hydrogenated to form feedstocks with good potential for producing olefins and aromatics. More severe hydrogenation of the coal liquid produces a product gas rich in ethane and benzene, each of which has value to a petrochemical producer.

When a typical coal liquid is hydrogenated and then cracked with steam under commercial conditions, the composition of the resulting furnace effluent is summarized in Table VI (Korosi *et al.*, 1976). A light gas oil of Kuwait origin, having a boiling range similar to a hydrogenated coal liquid and cracked under similar conditions, gave a product that is also shown in Table VI for comparative purposes. The hydrogenated coal liquid yielded somewhat less ethylene, about half the propylene, the same butadiene, and about twice the benzene plus other aromatics.

Liquid hydrocarbons produced by retorting of oil shale offer much promise in providing an alternative source of feedstocks for jet fuels and diesel oils, and for the manufacture of olefins. Table VII provides an insight as to the potential of a full-range shale oil boiling in the range of 300–1000°F (149–538°C). Data are shown for nontreated and hydrogenated liquid when pyrolyzed at commercial conditions in a coil reactor (Korosi *et al.*, 1980). Both samples were pyrolyzed under the same conditions. Table VII indicates that the hydrogenated shale oil gives an ethylene yield comparable to some full-range VGOs from crude-oil sources.

Schultz *et al.* (1955) reported the results of the pyrolysis of crude shale oils produced by the Bureau of Mines combustion process and by the Union Oil Company process. Variables that describe gasification results for petroleum oils characterize the pyrolysis behavior of the crude shale oils used in their study. Under the conditions of pyrolysis of the crude shale oils, gas yields varied from 33.5 to 52.9 wt% of the feed.

For both coal-derived feedstocks and shale oil, prices must be competitive with crude oil-derived feedstocks if they are to be considered seriously in the future.

TABLE VII

Pyrolysis Yields from Raw and Hydrogenated Shale Oil

	Raw shale oil		Hydrogenated shale oil	
Feedstock characteristics				
Specific gravity (60/60°F; 15.6/15.6°C)	0.920		0.867	
Distillation (°F)				
Initial b.p.	326	(163.3°C)	262	(127.8°C)
50% Evaporation	673	(356.1°C)	639	(337.2°C)
End point	1086	(585.6°C)	—	
H_2 (wt%)	12.00		13.16	
Furnace effluent (wt%)				
Hydrogen	0.3		0.5	
Methane	7.6		11.0	
Ethylene	13.1		19.8	
Ethane	2.7		3.0	
Propylene	11.0		13.9	
Propane	0.8		0.6	
1,3-Butadiene	3.5		5.0	
Other C_4's	5.7		5.3	
C_6–C_8 Aromatics	6.9		9.9	
Other C_5–400°F (204.4°C)	17.0		17.4	
Fuel oil	37.3		22.6	
$CO + CO_2$	0.7		—	
Total	100.0		100.0	

K. Ethylene from Ethanol and Methanol

When ethanol is heated and passed over a suitable catalyst such as alumina, silica, or phosphoric acid on a suitable support, high yields of ethylene are possible by dehydration. The chemistry of this reaction has been known for a long time and was practiced in Europe on a small scale. Ethylene from ethanol could not, however, economically compete with ethylene derived from hydrocarbon sources. With the cost of hydrocarbon feedstocks currently increasing rapidly, several nations with limited crude-oil supplies have considered producing ethanol by fermentation of carbohydrates and by dehydration of ethanol to ethylene. There is some evidence that ethylene can be produced directly from carbohydrates via fermentation using an ethylene-producing organism (Brandt, 1979). At the present time, ethylene from carbohydrates must be considered as a long-range development program.

Methanol can be used as a feedstock for the production of olefins. It is cracked primarily to ethylene and propylene over zeolite-type catalysts in accord with the expression:

$$2\ CH_3OH \xrightarrow{\text{catalyst}} C_2H_4 + 2\ H_2O$$

Although this route for production of ethylene can compete in some locations with more conventional hydrocarbon-cracking processes giving yields of ethylene of 50–60%, cheap methanol must be available from sources such as natural gas or coal-derived feedstocks.

VII. THE PYROLYSIS PROCESS AND PRODUCT DISTRIBUTION

A. General Considerations

The pyrolysis process by which ethylene and its co-products are produced has a profound influence on the economics of olefins manufacture. Details on furnace operation are cited in Chapter 14.

For any specific feedstock, the conditions of residence time, temperature level, degree of conversion, and hydrocarbon partial pressure under which the pyrolysis is carried out all affect the product distribution.

B. Plots of Product Distribution versus Severity

For light-hydrocarbon feedstocks having 2 to 4 carbon atoms, a convenient and widely accepted method of indicating the variation in effluent composition with severity is to plot the quantities of each major effluent constituent in moles per 100 mol of feed constituent charged against percentage conversion. For distillate feedstocks, it is customary to relate the quantities of effluent constituents in pounds per 100 pounds of furnace feed to some variable indicative of severity.

C. Distribution of Products from Naphthas, Gas Oils, and Other Distillates

Figure 11 shows qualitatively how the distribution of the major products from cracking a typical naphtha in a tubular reactor varies with cracking severity, expressed as kinetic severity function (KSF), which is defined as the integral of the reaction-velocity constant times the differential time (Zdonik *et al.*, 1968).

Constant flow rates of hydrocarbon and dilution-steam, and constant coil-outlet pressure are assumed. Therefore, the partial pressure of hydrocarbon at the coil outlet is approximately constant at the moderate- and high-severity levels, where this variable is important. Because the flow rate through the coil is unvarying, the equivalent time is also approximately

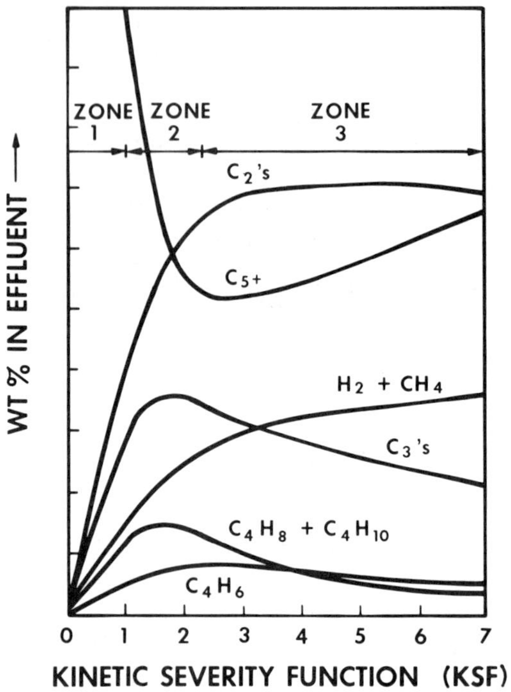

Fig. 11. Naphtha cracking—typical variation of product distribution with cracking severity.

constant. The equivalent time, which must be related to some specific reference temperature, is the time that, if the reactant could be raised instantaneously to the reference temperature and held there, would produce the same degree of disappearance as achieved in the actual variable-temperature apparatus. To afford some measure of comparison, the outlet temperature of the furnace radiant coil is often taken as the reference.

The severity diagram of Fig. 11 can be divided into three zones. In zone 1, the zone of low severity up to a KSF of about 1, the principal cracking reactions are those involving saturates. Because most of the butadiene produced is derived from secondary reactions, the slope of the curve for this product at the origin is close to zero.

In zone 2, primary reactions are still important as indicated by the still rapid decrease in C_5^+ material, but secondary reactions are becoming dominant. This zone, extending up to a KSF of about 2–2.5, yields hydrogen, methane, total C_2's, and butadiene at a rapid but gradually diminishing rate.

The rates of decomposition of propylene and butylenes catch up to, and then exceed, their rates of formation. Therefore, their concentrations in the reacting mixture pass through maxima. The propylene peak typically occurs at a KSF of about 1.7. The butylene peak usually appears at a slightly lower severity, although feedstocks rich in isoparaffins, which produce considerable isobutylene, tend to displace the peak to a severity level comparable to that of propylene.

In Zone 3, the primary reactions have virtually ceased, and secondary reactions become predominant. The C_5^+ passes through a minimum as the original saturates become exhausted, and the formation of stable aromatics from propylene and C_4's adds to the C_5^+ yield as the severity is increased.

The production of hydrogen plus methane rises in movement through the zones, but the C_2's and butadiene eventually pass through maxima. Butadiene yield typically peaks at a KSF of about 2.5, whereas the C_2's pass through a broad maximum at about KSF = 5, as the rate of ethylene degradation finally reaches and then surpasses its rate of formation from the higher olefins.

D. Maximum Practicable Severity

The practicable upper limit of severity that may be achieved in a tubular furnace is determined by either of two factors: (*a*) attainment of tube-metal temperatures so high as to be significantly detrimental to the life of the furnace components, or (*b*) excessive deposition of coke on the walls of furnace tubes and/or fouling of downstream piping and equipment. Coke deposition in pyrolysis coils is caused fundamentally by overcracking in the relatively stagnant gas film adjacent to the tube wall. In practice, the rapidity with which the coke deposit builds up is dependent on a number of factors in addition to the cracking severity. These include the type of feedstock being pyrolyzed, coil-design parameters, heat flux, hydrocarbon partial pressure, and mass velocity through the coil. Thus maximum severity criteria cannot be postulated without considering many variables.

By reference to Fig. 11 it can be seen that for distillate feedstocks at severities higher than those corresponding to minimum C_5^+ (maximum conversion to C_4 and lighter), the concentration of heavy aromatics, which are coke precursors, is increasing. Cracking at severities very far beyond this point is not feasible without resorting to uneconomical amounts of dilution steam. Even with cracking coils of sophisticated design operated under close supervision, severities of 3.5–4 appear at this time to be the maximum practicable for continuous commercial operation.

Fortunately, as Fig. 11 indicates, there is little, if any, incentive to push severity higher than a value for KSF of about 5 because ethylene yields are near a maximum and the other olefins normally sought have already passed through their maxima.

VIII. ETHYLENE UNIT PROCESS-FLOW DIAGRAM

A. General Concept of a Modern Plant

The process-flow diagram of a modern ethylene unit has undergone considerable alteration in the last 20 years. Because the feedstock spectrum has changed during this period, it has been necessary to make certain modifications to the flow sheet to accommodate the product spectrum and the more stringent product specifications.

Most cracking furnaces now operate at higher coil-outlet temperatures and at shorter residence times than previously, which requires more rapid quenching of the products of cracking to maximize ethylene yield. Overall efficiencies of furnaces have increased by improving the recovery of heat from the flue gases in economizer coils that generate steam. Transfer-line exchangers (TLEs) are now almost always included whenever they provide satisfactory performance. Quench-oil circuits have become more complicated. Ecological and economic considerations have promoted the recovery of condensed dilution steam for recycling back to the cracking furnaces. Previously, condensed dilution steam was discharged to the sewer. Designers of cracked-gas driers have a choice of alumina or molecular sieves for the desiccant. Selection is made on the basis of cost, moisture-adsorption capacity, structural stability of the desiccant, and proven performance. More selective acetylene catalysts have been developed with a trend to locating the catalyst-hydrogenation converters in the overhead stream of the de-ethanizer. The fractionation equipment for the product has been given more capability for separation to meet current product specifications. Refrigeration requirements per unit of ethylene production have been optimized to reduce operating costs.

B. General Processing Scheme

The recovery of components from the pyrolysis of feedstock in an ethylene unit into products such as hydrogen, residue gas, ethylene, propylenes, raw C_4's, cracked gasoline, and cracked fuel oil are normally separated by a combination of low-temperature fractionation and more conventional methods. Refrigerated reflux is employed to recover hydrogen, residue gas, and ethylene. The conventional fractionation methods are used to recover propylene, raw C_4's, and cracked gasoline. Although the cited products may be produced from a variety of feedstocks, there are similarities in the flow sheet for both a gaseous-feedstock plant and a liquid-feedstock plant.

C. Flow Sequence for Producing Ethylene from a Gaseous Feedstock

Figure 12 illustrates the flow sheet in a simplified form for producing ethylene and associated by-products from gaseous feedstocks such as ethane, propane, or a mixture of ethane and propane.

Ethane or propane is fed with dilution steam to the cracking furnace. The resulting effluent passes through TLEs for generation of high-pressure steam, which after superheating will be used as the motive steam for the cracked-gas and refrigeration compressors. The partially cooled, cracked-gas effluent is further cooled in a water-quench tower by a circulating water stream.

The cooled, cracked gas at about 105°F flows to the first stage of the cracked-gas compressor where it is compressed in either four or five stages depending on the lowest discharge temperature possible with the available cooling water. In between the third or fourth compressor stages, the acid gases are removed. Older types of plant designs removed the C_2 acetylene in the cracked-gas compressor circuit; however, modern plants convert or recover the C_2 acetylene in the de-ethanizer overhead.

After the cracked gas is compressed to approximately 500 psig (3.45×10^6 Pa), it is cooled to about 60°F (15.6°C), dehydrated, further cooled in a

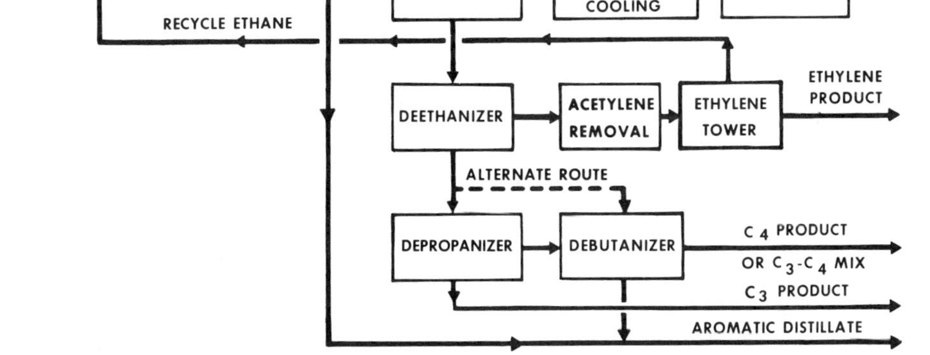

Fig. 12. Typical processing steps for producing ethylene from gas cracking.

series of exchangers, and fed to the demethanizer. In this system, methane is separated from the C_2 and heavier fractions. Hydrogen is also separated in this system from the methane either for hydrotreating or for delivery to battery limits.

The bottoms from the demethanizer are fed to the de-ethanizer where the C_2's are taken overhead. The C_2's are sent first to a catalytic-hydrogenation unit for conversion of C_2 acetylene and then to the ethylene tower. The bottoms from the de-ethanizer are sent to the depropanizer. Treated C_2's are fractionated in the ethylene tower into ethylene product and into ethane, which is recycled to cracking. The depropanizer produces an overhead of C_3 product and a bottoms of C_4 and heavier, which is often separated in a debutanizer into a C_4 product and an aromatic fraction.

When the feedstock is predominantly propane, the overhead of the depropanizer is often treated for removal of methylacetylene and propadiene by catalytic hydrogenation; this stream is then sent to a propylene fractionator to produce either chemical or polymer-grade propylene and a bottoms propane stream, which is generally recycled to cracking.

D. Flow Sequence for Producing Ethylene from a Liquid Feedstock

Figure 13 illustrates a flow sheet for an ethylene plant utilizing a liquid feedstock such as naphtha or gas oil. Feedstock is pumped from the storage area through preheat exchangers to the cracking furnaces. Superheated dilution steam is introduced into the process coil at an appropriate point in the convection section. The hydrocarbon is thermally cracked in the radiant coil to maximize the yields of the desired olefins.

The effluents from the furnaces pass through effluent TLEs and are then cooled further by quenching with a circulating oil. The quenched furnace effluent flows to a primary fractionator. This fractionation step removes the quench oil, dilution steam, and fuel-oil components. The fuel oil passes to a fuel-oil stripper for removal of gasoline components.

Usually the cracked gas is compressed in five stages in a steam turbine-driven centrifugal compressor to about 500 psig (3.45×10^6 Pa). Hydrocarbons condensed in the first three stages are flashed back to successively lower stages. The net hydrocarbon condensate from the first stage and the net gasoline from the primary fractionator are fed to a distillate stripper for debutanizing. The net bottoms from the stripper consist of C_5's and heavier, and are combined with the debutanizer net bottoms and sent to battery limits. The fifth-stage discharge condensate is fed to the condensate stripper for de-ethanizing, and the overhead vapors are sent to the fourth-stage suction. The bottoms from the condensate stripper are sent to the

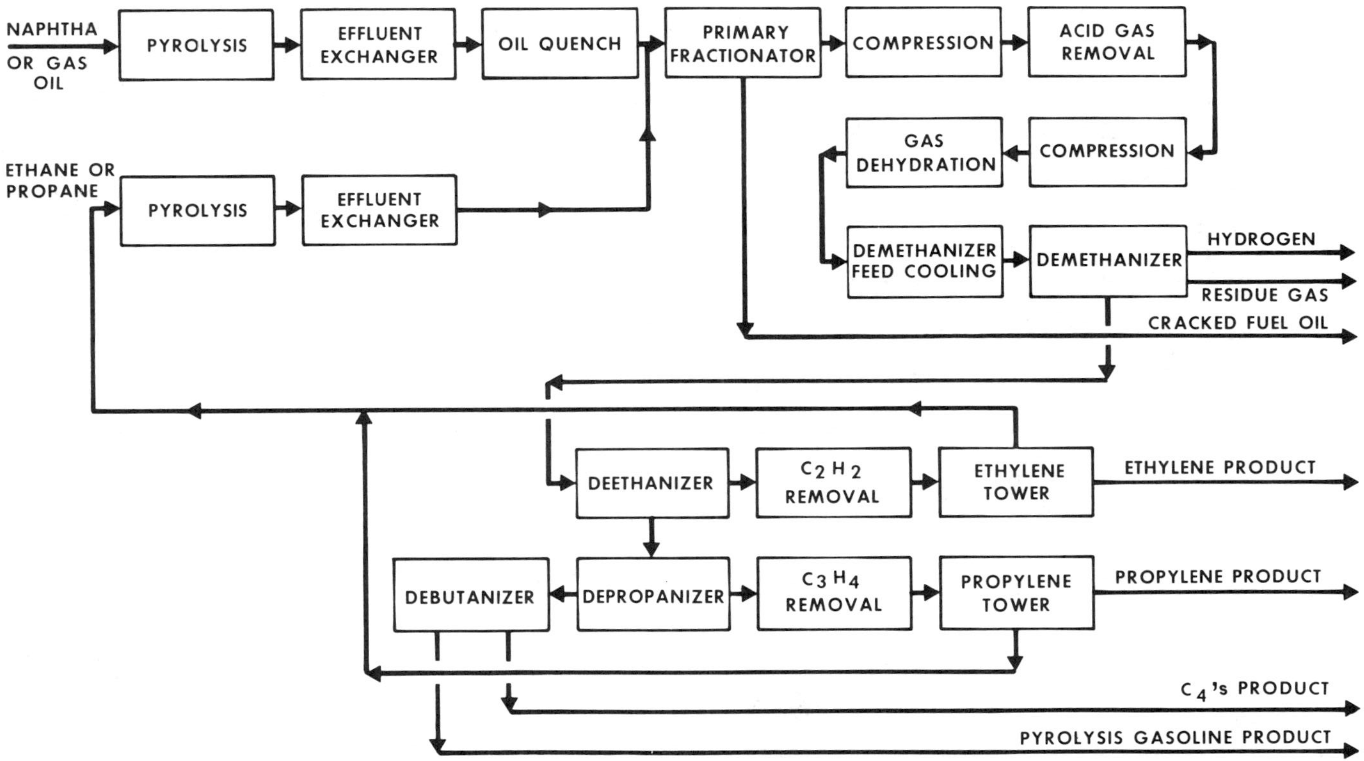

Fig. 13. Typical processing steps for producing ethylene from liquids cracking.

depropanizer. Acid gases are normally removed in a caustic scrubber between the fourth and fifth stages of compression. Cracked gas from the final stage of compression is dried in dehydrators charged with solid desiccant. This step prevents hydrate formation in the low-temperature fractionation equipment. The gas is then cooled by heat exchange with product streams and refrigeration, and fed to the demethanizer system where hydrogen and methane are separated from the C_2 and heavier components. In this system, hydrogen is produced for the hydrogenation of pyrolysis gasoline and acetylenes and for toluene hydrodealkylation.

Bottoms from the demethanizer are processed as described for the same fraction in the previous section on use of a gaseous feedstock. Bottoms from the debutanizer and distillate stripper are combined and pumped to storage or directly to the gasoline-hydrogenation unit.

Refrigeration for the low-temperature fractionation is furnished by an ethylene–propylene cascade system, which provides ethylene refrigeration at two levels and propylene refrigeration at four levels. Where possible, refrigerant vapors are condensed in reboilers, or heat is transferred from the refrigerant to the process, for increased refrigeration economy.

E. Description of Butadiene-Recovery Unit

The raw C_4 product from the ethylene unit is vaporized and sent into the first extraction tower. Lean solvent enters the tower several trays below the top tray, and absorbs butadiene and any residual acetylenes from the feed. Hydrocarbon reflux introduced at the top tray scrubs solvent from the butane–butene vapor overhead. The net butane–butene overhead stream is pumped to storage.

The rich solvent from the first extraction tower is pressured to the first stripper where hydrocarbons are stripped from the solvent. The net hydrocarbon overhead is cooled and compressed as feed to the second extraction tower.

The feed to the second extraction tower consists mainly of butadienes, heavy hydrocarbons, and acetylenes. Butadiene-rich overhead from the second extraction tower is condensed and pumped to the topping tower. The rich solvent bottoms are sent to the secondary stripper where acetylenes are stripped and sent to fuel. The lean solvent bottoms from both strippers are heat-exchanged and water-cooled before being recirculated.

In the topping tower, components lighter than butadiene, particularly methylacetylene, are separated overhead and recycled to the ethylene plant.

Topping-tower bottoms are pumped to the butadiene-product tower where the butadiene product is taken overhead and sent to storage. Heavy hydrocarbon bottoms are sent to fuel. A solvent-recovery and -refining system is included in the butadiene unit to maintain solvent purity and to remove dimers that are formed in the system.

All modern butadiene processes also use solvent additives to prevent foaming and fouling and to extend to onstream life of the plant.

F. Description of Pyrolysis-Gasoline Hydrogenation Unit

The hydrogenation unit for pyrolysis gasoline completely hydrogenates the olefins and diolefins in a two-stage catalytic reactor, and then separates the hydrogenated gasoline into fuel gas, a C_6–C_8 heart cut, and light- and heavy-gasoline fractions. Pyrolysis gasolines from the ethylene unit are pumped, after suitable heat exchange, to the first hydrogenation reactor. Methane and hydrogen are flashed from the effluent, which is then fractionated to produce a light gasoline consisting of C_5's and lighter, a heavy gasoline of C_9's +, and a C_6–C_8 heart cut. The heart cut is heated and then vaporized with hot hydrogen and sent to the second-stage reactor where the remaining olefins are completely hydrogenated. The cut is then stabilized and sent to the aromatics-extraction and -recovery unit.

G. Description of Aromatics-Recovery Unit

1. *Aromatics Extraction*

A hydrogenated C_6–C_8 heart cut from the pyrolysis-gasoline hydrogenation unit is pumped to a liquid–liquid extraction tower. Solvent pumped to the top of the tower passes through the tower as the continuous phase and absorbs the aromatics from the rising hydrocarbons, which are injected near the middle of the tower.

The undissolved paraffins leave as the overhead raffinate stream. The raffinate, being saturated with solvent, is water-washed to recover the solvent before recycling the raffinate to the ethylene unit for blending with the fresh feed to the pyrolysis furnaces.

The dissolved aromatics leave the base of the tower as the extract solution. A portion of the extract is separated from the solvent by normal distillation in the extract-stripper tower to serve as bottoms-reflux to the extraction tower. The extract stripper-tower bottoms are fed to the extract-recovery tower. The last traces of hydrocarbon are removed by steam stripping in the bottom part of the recovery column.

Overall economy is improved by using the overhead water from the recovery tower, which contains only small amounts of solvent, as wash water for the raffinate. The spent wash water is then reused as the stripping medium in the recovery tower.

2. *Purification*

The aromatic overhead from the recovery tower is heated from 175° to 250°F (79–121°C), depending on the composition of the aromatic overhead, and passed through two clay towers in series where objectionable color is removed. Piping is arranged to permit interchanging of flow through the towers, which allows the freshest bed to be the last in series. While clay is being changed, a single tower is used to maintain continuity of operation. The benzene–toluene–xylene stream is then separated by normal fractionation in three towers.

H. Description of Toluene-Hydrodealkylation Unit

Most of the nitration-grade toluene feed, together with recycle toluene, recycle gas, and makeup hydrogen, is preheated by exchange against hot-reactor effluent and then heated to reactor temperature in a fired heater. The heated mixture is passed through the first reactor, where some dealkylation occurs. The exothermic heat of reaction is removed by quenching the first reactor effluent with the remaining toluene feed, and the entire stream is then sent to the second reactor for additional dealkylation.

The effluent from the second reactor is cooled by exchange with reactor feed and stabilizer-tower feed, and by reboiling the stabilizer. The effluent is further cooled, usually in an air cooler, followed by a trim-water cooler, before entering the vapor–liquid separator drum. Net vapor from the separator drum is recycled to the cracked gas-compression system of the naphtha-pyrolysis unit for recovery of hydrogen plus small amounts of benzene and toluene present in this vent gas. Recycle gas is compressed and returned to the first reactor.

Separator-drum liquid is sent to the stabilizer tower where all components lighter than benzene are vaporized overhead. The net stabilizer overhead vapor is chilled for almost complete recovery of benzene and toluene, and the vent gas is sent to the fuel-gas system.

The bottoms from the stabilizer are sent to a clay treater for removal of trace quantities of polymer. They then are transferred to the benzene tower for separation of benzene overhead product from recycle toluene bottoms.

IX. ETHYLENE-PRODUCT DELIVERY

The ethylene product is supplied to the consuming plant in various ways. When the ethylene consumer produces his own ethylene, it may be fed directly to the satellite operational unit with the option to send some to intermediate storage. This storage provides continuity of operation of the satellite unit when the ethylene plant is operating at less than desired capacity or is shut down for unscheduled or planned maintenance.

If the ethylene producer sells his product to many users, he can supply a pipeline from which ethylene may be withdrawn on demand. This arrangement has many advantages because the pipeline supply can be backed up with many days of storage to fulfill contractual obligations.

X. ETHYLENE-PRODUCT STORAGE

Ethylene is normally produced as a liquid from the ethylene fractionator at about 275 psig (1.90×10^6 Pa) and −20°F (−28.9°C). It is often pumped to a pressure of 325 to 2000 psig (2.24×10^6 to 13.79×10^6 Pa), depending on its destination. When the ethylene product is to enter a pipeline or cavern storage, the ethylene is first pumped and vaporized against process streams.

When the ethylene is sent to liquid storage, it may either be subcooled and allowed to flow to pressure storage, or be subcooled, partially flashed, and sent to low-pressure storage. If super pressures are required, such as for a high-pressure polyethylene plant, the compression is furnished by compressors within the polyethylene plant.

Ethylene may be stored in several ways, depending on the quantity to be stored. Some of the storage systems are listed next.

1. Cylindrical steel pressure vessels at about 300 psig (2.07×10^6 Pa) and −20°F (−28.9°C). This method uses many vessels and is normally used for total capacities under 500,000 gallons (1893 m^3).
2. Subsurface steel pipe or pipe sections at pressures up to 1500 psig (10.34×10^6 Pa) and about 60°F (15.6°C). This method is not normally used for large capacities.
3. Product pipeline at pressures of 500 to 1800 psig (3.45×10^6 to 12.41×10^6 Pa) and ambient temperature. This method is limited in capacity to the length of the pipeline system.
4. Spheres at low pressures and temperatures in the range of −100°F to −150°F (−73.3 to −101.1°C). This method is more economical than storage in steel cylindrical vessels of capacities over about 800,000 gallons (3028 m^3).
5. Storage in atmospheric double-walled tanks at a temperature of −150°F (−101.1°C). In storing large volumes of about 2 million gallons (7571 m^3), this method is more economical than storage in spheres.
6. Available underground salt caverns at 1500 to 2000 psig (10.34×10^6 to 13.79×10^6 Pa) and about 110°F (43.3°C). For large-volume storage, in the range of 100,000 bbl (15,899 m^3), this method is the most economical.

XI. PLANNING A PETROCHEMICAL PROJECT

In the present worldwide economic climate involving supply and demand, inflation, recession, high interest rates, governmental controls, profits, losses, energy shortages, feedstock availability, and high prices, it is very difficult to make decisions and recommendations concerning new petrochemical projects. Today most companies are asking more questions about the viability of new ventures and are more critical than in the past of recommendations made by project planners. Management must in general be carefully assured that the proposed project has been tested for sensitivity of the products market, feedstock availability, pricing structure for feedstock and products, environmental effects, and capital-investment requirements.

To meet the needs of petrochemical producers, planning or investment models have been developed that consist of computer capabilities utilizing linear-programming techniques. Objectives are established within economic, technical, and marketing restraints to determine the best operational and processing schemes. Although the most common objective is to maximize profit, other objectives might include minimal use of feedstock, minimal capital investment, and maximal production of petrochemicals.

Complex processing schemes are evaluated to select the optimum process configuration and operating strategy based on specified input parameters that define feedstock availability, product requirements, and investment objectives. Severity levels of operation, crude-oil cut points, product price–volume relationships, and availability of utilities can be quickly evaluated. Such evaluations can be restricted to the olefins unit or can also include downstream processing.

The olefins plant has become a versatile unit for upgrading to premium value what would otherwise be lower valued refinery streams. The scope of some studies may start with crude oil to identify what units should be used, and what products should be produced for sales or as feedstocks for the olefins unit or to the aromatics-conversion units. With heavier crude oils, there is a greater need to process lighter fractions more prudently and to convert the bottom of the barrel into usable feedstocks and fuels. The economics of hydroprocessing of these refinery streams can also be evaluated with respect to conversion of upgrading for olefin-plant feedstocks.

Figure 14 is a typical scheme of processing units that might be used in a planning study. Based on the objectives presented, the study would include those process units necessary for the complex that is being considered, as well as complementary units that might affect the results to be achieved.

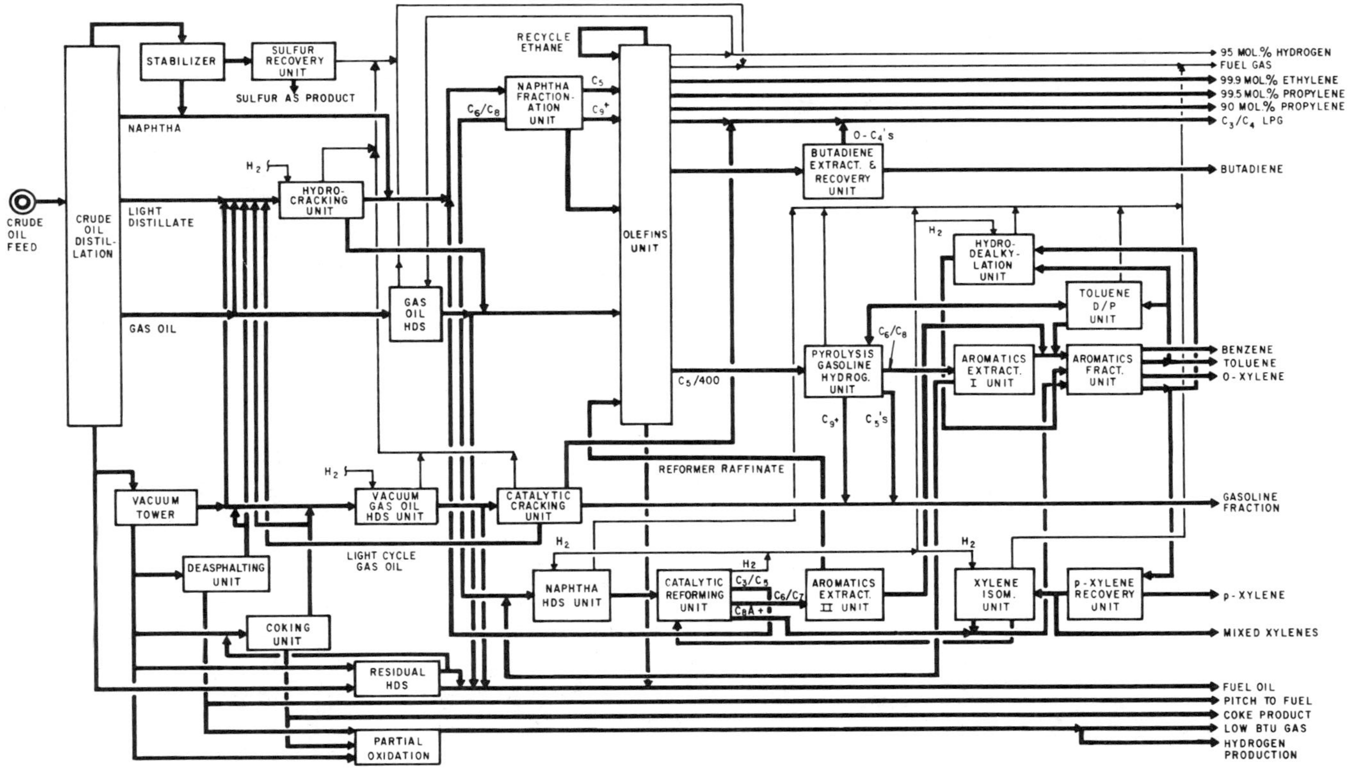

Fig. 14. Petrochemical refinery study flowchart.

At the beginning of the study, there must be agreement among all interested parties as to the general configuration of the system to be used. Feedstock compositions and availability, yields, product battery-limit conditions, product and raw-material costs, process-unit investments, royalties, off-site requirements, catalyst and chemical consumption, utility requirements, and viable market analyses are among the key factors.

Figure 15 is a representation of a stand-alone olefins unit receiving feeds from independent sources.

Figure 16 shows an arrangement of processing units that could be used in studying aromatics production and recovery from pyrolysis systems.

The more common feedstocks for olefins production can be provided by simple distillation of crude oil, but the bottom of the barrel is too large a portion of the crude to be disposed of in the fuels market by the olefins producer. Figure 17 shows alternatives available either for conversion of crude to petrochemical feedstock or for minimizing fuel products.

Because primary olefins and aromatics are items of commerce, many studies stop at that point. When liquid feedstocks are to be cracked, however, the other olefins and aromatic derivatives such as ethylbenzene, styrene, cyclohexane and propylene for acrylonitrile, may be included. Even ammonia production can be included in the study because of the hydrogen produced in the olefins unit, hydrogen produced by steam reforming, or partial oxidation of residual materials.

Independently, or in connection with the olefins study, the economics of the derivatives units can also be included, because recycle streams are often involved. For example, unconverted ethylene from polymerization processes may be recycled to the cracked-gas compressor or to the fractionation train.

Similarly, the hydrogen balance in the ethylene plant may be incorporated into recycle and recovery from degraded hydrogen streams in other units. Because the ethylene plant is both a steam generator and consumer, overall utility balances with derivative plants may have positive, synergistic effects. Interrelationships of fuel consumption among units, for both gaseous and liquid fuels, may develop integrated benefits that would not otherwise result.

Computer models, usually linear, can answer questions presented by a proposed set of conditions, and can also uncover further problems that must be treated before final conclusions can be reached. The initial analysis shows throughput quantities and costs under ideal conditions, displays, and the application of necessary restraining factors. Further computer runs will show more refined results after further restraints, availability of critical materials, and marketability of specific products have all been factored into the study. Subsequent computer results, obtained by adjusting

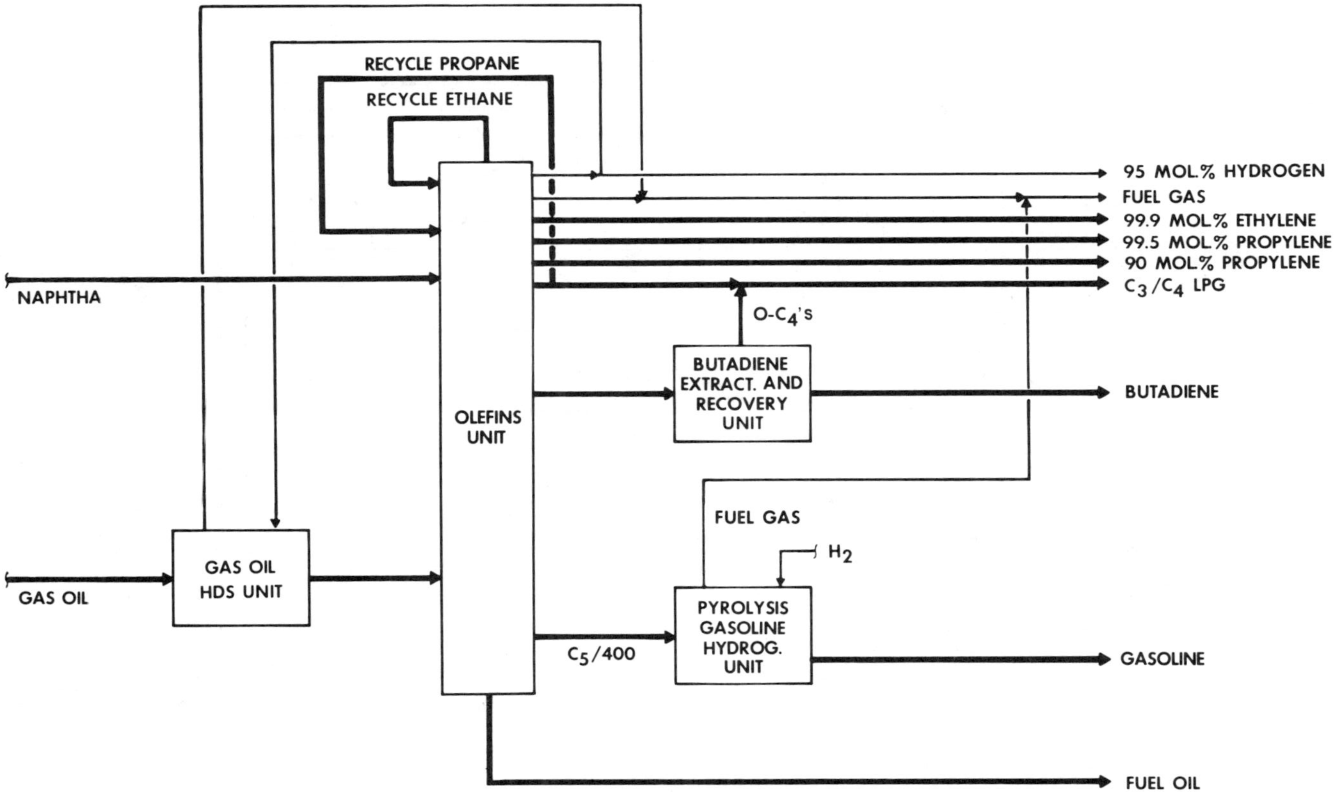

Fig. 15. Flowchart for olefins processing.

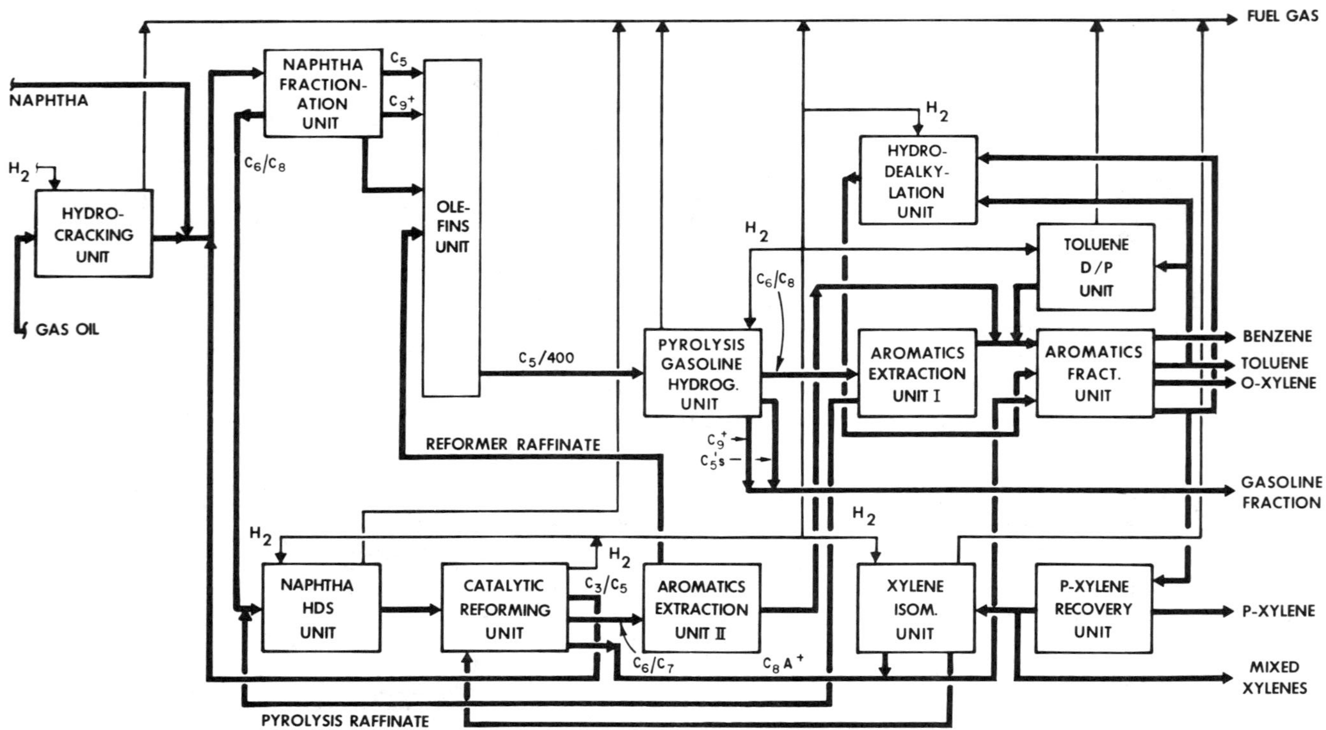

Fig. 16. Flowchart for aromatics processing.

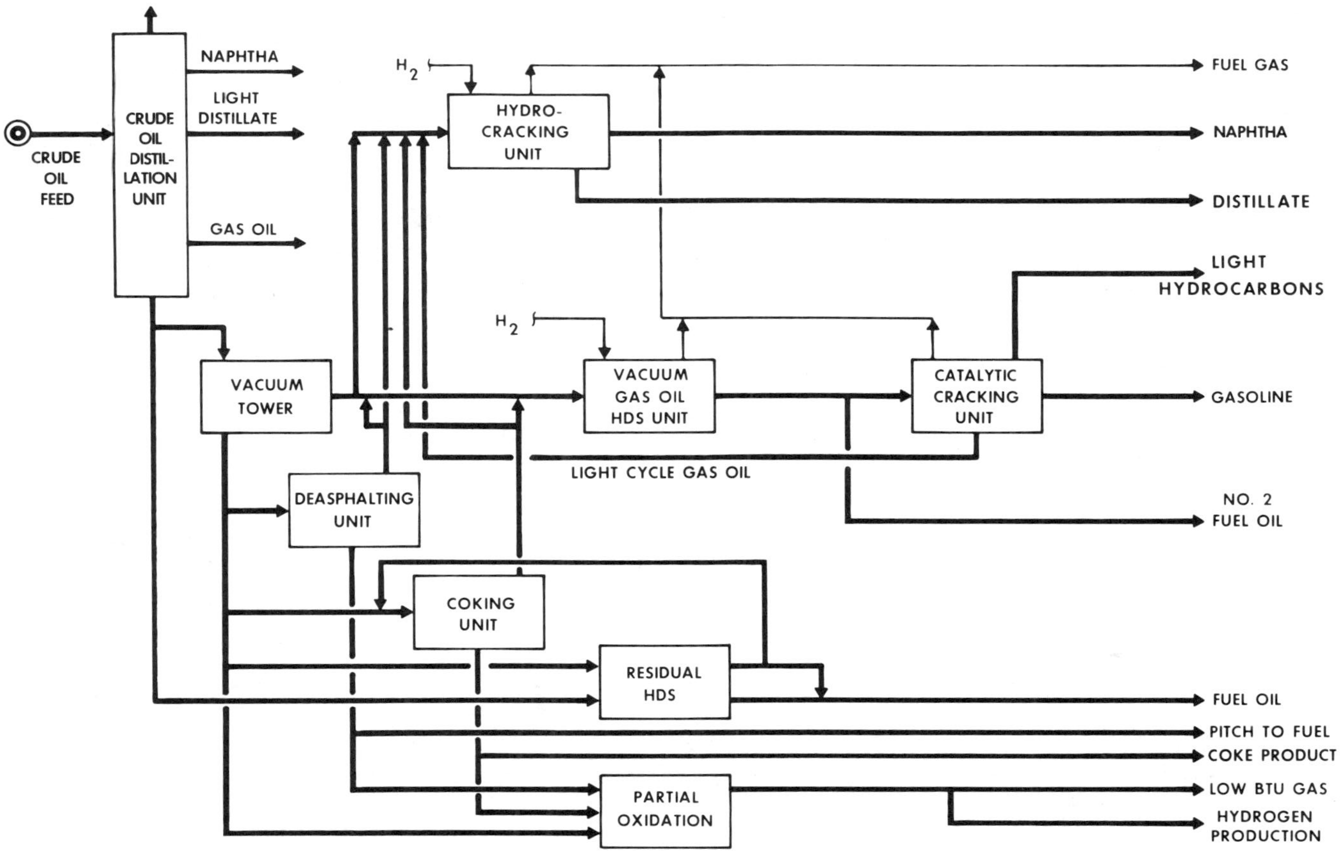

Fig. 17. Flowchart for residual processing.

inputs and objectives and comparing the interim information that has been generated, provide the investor with custom-formulated data. Computer printouts can summarize extensive data into clear statements by means of which intelligent and reliable investment decisions can be made. Operating costs for the production complex, including both variable and fixed costs, may also be clearly presented for evaluation. Sensitivity of optimal solutions can be evaluated with regard to specified feed and product values, unit investments, and volume constraints placed on feeds and/or products.

XII. APPROXIMATE COST OF ETHYLENE PLANTS

Cost estimates for ethylene plants vary widely depending on the scope and definition provided the cost estimator. The preparation of a "definitive cost estimate" is expensive and time-consuming, requiring firm quotations from vendors, knowledge of materials and utility costs, and, most important, an accurate estimate of the field-construction manhours and expenses. To make the cost estimator's problem more difficult, forecasts of the rates of escalation of costs over the purchasing and construction period must be made. In 1973, plant-construction costs increased by about 20% and in 1974, approximately 40% (*Chemical Week*, 1975). Between 1975 and January 1980, escalation has increased at an average rate of approximately 11% per year.

Figure 18 shows a range of inside-battery-limits plant costs for a 1 billion pounds/year (0.4359 billion kg/year) ethylene unit for ethane, propane, normal butane, full-range naphtha, AGO, and VGO feedstocks. A band of plant costs is shown for each feedstock to indicate that variations can occur due to specific plant location, vendor quotations at the time of preparing the cost estimate, client requirements, client product and mechanical specifications, extended construction schedules (which may be affected by equipment deliveries), and expected labor productivity.

The battery limits–approximate plant-cost estimates were developed for a Houston location at 1980 levels, and include all materials and equipment required for the actual processing operations, including the refrigeration systems. Not included in the plant-cost estimates are site preparation, piling (if required), major storage capacity, off-site steam boilers, water-cooling tower, water-treating facilities, flare, waste-disposal systems, product pipelines, office and warehouse buildings, any other facilities normally considered as off-sites, and the cost of land. The plant-cost estimates include all materials and equipment, construction labor, tools, temporary structures, fixed expenses, and engineer–contractor costs for engineering, design, supervision of construction, overhead, and profit. The

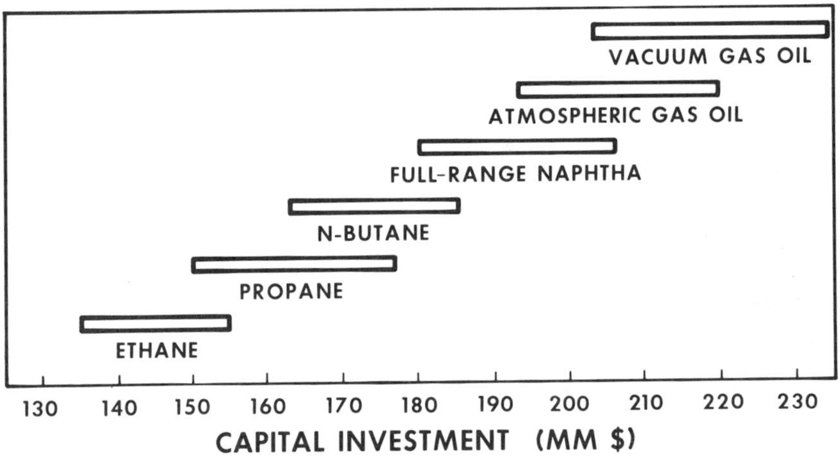

Fig. 18. Bar graph showing approximate comparative ethylene-plant costs in billions (MM) of dollars, for plants using six different feedstocks. Plant capacity: 1 billion pounds/year ethylene. (Based on U.S. Gulf Coast, I.S.B.L., December 1980.)

expenses for construction in the figures reflect "normal" productivity and a 40-hour working week. No allowance is included for premium pay or for diminished productivity deriving from an extended work week and/or shortage of craftsmen.

Many factors affect plant cost, including escalation as discussed earlier. Escalation rates are difficult to apply because the rate is not constant for each item. Other factors that might affect plant cost are feedstock flexibility (2–4%), product-slate flexibility (3–5%), labor costs—union versus nonunion (±10%), environmental factors (10%), energy-saving factors (2–5%), client specifications (2–4%), plant location—site conditions (2–5%), and schedule (2–5%). The indicated factors are typical and they explain why variations in plant-cost estimates are experienced.

In the 1960s, the unit cost of an ethylene plant per ton of ethylene decreased as plant size increased because technology improved at a considerable rate. By the early 1970s, this trend was changing. The economies of scale did not appear to improve materially, and, in fact, gave a negative effect. This result was caused by special problems encountered in large plants, such as the need for more prefabricated large pipe, special valves and fittings, the increased need for field equipment, and the increased construction costs. The increase in the size of ethylene plants is still technically feasible. The technological improvements and large size that helped to reduce ethylene-production costs in the past are likely to be less important in the future than feedstock and capital costs, energy costs, and inflation.

If the off-sites are not defined in detail for developing a cost estimate, an indication of the probable order of magnitude may be of sufficiently accuracy for the preliminary evaluation of a petrochemical project. For a gaseous-feedstock ethylene plant, the installed off-site cost could vary from 15 to 30% of the inside battery-limits plant cost, and for the liquid-feedstock plant, from 25 to 50%. The wide variation in the percentages can be explained, for example, by the type and size of the steam-boiler installation, the size of the water-treatment and cooling-water system, the capacity of the feed and product storage, the degree to which waste-effluent treatment is necessary, the quality and extent of the warehouse, maintenance, office, and laboratory facilities, need for railroad sidings, and need for product pipelines.

XIII. ETHYLENE-PRODUCTION COSTS

Ethylene-production costs are affected by many factors, such as the feedstock and the values of the by-products. In the past it has been easy to demonstrate that with liquid feedstocks (e.g., naphtha), ethylene could be produced at lower cost when cracking at low or moderate severity than when cracking at high severity. This result could easily be accounted for because at the lower severities more feedstock was required; moreover, because the by-products were of a greater value than the feedstock, the ethylene-production cost for a given ethylene-production capacity resulted in attractive economics. This situation was the basis of many early ethylene plants in western Europe and Japan. As naphtha costs increased, cracking furnaces were developed to give higher ethylene yields with less feedstock for a given ethylene production, and still produce ethylene at an attractive price. In the United States, ethylene could be produced at attractive prices because the ethane and propane feedstocks were available at low cost relative to liquid feedstocks, which gave American ethylene producers an advantage over western European producers.

When American ethylene producers were forced to turn to naphtha feedstocks to produce a wider range of products, European plants became more competitive because naphtha feedstocks were relatively less costly than in the United States. This situation, however, has changed because naphtha has become less available in Europe, and its cost is rising rapidly. In the first quarter of 1980, ethylene along the Gulf Coast in the United States was priced at about 24.5¢/pound (54¢/kg), while in western Europe it was approximately 36¢/pound (79.4¢/kg) (*Chemical Week*, 1980).

To provide a comparison of the economics of producing ethylene from two widely different feedstocks, Table VIII tabulates the overall material

TABLE VIII

Overall Material Balances for Ethane and Naphtha Feedstock Ethylene Plants[a]

	MM pounds/year[b]
Ethane feedstock	1244
Products	
Residue gas	153
Ethylene product	1000
C_3 and heavier	91
Total	1244
Full-range naphtha feedstock	3126
Products	
Residue gas	521
Ethylene product	1000
Propylene product	420
C_4 Product	
Butadiene	141
Other C_4's	108
C_5–400°F	741
Cracked fuel oil	195
Total	3126

[a] Capacity: 1 billion pounds/year ethylene.
[b] Abbreviation: MM, billions.

balance for an ethylene plant rated at 1 billion pounds/year (0.454 billion kg/year) of ethylene from ethane at 60% conversion and full-range naphtha at high severity. On a weight basis, about 2.5 times as much naphtha feedstock is required as compared to ethane to produce 1 billion pounds/year (0.454 billion kg/year) ethylene. The by-products also obviously play an important role in the economics of a naphtha-feedstock plant.

Table IX provides the values used in calculating the ethylene-production costs from ethane and naphtha feedstocks. These data are considered as average values, and the results of the calculations, presented in Tables X and XI, show that ethylene in the United States can be produced at a lower cost today from ethane than from naphtha. The ethane-feedstock plant has a production-cost advantage of 5.8¢/pound (12.5¢/kg), of which 4.9¢/pound (10.8¢/kg) is derived from a lower net raw-material and utilities cost. It is now understandable why many naphtha-feedstock plants that can obtain ethane, propane, or *n*-butane at artificial prices relative to crude-oil prices are modifying their operations to use these lighter feedstocks to a maximum.

TABLE IX

Data Used in the Development of Ethylene-Production Costs

	Costs	
Ethane		
¢/pound	8.0	(17.6¢/kg)
¢/gallon (3.11 pounds/gallon)	24.9	($65.78/$m^3$)
Full-range naphtha		
¢/pound	13.0	(28.7¢/kg)
¢/gallon (5.86 pounds/gallon)	76.2	($201.30/$m^3$)
AGO		
¢/pound	10.82	(23.9¢/kg)
¢/gallon (7.04 pounds/gallon)	76.2	($201.30/$m^3$)
By-product value		
Residue gas	$3.50/MM BTUs	($3.31/GJ) (LHV)
Propylene	16¢/pound	(35.3¢/kg)
Butadiene	28¢/pound	(61.7¢/kg)
Other C_4's	15¢/pound	(33.1¢/kg)
C_5–400°F (204.4°C)	16¢/pound	(35.3¢/kg)
Cracked fuel oil	8¢/pound	(17.6¢/kg)
Utilities		
Fuel gas	$3.50/MM BTUs	($3.31/GJ) (LHV)
High-pressure steam	$6.00/M pounds	($13.2/M kg)
Electricity	3.8¢/kw	
Circulating cooling water	5¢/M gallons	(1.3¢/m^3)
Operating cost		
Labor	$20,000/year/person	
Ethane	7 persons/shift	
Full-Range Naphtha	9 persons/shift	
Atmospheric Gas Oil	9 persons/shift	
Supervision	$30,000/year	
Maintenance (Material and labor)	4% Battery-limits plant cost	
Overhead expenses		
Direct overhead	30% (Labor + supervision)	
General plant overhead	65% Operating cost	
Insurance and property taxes	2% Battery-limits plant cost	
Depreciation	10% Battery-limits plant cost	
Interest	12% Working capital	
Ethane battery-limits plant cost	$135 MM	
Full-range naphtha battery-limits plant cost	$180 MM	
AGO battery-limits plant cost	$193 MM	
Working capital	10% Battery-limits plant cost	

TABLE X

Ethylene-Production Cost for an Ethane-Feedstock Ethylene Plant[a]

	¢/pound of C_2H_4	¢/kg of C_2H_4	% of C_2H_4 Production cost
Raw-materials cost: feedstock, catalysts, and chemicals	10.04	22.13	
By-product credit	3.33	7.34	
Net raw-materials cost	6.71	14.79	52.17
Utilities cost: fuel, steam, power, and cooling water	3.36	7.40	26.09
Operating cost: labor, supervision, and maintenance	0.60	1.32	4.66
Overhead expenses: direct, general plant, insurance, taxes, depreciation, and interest	2.20	4.85	17.08
Total C_2H_4-production cost	12.87	28.36	100.00
20% Return on battery-limits plant cost	2.70	5.95	
C_2H_4-Transfer price	15.57	34.31	

[a] Capacity: 1 billion pounds/year (0.454 billion kg/year) ethylene. Stream factor: 8000 h/year.

TABLE XI

Ethylene-Production Cost for a Naphtha-Feedstock Ethylene Plant[a]

	¢/pound of C_2H_4	¢/kg of C_2H_4	% of C_2H_4 Production cost
Raw-materials cost: feedstock, catalysts, and chemicals	40.75	89.84	
By-product credit	29.93	65.98	
Net raw-materials cost	10.82	23.86	57.83
Utilities cost: fuel, steam, power, and cooling water	4.17	9.19	22.29
Operating cost: labor, supervision, and maintenance	0.80	1.76	4.28
Overhead expenses: direct, general plant, insurance, taxes, depreciation, and interest	2.92	6.44	15.60
Total C_2H_4-production cost	18.71	41.25	100.00
20% Return on battery-limits plant cost	3.60	7.94	
C_2H_4-Transfer price	22.31	49.19	

[a] Feedstock: full-range naphtha. Capacity: 1 billion pounds/year (0.454 billion kg/year) ethylene. Stream factor: 8000 h/year.

Atmospheric gas oil is now recognized as a viable ethylene-plant feedstock and can be competitive with naphtha. In the mid-1970s, when naphtha was quoted at 28.4¢/gallon (4.85¢/pound; 10.7¢/kg) and AGO at 31.6¢/gallon (4.49¢/pound; 9.9¢/kg), ethylene could be produced from naphtha at 9.90¢/pound (21.8¢/kg) and from gas oil at 10.31¢/pound (22.73¢/kg). For this period, by dropping the gas-oil price by 0.8¢/gallon or 0.11¢/pound (0.243¢/kg), ethylene-production cost was made equal for naphtha and AGO. Clearly, when naphtha prices rise at a faster rate than gas-oil prices, those ethylene plants that have been designed to accommodate gas-oil feedstocks may be capable of attaining lower ethylene costs.

Table XII tabulates the overall material balance for an ethylene plant rated at 1 billion pounds/year (0.454 billion kg/year) of ethylene when using AGO pyrolyzed at high severity. Table XIII, which was developed assuming that the AGO price was equal to the naphtha price on a volumetric basis, shows an ethylene-production price of 14.02¢/pound (30.91¢/kg), which is lower than that produced from naphtha. The present results are different from the comparison made in the mid-1970s because even though the volumetric-feedstock prices are equal, the cents per pound of feedstock are quite different. Atmospheric gas oil costs 2.18¢/pound (4.81¢/kg) less than naphtha. Another difference that favors the AGO feedstock is the value selected for each by-product. Because the quantity of by-products is greater

TABLE XII

Overall Material Balance for an AGO-Feedstock Ethylene Plant[a]

	MM pounds/year[b]	MM kg/year
AGO feedstock	3646	1654
Products		
Residue gas	479	217
Ethylene product	1000	454
Propylene product	474	215
C_4 Product		
Butadiene	161	73
Other C_4's	137	62
C_5–400°F	709	322
Cracked fuel oil	686	311
Total	3646	1654

[a] Capacity: 1 billion pounds/year (0.454 billion kg/year) ethylene.

[b] Abbreviation: MM, billions.

TABLE XIII

Ethylene-Production Cost for an AGO-Feedstock Ethylene Plant

	¢/pound of C_2H_4	¢/kg of C_2H_4	% of C_2H_4 Production cost
Raw-materials cost: feedstock, catalysts, and chemicals	39.61	87.32	
By-product credit	34.84	76.81	
Net raw-materials cost	4.77	10.51	34.04
Utilities cost: fuel, steam, power, and cooling water	5.27	11.62	37.62
Operating cost: labor, supervision, and maintenance	0.85	1.87	6.07
Overhead expenses: direct, general plant, insurance, taxes, depreciation, and interest	3.12	6.88	22.27
Total C_2H_4-production cost	14.01	30.88	100.00
20% Return on battery-limits plant cost	3.86	8.52	
C_2H_4-Transfer price	17.87	39.40	

[a] Capacity: 1 billion pounds/year (0.454 billion kg/year) ethylene. Stream factor: 8000 h/year.

for the AGO, the by-product credit is larger and results in a cost advantage of 4.91¢/pound (10.82¢/kg) for ethylene. The example assumed equal volumetric-feedstock prices for naphtha and AGO, but it does not reflect the fact that current naphtha prices are higher than those of AGO.

XIV. FACTORS INFLUENCING COST OF ETHYLENE PRODUCTION

As mentioned in the previous section, several factors affect the cost of ethylene production. The type of feedstock, its price, and the selling price for by-products all have an important influence on cost of ethylene production; however, plant cost, utilities cost, and plant-operating capacity are also important in determining the cost. The various cost factors are worthy of brief comment.

A. Effect of Feedstock Cost

The largest component in the cost of ethylene production is the net raw-material cost, which is essentially the feedstock cost minus credits for by-products. Table XI was recalculated by increasing the price of a given naphtha feedstock from 13¢ to 14¢/pound (28.7¢ to 30.9¢/kg). This increase of 1¢/pound (2.2¢/kg) gave an ethylene-production cost of 21.84¢/pound (48.15¢/kg), which is 3.13¢/pound (6.90¢/kg) greater than the production cost shown in Table XI. This same differential in production cost, but negative, applies if the naphtha price is reduced by 1¢/pound (2.2¢/kg). This calculation emphasizes that the cost of feedstock is very important. This example was based, however, on changes in feedstock prices for feedstocks with identical characteristics. It could prove economical, for example, to buy a feedstock at a slight increase in cost if it has the potential to produce a greater yield of valuable olefins. The major point is that not all liquid feedstocks are of the same composition and some are more suited than others for production of olefins.

B. Effect of By-Product Prices

The production of ethylene and by-products formed by pyrolyzing a naphtha feedstock can be viewed as a means of upgrading a common hydrocarbon resource into more valuable products. Normally, the production cost for ethylene is greatly influenced by the amount and pricing of the by-products.

Although the following calculations are artificial, they were made by pricing all by-products such as propylene, butadiene, other C_4's, and the C_5–400°F pyrolysis gasoline at a fuel value of $3.50 per million BTUs ($3.31 per GJ) (LHV) and developing a new cost for production of ethylene. These calculations yield a disastrous set of economics because the ethylene-production cost in Table XI increases by 14.79¢/pound (32.61¢/kg) of ethylene, to give a production cost of 33.50¢/pound (73.86¢/kg). No current plant could afford to operate with such economics. The result emphasizes that the marketing department of any ethylene plant must find an assured market for both the ethylene and its by-products.

C. Effect of Plant Cost

Consider the naphtha case of Table XI. Increasing the plant cost by 10% from $180MM to $198MM, correcting those items that are a function of plant cost, and leaving all other items untouched, increase the ethylene-production cost from 18.71¢/pound (41.25¢/kg) to 19.07¢/pound (42.0¢/kg). The incremental increase is 0.36¢/pound (0.79¢/kg), which gives a cost of about 1.9% greater than the base case. If the ethylene selling price were fixed at 22.31¢/pound (49.19¢/kg), the increase in production cost would result in a decrease in revenue of $3.6MM/year.

D. Effect of Utilities Cost

If the naphtha-feedstock case is used as an example, implementation of various energy-saving features into the plant design to reduce the energy consumption by 20% and simultaneously to increase the capital cost by $9.0MM gives a production cost for ethylene of 18.03¢/pound (39.75¢/kg). This figure is a reduction of 0.69¢/pound (4.52¢/kg) for ethylene over the base case of 18.71¢/pound (41.25¢/kg) in Table XI. Such a reduction is attractive because it represents an annual savings of $6.9MM/year and a return of the capital expenditure in 1.3 year.

E. Effect of Plant Operating Capacity

If a process plant operates at partial capacity, it will not generate revenue at its design potential. When the ethylene plant is operated at reduced capacity because of reduced sales, certain costs remain constant and are independent of operational capacity. Certain utility usages are not proportional to the capacity of the plant's operation. Therefore, the utility costs would vary. The following tabulation shows that with an ethylene-transfer price of 22.31¢/pound (49.19¢/kg) taken from Table XI, the cost of ethylene production is at break-even when the ethylene plant is operated at about 60% of design capacity.

Operating capacity (% of design)	C_2H_4 Production cost	
	(¢/pound ethylene)	(¢/kg ethylene)
100	18.71	41.25
70	20.30	44.75
60	21.90	48.28
50	24.10	53.13

This particular exercise also illustrates that when future overcapacity is included in the design of an ethylene plant, the period to absorb the overcapacity should not be of too long duration.

REFERENCES

Brandt, D. E. (1979). *Oil Gas J.* **77**, 51–56.
Chemical Week (1975). **117**, 37.
Chemical Week (1980). **126**, 44–51.
Green, R. B. (1969). "Vacuum Gas Oil Cracking." Exxon Chem. Co., Florham Park, New Jersey.

Grimes, C. R. (1980). *ROC–USA Econ. Conf. Petrochem. Sect. Meet., Taipei.*

Korosi, A., Woebcke, H. N., and Virk, P. S. (1976). *Prepr. Pap.—Am. Chem. Soc., Div. Fuel Chem.* **21**(6), 190–197.

Korosi, A., Hallee, L. P., and Virk, P. S. (1980). *Prepr., Div. Pet. Chem., Am. Chem. Soc.* **25**, 621–631.

Schultz, E. B., Jr., Guyer, J. J., and Linden, H. R. (1955). *Ind. Eng. Chem.* **47**, 2479–2482.

Zdonik, S. B., Green, E. J., and Hallee, L. P. (1968). *Oil Gas J.* **66**, 91–94.

16

Materials of Construction in Ethylene Pyrolysis-Heater Service

S. IBARRA

I. INTRODUCTION

Operating temperatures of pyrolysis furnaces are sometimes in excess of 2000°F (1100°C). These temperatures are very detrimental to the life of the typical HK-40 furnace tubes, which normally have a 3- to 5-year life in the hot section of these furnaces. Short life is attributed to rapid carburization of internal-diameter (ID) surfaces, which subjects tubes to higher than normal stresses and results in creep cracking of furnace tubes.

As an aid to understanding the materials problems, the ethylene process will be presented, along with data on the carburization of furnace tubes.

II. DESCRIPTION OF THE PROCESS

Material problems encountered in petrochemical plants usually center on the ethylene-pyrolysis heater, which is the most important component of the thermal-cracking process. Figure 1 shows a typical upper convection

Pyrolysis: Theory and Industrial Practice

ISBN 0-12-048880-9

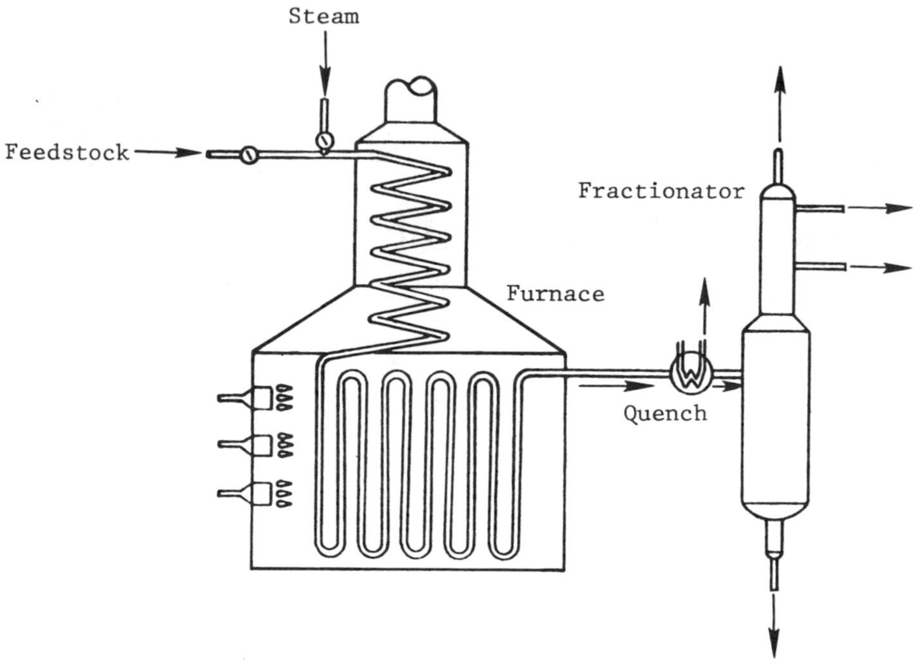

Fig. 1. Typical flow path for steam-cracking process.

or preheat section, and a lower radiant or cracking zone. Burners are provided on the sidewalls or on the bottom of the furnace to supply the heat. The number of burners is dependent on the heat required and may vary considerably.

The furnace contains several passes in parallel, and each pass contains a number of connected tubular members or furnace tubes that provide a flow path through the convection section and into the cracking section. Petroleum feedstock such as naphtha or ethane is supplied to the steam cracker; steam is mixed with the feed through an inlet line before entering the convection section.

Radiant coils in the furnaces are subjected to metal temperatures of approximately 2000°F (1100°C) and usually have about a 3-year life in the hottest section. The usual feedstocks are highly carburizing and cyclic—conditions that exist contribute to a decrease in coil life.

The high-temperature applications create problems with creep, which is accelerated by the carburization and low-cycle fatigue brought on by the cyclic conditions from temperature variations. Residence times in the section are carefully controlled to minimize polymerization and other undesirable reactions. Residence times in the cracking zone will range from about 0.1 to 10 s. Pressures within the tubes may range from 0 to 50 psi (0 to 345 kPa), but are not critical.

Reaction products are discharged from the coil into a header, from which they are transferred into a quench area to stop the cracking reaction promptly, and to prevent or minimize side reactions. Quenching agents such as higher boiling hydrocarbons and/or water can be supplied through a control valve. Alternately, the reaction may be stopped by cooling the furnace effluent in an exchanger while generating steam directly or indirectly. The mixture of quench-reaction products and quenching agent is discharged into the fractionating tower. An aromatic tar is withdrawn from the bottom of this tower, and the product is taken overhead (Ibarra, 1981).

Probably the most objectionable drawback relating to carbon formation is the deposition of coke on the interior of the tubes through which the cracking mixture flows. The coking rate, which varies with the type of feed employed, is continuous. As the coke builds up, it reduces the effective cross-sectional area of the tube, necessitating higher pressures to maintain a constant throughput.

Also, coke is an excellent thermal insulator, and its formation on tube walls must be accompanied by a sharp increase in furnace-tube temperatures to maintain cracking efficiency. Periodically, a furnace must be shut down for decoking. This involves elimination of the feedstock and application of a steam–air mixture. The result is a flame front in the coil, which spalls and burns off the coke, and prepares the tube for another cracking cycle.

It is believed that the decoking reaction results from an interaction of steam and coke, according to the following equation known as the water–gas reaction:

$$H_2O + C \longrightarrow CO + H_2$$

At steam-cracking temperatures (1650°F), this reaction is quite slow and would not be expected to be economically feasible as a decoking method. The decoking temperature is usually raised above 1850°F, and the reaction causes tube-metal temperatures of about 2000°F. It is believed that as coke forms on the tube surfaces there is some diffusion between the coke and the furnace tube. Some coke goes into solid solution in the metal tube, while trace amounts of metals diffuse into the coke layer. Several samples of coke were examined in earlier experiments and the results showed evidence of chromium, nickel, and iron in the coke deposits. These trace amounts of metals are believed to catalyze the water–gas reaction, allowing it to proceed at favorable rates. Our decoking procedures were conducted using steam and air (Ibarra, 1980). This allowed better burnoff without leaving fine particles of coke in the system. Steam decoking without air has been used in the field, but fine coke particles are not burned off. These particles travel through the tube and cause erosion at the U-bends.

III. MATERIALS FOR ETHYLENE-PYROLYSIS SERVICE

Heater coils, headers, and other components for pyrolysis furnaces are designed for cast heat-resistant alloys containing basically chromium, nickel, and iron. Practically all commercial methods of fabrication are employed in making these coil-system components—extrusion, static, casting, and centrifugal casting.

For straight tubular portions of the coils, tubes are usually made from centrifugal castings. Both static and centrifugal castings contain a substantial amount of carbon (>0.3%), which imparts higher temperature properties. Centrifugal castings have greater creep-rupture strength than static castings because they have a sounder structure that is more uniform and free from porosity. The centrifugal-cast tubes are produced in a metal mold spinning at speeds sufficient to impart a force of about 75 to 100 times that of gravity to the molten metal in the mold. The high spinning speed and the casting conditions employed result in almost unidirectional solidification and a high-quality product. In centrifugal castings, the ID region is the last to solidify, and it contains shrinkage porosity. This means that low-density inclusions and dissolved gases are rejected into the last metal to solidify near the ID, leaving the remainder of the casting sound and of good quality. It is important to machine the ID to remove about 3/32 in. (2.4 mm) of metal. Defects here act as a catalyst for carburization and reduce tube life by subjecting the remainder of the tube to higher than normal stresses, which causes creep cracks.

At the present time, HK-40 is the most commonly used alloy for ethylene-pyrolysis heater-tube application. It is a 25% Cr, 20% Ni stainless-steel casting having a carbon content of 0.35 to 0.45%. HP-50 and some of its variations are being used when higher process temperatures require a higher carburization resistance. HP-50 is a 25% Cr, 35% Ni stainless-steel casting with 0.5% carbon. Thermalloy 63CW is also being used for this application. This alloy is known as Supertherm and contains 28% Cr, 35% Ni, 5% W, 15% Co, and 0.50% carbon. Thermalloy 63CW has exhibited better carburization resistance than HK-40 and HP-50, but its cost is high.

The microstructure of the as-cast HK-40 alloy consists of an austenitic matrix with eutectic carbides in the grain boundaries. The austenitic matrix contains appreciable dissolved carbons, and during heating at temperatures above 900°F, chromium carbides precipitate within the austenitic matrix. The precipitation, occurring first adjacent to the eutectic carbides, is exceedingly fine. During long-term holding at about 1600°F, carbide precipitation becomes general throughout the matrix and the precipitated carbides coalesce. Above 1800°F, precipitation and coalescence occur more rapidly and, if the temperature is sufficiently high, some solutioning of the coalesced carbide particles occurs. Carbon in the form of carbides is the principal strengthener and is responsible for good creep and rupture strength at temperatures above 1200°F (650°C). These properties are attributed to the presence of very fine stable precipitated carbides within the matrix of the alloys. When these fine particles coalesce and dissolve, the material's high-temperature strength decreases (Jackson, 1973).

Although HK-40 is the most common material used, selection for this service should be based on carburization resistance, creep-rupture strength, ductility, weldability, and cost.

IV. CARBURIZATION RESISTANCE

Pyrolysis heater tubes are subjected to a variety of high-temperature corrosion environments that can cause carburization. Corrosion weakens structural components by consuming the stress-carrying ability of the material and elevating the stresses present in the process. It also creates surface discontinuities that act as stress raisers under cyclic loading or thermal shock.

Carburization has been a major corrosion problem in high-temperature pyrolysis service. The phenomenon occurs through a reaction between the gaseous phase and the base metal if the protective films normally found on these stainless steels are destroyed and cannot regenerate. The presence of stable oxide scales has been shown to retard the carburization of these heat-resisting alloys at high temperatures. The chromium oxide produced on an alloy exposed to a carburizing–oxidizing environment provides a barrier to carbon transport while the scale remains intact. Because these oxide films are unable to exert a protective effect, there is a diffusion of carbon into the base metal and metal diffuses toward the surface. Carbon combines with chromium and forms chromium carbides in the matrix. Loosening of the grain structure also permits penetration of the carbon into grain boundaries (Muller, 1968).

Research has been done at Gulf Science and Technology Company to analyze the carburization characteristics of HK-40 and to determine how the material reacts to simulated furnace environments. The tests were carried out in a laboratory gas-pyrolysis cracker as shown in Fig. 2

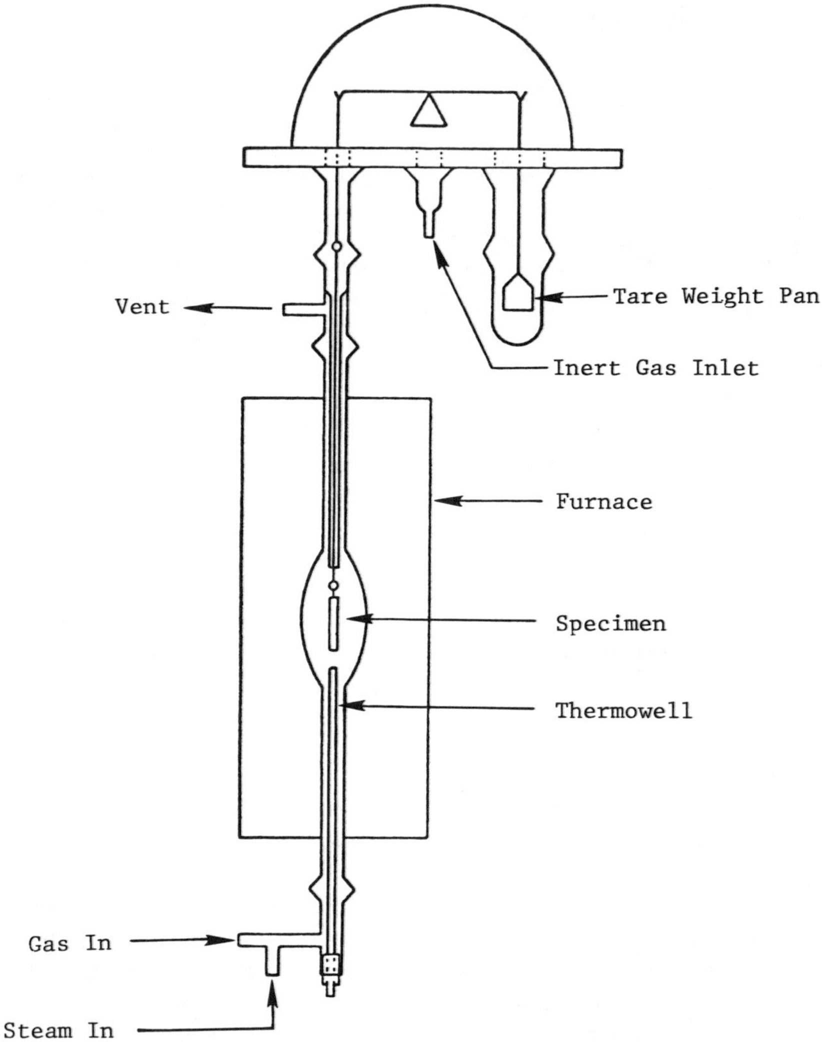

Fig. 2. Schematic diagram of a laboratory gas-pyrolysis cracker. This material is suspended from one end of a microbalance arm in the cracking tube so that changes in weight can be monitored.

(Ibarra, 1977). The furnace-tube material was placed in the reaction chamber to determine how it would react to simulated furnace environments. The tube material was machined to form a $1/4 \times 1/4 \times 1\frac{1}{2}$-in. sample with a 1/8-in. diameter hole drilled through the length. The tube material was suspended from one end of the microbalance arm in the cracking tube so that weight gains or losses could be monitored. During the experiment, weight gains were the result of surface coke formation that occurred during the cracking cycle. During the steam–air decoking cycles, weight losses due to the coke burnoff could be observed. Carburization of the tube material was detected by microstructural observation.

A typical cracking cycle was performed with 75% ethane containing 200 ppm H_2S plus 25% steam. A cracking cycle operating at 1950°F usually lasted 30 h and resulted in weight gains of 200–600 mg. The typical decoking cycle lasted 4 h or until the weight loss had stabilized. From our experiments we have drawn the following conclusions.

1. Carburization occurs more severely during the decoking cycle. Samples examined metallographically after a single cracking cycle had a small amount of carburization. Samples that had been through both the cracking and decoking cycles displayed a dramatic increase in carburization.
2. Furnace-tube temperature affects rate of carburization, but microstructural observation indicated that thermal aging of the material did not appear to affect the rate of carburization.
3. Reduction in the percentage of air during steam–air decoking will reduce the amount of carburization, but further tests are needed to establish optimum steam/air ratios.
4. Aluminized (diffusion) coatings develop a protective oxide film that can reduce carburization, but it does not eliminate all carburization.
5. Silicon content appears to have a strong influence on resistance to carburization. It may be beneficial to raise the lower limit of silicon in our specifications to improve the carburization resistance of HK-40.

V. CREEP-RUPTURE STRENGTH OF MATERIALS IN SERVICE

Creep can affect coil systems in a pyrolysis furnace by relaxing induced stresses (stress relaxation), by distorting the piping configuration, and by causing creep rupture. Stress relaxation by creep is beneficial in that it prolongs service life by enabling very high initial stress levels to be safely tolerated. However, if creep causes distortion in the piping configuration, it can have a deleterious effect because it can upset the balance between the support forces and the pipe weight.

Many commonly used heat-resistant alloys have a creep-rupture strength of a few thousand pounds per square inch at 1700°F. This usually allows a margin of safety, and many stress problems can be tolerated or ignored. However, when the temperature increases to 2000°F, creep-rupture strength diminishes to a few hundred pounds per square inch, so stress problems have to be considered more carefully.

TABLE I

Comparative Creep-Rupture Strengths for Three Different Pyrolysis-Tube Alloys[a]

Alloy	Time to rupture (h)
HK-40	
1800°F	200
2000°F	10
HP-50	
1800°F	350
2000°F	10
Supertherm (Thermalloy 63CW)	
1800°F	10,000
2000°F	80

[a] Two different temperature conditions were tested; a constant pressure of 4000 psi was maintained.

As mentioned previously, carburization attacks the ID of pyrolysis tubes, and the depth of carburization can be as much as one half the wall thickness. Carburized metal will tend to increase in volume and has a higher thermal-expansion coefficient. This results in tensile stresses on the uncarburized layer, making the tube more susceptible to creep-rupture failure. It is important to have a material that not only is resistant to carburization but also has sufficient creep-rupture strength to withstand the applied stresses when carburization does occur.

Table I is a comparison of the creep-rupture strengths of HK-40, HP-50, and Supertherm (Thermalloy 63CW); it shows the latter to be superior.

These properties are important because they show that even though a material is carburized, it can remain in service longer even under the high tensile stresses imposed by the carbon's volumetric expansion if the remaining uncarburized wall has strong creep-rupture properties (Jackson, 1973).

VI. DESIGN FOR IN-SERVICE DUCTILITY

Although carbon is a potent strengthener, it reduces room-temperature ductility and toughness. *Ductility* can be defined as the ability to strain before fracture and is commonly measured in terms of tensile elongation, reduction in area, or impact energy. In the as-cast condition, these alloys usually exhibit at least 15% tensile elongation and about 5 to 10 ft/pound in the Charpy V-notch impact test.

After exposure at temperatures of 1200 to 1800°F, the values for these properties fall to 2% tensile elongation and 2 to 3 ft/pound Charpy V-notch impact energy. Ductility is important in designing pyrolysis coils because after high-temperature service the coils will have low ductility either when cold or when relatively hot. These low ductilities are important to consider because they indicate low resistance to thermal shock and to cyclic thermal-stress fatigue.

The ductility of these heat-resistant alloys is strongly influenced by carbon content. Ductility could be increased by lowering carbon content and by using HK-20 instead of HK-40. However, this would result in a reduction in the creep-rupture strength and is not recommended.

VII. WELDABILITY PROBLEMS

Welding of as-cast HK-40 and HP-50 is considered routine. The welding of Supertherm (Thermalloy 63CW) is a little more difficult but is being performed successfully in various plants (Krysiak, 1979). The weldability of these cast heat-resistant alloys is particularly important in pyrolysis service because the materials are subjected to such severe carburization and because weldments are targets of a combination of stress concentrations, thermal stress, low ductility, and residual stresses.

After these alloys have been in service, welding is more difficult because the thermal aging has reduced the ductility of the casting, thus making it more crack sensitive. The problems are compounded when welding a new as-cast tube section to an aged and carburized tube section, as is done frequently in repairing a failed section of a coil. Welding is usually with a gas tungsten-arc root pass to fuse the lands and subsequent cover passes of matching filler metal. It is necessary to remove the carburized ID surface of the older section of the coil adjacent to the proposed weld area. It may be necessary to perform a solution anneal at this proposed weld area if welding is not successful. This solution anneal will dissolve the carbides and thus increase the ductility.

Solution annealing is not a simple operation because slow cooling from the solution temperature of 2200°F (1205°C) will produce a similar low-ductility alloy. The tubes must be allowed to air cool to 1100°F (595°C) after solution annealing to retain the needed ductility.

Good-quality welds can be assured by taking the following precautions:

1. Removing damaged material and grinding the carburized ID around the weldment area
2. Solution annealing to improve ductility, especially for the more difficult Supertherm (Thermalloy 63CW)

3. Using thoroughly qualified welders and specifying welding procedures
4. Providing gradual transitions at welded joints to avoid sharp changes in the cross section that could increase residual stresses
5. Using gas tungsten-arc welding for the root pass with inert gas backup, and completing with matching filler wires
6. Inspecting the root pass, using dye-penetrant techniques
7. Smoothing contours and removing surface irregularities by grinding

VIII. SOME CONCLUSIONS REGARDING MATERIAL SELECTION WITH THESE ALLOYS

The heating-coil system in an ethylene-pyrolysis furnace can be regarded as a complex dynamic piping system made of heater tubes, return bends, inlet and outlet headers, and supports, For pyrolysis service, these systems must be able to cope with a variety of operating problems, such as (*a*) weakening of material due to carburization and subsequent high stresses induced in coils; (*b*) thermal fatigue resulting from aging, from cycling between pyrolysis cracking and decoking operations, and from startup shutdown; (*c*) stresses induced by uneven temperature distribution along the length of the heater coils, or by coke buildup inside heater coils and dead pockets; and (*d*) localized overheating such as in heavily coked tubes.

With these problems, it is essential that the material selection be specifically designed for each application. For instance, it may seem that Supertherm (Thermalloy 63CW) is the best alloy for a particular case because of its high creep-rupture strength. However, weldability after long-term service must be taken into consideration. The cost of using such an alloy must also be taken into account because one must determine whether to provide a longer lasting but more expensive material, or to use a less expensive material with a more costly maintenance problem.

REFERENCES

Ibarra, S. (1977). Rep. No. JF00. Technol. Mater. Dep., Gulf Sci. Technol. Co., Pittsburgh, Pennsylvania.

Ibarra, S. (1980). *Met. Prog.* 62–66.

Ibarra, S. (1981). Rep. No. 560RM151, File JD00. Technol. Mater. Dep., Gulf Sci. Technol. Co., Pittsburgh, Pennsylvania.

Jackson, R. J., Van Echo, J. A., and Roach, D. B. (1973). Report, July. Battelle Meml. Inst., Columbus, Ohio.

Krysiak, K. F. (1979). *Corrosion/79, Atlanta, Ga.* Pap. No. 127.

Muller, K. (1968). *Nickel-Ber.* by International Nickel Co..

17

Process Control of Pyrolysis Systems

FRANK W. TSAI

I. INTRODUCTION

A hydrocarbon-pyrolysis system usually consists of a cracking furnace where feed is heated and converted into a mixture containing mostly olefins, and aromatics. The cracking furnace is followed by a quenching device consisting of a transfer-line exchanger (TLE) and/or oil quench where the furnace effluent is quickly cooled below the cracking temperature. The control of such a system generally satisfies certain objectives such as maximizing production and minimizing cost. The operation of a modern pyrolysis system is sufficiently complicated to justify an on-line computer-control system.

Pyrolysis: Theory and Industrial Practice

ISBN 0-12-048880-9

II. CONVERSION AND SELECTIVITY

In the cracking of hydrocarbons, the two principal dependent variables are conversion and selectivity. *Conversion* is a measure of the depth of cracking as indicated by the percentage disappearance of the feedstock. *Selectivity* indicates the quantity of desired products, such as ethylene and propylene, that are obtained per unit of feedstock material converted.

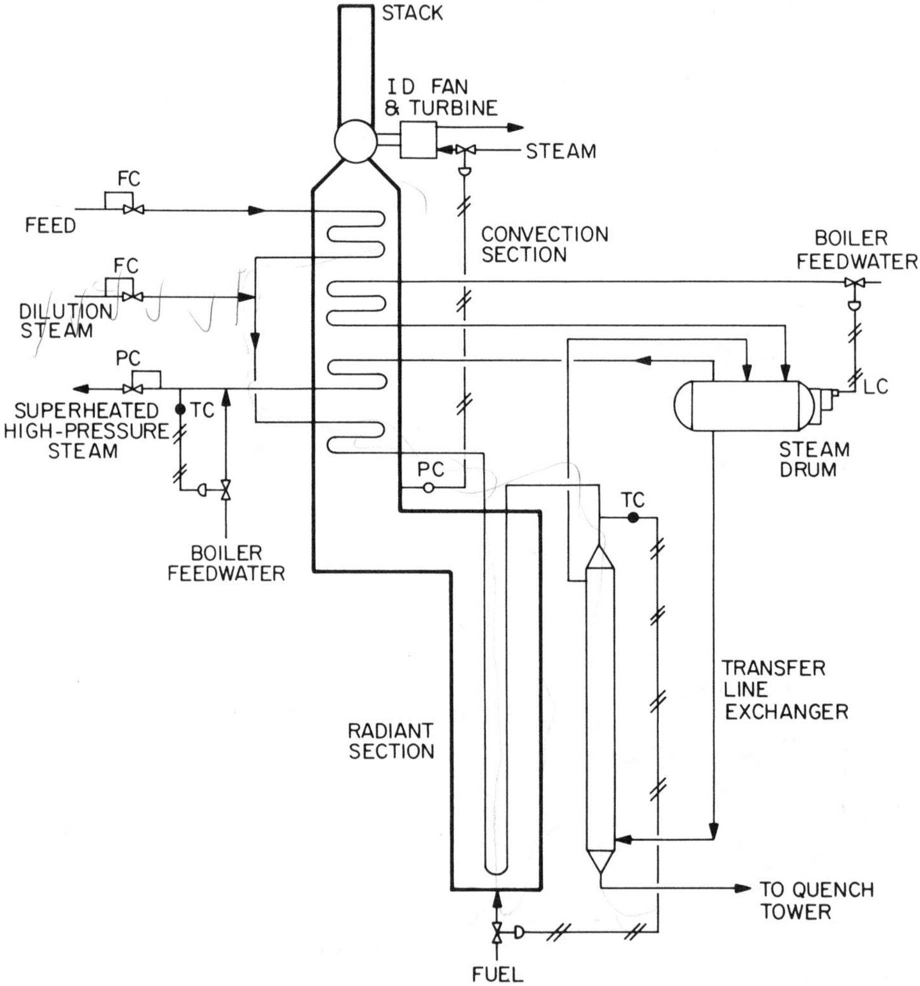

Fig. 1. Furnace-control scheme. ID, Induced draft; FC, flow controller; LC, level controller; PC, pressure controller; TC, temperature controller.

Conversion is generally controlled by reaction temperature, residence time, and partial pressure of hydrocarbon. Selectivity is also related to the reaction temperature and residence time; in addition, it is particularly sensitive to the partial pressure of the hydrocarbon. That pressure along the coil is affected by the flow rates of hydrocarbon and dilution steam, and by fouling of radiant coils and TLEs. The outlet pressure of the TLE also affects the partial pressure of the hydrocarbon, but it is largely fixed by the design of the cracked-gas compressor and the quench tower. As the cost of feedstock increases, the selectivity—that is, the quantity of feedstock required for a given level of olefin production—becomes a more important consideration.

III. PROCESS VARIABLES AND NORMAL OPERATION

The conventional control system for a pyrolysis furnace uses individual flow controllers for individual hydrocarbon and dilution-steam flows to each pass. The outlet temperature of the furnace-process stream is used to control the firing of fuel in the firebox. A typical control scheme is shown in Fig. 1.

A. Furnace Feed and Steam Control

In modern cracking furnaces each firebox contains several radiant coils. In design of large-diameter coils ($\geq$4 in.; $\geq$10.2 cm), there are 4 to 12 parallel passes. With small-diameter coils ($2\frac{1}{2}$ to 3 in.; 6.4 to 7.6 cm), there are 16 to 32 parallel passes. One unit with extremely small-diameter coils (1 in.; 2.54 cm) has 84 parallel passes. The flows of hydrocarbon and dilution steam are controlled using one of the three schemes (designated as 1, 2, and decoking. It is costly to install and maintain, however.

Scheme 1 (Fig. 2) is the most commonly used control method. Flows of hydrocarbon and dilution steam to each individual pass are under direct control. This scheme provides wide flexibility during normal operation and decoking. It is costly to install and maintain, however.

Another disadvantage of this scheme is that it is difficult to have a constant dilution steam/hydrocarbon ratio to each pass in the radiant box. For a split-coil design where several parallel inlet passes are combined into one pass near the outlet, the number of flow controllers of hydrocarbon and dilution steam can be excessive.

Scheme 2 (Fig. 3) has fewer controllers than Scheme 1 (Fig. 2) but also has less flexibility in operation. This scheme provides a constant dilution

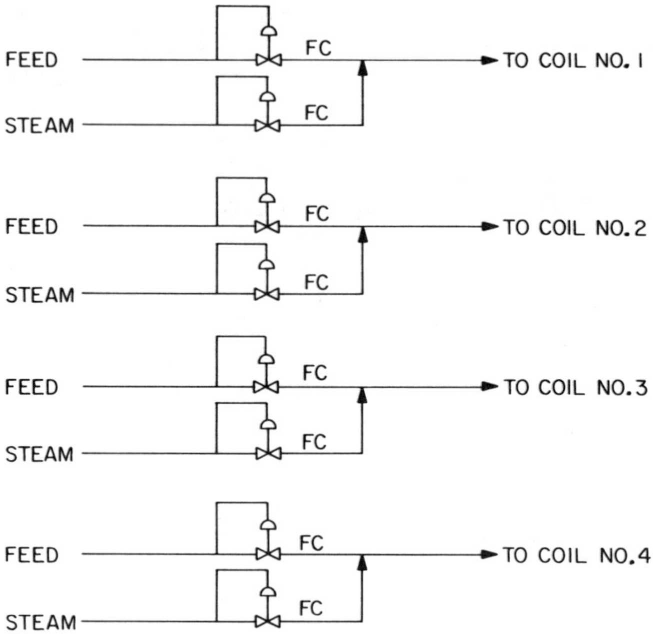

Fig. 2. Furnace feed and steam control (scheme 1). FC, Flow controller.

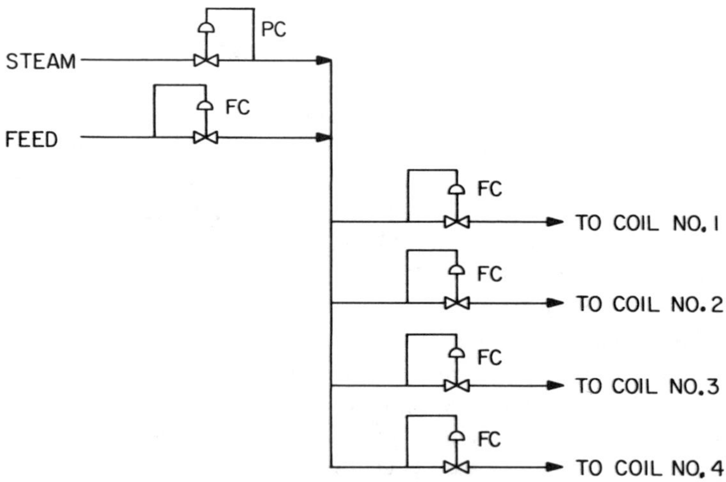

Fig. 3. Furnace feed and steam control (scheme 2). FC, Flow controller; PC, pressure controller.

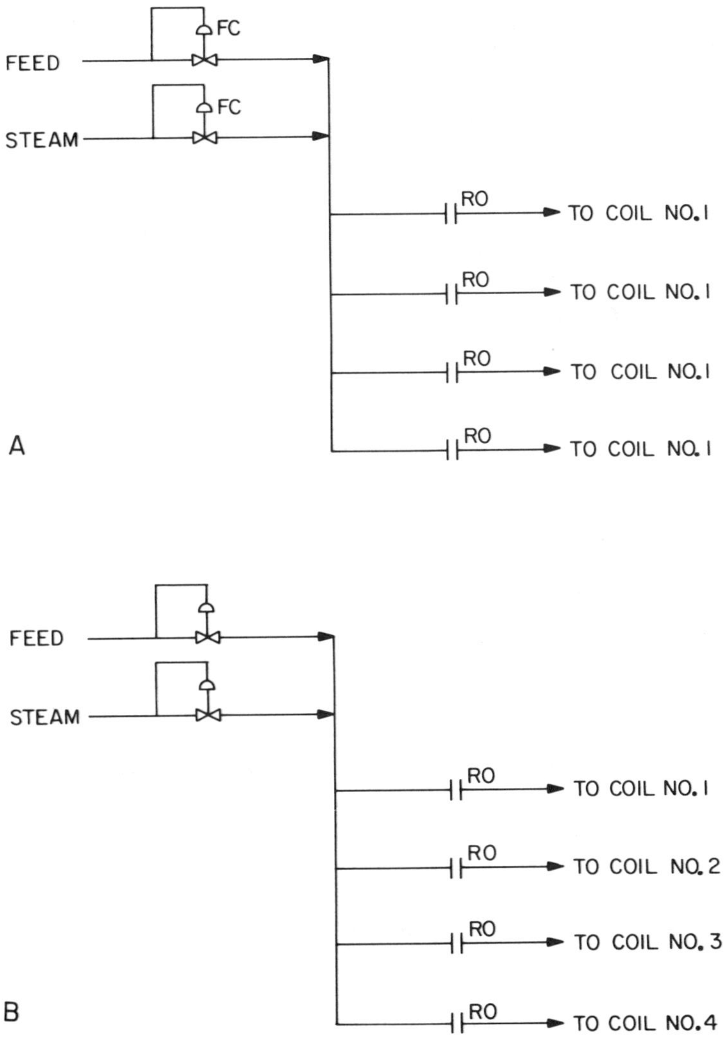

Fig. 4. Furnace feed and steam control (scheme 3). (A) Split-coil design; (B) Straight-coil design, FC, Flow controller; RO, restriction orifice.

steam/hydrocarbon ratio in all parallel passes. Some operating personnel believe that this scheme gives better control, even though the dilution steam/hydrocarbon ratio to each pass is removed from furnace control.

Scheme 3 (Fig. 4) is very similar to Scheme 2 (Fig. 3); however, restriction orifices or venturis replace flow controllers to each parallel pass. This design has been used in both small-diameter tube and split-coil designs.

Again, the flexibility to change the dilution steam/hydrocarbon ratio for each pass is not available. This scheme has the lowest investment and maintenance costs of the three shown.

B. Crossover-Temperature Control

Control of the crossover temperature—the temperature of the process stream leaving the convection section and entering the radiant tubes—is an important factor in design. Once a furnace is built, there is very little one can do to change the crossover temperature. For short-residence-time furnaces, equal crossover temperatures are very important in order to obtain equal severity of cracking per pass. Various designs have been used to equalize the exit temperature of each parallel pass from the convection section. These include using larger-than-average spacing in the tube layout of the lower convection coil, cross-linking of parallel passes, and combining parallel passes into a single mixing header and then separating them again into numerous parallel passes.

C. Superheated Steam-Temperature Control

Steam produced in the TLE is normally superheated by heat exchange with hot flue gas in the convection section. The saturated steam from the steam drum is superheated from 750 to 950°F (399 to 510°C). The amount of steam produced in the TLE decreases as fouling increases.

Furthermore, in order to sustain the rate of pyrolysis, firing of the furnace will be increased when radiant tubes are fouled. Then, because of the increase of flue-gas volume and temperature, and the decrease in the production rate of steam, the temperature of the steam will be higher than the initial value at clean furnace and TLE conditions. If the steam temperature is high, the tube-metal temperature will also be high. The high tube-metal temperature could damage the tubes in the superheating coil. Many designs of temperature control to minimize this problem are in service. The simplest control is to inject boiler-feed water to lower the steam temperature. In this design, the flow from the superheating coil must have an exit temperature higher than the required final superheat temperature at all times. Finchley (1979) described a more complicated design involving an attemperator.

D. Draft Control

A furnace with a high thermal efficiency usually has an induced-draft fan. The fan can be motor driven or steam-turbine driven, and it sustains flue-gas flow with maximum heat recovery from the flue gas. The draft at the bridge-wall, located between the radiant and convection sections on the flue-gas side, is controlled at a given value by use of the vanes at the suction inlet in the motor-driven case, or by fan speed with the turbine drive. Various combustion-control systems have been described by Kane (1980).

E. Furnace-Firing Control

The rate of furnace firing is often used to control cracking severity so that a given objective can be achieved. There are numerous ways to manipulate the firing rate. The most commonly used control point is the outlet temperature of the furnace. By adjusting the fuel-firing rate, a constant furnace-outlet temperature can be obtained. Another control point for the firing rate can be the composition of the furnace effluent as measured by a given key component or ratio of two components. This approach requires analysis of furnace effluent by a gas chromatograph or a mass spectrometer.

Actual plant data, as well as SPYRO (Dente *et al.*, 1979) computer simulations, indicate that in production of ethylene the selectivity decreases when the pyrolysis system is fouled. This effect is shown in Fig. 5. The solid lines are from computer simulations, and the connected points show operating data for two plants. Coke thickness is calculated from the measured tube-wall temperature near the furnace outlet. The operating data from ethane cracking confirm that selectivity decreases when the pyrolysis system is fouled, as well as when the furnace outlet pressure is increased.

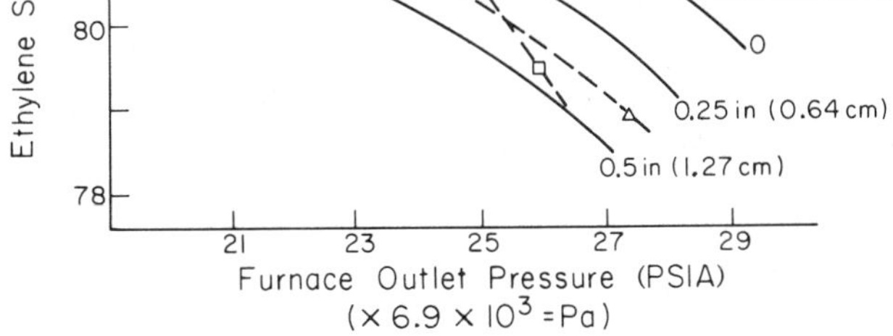

Fig. 5. Plot of ethylene selectivity versus furnace-outlet pressure. △, Plant A; □, plant B.

F. Outlet-Temperature Control

Control of the outlet temperature is the simplest method used to control the firing rate of the furnace. It is also the least costly approach. As can be seen in Fig. 5, however, when the pyrolysis system is fouled, the selectivity cannot be maintained at constant outlet temperature for the coil. Furthermore, in modern cracking furnaces, there are usually multiple-radiant coils (e.g., up to 84). Only the temperature from one coil or a combination of several coils is used as the control point. If the control is by the outlet temperature of one coil, then the other coils may have different temperatures. Plant data indicate that outlet temperatures from coils not under control can vary as much as ±20°F (±11°C) away from the control point. This variability not only presents a problem in control of selectivity but also causes heavy coking among coils with high outlet temperatures. That often causes premature shutdown of furnaces with only one coil heavily fouled.

G. Component Control

In gas cracking, conversion can be measured by the disappearance of feedstock. Measurement of the key component in the effluent can be used for conversion control. Because of fouling of the furnace and the TLE, the selectivity will drop off at constant conversion. This effect can be seen in Table I, which shows data for computer simulation of an ethane cracker. For constant ethane content in the furnace effluent, the ethylene yield drops off with increased coking, but the methane yield increases. If the furnace firing is controlled at constant conversion of ethane, the selectivity cannot be held constant. Therefore, control of selectivity by using only one

TABLE I

Production Composition in Ethane Cracking (Computer Simulation)

	Coking		
Yield (wt%)	Clean 23.3 psia (0.14×10^6 Pa)[a]	Average 26.2 psia (0.18×10^6 Pa)	Heavy 31.3 psia (0.22×10^6 Pa)
Methane	4.6	4.8	5.3
Ethylene	47.3	46.6	45.5
Ethane	38.1	38.1	38.1
Propylene	1.2	1.3	1.6

[a] Coil-outlet pressure.

effluent component is not sufficient to reach the control objective. Use of the ratio of two effluent compounds can improve the control of selectivity. In an ethane-cracking furnace, the most sensitive ratio is methane/propylene. The methane/propylene ratio proves to be useful for controlling selectivity not only for gas cracking but also for naphtha/gas-oil cracking.

At a given selectivity, the pyrolysis furnace can be controlled during the course of a run using the methane/propylene ratio in the effluent. The methane/propylene ratio is controlled not by fuel firing but by the flow of hydrocarbon and/or steam with fuel firing held constant.

IV. OPTIMAL CONTROL

The optimal control of the pyrolysis system requires adjustment of the pertinent variables so that certain performance objectives are achieved without violating constraints on the system. In addition, to meet a given production demand, constant selectivity is often used as an objective. As the pyrolysis system begins to foul, a given selectivity can be maintained by adjusting the total flow of hydrocarbon and dilution steam and by adjusting the dilution-steam ratio. The precise control strategy requires the solution of a mathematical model and the use of an on-line computer.

For a given pyrolysis system, the selectivity is affected by (*a*) TLE fouling, (*b*) radiant-coil fouling, (*c*) total flow of hydrocarbon and dilution steam, and (*d*) dilution-steam ratio. The TLE fouling and radiant-coil fouling are disturbances in the system, although proper operation and design can minimize this fouling. The total flow of hydrocarbon and dilution steam and the dilution-steam ratio can be manipulated. They can be adjusted to achieve a given partial pressure of hydrocarbon and residence time. When a pyrolysis system is fouling, not only will the total feed rates of hydrocarbon and dilution steam have to be reduced but the dilution-steam ratio may have to be increased. Note should be made that the controlling partial pressure of hydrocarbon is not at the coil outlet. It is better defined as the partial-pressure profile along the coil. Therefore, the solution is not straightforward and obvious.

The optimal control strategy is to operate the pyrolysis system near system constraints. The most important constraint in a pyrolysis system is the tube-metal temperature of the radiant coil.

As the radiant coil and the TLE are fouled, the inside surfaces of the radiant coil and the tube of the TLE are coated with a layer of coke. This coke retards heat transfer and reduces the free cross-section of flow area. During a run, as coke is deposited, the tube-metal temperature

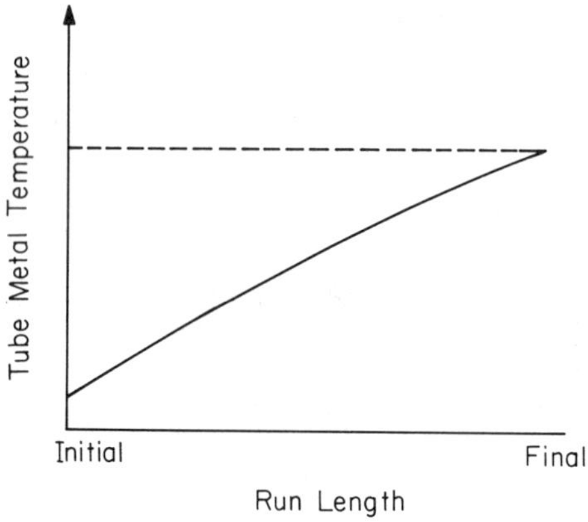

Fig. 6. Plot of tube-metal temperature versus run length. ——, Constant feed; ----, decreasing feed at high initial rates of feed and dilution steam.

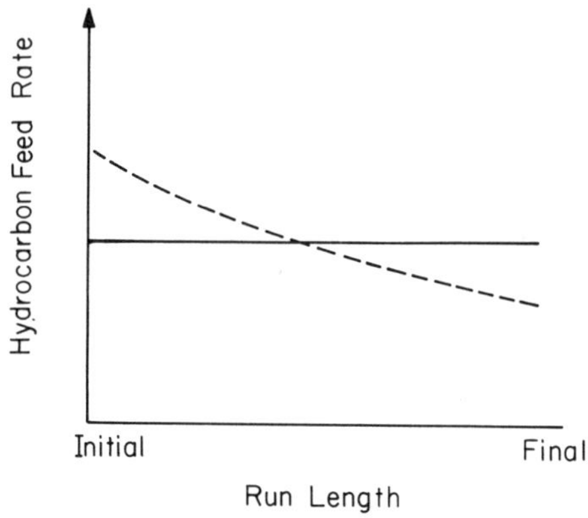

Fig. 7. Plot of hydrocarbon feed rate versus run length. ——, Rising tube-metal temperature; ----, constant tube-metal temperature.

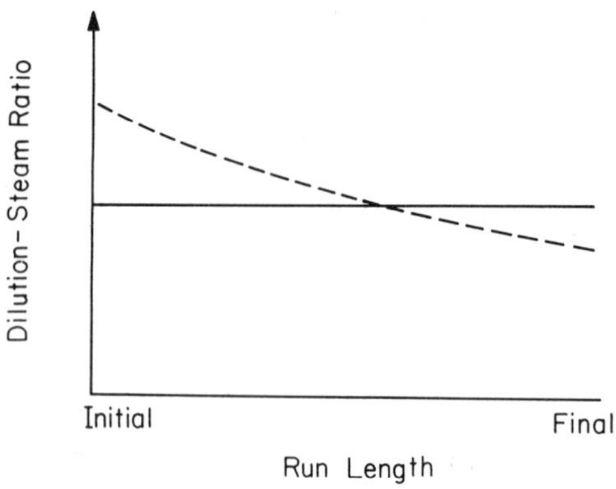

Fig. 8. Plot of dilution-steam ratio versus run length. ——, Rising tube-metal temperature; ----, constant tube-metal temperature.

of the radiant coil will increase at a given cracking severity and feed rates of hydrocarbon and dilution steam. When the maximum permissible tube-metal temperature is reached, the pyrolysis system will be taken off line for decoking. In order to operate near the maximum tube-metal temperature during a run, as shown schematically in Fig. 6, the feed rates of hydrocarbon and dilution steam are increased above the average feed rate. The effects can be seen in Figs. 7 and 8. As the feed rate is increased at a given severity, more fuel will be fired. This result can be seen in Fig. 9. Therefore, the tube-wall temperature will approach the maximum permitted

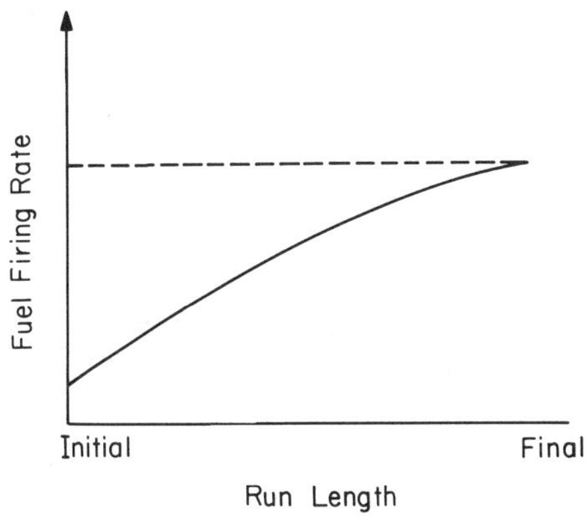

Fig. 9. Plot of fuel-firing rate versus run length. ——, Rising tube-metal temperature; ----, constant tube-metal temperature.

at all times. In order to maintain a particular hydrocarbon partial pressure profile in the presence of a higher pressure drop through the radiant coil caused by the increased feed rate, the dilution-steam ratio is also increased to compensate for the increased total pressure. The partial-pressure profile of the hydrocarbon remains essentially unchanged (Fig. 10). This result is different from the conventional mode of operation where the partial-pressure profile of the hydrocarbon increases with fouling of the pyrolysis system. By adjusting the total flow of hydrocarbon and dilution steam, and the dilution-steam ratio, the partial-pressure profile of the hydrocarbon can be maintained in the presence of coking as shown in Fig. 11. The solid line is based on a clean pyrolysis system, and the dotted line is based on a fouled system with reduced flow of hydrocarbon and dilution steam. Although the two curves are not exactly the same, nevertheless there is almost no difference in selectivity between these two operations.

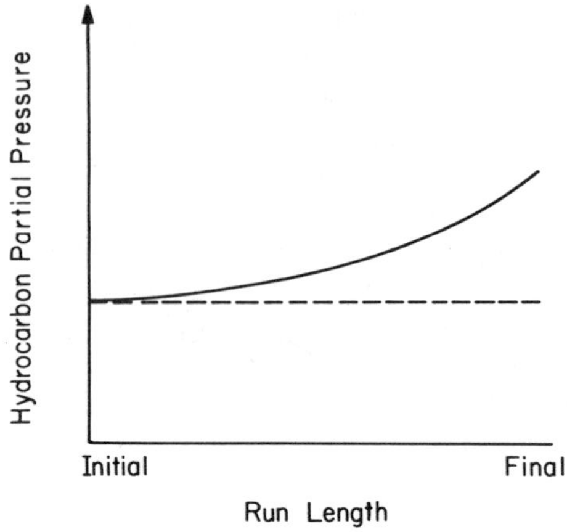

Fig. 10. Plot of hydrocarbon partial pressure versus run length. ——, Constant feed and rising tube-metal temperature; ----, decreasing feed and constant tube-metal temperature.

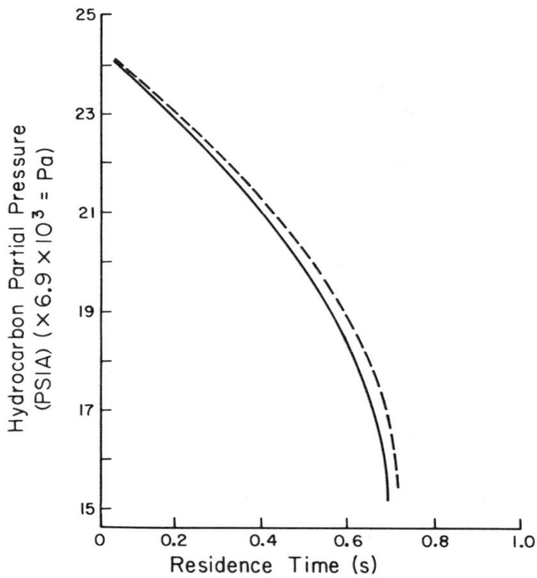

Fig. 11. Plot of hydrocarbon partial pressure versus residence time for fouled (----) and clean (——) conditions.

Therefore, by increasing the flow of hydocarbon and dilution steam when the pyrolysis system is clean, production can be maximized as suggested in Fig. 12. As the system is fouled, adjustment of the total flow of hydrocarbon and dilution steam, as well as the dilution-steam ratio, will

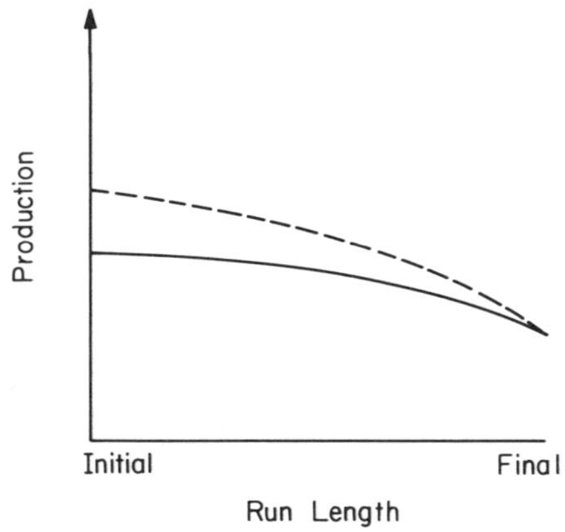

Fig. 12. Plot of production versus run length. ——, Constant feed and rising tube-metal temperature; ----, decreasing feed and constant tube-metal temperature.

allow the selectivity to be maintained for maximizing profit and/or minimizing production cost as implied in Fig. 13. Development of the optimal operating conditions is not simple because of the complicated relationships among conversion, residence time, and the partial-pressure profile of the hydrocarbon. An on-line computer can speed establishment of the optimal solution without violation of constraints at any time. This discussion of optimal operation has been based on an ethane cracker for simplicity. A pyrolysis system for naphtha or gas oil could be adapted to this same approach.

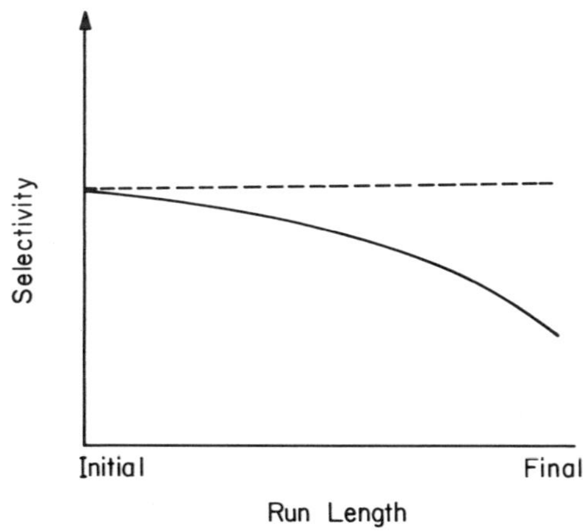

Fig. 13. Plot of selectivity versus run length. ——, Constant feed and rising tube-metal temperature; ----, decreasing feed and constant tube-metal temperature.

REFERENCES

Dente, M., Ranzi, E., Barendregt, S., and Goossens, A. G. (1979). *AIChE Annv. Meet., 72nd, San Francisco, Calif.*

Finchley, P. (1979). *Chem. Eng. (N.Y.)* **86**(17), 120–134.

Kane, L. A. (1980). *Hydrocarbon Process.* **60**(6), 65–68.

18

Transfer-Line Exchangers

ALFRED MOL

I. QUENCH COOLERS

The gases that exit pyrolysis-furnace coils contain significant usable heat. Much of this heat can be recovered by use of transfer-line exchangers (TLEs) or quenchers. At the cracking-coil outlet, the primary and secondary

Pyrolysis: Theory and Industrial Practice

ISBN 0-12-048880-9

TABLE I

Impact of Adiabatic Cracking in Transfer Line and TLE-Inlet Cone[a]

	Feedstock		
	Ethane	Naphtha	AGO
Coil design	Conventional (115 m)	Conventional (57 m)	Split
Coil			
Residence time (s)	0.916	0.456	0.357
Feed conversion (%)	62.8	89.8[b]	98.1[c]
Outlet temperature (°C)	836	844	797
Yields (wt%)			
Ethylene	47.3	26.0	20.1
Propylene	1.7	16.1	13.6
C_4's	1.7	11.1	11.0
TL/TLE cone			
Residence time (s)	0.030	0.062	0.042
Feed conversion (%)	65.4	96.7	99.3
Outlet temperature (°C)	818.5	832	789
Yields (wt%)			
Ethylene	48.6	28.6	21.3
Propylene	1.7	13.7	13.4
C_4's	1.8	8.5	10.2

[a] Basis: PHENICS (phenomenologic cracking simulation) model.
[b] n-C_5.
[c] n-C_{15}.

reactions continue in the transfer line under adiabatic conditions, as shown in Table I. The table depicts the considerable impact of adiabatic cracking on yields, conversion, and process temperature. In case the residence time in the transfer line is excessive (say more than 10–15% of the residence time in the cracking coils), the yields of valuable primary products such as acetylene, butadiene, and ethylene decrease. In order to avoid excessive transfer-line cracking, there is either a direct or an indirect quench downstream of the transfer line. Direct quenching features a more rapid temperature decrease than indirect quenching (Sundstrom and DeMiciell, 1971). However, indirect quenching permits heat recovery at a high temperature level suitable for generation of high-pressure (HP) steam.

In older energy-inefficient steam crackers, the furnace effluent was quenched directly with water (in case of ethane or propane feed) or with a suitable oil (in case of liquid feedstocks). Transfer-line exchangers, also called quench coolers, often utilized a heating medium such as Dowtherm until the late 1960s. The first TLE generating HP steam used for plant turbocompressors was developed by Schmidt'sche GmbH, Kassel, Federal Republic of Germany (Knuelle, 1972, 1973). Since then, quench coolers of other designs generating HP steam suitable for feedstocks ranging from ethane through atmospheric gas oil (AGO) have become an integral part of steam crackers and have resulted in a substantial improvement in operating cost.

II. REQUIREMENTS FOR QUENCH COOLERS

A. Rapid and Uniform Cooling

In order to avoid secondary reactions at lower temperatures leading to less valuable products, the residence time between furnace outlet and quench-cooler inlet has to be limited to some 10–15% of the residence time in the cracking coils. A guide cone in the inlet chamber of the TLE is sometimes incorporated for proper flow distribution. Because of poor flow distribution for multiple-tube TLEs, the mass velocity is higher for the central tubes than for the outer tubes. Consequently, an increase of TLE capacity has a negative effect on the flow distribution over the TLE hot tubesheet.

B. Small Pressure Drop

Because the cracking selectivity is directly dependent on the cracking-coil outlet pressure, the pressure drop over the quench cooler must be limited to a few tenths of an atmosphere. Figure 2 depicts that for a multitube TLE design, there is a substantial pressure drop in the inlet chamber and at the hot tubesheet. It is of considerable importance that no major increase of pressure drop should occur during the run. Any coke spalled off the cracking coils may accumulate at the TLE hot tubesheet and increase the pressure drop. Therefore, large-bore tubes are preferred, particularly for cracking of light-hydrocarbon gases. Conventional TLE-tube inner diameter (ID) is 23 mm, and for multitube TLEs some 35-mm ID tubes are used. Some quench coolers (USX-D; see Fig. 6) may have an ID as large as 90 mm.

C. Desired Heat Recovery

Because quench-cooler heat recovery has a substantial effect on plant-operating cost (Mol and Westenbrink, 1974), this heat recovery should be reviewed carefully. For cracking of light-hydrocarbon gases, there is no problem of TLE-tube fouling by condensation, and therefore, it is economical to have two quench coolers in series—that is, one HP steam-generating TLE (or Ultra Selective exchanger) in series with a secondary (also called "postquench cooler") TLE for further heat recovery. The secondary TLE may serve for generation of dilution steam or for preheating. For naphtha feed the HP steam-pressure level should be more than 80 atm to avoid too low tube-skin temperatures and consequent rapid fouling by condensation. Present economics dictate high (125 atm or so) steam-pressure level, which is favorable to a low TLE tube-fouling tendency.

High heat recovery can be ensured by low TLE-outlet temperature. A sound economical design for naphtha feed is a TLE-outlet temperature at the start of run (SOR) that is about 30°C above the HP steam-generating temperature.

With a gas-oil feed and because of a larger fouling rate, a higher TLE-outlet temperature is justified. Conventional TLE designs for gas-oil feed frequently have a SOR outlet temperature as high as 550–600°C. However, because the minimum TLE tube-wall temperature is only slightly dependent on TLE-outlet temperature at SOR, recent TLE designs for gas-oil feed have SOR TLE-outlet temperatures as low as 430–450°C. As a result there has been a substantial improvement in the economics of gas-oil processing.

D. Long Continuous Run Length

Fouling of the TLE tubes by condensation and possible blocking of the hot tubesheet by coke spalled off the cracking coils will cause an increase of TLE-outlet temperature and of pressure drop. During the first days of the run for a liquid feedstock, increase of TLE-outlet temperature due to condensation is substantial (for gas oil as high as 30°C on the first day), and then stabilizes at the "plateau temperature." Then a typical TLE-outlet temperature rise is 1–2°C per day for AGO and 0.2–0.4°C per day for a full-range naphtha feed.

Reasonable run lengths can be ensured by designing for a proper margin between TLE-outlet temperature at start and end of run. With a suitable TLE-tube ID, the increase in TLE-outlet temperature during a run with gas feed will be less than 20°C. For a naphtha feed, the TLE-outlet temperature increase is a function of feed characteristics and cracking severity. For light naphthas, the increase during a run is 20–30°C; for heavy naphtha, 100–120°C. The TLE run length is usually the limiting factor for the integral cracking heater–TLE combination in the case of a gas-oil feedstock.

However, with an AGO feedstock also, a TLE run length of 35–40 days at normal cracking severity is feasible (Mol, 1974).

E. Low Maintenance

The maintenance costs for periodic cleaning of TLEs may be excessive and jeopardize the plant-service factor in case of (*a*) an extremely short TLE cycle length and/or (*b*) a long cleaning time per run.

Conventionally, the heater is cooled down after decoking, the TLE cones are removed, and the TLE tubes are mechanically cleaned with a high-pressure (500–700 atm) water jet, or sometimes—in case of hard coke—by hydro-sandblasting. Therefore, proper accessibility to the TLE, preferably from grade level, is recommended.

Pseudo on-line decoking of the cracking coils through the TLE results in some cleaning effect for the TLE too, particularly for a gas-oil feed. Because the TLE run length is rather short for an AGO feed, however, with pseudo on-line decoking the TLE has only to be hydrojetted after every third decoking operation. Numerous TLEs have suffered from water-side corrosion, particularly if the boiler-feed water is of poor quality. For the shell-and-tube types of TLEs, installation of a discontinuous-blowdown facility is recommended for removing the water above the critical hot tubesheet. Because the hot tubesheet is most vulnerable to water-side corrosion, a horizontal TLE is preferred to a vertical TLE.

III. SCHMIDT'SCHE QUENCH SYSTEM

The Schmidt'sche (SHG) indirect-quench system consists of separate TLEs, steam drum(s), risers, and downcomers. The basic element of the SHG TLE (see Figs. 1 and 2) is the double-tube register consisting of several concentric tubes (1 and 2) welded at their ends (4) into an oval tube (3). In this way the internal tube with product stream is surrounded by a water jacket, consisting of an external tube and the two oval headers.

Several double-tube registers with different tube numbers are connected in one cooler body to the oval headers to produce gas-tight welds. The ends of the oval headers are connected to the downcomer distributors and riser headers (6) by means of transition pieces (5). The tubesheets are formed by welding together the oval headers. Connecting cylinders (7) with flanges (8) will be welded for the attachment of the gas-inlet (9) and gas-outlet (10) chambers. (Refer to Fig. 2.)

Fig. 1. Schmidt'sche TLEs. This photograph depicts the absence of TLE shell. (Reprinted with permission of Schmidt'sche GmbH.)

The original SHG units were provided with a conical hot tube-sheet, but the present designs feature a flat hot tubesheet. The flat tubesheet is preferred with respect to water-side fouling. The product-stream flow is normally upward. This design has been successfully applied worldwide for feedstocks ranging from ethane to gas oil.

IV. BORSIG QUENCH SYSTEM

The Borsig quench system also consists of separate TLEs, steam drum(s), risers, and downcomers (Deuse, 1971). This unit is of the shell-and-tube type (Fig. 3). The mechanical problems related to the large temperature and pressure differences over the hot tubesheet are solved by

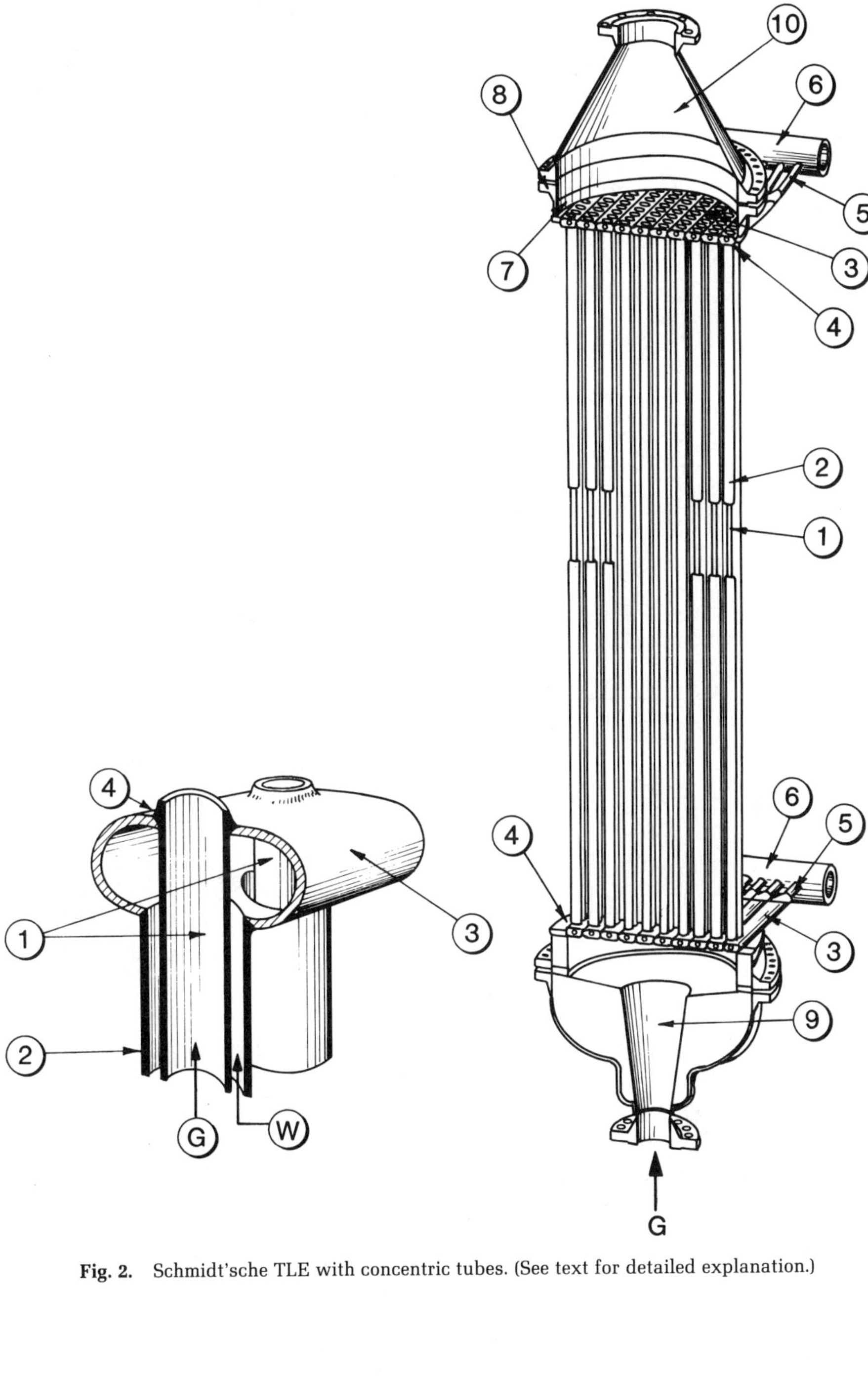

Fig. 2. Schmidt'sche TLE with concentric tubes. (See text for detailed explanation.)

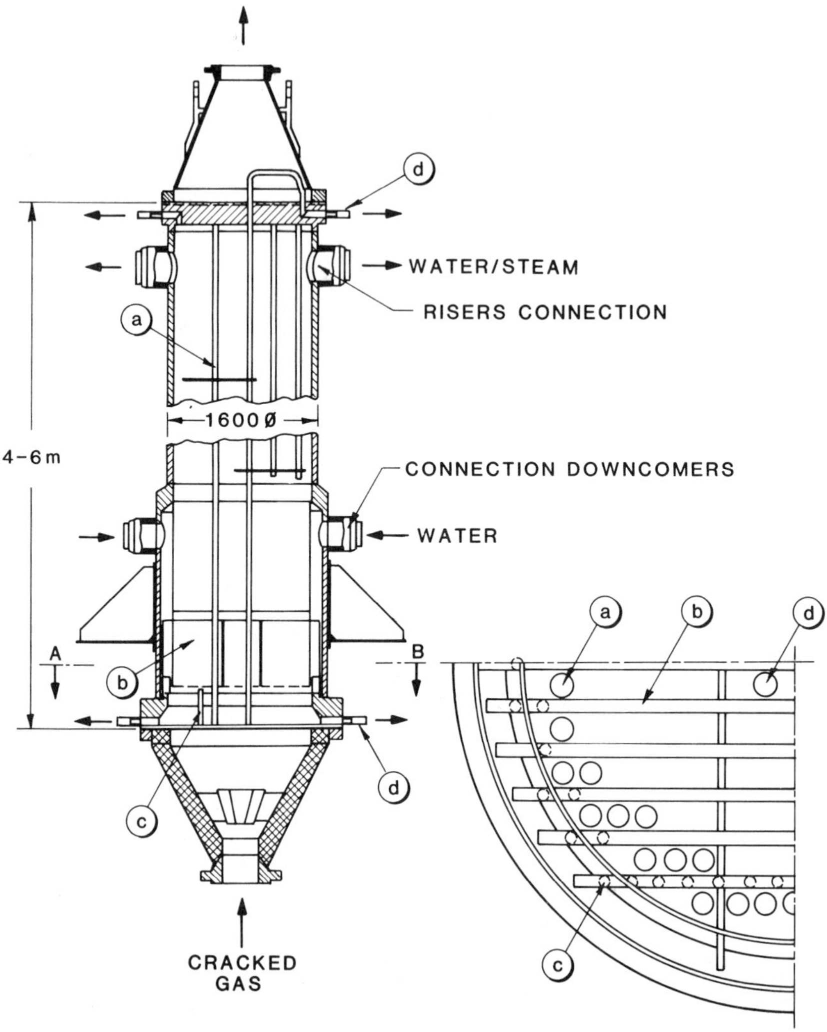

Fig. 3. Borsig TLE with thin, studded tubesheet. (Reprinted with permission from Ullmanns Enzykl. Copyright Verlag Chemie GmbH.)

application of a relatively thin (15 mm or so) tubesheet; however, the tubesheet is studded on the steam–water side. The downcomers from the steam drum are welded directly to the TLE shell just above the hot tubesheet. Current TLE-shell designs often are conical near the downcomers to ensure better water distribution. The present vertical TLEs are provided with a facility to draw discontinuously a blowdown from near the critical hot tubesheet. The Borsig TLE can be arranged vertically as well as horizontally. The vertical arrangement is more common because of overall plot plan requirements. For a horizontal TLE, a larger TLE-shell diameter is required to (*a*) ensure proper water distribution and steam–water removal, and (*b*) to meet the requirements of larger tube-to-tube spacing.

The type of TLE-outlet tubesheet depends on the design outlet temperature at the end of run (EOR). For gaseous and naphtha feedstocks, the outlet tubesheet is conventionally thick and uncooled. For a gas-oil feed requiring higher TLE-outlet temperatures, a more expensive, thin, studded tubesheet design may be necessary for the hot tubesheet.

Borsig TLEs with the thin, hot tubesheet have gained a considerable market share for feedstocks ranging from ethane to AGO.

V. MITSUBISHI QUENCH SYSTEM

This system, often called M-TLX, consists of an integrated quench cooler with a steam drum located on top of the M-TLX. Actually part of the cooling tubes are in the steam drum, as shown in Fig. 4 (Sato and Ohnishi, 1971). The lower parts of the cooling tubes are straight and concentric, with the product in the inner tube, like the Schmidt'sche TLE. The upper parts of the cooling tubes inside the shell have a spiral shape, which makes it possible to dissipate the thermal stresses during on-line decoking. The inside of the shell contains many helical cooling tubes arranged in concentric circles, and a descending pipe for boiler-water circulation in the central region.

The design is particularly applicable for on-line decoking because of the helical cooling tubes in which the M-TLX can be decoked without boiler water present. The schedule for steam reduction and air addition is shown in Fig. 5, along with the resultant furnace-outlet temperature, shell pressure, and water level. The M-TLX and the unique decoking procedure are most convenient for gas-oil feedstock, improving the heater–quench service factor. The M-TLX was introduced around 1970 and is presently used in about eight different steam crackers in Japan and the People's Republic of China.

The present M-TLX II consists of individual quench cooler(s), steam drum, risers, and downcomers. The M-TLX II no longer has helical tubes but has straight concentric tubes featuring lower investment than the M-TLX I (Sakaguchi *et al.*, 1980).

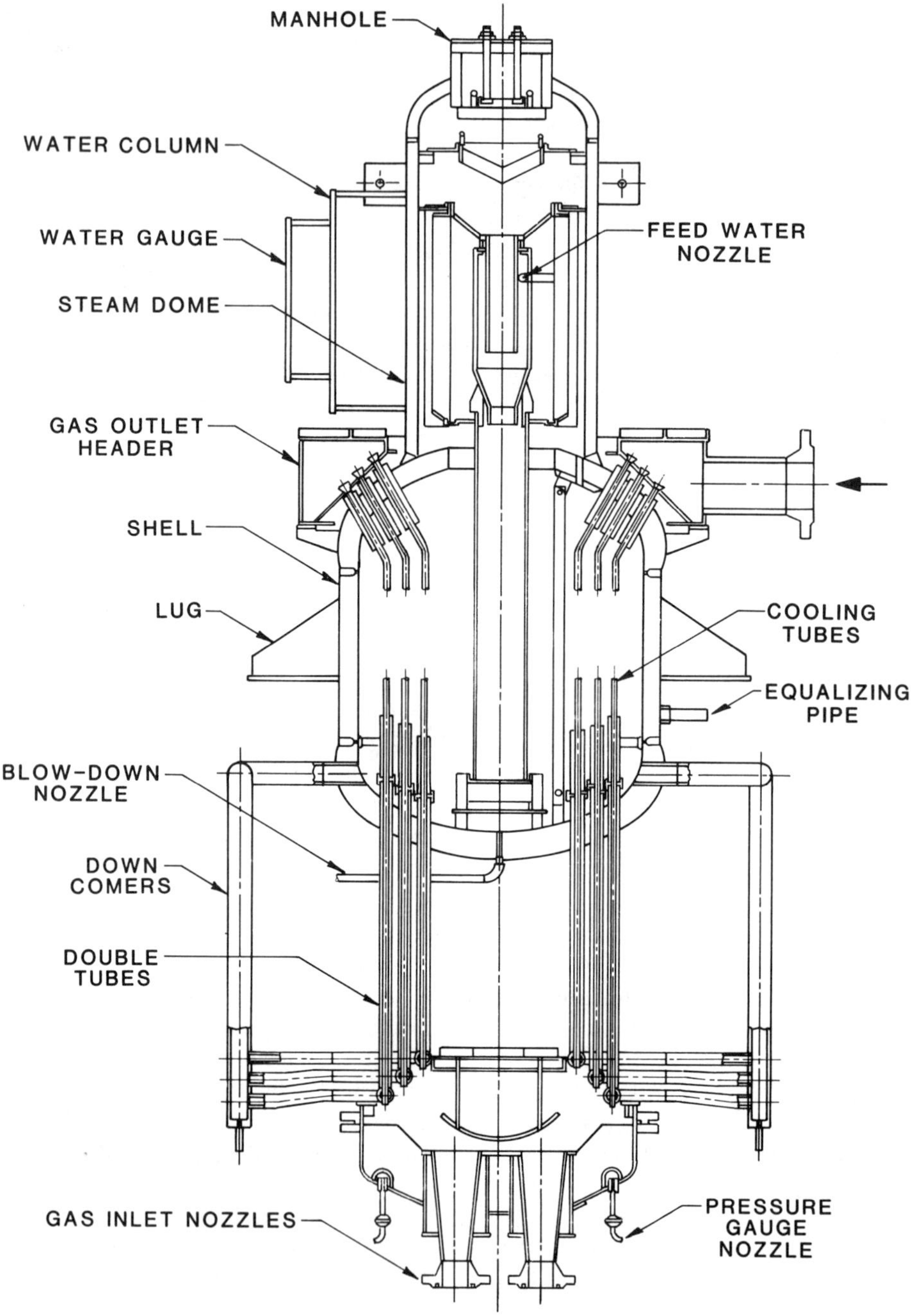

Fig. 4. Mitsubishi M-TLE with integral quench cooler and steam drum. Lower parts of cooling tubes are straight and concentric; upper parts of cooling tubes are helical. (Reprinted with permission of Mitsubishi Petrochemical Co., Ltd.)

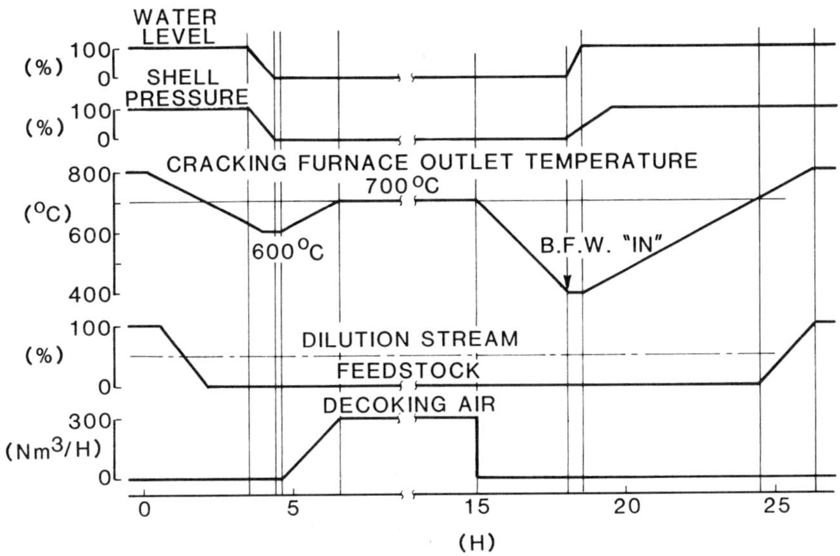

Fig. 5. Typical schedule of steam–air decoking for cracking furnace with M-TLX. (Reprinted with permission of Mitsubishi Petrochemical Co., Ltd.)

VI. STONE AND WEBSTER QUENCH SYSTEMS

Stone and Webster (S & W) has developed several proprietary quench-cooler designs (Woebcke *et al.*, 1971). The basic characteristic of these quench coolers is the absence of a tubesheet. The S & W quench coolers are called USXs (UltraSelective exchangers). The original USX consisted of an annular bayonet-type design, whereby the bayonet was the combined riser/downcomer. Later S & W steam crackers have been equipped with the USX-D type of exchanger, which is a double-tube exchanger with the inner tube of somewhat larger diameter (90 mm or so) than the transfer line between the heater outlet and USX inlet. This is shown in Fig. 6. Because the USX-D outlet temperature is relatively high (500–550°C), there is usually downstream of the USX-D one conventional multitube TLE (16 or so tubes per heater Fig. 7) for further waste-heat recovery. The USX-D has a proven record for successful performances with ethane, naphtha, and gas-oil feedstocks (Howard and Hoggett, 1975).

For ethane crackers, with the cracking-coil outlet at the base of the firebox and where coking has been a traditional problem, the USX-Omega type has been found to be satisfactory. The USX-Omega also has the large-diameter double-tube design, but is in the shape of an inverted hairpin.

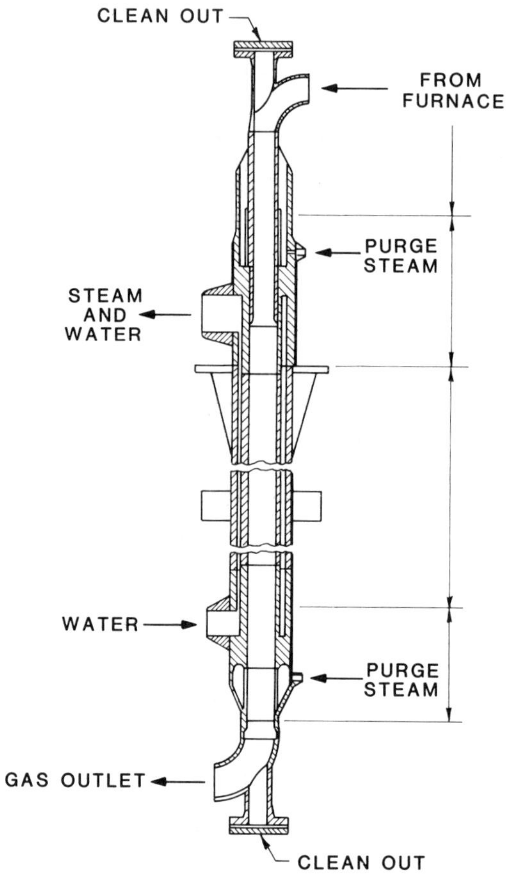

Fig. 6. USX Type "D" exchanger. (Reprinted with permission of Stone and Webster Engineering Corp.)

Key features of the various USX designs are as follows.

1. The gas flows are at low Reynolds numbers and have a streamlined pattern from the coil outlet to the cooling surface, thus eliminating turbulent eddies, which create backmixing and promote fouling because of secondary reactions.
2. The ID is larger than the upstream radiant-coil/transfer line, so the pressure drop is low at relatively high mass velocities, thereby minimizing coke and tar deposition.
3. Because the USX quench cooler does not have a tubesheet, any coke spalled from the radiant coils will travel through the exchanger without causing plugging and subsequent increase of pressure drop.

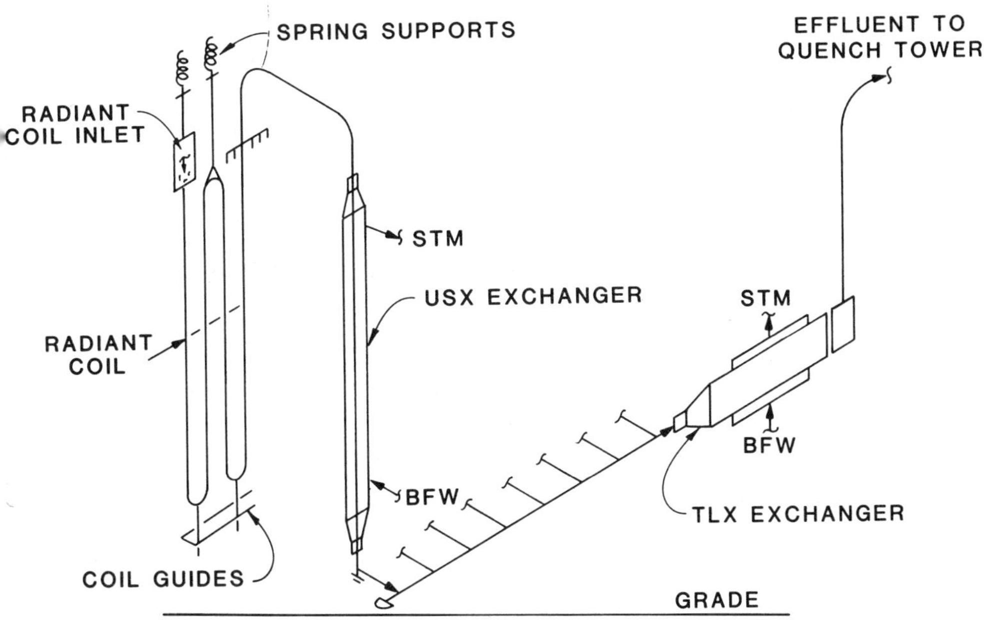

Fig. 7. Stone and Webster USC-type cracking furnace radiant-coil and effluent cooling system. (Reprinted with permission of Stone and Webster Engineering Corp.)

The USX quench coolers have been used since the mid-1960s in S & W crackers in the United States, Canada, Europe, Japan, Taiwan, Turkey, People's Republic of China, Singapore, and other countries.

VII. OTHER QUENCH SYSTEMS

Other quench systems have been developed by engineering contractors and specialized vendors. Conventional shell-and-tube TLEs of proprietary design for the generation of steam at up to 45 atm pressure have been developed by Lummus, Kellogg, Linde, Foster Wheeler, and others, mainly for ethane crackers. The best-known conventional TLE used for ethane crackers is the Lummus horizontal TLE. In this design, the TLE is located under the firebox with the steam drum on top of the firebox. This design is used in ethane crackers in the United States, Mexico, and Algeria. However, with present high energy costs, higher steam-pressure levels are used, eliminating these conventional TLEs.

Other quench systems of importance include the following.

1. The Mitsui innovative quencher (Fig. 8). This design of integral TLE and steam drum features concentric tube passes, with steam–water in the

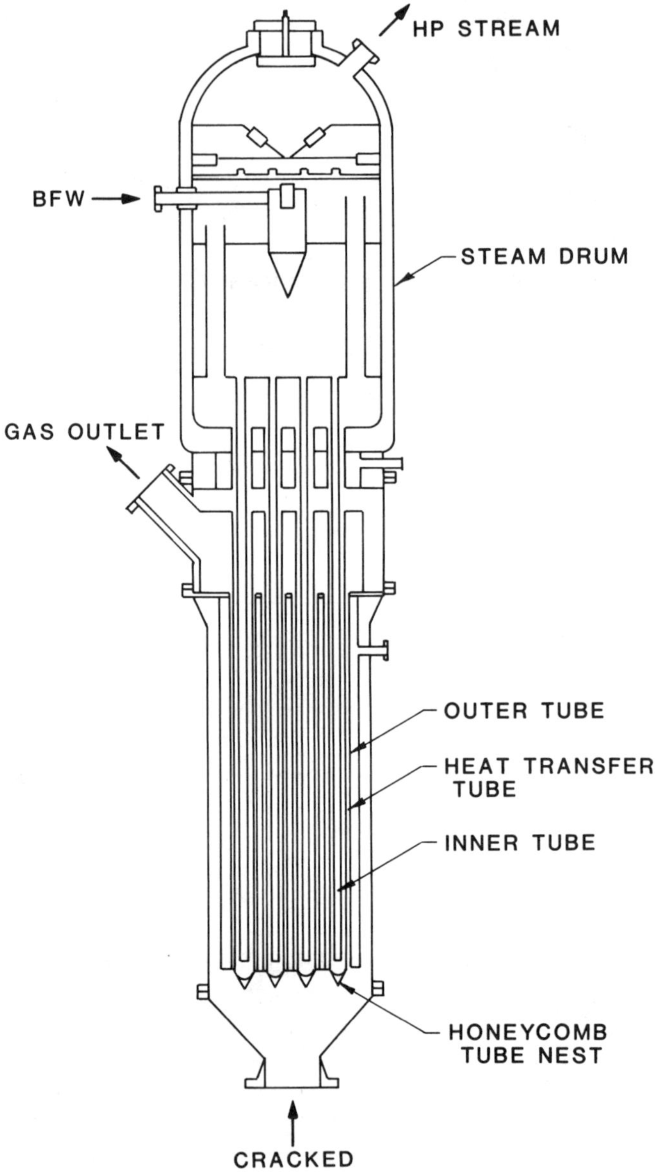

Fig. 8. Mitsui innovative quencher with integral quench cooler and steam drum. (Reprinted with permission of Mitsui Engineering and Shipbuilding Co.)

center tubes and cracked gas in the annular pass. The flow pattern, compared with conventional TLEs, is improved by use of hexagonal openings of the outer tubes at the hot tubesheet (Shimizu and Nonaka, 1982).

2. Lead-bath quench systems. There have been at least two independent efforts to commercialize a molten-lead-bath quench generating HP steam. A lead-bath quench design may have some advantageous features with respect to periodic mechanical cleaning compared to conventional multitube TLEs. However, these systems have not been successfully designed nor operated.

3. Lent jes-Rekuperator TLE. The shell of each TLE contains both the cooling tubes and the centrally located downcomer. The number of cooling

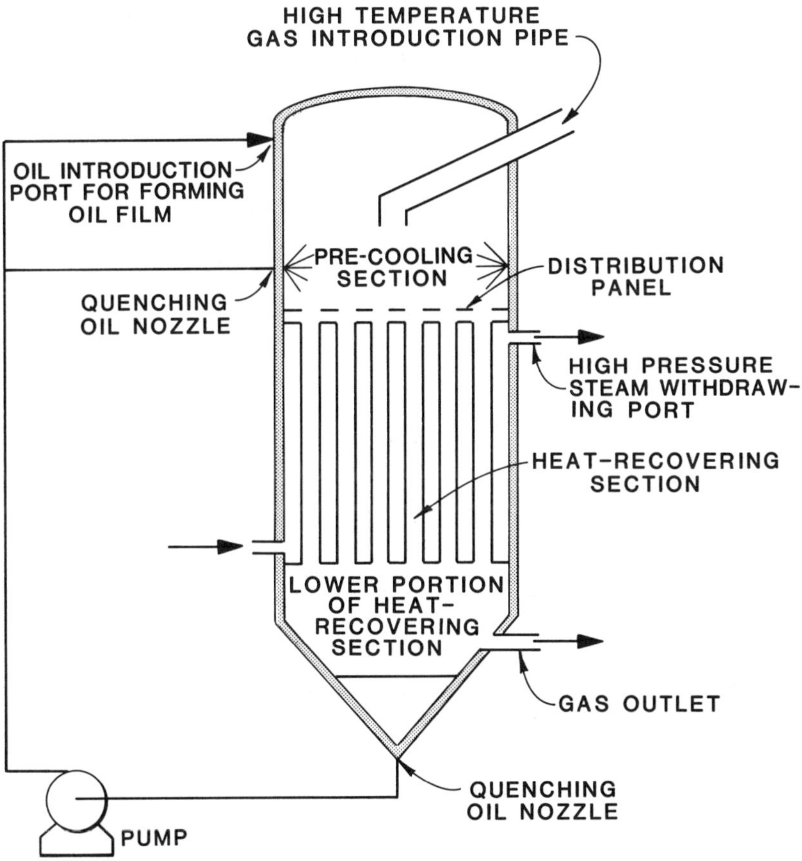

Fig. 9. Ozaki quench cooler. The OQC can process effluent of heavy-oil cracking, because of liquid oil film on cooler tubes. (Reprinted with permission of *Chemical Week*. Copyright 1977 by McGraw-Hill.)

tubes per cooler is limited to about six. As the cooler capacity is limited, several TLEs may have to be used for large-capacity cracking coils. Operating experience has been obtained for crackers in the Federal Republic of Germany and Poland.

4. Ozaki quench cooler (OQC). One of the latest developments in quenching technology is represented by the OQC as employed for the ACR (advanced cracking reactor) developed by Union Carbide/Kureha/Chiyoda. Because the ACR process is based on heavy-oil feedstocks, conventional quench coolers are impractical because of rapid fouling. In the OQC (see Fig. 9), the inlet-chamber quenching oil is jetted into the high-temperature reaction-product stream to precool the gas. The mixed phase—containing steam, hydrocarbon products, and quench oil—is introduced into a tubular exchanger, where heat is recovered to produce HP steam. Part of the liquid flows down the interior surface of the exchanger tubes and forms a liquid film that coats the walls and prevents coking (Anonymous, 1977).

The historical development of quench-cooler technology is shown in Table II, beginning with the Schmidt'sche cooler in 1960 and continuing to the Ozaki cooler in the late 1970s.

TABLE II

Historical Review of Quench-Cooler Developments

1960	Schmidt'sche double-tube TLE at ROW, Germany
1961–1962	Borsig thin-tubesheet TLE
Mid-1960s	USX Bayonet and USX-Omega
Late 1960s	USX-D for naphtha
1970	M-TLX
1970–1973	Development of AGO processing in quench coolers
1974	Postquench cooler for gas feed
1975	Larger TLE diameters
Late 1970s	Ozaki quench cooler for ACR process

VIII. TECHNICAL AND ECONOMIC DESIGN CRITERIA

A. TLE Capacity

Low investments can be realized when only one TLE per cracking heater is used. Presently the largest TLE capacity is 22 tons of naphtha per hour. In general the preferred flow distribution and process control decrease at larger TLE capacity. Frequently only one, two, or three cracking coils per quench cooler are used.

B. Mass Velocity and Heat Flux

The mass velocity is taken as high as possible for optimal heat-transfer conditions, which are product side limiting. The maximum heat flux (at entry) must be limited to prevent film boiling. Entry-heat fluxes as high as 500,000 kcal/h m^2 have been successful. In view of pressure drop, heat flux, and tube size requirements, the recommended mass velocities for tubular, conventional TLE's are shown in the table.

Feedstock	Mass velocity (kg/s m^2)
Ethane	50–55
Naphtha	55–60
AGO	70–75

As for USXs, there is no critical hot tubesheet, consequently the mass velocity can be 20% higher.

The entry-heat flux (O_E) can be calculated as follows.

$$O_E = k_E(T_i - T_w)$$

where k_E = the overall heat-transfer coefficient at the entry, T_i = quench-cooler inlet temperature, and T_w = steam–water temperature.

k_E can be calculated as follows.

$$\frac{1}{k_E} = \frac{1}{\alpha_i} + \frac{1}{\alpha_w} + \frac{1}{f_i} + \frac{1}{f_w} + \frac{\lambda}{d}$$

where α_i = heat-transfer coefficient on the product side, α_w = heat-transfer coefficient on the water–steam side, f_i = fouling coefficient on the product side, f_w = fouling coefficient on the water–steam side, λ = thermal conductivity of the TLE tube, and d = tube thickness.

C. Tube Size

Larger diameter cooler-tubes avoid blockage by coke particles from the radiant coils, but they are less effective with respect to heat transfer. Conventional TLE IDs were usually 23 mm, but presently larger diameters of 35–38 mm are favored. The tube length for the conventional TLEs is usually 4–6 m. In contrast, the USX-D inner diameter is some 90 mm, with a tube length of 8–9 m.

D. Outlet Temperature and Steam Pressure

For gas cracking, the TLE-outlet temperature will be some 30°C above the steam-generating temperature. The postquench-cooler outlet temperature can be as low as 200°C because tube fouling will not occur. For older gas crackers, the steam-generating pressure is frequently only 45–50 atm. Present fuel prices require a higher thermodynamic efficiency and thus, higher steam-pressure levels (Mol, 1978).

For liquid feedstocks, high steam-pressure levels favor longer TLE cycle length. Common practice is to generate steam with a pressure of 125 atm. The TLE SOR outlet temperature for naphtha is some 30°C above the steam-generating temperature; for AGO, it is 110–125°C.

E. Pressure Drop

Figure 10 depicts the distribution of pressure drop at SOR for a conventional tubular TLE. The pressure drop over the cooling tubes appears to be about 70% of the pressure drop over the total TLE. The pressure drop at the inlet nozzle, tubesheet inlet, and so on, is considerable. In the past years there has been a trend toward a somewhat lower pressure drop featuring lower radiant-coil outlet pressure. An acceptable pressure drop at clean conditions for a conventional TLE is 0.2 kg/cm^2.

F. Circulation Rate

The circulation rate is the ratio of water to steam flow over the risers. A positive circulation rate must be maintained to avoid stagnant water spaces in the shell and resultant steam blanketing. Circulation is determined by the cooling and flow along the hot tubesheet to avoid coke deposits; circulation rates are higher for vertical than for horizontal TLEs. Coke deposition is less likely for concentric-tube design. Recommended rates are as follows.

TLE arrangement	Circulation ratio (Water/steam flow)
Vertical	12–14
Horizontal	10

G. Hold-up Time

Hold-up time is the run time available from the boiler-feed water in the steam drum—from normal level to empty—in case of feed-water pump or

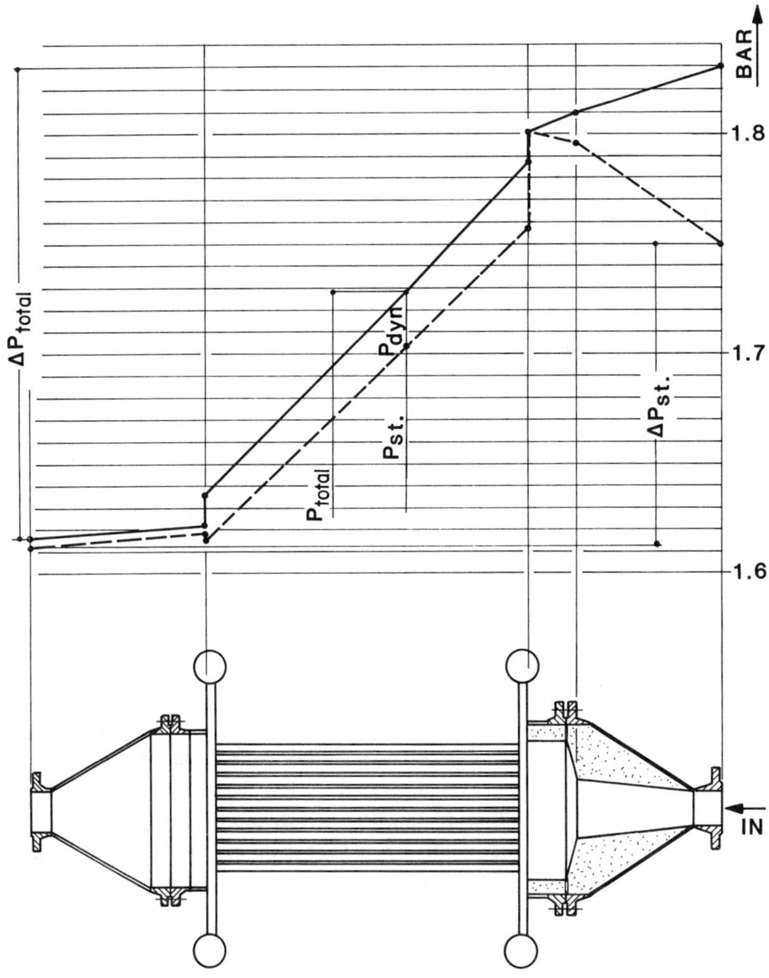

Fig. 10. Pressure drop (total and static) over Schmidt'sche TLE. This figure depicts relatively large pressure drop over TLE inlet. (Reprinted with permission of Schmidt'sche GmbH.)

controller failure. The time required for corrective action at low-level alarm depends on instrumentation, plot plan, and operators' experience.

Hot-tubesheet damage is less likely to occur in vertical quench coolers because the water in the cooler adds to the hold-up time. Whereas in the past a hold-up time could be as high as 15 min, the trend toward lower investment has resulted in a present practice of some 5 min of hold-up time.

H. Design Code

Noteworthy in the context of investment is the impact of design code on shell-wall thickness and consequently investment. The German code (i.e., AD-Merkblaetter) permits use of considerably smaller TLE and steam drum-shell thickness than for ASME Section 8 Division 1 or 2. The investment for steam drums (40% of total indirect-quench system) based on the AD-Merkblaetter code is some 10–12% less than those designed using the ASME Section 8 Division 1.

IX. QUENCH-COOLER FOULING

Basically, there are three different mechanisms for quench-cooler fouling (Lohr and Dittmann, 1978).

1. Coke produced in the cracking coils can spall off and block the TLE-inlet tubesheet. This effect of blocking the inlet tubesheet is called mushroom formation, and is most serious for ethane feedstock.
2. Nonideal flow distribution in the TLE-inlet chamber and at the hot tubesheet causes turbulent eddies and backmixing, resulting in overcracking and additional fouling.
3. Pyrolysis tars can condense on the cool tube wall, because the dew point is lower than the tube-wall temperature. This mechanism is most important for gas-oil and heavy-naphtha feedstocks. On the tube wall, the condensed high-boiling hydrocarbons are converted into a cokelike substance. Typically, fouling due to condensation is fast at SOR. TLE-outlet temperature rise during the first day onstream may be as much as 40–50°C for AGO feed. The coke layer reduces the heat-transfer rate, and the inner surface temperature becomes higher than the original tube-wall temperature at that location. Therefore, the fouling rate decreases to an equilibrium temperature level (called plateau temperature). Typical quench-cooler fouling for naphtha and gas-oil feeds is depicted in Fig. 11. The average TLE-outlet temperature depends on SOR outlet temperature, feedstock characteristics, cracking severity, steam dilution, process pressure, HP steam pressure, mass, and heat transfer.

TLE-outlet temperature at SOR has only a marginal influence on the minimum surface temperature and thus on the fouling rate. A rule of thumb for TLE fouling for AGO feed is that the maximum cracking severity has to be limited to a minimum of 7 wt% hydrogen in the C_5^+ fraction. Cycle length for an AGO feed at normal cracking severity can be 35–40 days for conventional TLEs as well as for the USX-D type of quench cooler.

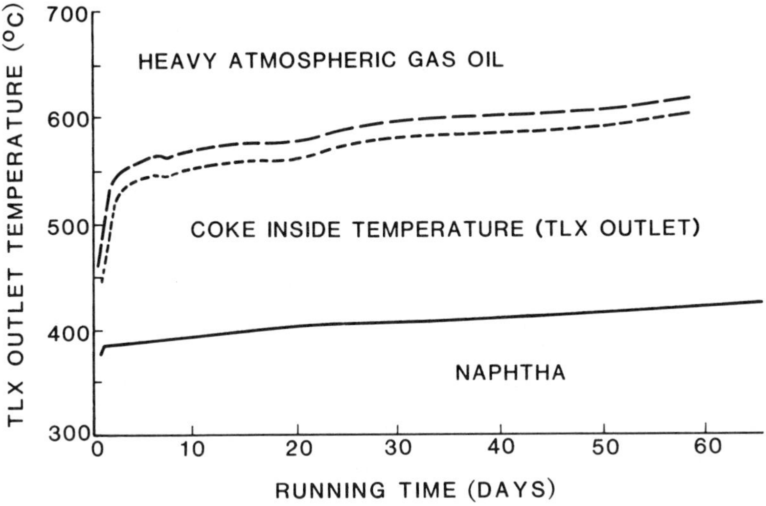

Fig. 11. TLE-outlet temperature versus run time. This figure depicts fast outlet-temperature rise over first days of run (from *Oil and Gas Journal*).

REFERENCES

Anonymous. (1977). *Chemical Week* **121**(13), 39.

Arisaki, K., Hiruta, T., Kawamura, S., and Yonezawa, M. (1976). *CEER. Chem. Econ. Eng. Rev.* **8**(97), 33–35.

Chen, J., and Maddock, M. J. (1973). *Hydrocarbon Process,* **52**(5), 147–150.

Chen, J., and Vogel, W. (1973). *AIChE Natl. Meet., 74th, New Orleans, La.*, p. 87b.

Deuse, K. H. (1971). *Ingenieursblad* **40**, 614–618.

Hinchley, P. (1979). *Chem. Eng. (N.Y.)* **86**(17), 120–134.

Howard, G. J., and Hoggett, P. N. (1975). *Oil Gas J.* **73**(30), 81–85.

Knuelle, H. R., (1972). *Chem. Eng. Prog.* **68**(7), 53–56.

Knuelle, H. R. (1973). *Ingenieursblad* **42**, 666–672.

Lohr, B., and Dittman, H. (1978). *Oil Gas J.* **76**(20), 63–68.

Mol, A. (1974). *Hydrocarbon Process.* **53**(7), 115–118.

Mol, A. (1978). *Hydrocarbon Process.* **57**(7), 179–184.

Mol, A., and Westenbrink, J. J. (1974). *Hydrocarbon Process.* **53**(2), 83–87.

Sakaguchi, N., Kawada, A., Kono, K., Hashimoto, R., Okamoto, T., Tamura, M., and Okuda, T. (1980). *Mitsubishi Juko Giho* **17**(5).

Sato, T., and Ohnishi, Y. (1971). *Bull. Jpn. Petrol. Inst.* **13**, 279–284.

Shimizu, S., and Nonaka, M. (1982). *CEER, Chem. Econ. Eng. Rev.* **14**(155), 35–40.

Sundstrom, D. W., and DeMiciell, R. L. (1971). *Ind. Eng. Chem., Process Des. Dev.* **10**, 114–122.

Woebcke, H., Dutkiewicz, B. K. and Espanola, W. R. (1971). U.S. Patent 3,583,476.

Index

B

C

D

E